T0234897

Optimization of Structural and
Mechanical Systems

Mechanical Systems

Optimization of
Structural and
Mechanical Systems

Editor

Jasbir S Arora

University of Iowa, USA

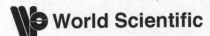

World Scientific

NEW JERSEY · LONDON · SINGAPORE · BEIJING · SHANGHAI · HONG KONG · TAIPEI · CHENNAI

Published by

World Scientific Publishing Co. Pte. Ltd.

5 Toh Tuck Link, Singapore 596224

USA office: 27 Warren Street, Suite 401-402, Hackensack, NJ 07601

UK office: 57 Shelton Street, Covent Garden, London WC2H 9HE

British Library Cataloguing-in-Publication Data
A catalogue record for this book is available from the British Library.

First published 2007 (Hardcover)
Reprinted 2016 (in paperback edition)
ISBN 978-981-3203-34-1

OPTIMIZATION OF STRUCTURAL AND MECHANICAL SYSTEMS

ISBN-13   978-981-256-962-2
ISBN-10   981-256-962-6

Printed in Singapore

PREFACE

This book covers several important topics on the subject of optimization of structural and mechanical systems. Computational optimization methods have matured over the last few years due to the extensive research by applied mathematicians and the engineering community. These methods are being applied to a variety of practical applications. Several general-purpose optimization programs as well as programs for specific engineering applications have become available recently. These are being used to solve practical and interesting optimization problems.

The book covers state-of-the-art in computational algorithms as well as applications of optimization to structural and mechanical systems. Formulations of the problems are covered and numerical solutions are presented and discussed. Topics requiring further research are identified. Leading researchers in the field of optimization and its applications have written the material and provided significant insights and experiences with the applications. The topics covered include:

- ❖ Optimization concepts and methods
- ❖ Optimization of large scale systems
- ❖ Optimization using evolutionary computations
- ❖ Multiobjective optimization
- ❖ Shape optimization
- ❖ Topology optimization
- ❖ Design sensitivity analysis of nonlinear structural systems
- ❖ Optimal control of structures
- ❖ Nonlinear optimal control
- ❖ Optimization of systems for acoustics
- ❖ Design optimization under uncertainty
- ❖ Optimization-based inverse kinematics of articulated mechanisms
- ❖ Multidisciplinary design optimization
- ❖ mesh free methods for optimization
- ❖ Kriging metamodel based optimization,

- ❖ Sensitivity-free formulations for structural and mechanical system optimization
- ❖ Robust design based on optimization
- ❖ Parallel computations for design optimization
- ❖ Semidefinite programming for structural optimization.

The book is suitable for advanced courses on optimization of structural and mechanical systems. It is also an invaluable resource for researchers, graduate students, and practitioners of optimization.

I would like to thank all the authors for their diligence and meticulous work in writing their chapters. Without their hard work this book would not be possible. I would also like to thank the staff at World Scientific Publishing Company for their patience and help in finalizing the material for the book.

Finally, I would like to thank all my family members for their unending support, patience and love.

Jasbir S. Arora
Iowa City, Iowa, USA
4 December 2006

LIST OF CONTRIBUTORS

Chapter 1
Introduction to Optimization
Jasbir S. Arora
Department of Civil and Environmental Engineering
Department of Mechanical and Industrial Engineering
Center for Computer Aided Design
The University of Iowa
Iowa City, Iowa, U.S.A.
E-mail: Jasbir-Arora@uiowa.edu

Chapter 2
Optimization of Large Scale Systems
José Herskovits, Evandro Goulart and Miguel Aroztegui
Mechanical Engineering Program
COPPE – Federal University of Rio de Janeiro
Caixa Postal 68503, 21945 970 Rio De Janeiro, Brazil
E-mail: jose@optimize.ufrj.br

Chapter 3
Structural Optimization Using Evolutionary Computation
Christopher M. Foley
Department of Civil & Environmental Engineering, Marquette University
P.O. Box 1881, Milwaukee, Wisconsin 53201
E-mail: chris.foley@marquette.edu

Chapter 4
Multiobjective Optimization: Concepts and Methods

Achille Messac
Department of Mechanical, Aerospace, and Nuclear Engineering
Rensselaer Polytechnic University
Troy, N.Y. 12180 USA
E-mail: messac@rpi.edu

Anoop A. Mullur
Department of Mechanical, Aerospace, and Nuclear Engineering
Rensselaer Polytechnic University
Troy, N.Y. 12180 USA
E-mail: mullua@rpi.edu

Chapter 5
Shape Optimization

Tae Hee Lee
School of Mechanical Engineering
Hanyang University, Seoul 133-791, Korea
E-mail: thlee@hanyang.ac.kr

Chapter 6
Topology Optimization

Martin P. Bendsoe
Department of Mathematics
Technical University of Denmark
DK-2800 Lyngby, Denmark
E-mail: M.P.Bendsoe@mat.dtu.dk

Ole Sigmund
Department of Mechanical Engineering,
Technical University of Denmark
DK-2800 Lyngby, Denmark
E-mail: sigmund@mek.dtu.dk

Chapter 7
Shape Design Sensitivity Analysis of Nonlinear Structural Systems

Nam Ho Kim
Department of Mechanical and Aerospace Engineering
University of Florida, P.O. Box 116250
Gainesville, Florida 32611, USA
E-mail: nkim@ufl.edu

Chapter 8
Optimal Control of Structures

Satish Nagarajaiah
Department of Civil & Environmental Engineering
Department of Mechanical Engineering & Material Science
Rice University, Houston, Texas, USA
E-mail: nagaraja@rice.edu

Sriram Narasimhan
Department of Civil Engineering
University of Waterloo, Waterloo, Canada
E-mail: snarasim@uwaterloo.ca

Chapter 9
Optimization of Systems for Acoustics

Ashok D. Belegundu and Michael D. Grissom
The Pennsylvania State University, University Park, PA 16802
E-mail: abelegundu@psu.edu

Chapter 10
Design Optimization Under Uncertainty

Sankaran Mahadevan
Department of Civil and Environmental Engineering, Vanderbilt University
VU Station B # 351831, 2301 Vanderbilt Place
Nashville, Tennessee, USA
E-mail: sankaran.mahadevan@vanderbilt.edu

Chapter 11
Design Optimization with Uncertainty, Life-cycle Performance and Cost Considerations

Dan M. Frangopol
Department of Civil, Environmental, and Architectural Engineering
University of Colorado, Boulder, CO 80309-0428, USA
E-mail: dan.frangopol@colorado.edu

Kurt Maute
Department of Aerospace Engineering Sciences
University of Colorado, Boulder, CO 80309-0429, USA
Email: maute@colorado.edu

Min Liu
Formerly, Department of Civil, Environmental, and Architectural
Engineering, University of Colorado, Boulder, CO 80309-0428, USA
Email: minliu@illinoisalumni.org

Chapter 12
Optimization-based Inverse Kinematics of Articulated Linkages

Karim Abdel-Malek and Jingzhou Yang
Center for Computer-Aided Design
The University of Iowa, Iowa City, IA 52242, USA
E-mail: jyang@engineering.uiowa.edu

Chapter 13
Multidisciplinary Design Optimization

Gyung-Jin Park
Department of Mechanical Engineering, Hanyang University
1271 Sa Dong, Ansan City, Gyeonggi Do, Korea 426-791
E-mail: gjpark@hanyang.ac.kr

Chapter 14
Meshfree Method and Application to Shape Optimization

J. S. Chen
Civil & Environmental Engineering Department
University of California, Los Angeles
Los Angeles, CA 90095-1593
E-mail: jschen@seas.ucla.edu

Nam Ho Kim
Mechanical & Aerospace Engineering Department
University of Florida
Gainesville, Florida 32611-6250
E-mail: nkim@ufl.edu

Chapter 15
Sensitivity-free Formulations for Structural and Mechanical System Optimization

Jasbir S. Arora and Qian Wang
Department of Civil and Environmental Engineering
Department of Mechanical and Industrial Engineering
Center for Computer Aided Design
The University of Iowa
Iowa City, Iowa, U.S.A.
E-mail: Jasbir-Arora@uiowa.edu

Chapter 16
Kriging Metamodel Based Optimization

Tae Hee Lee and Jae Jun Jung
School of Mechanical Engineering
Hanyang University, Seoul 133-791, Korea
E-mail: thlee@hanyang.ac.kr

Chapter 17
Robust Design Based on Optimization

Byung Man Kwak
Department of Mechanical Engineering
Korea Advanced Institute of Science and Technology
373-1 Guseong-dong, Yuseong-gu, Daejeon 305-701, South Korea
E-mail: bmkwak@khp.kaist.ac.kr

Chapter 18
Parallel Computations for Design Optimization

S. D. Rajan and A. Damle
Department of Civil Engineering
Arizona State University, Tempe, AZ 85287
E-mail: S.Rajan@asu.edu

Chapter 19
Semidefinite Programming for Structural Optimization

Makoto Ohsaki
Department of Architecture and Architectural Engineering, Kyoto University
Kyotodaigaku-Katsura, Nishikyo, Kyoto 615-8540 Japan
E-mail: ohsaki@archi.kyoto-u.ac.jp

Yoshihiro Kanno
Department of Urban and Environmental Engineering, Kyoto University
Sakyo, Kyoto 606-8501 Japan
E-mail: Kanno@archi.kyoto-u.ac.jp

Chapter 20
Nonlinear Optimal Control

Soura Dasgupta
Department of Electrical and Computer Engineering
The University of Iowa, Iowa City, Iowa, USA
E-mail: soura-dasgupta@uiowa.edu

CONTENTS

CHAPTER 1

INTRODUCTION TO OPTIMIZATION

Jasbir S. Arora

Department of Civil and Environmental Engineering
Department of Mechanical and Industrial Engineering
Center for Computer Aided Design
The University of Iowa
Iowa City, Iowa, U.S.A.
E-mail: Jasbir-Arora@uiowa.edu

Basic concepts of optimization are described in this chapter. Optimization models for engineering and other applications are described and discussed. These include continuous variable and discrete variable problems. Optimality conditions for the continuous unconstrained and constrained problems are presented. Basic concepts of algorithms for continuous and discrete variable problems are described. An introduction to the topics of multiobjective and global optimization is also presented.

1. Introduction

Optimization is a mature field due to the extensive research that has been conducted over the last about 60 years. Many types of problems have been addressed and many different types of algorithms have been investigated. The methodology has been used in different practical applications and the range of applications is continuously growing. Some of the applications are described in various chapters of this book. *The purpose of this chapter is to give an overview of the basic concepts and methods for optimization of structural and mechanical systems.* Various optimization models are defined and discussed. Optimality conditions for continuous variable optimization problems are presented and discussed. Basic concepts of algorithms for continuous variable and discrete variable optimization problems are described. Topics of multiobjective and global optimization are also introduced. The material of the chapter is available in many textbooks on optimization.[1-7] It is derived from several recent publications of the author and his co-workers.[7-34]

1

2. Optimization Models

Transcription of an optimization problem into a mathematical formulation is a critical step in the process of solving the problem. If the formulation of the problem as an optimization problem is improper, the solution for the problem is most likely going to be unacceptable. For example, if a critical constraint is not included in the formulation, then most likely, that constraint is going to be violated at the optimum point. Therefore special attention needs to be given to the formulation of the optimization problem.

Any optimization problem has three basic ingredients:
- *Optimization variables*, also called *design variables* denoted as vector **x**.
- *Cost function*, also called the objective function, denoted as $f(\mathbf{x})$.
- *Constraints* expressed as equalities or inequalities denoted as $g_i(\mathbf{x})$.

The variables for the problem can be continuous or discrete. Depending on the types of variables and functions, we obtain continuous variable, discrete variable, differentiable and nondifferentiable problems. These models are described next; for more details and practical applications of the models, various references can be consulted.[7,9-12,14,16,25-30]

2.1. *Optimization Models: Continuous Variables*

Any continuous variables optimization problem can be transcribed into a *standard nonlinear programming (NLP) model* defined as minimization of a cost function subject to equality constraints and inequality constraints expressed in a "≤" form as Problem P.[7]

Problem P. Find the optimization variable vector $\mathbf{x} = [x_1\ x_2\ \ x_n]^T$ to minimize a cost function $f(\mathbf{x})$ subject to equality and inequality constraints:

$$g_j(\mathbf{x}) = 0,\ j = 1 \text{ to } p \tag{1}$$

$$g_j(\mathbf{x}) \leq 0,\ j = p+1 \text{ to } m \tag{2}$$

where n is the number of variables, p is the number of equality constraints, and m is the total number of constraints. Note that the *explicit lower* and *upper bounds* on the variables are included in Eq. (2). However, for efficient numerical calculations the simple form of these constraints is exploited.

The *feasible set* for the problem is defined as a collection of all the points that satisfy the constraints of Eqs. (1) and (2). It is also called the *constraint set*, and is denoted as S:

$$S = \left\{ \mathbf{x} \middle| \, g_j(\mathbf{x}) = 0, \, j = 1 \text{ to } p; \, g_j(\mathbf{x}) \leq 0, \, j = p+1 \text{ to } m \right\} \tag{3}$$

Thus the Problem P can be written simply as

$$\underset{\mathbf{x} \in S}{\text{minimize}} \, f(\mathbf{x}) \tag{4}$$

It is important to note that the feasible set for a problem may be empty if there are too many constraints on the problem or if there are conflicting constraints. In general, this is difficult to determine before the problem is solved. Only after a numerical algorithm fails to find a feasible point for the problem, we can conclude that the set S is empty.[21] In that case the problem formulation needs to be examined to relax some of the constraints, or eliminate conflict in the constraints. In addition, it is difficult to know, in general, if there is a solution to the Problem P. However, the question of *existence of a solution* can be answered with certain assumptions about the problem functions. It turns out that if $f(\mathbf{x})$ is *continuous* on a *nonempty feasible set S*, all constraint functions are continuous, and all inequalities contain their boundary points (i.e., expressed as "$\leq$" and not simply as "$<$"), then there is a solution for Problem P. When these requirements are satisfied, a robust numerical algorithm is guaranteed to converge to a solution point.

If there are no constraints on the variables, the set S is the entire design space and the problem is called an *unconstrained optimization problem*. If all the functions are linear in terms of the variables, the Problem P is called a *linear programming* (LP) problem. If the cost function is quadratic and the constraints are linear, the problem is called a *quadratic programming* (QP) problem.

An inequality constraint $g_i(\mathbf{x}) \leq 0$ is said to be *active* at a point $\mathbf{x}$ if it is satisfied as an equality at that point, i.e., $g_i(\mathbf{x}) = 0$. It is said to be *inactive* if it has negative value at that point, and *violated* if it has positive value. An *equality* constraint is always either *active* or *violated* at any point.

In some applications, several objective functions need to be optimized simultaneously. These are called *multiobjective* optimization problems. They are usually transformed into Problem P by combining all the objective functions to form a composite scalar objective function. Several approaches to accomplish this objective are summarized in a later section.[7,32,35-37]

When a gradient-based optimization method (discussed in a later section) is used to solve Problem P, the cost and constraint functions are assumed to be twice differentiable.

2.2. *Optimization Models: Mixed Variables*

In many practical applications of optimization, discrete variables occur naturally in the problem formulation. For example,
- plate thickness must be selected from the available dimensions,[7]
- material properties must correspond to the available materials,[7,25]
- structural members must be selected from a catalog,[14,26,29]
- number of reinforcing bars in a concrete member must be an integer,[28]
- diameter of rods must be selected from the available sizes,[7,28]
- number of bolts must be an integer,[27]
- number of strands in a prestressed member must be an integer.[28]

Discrete variables must be treated properly in numerical optimization procedures. A mixed continuous-discrete variable optimization problem is defined next as Problem MP.

Problem MP. A general mixed discrete-continuous variable nonlinear optimization problem is defined by modifying Problem P to minimize the cost function $f(\mathbf{x})$ subject to the constraints of Eqs. (1) and (2) with the additional requirement that each discrete variable be selected from a specified set:

$$x_i \in D_i, \; D_i = (d_{i1}, d_{i2}, \ldots, d_{iq_i}); \; i = 1 \text{ to } n_d \tag{5}$$

where n_d is the number of *discrete design variables*, D_i is the set of discrete values for the ith variable, q_i is the number of available discrete values for the ith variable, and d_{ik} is the kth discrete value for the ith variable. Note that the foregoing problem definition includes *integer variable* as well as 0-1 *variable* (on-off variables, binary variables) problems. If the problem has only *continuous variables*, and the functions f and g_j are twice continuously differentiable, we obtain the Problem P. Many discrete variable optimization problems have nondifferentiable functions; therefore gradient-based methods cannot be used to solve such problems. However, methods that do not require gradients of functions are available to solve such problems.

It is also important to note that the discrete variable optimization problems usually require considerably more computational effort compared to the continuous variable problems. This is true even though the number of feasible points with discrete variables is finite and they are infinite with continuous variables.

3. Optimality Conditions for Problem P

3.1. *Definitions and General Concepts*

The optimality conditions are the mathematical conditions that characterize a minimum point for the problem. Let us first define what is meant by a *minimum* point for the cost function $f(\mathbf{x})$ before discussing the optimality conditions.

Local Minimum. The cost function $f(\mathbf{x})$ has a *local minimum* (*relative minimum*) at a point $\mathbf{x}^*$ in the *feasible set S* if the function value is the smallest at the point $\mathbf{x}^*$ compared to all other points $\mathbf{x}$ in a small *feasible neighborhood* of $\mathbf{x}^*$, i.e.,

$$f(\mathbf{x}^*) \leq f(\mathbf{x}) \tag{6}$$

If *strict inequality* (i.e., $f(\mathbf{x}^*) < f(\mathbf{x})$) holds, then $\mathbf{x}^*$ is called the *strict* or *isolated local minimum.*

Global Minimum. The cost function $f(\mathbf{x})$ has a *global minimum* (also called an *absolute minimum*) at a point $\mathbf{x}^*$ if Inequality (6) holds for *all* $\mathbf{x}$ in the feasible set S. If strict inequality holds, then $\mathbf{x}^*$ is called the *strict* or *unique global minimum*.

These definitions show that for the local minimum, we test the inequality in Eq. (6) only for a small *feasible domain* around the point $\mathbf{x}^*$, and test it over the *entire* feasible set S for the global minimum point. Note that the cost function $f(\mathbf{x})$ can have many global minimum points as long as the function values are the same at all the points. Similarly, there can be multiple local minima in the small feasible domain.

The foregoing definitions of local and global minima cannot be used directly to find the minimum points for the Problem P. However, they can be used to derive the *optimality conditions* that characterize a local minimum point. Note that they cannot be used to derive optimality conditions for a global minimum point for the function $f(\mathbf{x})$. The reason is that the global optimality conditions require knowledge about the global behavior of $f(\mathbf{x})$. For a discrete variable problem, the definitions are useful because there are only a finite number of points to be checked for optimality of a given point. In fact most stochastic methods for optimization of discrete variable problems, described in a later section, use the definitions to check optimality of a point in its neighborhood.

The optimality conditions can be divided into two categories: *necessary* and *sufficient*. The necessary conditions must be satisfied for a point to be a candidate minimum point. The points that satisfy the necessary conditions are called *stationary points*. Note however that a point satisfying the necessary conditions need not be a minimum point, i.e., points that are not minima may also satisfy the necessary conditions. The *sufficient condition* if satisfied determines the stationary point to be a local minimum point. If the sufficient condition is not satisfied, no conclusion about the optimality of the stationary point can be drawn. We shall describe both the necessary and sufficient conditions. Sample problems showing the use of these conditions can be found in many textbooks.[2-7,38]

The optimality conditions are used in two ways: (i) they are used to develop numerical methods for finding minimum points, and (ii) they are used to check optimality of a given point; i.e., using them, a stopping criterion for the iterative numerical algorithm can be defined. We shall first present the optimality conditions for the unconstrained problem and then for the general constrained Problem P.

3.2. *Optimality Conditions for the Unconstrained Problem*

When there are no constraints, the problem is to minimize just the cost function $f(x)$. The conditions for x^* to be a minimum point for the function $f(x)$ are derived by analyzing the local behavior of the function at the point x^*; i.e., Taylor's expansion for the function.

First Order Necessary Condition. If x^* is a local minimum for the cost function $f(x)$, then the gradient (*first derivatives*) of $f(x)$ at x^* must vanish, that is, $\partial f/\partial x_i = 0$, $i = 1$ to n.

Second Order Necessary Condition. If x^* is a local minimum for the function $f(x)$, then its Hessian $H = [\partial^2 f/\partial x_i \partial x_j]$ at x^* must be at least *positive semidefinite*; i.e., all its eigenvalues must be nonnegative.

Second Order Sufficient Condition. If the matrix $H(x^*)$ is *positive definite* at the stationary point x^*, then x^* is an isolated local minimum point. (A matrix is called positive definite if all its eigenvalues are positive).

Any point x^* satisfying the necessary conditions of optimality is called a *stationary point*. If a stationary point is neither a minimum nor a maximum, then it is called an *inflection point*. It should be noted that the optimality conditions are based on derivatives of $f(x)$ and not the function value. Therefore, the minimum point is not changed if a constant is added to the function, or the function is scaled by a positive constant. The optimum value of the cost function does, however, change in the process.

3.3. *Optimality Conditions for the Constrained Problem*

We now present the optimality conditions for a constrained problem which involve constraints in addition to the cost function. Although a constrained problem can have minimum points where no constraints are *active*, this usually does not happen in practical applications. The case where no constraints are active, the optimum point is inside the feasible set S and the foregoing optimality conditions for the unconstrained problem apply; i.e., the optimality conditions for the unconstrained problem are a special case of those for the constrained problem. The conditions for the constrained problem can be expressed in several alternate but equivalent ways. We shall present the conditions that are most commonly used in the modern literature. These are known as *Karush-Kuhn-Tucker* or *KKT conditions*.

Regular Point. An assumption in the derivation of the KKT necessary conditions is that the minimum point be a regular point of the feasible set S. A point $\mathbf{x}$ is called a *regular point* of the feasible set S if the cost function is continuous and the gradients of all the active constraints are linearly independent at the point. The number of linearly independent vectors cannot be more than n, the number of variables, i.e., dimension of each vector. Thus the total number of active constraints cannot be more than the number of variables at the regular point; i.e., at a minimum point.

Karush-Kuhn-Tucker Necessary Conditions. Let the *Lagrangian* for the Problem P be defined as

$$L(\mathbf{x},\mathbf{u}) = f(\mathbf{x}) + \mathbf{u} \bullet \mathbf{g}(\mathbf{x}) \qquad (7)$$

where $\mathbf{u}$ is a vector of *Lagrange multipliers* for the constraints $\mathbf{g}$ that needs to be determined and a "$\bullet$" implies scalar product of vectors. Let $\mathbf{x}^* \in S$ be a local minimum for $f(\mathbf{x})$. Also, let the gradients of the active constraints at $\mathbf{x}^*$ be linearly independent (i.e., point $\mathbf{x}^*$ is a *regular point* of the feasible set). Then there exist *unique* Lagrange multipliers u_i^* such that

$$\nabla L(\mathbf{x}^*) = \mathbf{0}, \quad \text{or} \quad \nabla f(\mathbf{x}^*) + \nabla \mathbf{g}(\mathbf{x}^*)\mathbf{u}^* = \mathbf{0} \qquad (8)$$

$$u_i^* g_i(\mathbf{x}^*) = 0, \ i = (p+1) \text{ to } m \qquad (9)$$

$$u_i^* \geq 0, \ i = (p+1) \text{ to } m \qquad (10)$$

where $\nabla \mathbf{g}(\mathbf{x}^*)$ is an $n \times m$ matrix. Equations (8) show that the Lagrangian L is stationary with respect to $\mathbf{x}$ since the gradient of the Lagrangian is zero, Eq. (9) shows that either the Lagrange multiplier for the ith inequality is zero or the

J. S. Arora

constraint is active at the minimum point (if $u_i = 0$, g_i must be ≤ 0 for feasibility of $\mathbf{x}^*$), and Eq. (10) shows that the Lagrange multipliers for the inequality constraints must be nonnegative. In addition, $\mathbf{x}^*$ must be a feasible point. Equation (9) is called the *switching condition* or *complementary slackness* condition because it identifies active and inactive inequality constraints. Note that there are n variables and m Lagrange multipliers, and thus $(n+m)$ unknowns. There are $(n+m)$ equations (n equations in conditions (8), p equalities, and m-p equations in condition (9)). Therefore, the necessary conditions give a determinate system of equations, though it is usually nonlinear.

The gradient condition of Eq. (8) can be re-arranged as:

$$-\nabla f(\mathbf{x}^*) = \nabla \mathbf{g}(\mathbf{x}^*)\mathbf{u}^* \tag{11}$$

This form of the equation brings out the physical meaning of the gradient condition. It shows that at the minimum point, the steepest descent direction (negative of the cost function gradient) is in the range of gradients of the active constraints; i.e., a linear combination of them with Lagrange multipliers as the scalars of the linear combination.

The regularity check for $\mathbf{x}^*$ is an important part of KKT conditions. If this check is not satisfied, all other KKT conditions may or may not be satisfied at $\mathbf{x}^*$. For example, the Lagrange multipliers may not be unique at $\mathbf{x}^*$. However, the irregular points can also be local minimum points where KKT conditions may be actually violated.

Second order optimality conditions can be used to distinguish the minimum points from others. These conditions also involve Hessians of the functions, as for the unconstrained problems; e.g., Hessians of the active constraints at $\mathbf{x}^*$. We briefly discuss these conditions next.[1-7]

Second-order Necessary Condition. Let $\mathbf{x}^*$ satisfy the first order KKT necessary conditions for Problem P. Let the Hessian of the Lagrange function L at $\mathbf{x}^*$ be defined as

$$\nabla^2 L = \nabla^2 f + \sum_{i=1}^{m} u_i^* \nabla^2 g_i \tag{12}$$

Let there be nonzero feasible directions, $\mathbf{d} \neq \mathbf{0}$, as solutions of the following linear system at $\mathbf{x}^*$:

$$(\nabla g_i \bullet \mathbf{d}) = 0, \, i = 1 \text{ to } p \text{ and for those } i > p \text{ with } g_i(\mathbf{x}^*) = 0 \tag{13}$$

That is, the vectors $\mathbf{d}$ are in the null space of the gradients of the active constraints. Then if $\mathbf{x}^*$ is a local minimum point for the optimum design problem, it must be true that

$$Q \geq 0 \text{ where } Q = \left(\mathbf{d} \bullet \nabla^2 L(\mathbf{x}*)\mathbf{d}\right) \tag{14}$$

Note that any point that does not satisfy the second-order necessary condition cannot be a local minimum point.

Second-order Sufficient Condition. Let $\mathbf{x}*$ satisfy the first-order KKT necessary conditions for Problem P. Let the Hessian of the Lagrange function L be defined at $\mathbf{x}*$ as in Eq. (12). Let there be nonzero feasible directions, $\mathbf{d} \neq \mathbf{0}$, as solutions of the following linear system at $\mathbf{x}*$:

$$(\nabla g_i \bullet \mathbf{d}) = 0, \, i = 1 \text{ to } p \text{ and for those } i > p \text{ with } g_i(\mathbf{x}*) = 0 \text{ and } u_i^* > 0 \tag{15}$$

That is, the vectors $\mathbf{d}$ are in the null space of the gradients of the active constraints with $u_i^* > 0$ for $i > p$. Also let $(\nabla g_i \bullet \mathbf{d}) \leq 0$ for those active inequalities with $u_i^* = 0$. If

$$Q \geq 0 \text{ where } Q = \left(\mathbf{d} \bullet \nabla^2 L(\mathbf{x}*)\mathbf{d}\right) \tag{16}$$

then $\mathbf{x}*$ is an isolated local minimum point (isolated means that there are no other local minima in the neighborhood of $\mathbf{x}*$).

Equations (13) and (15) define vectors that are in the null space of the gradients of the active constraints. There is slight difference in the two null spaces defined by these equations. In Eq. (13), all the active inequalities are included. However in Eq. (15), only the active inequalities with positive multipliers are included. Note that if the Hessian $\nabla^2 L$ is positive definite at $\mathbf{x}*$ then both the second order necessary and sufficient conditions for a local minimum are satisfied. Conversely, if it is negative definite or negative semidefinite, then the second order necessary condition is violated, and the point $\mathbf{x}*$ cannot be a minimum point.

3.4. *Global Optimality and Convexity*

Often a question is asked - is the optimum solution a global minimum? Usually the answer to this question is that the solution is only a local minimum. The global solution for the problem can be found by either an exhaustive search of the feasible set S, or by showing the problem to be convex. Both procedures require extensive computations. If the problem is convex, then any local minimum is also a global minimum and the KKT first order conditions are necessary as well as sufficient. The question of convexity of a problem is briefly addressed here. Methods for finding a global solution are described in a later section.

Problem P is called a *convex programming problem*, if the cost function $f(\mathbf{x})$ is convex over the convex feasible set S. Therefore, we need to discuss convexity of the feasible set S and the function $f(\mathbf{x})$. A set of points S (vectors $\mathbf{x}$) is called a *convex set* if and only if for any two points A and B in the set S, the entire line segment AB is also in the set. Graphically this means that a convex set has no re-entrant corners or holes. By this definition, we see that linear equalities and inequalities always define a convex feasible set. Also, a nonlinear equality always defines a nonconvex feasible set. However, in general, the graphical definition is difficult to use to check convexity of a set because an infinite pair of points will have to be considered. Therefore, a better computational procedure is needed to check convexity of a function.

The feasible set for the Problem P is defined by the functions $g_i(\mathbf{x})$, $i = 1$ to m. It turns out that if all the functions are convex, then the feasible set S is convex. Thus we need to know how to check convexity of a function.

A function of n variables is convex if and only if its Hessian is at least positive semidefinite everywhere over its domain of definition. If a function $g_i(\mathbf{x})$ is convex, then the set defined by the inequality $g_i(\mathbf{x}) \leq e_i$ is convex, where e_i is any constant. Note that this is not an "if and only if" condition; that is if $g_i(\mathbf{x})$ fails the convexity set, the feasible set defined by it may still be convex. In other words this is only a sufficient condition but it is not a necessary condition. Note that convexity checks for a problem are quite extensive. The Hessian of each nonlinear problem function needs to be evaluated and its form needs to be checked over the entire feasible domain.

The following points should be noted for *convex programming problems*:

(i) A convex programming problem can have several global minimum points where the cost function has the same numerical value.

(ii) The convexity check for a constraint function can sometimes fail if the form of the constraint is altered; however, the feasible set defined by the constraint may still be convex. Ref. 7 contains an example that illustrates this point.

(iii) If the convexity checks fail, the problem can still have a global minimum point in the feasible set. However, it is difficult to conclude that a global solution has been reached.

3.5. *Lagrange Multipliers*

It turns out that the optimum values of the Lagrange multipliers for the constraints represent relative importance of the constraints with respect to the cost function. We discuss this importance here. Also, many times it is useful in practical applications to scale the cost function and constraints to avoid numerical instabilities. We discuss the affect of this scaling on the Lagrange multipliers for the constraints.

3.5.1. *Changes in Constraint Limit*

Let us first study how the optimum value of the cost function is affected if a constraint limit is changed; i.e., a constraint is relaxed or tightened. Assume that Problem P has been solved with the current limit values for the constraints as zero. Let e_i be a small variation in the right hand side of the ith constraint. It is clear that the optimum point for the perturbed problem is a function of the vector $\mathbf{e}$, i.e., $\mathbf{x}^* = \mathbf{x}^*(\mathbf{e})$. Also $f = f(\mathbf{e})$. However, these are implicit functions of $\mathbf{e}$, and the following result gives a way of calculating the implicit derivatives $\partial f/\partial e_i$[6,7]

Sensitivity to Constraint Variations. Let $\mathbf{x}^*$ be a regular point that, together with the multipliers u_i^*, satisfies both the KKT necessary conditions and the sufficient conditions for an isolated local minimum point for the Problem P. If for each $g_i(\mathbf{x}) = 0$ for $i > p$, it is true that $u_i^* > 0$, then the solution $\mathbf{x}^*(\mathbf{e})$ of the modified problem is a continuously differentiable function of $\mathbf{e}$ in some neighborhood of $\mathbf{e} = 0$. Furthermore,

$$\frac{\partial f(\mathbf{x}^*(0))}{\partial e_i} = -u_i^*; \quad i = 1 \text{ to } m \tag{17}$$

It is useful to note that if the conditions stated in this result are not satisfied, existence of the implicit derivative of Eq. (17) cannot be ruled out. That is, the derivatives may still exist but their existence cannot be guaranteed. Using Eq. (17), we can estimate a change in the cost function due to a change in the right hand side of the ith constraint. First order Taylor expansion for the cost function about the point $e_i = 0$ is given as

$$f(e_i) = f(0) + \frac{\partial f(0)}{\partial e_i} e_i \tag{18}$$

where $f(0)$ is the optimum cost function value obtained with $e_i = 0$. Substituting from Eq. (17), we obtain change in the cost function Δf due to the change e_i as

$$\Delta f = f(e_i) - f(0) = -u_i^* e_i \tag{19}$$

Using the result of Eq. (19), we can show that the Lagrange multiplier corresponding to a "$\leq$ type" constraint must be nonnegative. To see this, let us assume that we want to relax an active inequality constraint $g_i \leq 0$ by selecting $e_i > 0$. This way, the feasible set for the problem gets expanded. Thus the minimum value for the cost function should reduce or stay unchanged with the expanded feasible set. However, Eq. (19) shows that if $u_i^* < 0$, relaxation of the constraint ($e_i > 0$) results in an increase in cost (i.e., $\Delta f > 0$). This is not possible, and therefore, the Lagrange multiplier for a "$\leq$ type" constraint cannot be negative.

3.5.2. *Scaling of Cost Function.*

Some times in practical applications, the cost function for the problem is normalized by multiplying it with a positive constant. Although this scaling does not affect the optimum point, it does change the Lagrange multipliers for all the constraints. Using the KKT conditions of Eq. (8), it can be shown that all the Lagrange multipliers also get multiplied by the same scale factor.[7] Let u_i^* be the Lagrange multiplier for the ith constraint with the original cost function. Let the cost function be scaled as $f^{new} = \alpha f$, where $\alpha > 0$ is a given constant, and u_i^{new} be the new value of the Lagrange multiplier for the ith constraint at optimum. Then the new and old Lagrange multipliers are related as

$$u_i^{new} = \alpha u_i^*; \quad i = 1 \text{ to } m \tag{20}$$

3.5.3. *Scaling of Constraints*

In numerical calculations, it is useful to normalize all the constraints (normalization of constraints is discussed in the next section). This scaling of a constraint does not change its boundary, so it has no effect on the optimum point or the cost function. However, the Lagrange multiplier for the constraint is affected. Using the KKT conditions of Eq. (8), it can be shown that the Lagrange multiplier for the scaled constraint gets divided by the same scale factor.[7] Let the ith constraints g_i be divided by $\beta_i > 0$ as $g_i^{new} = g_i/\beta_i$ and u_i^* and u_i^{new} be the corresponding Lagrange multipliers for the original and the scaled constraints, respectively. The new and original Lagrange multipliers are related as

$$u_i^{new} = \beta_i u_i^* \tag{21}$$

4. Basic Concepts Related to Computational Algorithms

Optimization methods for structural and mechanical systems have matured to the point where they are being used routinely in many practical applications. Many journals dedicated to the field of optimization and many textbooks on the subject can be consulted for the range of applications. Various chapters of this book contain a good sample of practical applications.

Real-world problems are usually quite complex. Each application has its own requirements, simulation methods and constraints to meet. In addition, the desire to solve more complex and larger problems also grows as computer-based computational tools improve. Furthermore, since the methods have matured substantially during the last decade, more nonexperts of optimization techniques are beginning to use this new methodology in their routine work. These considerations dictate the use of a theoretically sound and numerically reliable algorithm. Use of such an algorithm can remove uncertainty about the algorithm behavior, allowing the users to concentrate on their application. Such theoretically sound algorithms, although computationally more expensive, are more cost-effective in the long run.

In the remaining sections, some basic concepts related to numerical algorithms for optimization of structural and mechanical systems are presented and discussed. Algorithms for continuous variable problems as well as discrete variable problems are outlined. The ideas of a descent function, constraint normalization, and potential constraint strategy are introduced. Convergence of an algorithm is discussed and attributes of a good algorithm are presented. It is important to note that the gradient-based algorithms converge only to a *local minimum* point for the Problem P. Algorithms for finding a *global solution* require extensive numerical calculations and are outlined in a later section. Multiobjective optimization algorithms are also discussed.

4.1. *A Basic Gradient-based Algorithm*

Gradient-based optimization algorithms use the following iterative prescription:

$$\mathbf{x}^{(k+1)} = \mathbf{x}^{(k)} + \alpha_k \mathbf{d}^{(k)}; \quad k = 0, 1, 2,.... \tag{22}$$

where the superscript k represents the iteration number, $\mathbf{x}^{(k)}$ is the current estimate of the optimum design, $\alpha_k \mathbf{d}^{(k)}$ is a change in design, $\alpha_k > 0$ is a step size, $\mathbf{d}^{(k)}$ is a search direction, and $\mathbf{x}^{(0)}$ is the starting point.

Gradient-based algorithms are broadly classified as *primal methods* and *transformation methods*. In the *primal methods* the direction vector $\mathbf{d}^{(k)}$ is calculated using the problem functions and their gradients at the point $\mathbf{x}^{(k)}$. Then the step size is calculated along $\mathbf{d}^{(k)}$ that needs only the function values. Different algorithms can be generated depending on how the direction $\mathbf{d}$ and step size α are calculated. In many algorithms, $\mathbf{d}$ is calculated by solving a linear or quadratic programming subproblem. Several philosophies have been used to develop various algorithms. For example, if an intermediate point or the starting point is infeasible, many methods iterate through the infeasible region to reach the final solution; many others correct the constraints to reach the feasible set first and then move along the boundary to reach the solution point. Still others make special calculations not to violate constraints during the iterative process. Some algorithms generate and use second order information for the problem as the iterations progress.

In the *transformation methods* the solution process for Problem P is transformed to a sequence of unconstrained minimization problems. Solutions of the unconstrained problems converge to solution of the original problem. They include *barrier* and *penalty function methods* as well as the *augmented Lagrangian* or *multiplier methods*.[7,15,19,20] In the transformation methods, a transformed function is constructed by adding a penalty term for the constraint violations to the cost function, as $\Phi(\mathbf{x},r) = f(\mathbf{x}) + P(\mathbf{g}(\mathbf{x}),r)$, where r is a scalar or vector of penalty parameters and P is a real valued function whose action of imposing the penalty is controlled by r.

Many methods have been developed and evaluated based on the strategies described in the foregoing. Robust and general algorithms are based on the following four basic steps:

 (i)   Linearization of cost and constraint functions about the current point.
 (ii) Definition of a search direction determination subproblem.
 (iii) Solution of the subproblem for the search direction.
 (iv) Calculation of a step size along the search direction.

4.2. *Constraint Normalization*

It is useful to normalize all the constraint functions in numerical calculations because it is not easy to determine which constraint is more severely violated if they are not normalized. Also, in numerical calculations, one value for the parameter to check feasibility of all the constraints cannot be used. As examples, consider a stress constraint as $\sigma \leq \sigma_a$ and a displacement constraint as $\delta \leq \delta_a$, where σ is the calculated stress, $\sigma_a > 0$ is an allowable stress, δ is the calculated deflection, and $\delta_a > 0$ is an allowable deflection. Since the units for the two constraints are different their values are of widely differing orders of magnitude. If they are violated during the iterative solution process, it is difficult to judge the severity of their violation. However, if they are normalized as $R - 1.0 \leq 0$, where $R = \sigma / \sigma_a$ for the stress constraint, and $R = \delta / \delta_a$ for the deflection constraint, then it is easy to compare their values.

4.3. *Potential Constraint Strategy*

The optimization methods solve a subproblem to determine the search direction at each iteration. The subproblem is defined using gradients of the constraints. A subproblem that uses gradients of only a subset of the constraints is said to use a *potential constraint strategy*. The potential constraint set is comprised of the indices of active, nearly active and violated constraints, such as the index set I_k

at the kth point $\mathbf{x}^{(k)}$:

$$I_k = \left\{ i \mid i = 1 \, to \, p \text{ and all } i > p \text{ with } (g_i + \varepsilon) \geq 0 \right\} \tag{23}$$

where $\varepsilon > 0$ is a small number used to determine nearly active inequalities. Note that the equality constraints are always included in the index set I_k.

4.4. *Descent Function*

It is important to monitor progress of the iterative optimization process towards the minimum point. This can be done if a function can be defined that decreases at every iteration. Such a function is called the *descent function* or the *merit function*. The cost function is a descent function for the unconstrained optimization problems because it is required to reduce at each iteration. For constrained problems, many descent functions have been used. These functions must include the effect of constraint violations. The descent function is used in the process of step size determination. The basic idea is to compute a step size along the search direction $\mathbf{d}^{(k)}$ such that the descent function is decreased. The descent function also has the property that its minimum value is the same as the cost function.

4.5. *Convergence of an Algorithm*

An algorithm that has been proven to converge starting from an arbitrary point is called a *globally convergent* method, and satisfies two requirements: (i) there is a descent function for the algorithm, and (ii) the search direction $\mathbf{d}^{(k)}$ is a continuous function of the variables. This requirement implies that the active constraints are not coming in-and-out of the active set. This is called "zigzagging" of constraints.

4.6. *Attributes of a Good Algorithm*

A good algorithm for practical applications should have the following attributes:
  (i)   *Robustness*: The algorithm must be convergent to a local minimum point starting from any initial estimate.
  (ii) *Generality*: The algorithm must be able to treat equality as well as inequality constraints.
  (iii) *Accuracy*: The algorithm must be able to converge to an optimum point as accurately as desired.
  (iv) *Ease of Use*: Implementation of the algorithm must be such that it requires

minimum of input for use of the algorithm by the experienced as well as inexperienced users.

(v) *Efficiency*: The algorithm must have a faster rate of convergence, i.e., at least superlinear. The algorithm should be able to treat linear constraints efficiently. It should be able to exploit sparsity structure of the problem functions, especially for large-scale problems.

5. Overview of Computational Algorithms

Many numerical methods have been developed and evaluated for constrained and unconstrained optimization problems.[1-7,20,33] In addition, algorithms for discrete variable and nondifferentiable problems have been discussed.[16] Many practical applications require optimization of several objective functions, and therefore, procedures to treat multiple objectives in an optimization problem have been developed. In this section, we describe the basic concepts of these algorithms.

5.1. *Gradient-based Algorithms*

The gradient-based methods are suitable for problems with continuous variables and differentiable functions because they utilize gradients of the problem functions. The methods have been thoroughly researched and a considerable body of literature is available on the subject. They include sequential quadratic programming and augmented Lagrangian methods. We discuss the basic concepts related to these methods. The interior point methods, developed initially for linear problems, have also been extended for nonlinear problems.

5.1.1. *Linearization and Sequential Linear Programming*

All search methods start with an initial estimate for the optimum point and iteratively improve it. The improvement is computed by solving an approximate subproblem which is obtained by writing linear Taylor's expansions for the cost and constraint functions. Let $\mathbf{x}^{(k)}$ be the estimate for the optimum point at the kth iteration and $\Delta\mathbf{x}^{(k)}$ be the desired change. Instead of using $\Delta\mathbf{x}^{(k)}$ as a change in the current point, usually it is taken as the search direction $\mathbf{d}^{(k)}$ and a step size is calculated along it to determine the new point. We write Taylor's expansion of the cost and constraint functions about the point $\mathbf{x}^{(k)}$ to obtain a linearized subproblem as

$$\text{minimize } (\mathbf{c} \bullet \mathbf{d}) \qquad (24)$$

subject to the linearized equality constraints

$$(\nabla g_j \bullet \mathbf{d}) = e_j; \quad j = 1 \text{ to } p \tag{25}$$

and the linearized inequality constraints

$$(\nabla g_j \bullet \mathbf{d}) \le e_j; \quad j > p \text{ and } j \quad I_k \tag{26}$$

where $e_j = -g_j(\mathbf{x}^{(k)})$, and $\mathbf{c} = \nabla f(\mathbf{x}^{(k)})$. Note that a potential constraint strategy for inequality constraints is used in Eq. (26). If it is not to be used, ε can be set to a very large number in defining the index set I_k in Eq. (23).

Since all the functions in Eqs. (24) to (26) are linear in the variables d_i, linear programming can be used to solve for d_i. Such procedures are called *Sequential Linear Programming* methods or in short *SLP*. Note, however, that the problem defined in Eqs. (24) to (26) may not have a bounded solution. Therefore, limits must be imposed on changes in the variables. These constraints are called *move limits* in the optimization literature and can be expressed as

$$-\Delta_i \le d_i \le \Delta_i; \quad i = 1 \text{ to } n \tag{27}$$

where Δ_i is the maximum allowed decrease or increase in the ith variable, respectively at the kth iteration. The problem is still linear in terms of d_i, so LP methods can still be used to solve it. Selection of the move limits at every iteration is important because success of the SLP algorithm depends on them. However, selection of proper move limits is quite difficult in ptactice.

5.1.2. *Sequential Quadratic Programming - SQP*

To overcome drawbacks of SLP, sequential quadratic programming methods (SQP) have been developed where a *quadratic programming (QP) subproblem* is solved to find a search direction and a descent function is used to calculate a step size in that direction.

Subproblem QP.

$$\text{Minimize } (\mathbf{c} \bullet \mathbf{d}) + \tfrac{1}{2}(\mathbf{d} \bullet \mathbf{H}\mathbf{d}) \tag{28}$$

subject to the linearized constraints in Eqs. (25) and (26) where $\mathbf{H}$ is an $n \times n$ matrix that is an approximation to the Hessian of the Lagrangian function.

Different definitions of the QP subproblem generate different search directions. Once a direction has been determined, a step size is calculated by minimizing a descent function along it. The descent function for the constrained problems is constructed by adding a penalty for constraint violations to the cost function. One of the properties of the descent function is that its value at the optimum point be the same as that for the cost function. Also, it must reduce along the search direction at each iteration. In other words, the search direction must be a *descent direction* for the function. Several descent functions have been developed and used with different algorithms. We shall introduce Pshenichny's descent function Φ due to its simplicity and success in solving a large number of problems.[4,18,24,33,34] It is the exact penalty function defined as

$$\Phi(\mathbf{x}) = f(\mathbf{x}) + RV(\mathbf{x}) \tag{29}$$

where $R > 0$ is a penalty parameter and $V(\mathbf{x}) \geq 0$ is the maximum constraint violation among all the constraints. Note that R is required to be finite but larger than the sum of the magnitude of all the Lagrange multipliers.

It is important to note that calculation of an exact minimum point for the descent function along the search direction is quite costly. Therefore in most practical implementations of any optimization algorithm, only an approximate step size is determined. This is done using the so-called *inaccurate* or *inexact line search*. In the inaccurate line search procedure, one starts with the trial step size as one. If the descent condition is not satisfied, the trial step is taken as half of the previous trial. If the descent condition is still not satisfied, the trial step size is bisected again. The procedure is continued, until the descent condition is satisfied; i.e., a sufficient reduction in the descent function has been achieved. Performance of several SQP algorithms has been evaluated in Ref. 33.

5.1.3. *Augmented Lagrangian Method*

There is a class of computational methods that transform the constrained problem to an unconstrained problem and solve it by using unconstrained optimization methods. These are called *sequential unconstrained minimization techniques*.[1] The basic idea of these methods is to define an augmented functional by adding a penalty term to the cost function. The penalty term consists of the constraint functions multiplied by the penalty parameters. The penalty parameters are selected and the unconstrained function is minimized. Then the penalty parameters are increased and the unconstrained function is minimized again. The procedure is repeated until there is very little change in the solution. An advantage of the methods is that the unconstrained optimization algorithms and the associated software can be used to solve constrained problems. One drawback of the methods is that the penalty parameters are required to go to infinity to obtain an optimum solution. This can cause instability in numerical calculations.

To overcome difficulty of the foregoing methods, a different class of methods has been developed that do not require the penalty parameters to become infinite. The penalty parameters are required to be sufficiently large but finite. These are called the augmented Lagrangian methods or the *multiplier methods*. The augmented functional is defined as

$$\Phi = f(\mathbf{x}) + \frac{1}{2}\sum_{i=1}^{p} r_i (g_i + \theta_i)^2 + \frac{1}{2}\sum_{i=p+1}^{m} r_i \left[(g_i + \theta_i)_+\right]^2 \tag{30}$$

where $r_i > 0$ are the penalty parameters, θ_i for $i = 1$ to p are the multipliers for the equality constraints, $\theta_i \geq 0$ for $i > p$ are the multipliers for the inequality constraints, and $(x)_+ = x$ if $x > 0$, and $(x)_+ = 0$ if $x \leq 0$. The idea of multiplier methods is to start with some values for the parameters r_i and θ_i and minimize the augmented function of Eq. (30). These parameters are then adjusted using some procedure and the process is repeated until optimality conditions are satisfied. For more detailed discussion and applications of the methods, Refs. 10, 15, 19 and many works cited therein should be consulted.

It is important to note that the augmented functional, such as the one in Eq. (30), have been used as descent functions for many SQP methods to determine an appropriate step size along the search direction.[34]

5. 2. *Algorithms for Discrete Variable Problems*

The continuous variable optimization problem has infinite feasible points when the feasible set is nonempty. In contrast, the discrete variable problem has only a finite number of feasible points from which the optimum solution needs to be determined. However, it is more difficult and time consuming to find an optimum solution for the discrete variable problem compared to the continuous variable problem. The reason is that there are no optimality conditions to guide the numerical search process. We usually need to enumerate on the discrete points and use the definition of the minimum point in Eq. (6) to find the best solution. Many methods try to reduce this computational burden by using stochastic ideas or heuristic rules.

The solution algorithm for a mixed-discrete variable optimization problem depends on the type of problem. Five types of mixed variable problems are defined in Refs. 7, 11, 12 and 16 based on the characteristics of variables and problem functions. Also methods to solve the problems are identified. For example, if the problem functions are continuous and the discrete variables can have non-discrete values during the solution process, then gradient-based algorithms can be used to guide the search for a discrete optimum solution. If the problem functions are nondifferentiable and discrete variables must have only discrete values, then implicit or explicit enumeration methods or stochastic methods can be used to solve the problem.

There are basically two classes of methods for solving discrete variable problems: (i) enumeration methods, either implicit or explicit, such as the branch and bound algorithm, and (ii) stochastic or evolutionary methods, such as genetic algorithms and simulated annealing. Detailed review of the methods and their applications are presented in Refs. 7, 11, 12, 14, 16, 25-30. Here we summarize basic concepts and ideas of the methods from these references.

Branch and Bound Method. This is one of the most commonly used methods to solve discrete variable problems.[7,12,16] It is also called an implicit enumeration method because one systematically tries to reduce the entire enumeration. It was initially developed for LP problems for which a global solution is obtained. The method has also been applied to nonlinear problems for which there is no guarantee of optimum or even a feasible solution. The method uses the concepts of *branching, bounding* and *fathoming* to perform the search for the optimum solution. The solution space for the problem is represented as branches of an inverted tree. Each node of the tree represents a possible discrete solution. If the solution is infeasible, then either the branch is truncated if the cost function is higher than a previously established upper bound, or other branches are searched for a better solution from that node. A node is said to be *fathomed* if no better solution is possible with further branching from that node. When the solution at a node is feasible, it either represents a new upper bound for the optimum if the cost function is smaller than a previously established bound, or the node can be fathomed if no better solution is possible with further branching. The method can be implemented in two different ways. In the first one, non-discrete values for the discrete variables are not allowed during the solution process. Therefore enumeration on the discrete variables needs to be done as explained above. In the second implementation, non-discrete values for the variables are allowed. Forcing a variable to have a discrete value generates each node of the tree. This is done by defining a subproblem with appropriate constraints on the variable to force out a discrete value for the variable. The subproblem is solved using either LP or NLP methods.

Simulated Annealing. Simulated annealing (SA) is a stochastic method that can be used to find the global minimum for a mixed variable nonlinear problem.[7] The method does not require continuity or differentiability of the problem functions. The basic idea is to generate random points in a neighborhood of the current best point and evaluate the problem functions there. If the cost function (penalty function for constrained problems) value at any of those points is smaller than the current best value, then the point is accepted, and the best cost function value is updated. If it is not, then the point is sometimes accepted and sometimes rejected. The acceptance is based on the value of the probability density function of Bolzman-Gibbs distribution. If this probability density function has a value greater than a random number, then the trial point is accepted as the best solution. The probability density function uses a parameter called the temperature. For the optimization problem, this temperature can be the target value for the cost function. Initially, a larger target value is selected. As the trials progress, the target value is reduced (this is called the cooling schedule), and the process is terminated after a large number of trials. The acceptance probability steadily decreases to zero as the temperature is reduced. Thus in the initial stages, the method is likely to accept worse points while in the final stages, the worse points are usually rejected. This strategy avoids getting trapped at local minimizers. The main deficiencies of the method are the unknown rate at which the target level is to be reduced and uncertainty in the total number of trials.

Genetic Algorithms. As simulated annealing, these methods are also in the category of stochastic search methods.[35,37,40-44] In the methods, a set of alternative points (called the *population*) at an iteration (called *generation*) is used to generate a new set of points. In this process, combinations of the most desirable characteristics of the current members of the population are used that results in points that are better than the current ones. Thus, the average fitness of successive sets of points improves giving better values for the fitness function. Here fitness is defined using the cost function or the penalty function for constrained problems. The fitness value is calculated for each member of the population. An advantage of this approach is that derivatives of the functions are not needed. One starts with a set of randomly generated points. A finite length string, such as a binary string of 0's and 1's, is usually used to represent each point. Three operators are needed to implement the algorithm: (i) reproduction; (ii) crossover; and (iii) mutation. *Reproduction* is an operator where an old string (point) is copied into the new population according to its fitness. More highly fit strings (those points with smaller fitness values) receive higher numbers of offspring (new points). The crossover operator corresponds to allowing selected members (points) of the population to exchange characteristics of the points among themselves. Crossover entails selection of starting and ending positions on a pair of mating strings (points) at random and simply exchanging the string of 0's and 1's between these positions. *Mutation* corresponds to selection of a few members (points) of the population, determining a location on the strings at random, and switching the 0 to 1 or vice versa. The foregoing three steps are repeated for successive generations of the population until no further improvement in the fitness is attainable, or the number of generations reaches a specified limit. The member in this generation with the highest level of fitness is taken as the optimum point.

Integer Programming. The problem is called an integer programming (IP) problem when the variables are required to take on integer values. If all the functions are linear, an integer linear programming (ILP) problem is obtained, otherwise it is nonlinear. The ILP problem can be converted to 0-1 programming problem. Linear problems with discrete variables can also be converted to 0-1 programming problems. Many algorithms are available to solve such problems,[45] such as the branch and bound method discussed earlier.

Sequential Linearization Methods. Nonlinear discrete optimization problems can also be solved by sequential linearization procedures. The functions of the problem must be differentiable to use such a procedure. The nonlinear problem is first linearized at the current point. Then an ILP method is used to solve the linearized subproblem. A modification of this approach is to obtain a continuous optimum point first, and then linearize and use IP methods. This process can reduce the number of ILP problems to be solved. Restricting the number of discrete values to a neighborhood of the continuous solution can also reduce the size of the ILP problem.

Rounding-off Techniques. Rounding-off is a simple approach where an optimum solution is first obtained by assuming all the variables to be continuous. Then using heuristics, the variables are rounded-off to the nearest available discrete values to obtain a discrete solution. The procedure is applicable to a restricted class of problems where discrete variables can have non-discrete values during the solution process. The process may not result in a feasible point for the discrete variable problem. Note that it is not necessary to round-up all variables to their nearest discrete neighbors. Some of them could be rounded-down while others could be increased. The difficulty with this approach is in the selection of variables to be increased and the variables to be decreased. The strategy may not converge, especially in case of high nonlinearity and widely separated allowable discrete values. In that case, the discrete minimizer need not be in a neighborhood of the continuous solution. As an alternative, a dynamic rounding-off strategy has been used where only one variable is rounded-off to its discrete neighbor at a time. The selected variable is then fixed at the discrete value and the problem is optimized again. This process is repeated until all variables are selected and fixed to discrete values.

Neighborhood Search Method. Some times it is reasonable to enumerate on the discrete variables, especially when the number of variables is small. With all the discrete variables fixed at their chosen values, the problem is then optimized for the continuous variables. This approach has some advantages over BBM: it can be implemented easily with an existing NLP solver, the problem to be solved is smaller and the gradient information with respect to the discrete variables is not needed. However, in general, the approach is less efficient than an implicit enumeration method, such as the BBM, as the number of discrete variables and size of the discrete set of values become large. To reduce the number of enumerated cases, a neighborhood search method has been used which first obtains a continuous solution with all the discrete variables considered as continuous. Then only a few discrete values near the continuous solution are selected for explicit enumeration.

5.3. *Multiobjective Optimization*

There are many practical applications where we need to optimize two or more objective functions simultaneously. These are called *multiobjective, multi-criteria*, or *vector optimization* problems. Here, we give a brief introduction to the subject by describing some basic concepts, terminology and solution methods. Material for this section is derived from Refs. 7 and 32; for more details, references cited in there and many other sources can be consulted, such as Refs. 35-37, 44.

5.3.1. *Terminology and Basic Concepts*

The Problem P defined earlier is modified to multiobjective optimization problems as follows: find $\mathbf{x}$ S to minimize

$$\mathbf{f}(\mathbf{x}) = \left(f_1(\mathbf{x}), f_2(\mathbf{x}), \ldots, f_k(\mathbf{x})\right) \tag{31}$$

where k is the number of objective functions in the vector $\mathbf{f}(\mathbf{x})$. A collection of all the objective function vectors is called the *criterion space*. The feasible criterion space Z is defined as the set of objective function values corresponding to the feasible points in the variable space; i.e.,

$$Z = \{\mathbf{f}(\mathbf{x}) \,|\, \mathbf{x} \in S\} \tag{32}$$

Algorithms for solution of a single-objective optimization problem give local minima for the cost function in the feasible set. If all local minima are found, then a global minimum point can be identified. In contrast, the process of solving a multiobjective optimization problem is less definite. Usually this problem does not have a unique solution; i.e., there is no point **x** that minimizes all the objectives simultaneously. Therefore, it is not clear what is meant by the minimum of multiple objective functions. Usually, the objectives have opposing characteristics, since a point that decreases the value of one function may increase the value of another. However, there can be infinite solution points for the problem in the sense of *Pareto optimality*. This is the predominant concept in defining solutions for multiobjective optimization problems that is discussed next.

Pareto Optimal Points. A point $x^* \in S$ is *Pareto optimal* if and only if there does not exist another point $x \in S$ such that $f(x) \leq f(x^*)$ with at least one $f_i(x) < f_i(x^*)$. In other words, a point $x^* \in S$ is called Pareto optimal if there is no other point $x \in S$ that reduces at least one objective function without increasing another one. Pareto optimal points are also called *efficient points* of the feasible set S.

Non-dominated Points. Another common concept is that of *non-dominated* and *dominated* points of the feasible criterion space Z. A vector of objective functions $f^* = f(x^*) \in Z$ is *non-dominated* if and only if there does not exist another vector $f \in Z$ such that $f \leq f^*$ with at least one $f_i < f_i^*$. Otherwise, f^* is *dominated*.

Utopia Point. A vector of objective function values f° in the criterion space is called the utopia point if $f_i^\circ = \min\{f_i(x)| \text{ for all } x \in S\}$, $i = 1$ to k. It is also called the *ideal point*. Utopia point is a unique point in the criterion space that is obtained by minimizing each objective function without regard for other objective functions. Each minimization yields a point in the variable space and the corresponding value for the objective function. It is rare that each minimization will end up at the same point. That is, one point cannot simultaneously minimize all the objective functions. Thus, the utopia point exists only in the criterion space and, in general, is not attainable in the variable space.

Compromise Solution. Since the utopia point is not attainable, the next best thing is a solution that is as close as possible to the utopia point. Such a solution is called a *compromise solution*. The methods that seek different compromise solutions are collectively called *compromise programming*.

5.3.2. *Solution Methods*

Since the multiobjective optimization problem has infinite solutions (the *Pareto optimal set*), the user needs to select a solution that suits the requirements of the application. Therefore we may need to generate the entire Pareto set or at least a good representation of it so that the user can select the desired solution. Most solution methods for multiobjective optimization problems combine various objective functions to define a composite scalar function for the problem. This way, a single-objective optimization method can be used to solve the problem. By varying parameters of the composite function, different optimum solutions for the problem can be generated. Some methods always yield Pareto optimal solutions but may skip certain points in the Pareto optimal set; i.e., they may not be able to capture all of the Pareto optimal points. Alternatively, other methods are able to capture all of the points in the Pareto optimal set but may also provide non-Pareto optimal points as well. The former quality is beneficial when one is interested in using a method to obtain just one solution point. The latter quality is useful when the complete Pareto optimal set needs to be generated.

Weighted Sum Method. The weighted sum method is the most common approach to multiobjective optimization. Each objective function is scaled by a weighting factor $w_i > 0$ as $w_i f_i(\mathbf{x})$. Then all the objective functions are added together to form a composite objective function to be optimized:

$$U = \sum_{i=1}^{k} w_i f_i(\mathbf{x}) \tag{33}$$

The objective functions are usually normalized before the weights are assigned to them. The relative value of the weights generally reflects the relative importance of the objectives. This is another common characteristic of the weighted sum methods. If all of the weights are omitted or are set to one, then all objectives are treated equally. The weights can be used in two ways. The user may either set w_i to reflect preferences before the problem is solved, or systematically alter them to yield different Pareto optimal points (generate the Pareto optimal set). The method is quite easy to use; selection of proper weights is the most difficult part that requires thorough knowledge of the objective functions and their relative importance.

Weighted Global Criterion. A broader class of weighted sum methods is based on weighted global criterion which is defined as:

$$U = \left\{ \sum_{i=1}^{k} \left[w_i \left(f_i(\mathbf{x}) - f_i^\circ \right) \right]^p \right\}^{1/p} \tag{34}$$

The root $1/p$ may be omitted because the formulations with and without the root theoretically provide the same solution. The solution with this formulation depends on the values of both w_i and p. Generally, p is proportional to the amount of emphasis placed on minimizing the function with the largest difference between $f_i(\mathbf{x})$ and f_i°. Larger p puts more emphasis on minimizing the largest difference. p and w_i typically are not varied or determined in unison. Rather, a fixed value for p is selected, and then, either w_i is selected to reflect preferences before the problem is solved, or it is systematically altered to yield different Pareto optimal points. For computational efficiency or in cases where the utopia point f_i° may be difficult to determine, one may replace f_i° with an approximate value for it in Eq. (34). The approximation for f_i° is called an *aspiration point, reference point, goal,* or *target point*. When this is done, U is called an *achievement function*.

The global criterion reduces to other common methods with different values of p. For instance, when $p = 1$, Eq. (34) is similar to a weighted sum with the objective functions adjusted with the utopia point. When $p = 2$ and weights equal to 1, Eq. (34) represents the distance of the current point $f_i(\mathbf{x})$ from the utopia point, and the solution usually is called *compromise solution* as mentioned earlier. When $p = \infty$, Eq. (34) reduces to the well known min-max method.

Other Methods. There are other useful methods that reduce the multiobjective optimization problem to a single-objective optimization problem:

- *Lexicographic method* where the objective functions are arranged in the order of their importance and a sequence of optimization problems is solved: minimize $f_i(\mathbf{x})$ subject to $f_j(\mathbf{x}) \le f_j(x_j^*)$; $j = 1$ to $(i-1)$; $i > 1$; $i = 1$ to k. The process is stopped when two consecutive problems have same solution.

- The *ε-constraint method* minimizes a single, most important objective function $f_s(\mathbf{x})$ with other objective functions treated as constraints: $f_i(\mathbf{x}) \le \varepsilon_i$; $i = 1$ to $k; i \ne s$, where ε_i is the upper limit for the objective function $f_i(\mathbf{x})$. A systematic variation of ε_i yields a set of Pareto optimal solutions.

- *Goal programming* approaches set goals b_j for each objective function $f_j(\mathbf{x})$. Then, the total deviation from the goals is minimized. In the absence of any other information, goals may be set to the utopia point, i.e., $b_j = f_j^\circ$. In that case, the method becomes a special case of the global criterion method.

Besides the scalarization methods discussed in the foregoing paragraphs, there are methods that treat all the objective functions at the same time and generate the Pareto optimal set for the problem. A prominent method in this class is the *genetic algorithm* for multiobjective optimization problems.[35-37,44] This method is an extension of the genetic algorithms described earlier for single objective problems. Additional genetic operators are used to generate the new population for the next generation. For each generation, a possible set of Pareto optimal points for the problem is identified. These play a major role in generating new points for the next generation. The iterative process is repeated for a long period of time. At the end, an approximation to the Pareto optimal set is obtained. Since genetic algorithms do not require gradient information, they can be effective regardless of the nature of the problem functions.

5.4. *Algorithms for Global Solution*

Thus far, we have addressed mainly the problem of finding a local minimum for the cost function. However, in some practical applications, it is important to find globally optimum solutions as opposed to the local ones. The question of when a local solution is also a global optimum is quite difficult to answer because there are no mathematical conditions that characterize a global solution, except for convex programming problems, as discussed earlier. Therefore even when a global solution has been found, it is not possible to recognize it. Due to this reason, it is impossible to define a precise stopping criterion for a computational algorithm for global optimization. Usually, the best solution obtained by an algorithm after it is allowed to run for a long time is accepted as the global solution for the problem. In general, the quality of the solution depends on how long the algorithm is allowed to run. It is important to note that the computational effort to solve a global optimization problem increases enormously as the number of design variables increase. Thus, it remains a challenge to solve the global optimization problem efficiently.

In this section, we present some basic concepts of procedures that can be used to calculate a global solution. We consider the problem with continuous variables and functions. For discrete and nondifferentiable problems, the simulated annealing and genetic algorithms, described earlier, can be used for global optimization. In general, global optimization methods can be divided into two major categories: *deterministic* and *stochastic*. This classification is based on whether or not they incorporate any stochastic procedures to solve the global optimization problem. In the following subsections, we describe basic concepts of some of the methods in both of these categories. The material is derived from the work of the author and his co-workers.[7,17,22,] Numerous other references cited in these articles can be consulted for more details; e.g., Refs. 46-52.

5.4.1. *Deterministic Methods*

An exhaustive search of the feasible set S is performed in these methods to find the global minimum. The success of the method can be guaranteed for only the functions that satisfy certain conditions. We shall describe basic ideas of four deterministic methods: covering, zooming, generalized descent and tunneling methods.

Covering Methods. The basic idea of these methods is to cover the entire feasible set S by evaluating the cost function at all the points in order to search for a global minimum.[46] This is an enormous calculation and therefore all the covering methods try to implicitly cover the entire set by evaluating the functions at some selected points. Some methods exploit certain properties of the cost function to accomplish this objective. Covering methods have been used mainly to solve two variable problems because for 3 and more variables, the number of computations becomes very large.

Zooming Method. This method uses a target value for the global minimum of the cost function which is imposed as a constraint in the solution process.[7,22] Once the target is achieved, it is reduced further to zoom-in on the global minimum. The method combines a local minimization method with successive truncation of the feasible set S. The basic idea is that once a local minimum point has been found, the problem is redefined in such a way that the current solution is eliminated from any further search by adding the constraint $f(\mathbf{x}) \leq rf(\mathbf{x}^*)$, where $f(\mathbf{x}^*)$ is the cost function value at the current minimum point and $0 < r < 1$ if $f(\mathbf{x}^*) > 0$, and $r > 1$ if $f(\mathbf{x}^*) < 0$. The redefined problem is solved again and the process is continued until no more minimum points can be found. The method has a drawback in that as the target level for the global minimum is lowered, the feasible set for the problem shrinks and may even become disjointed. Therefore as the global minimum is approached, finding even a feasible point for the re-defined problem becomes time consuming.[21]

Methods of Generalized Descent. These methods are generalization of the descent methods where finite descent steps are taken along the search directions (i.e., straight lines). In those methods, it is sometimes difficult to find a suitable step size along the search direction. Therefore, it may be more effective if we deliberately follow a curvilinear path (trajectory) in the design space. The curvilinear paths are generated by integrating certain first or second order differential equations. The differential equations use the function values and its gradient along the trajectories. The search for the global minimum is based on solution properties of these differential equations. An important property is that their trajectories pass through majority of the stationary points for the cost function. There are conditions that can determine whether or not the trajectory will pass through all the local minimum points. In that case, the global minimum is guaranteed to be found. The methods have been used for problems with only a few variables.

Tunneling Method. The basic idea of the tunneling method is to execute the following two phases iteratively until some stopping criterion is satisfied: the local minimization phase and the tunneling phase. The method was initially developed for unconstrained problems and then extended for constrained problems.[48] A local minimum $x*$ for the problem is calculated in phase one. The tunneling phase determines a new starting point for phase one that is different from $x*$ but has cost function value smaller than or equal to the known minimum value. The tunneling phase is accomplished by finding a root of the nonlinear *tunneling function*, $T(x)$. This function is defined in such a way that it avoids previously determined local minima and the starting points. The two phases are repeated until no suitable roots of the tunneling function can be found. This is realized numerically when $T(x) \geq 0$ for all x. This problem is difficult to solve efficiently because finding a suitable point in the tunneling phase is in itself a global optimization problem.

5.4.2. *Stochastic Methods*

Most stochastic methods depend on random processes to search for the global minimum point. Some methods are useful for only continuous variable problems while others can be used for all types of problems. These methods are some variation of the pure random search. They try to reduce its computational burden. Pure random search evaluates $f(x)$ at N sample points drawn from a random uniform distribution over the feasible set. The smallest function value found is the candidate global minimum for $f(x)$. The sample size N must be quite large in order to get a good estimate of the global solution. Therefore the method is quite inefficient due to the large number of function evaluations. *Single start method* is a simple extension of the method in which a single local search is performed starting from the best point found in the random search.

The stochastic ideas are used in two ways in these methods: (i) to decide stopping criteria for the methods, and (ii) to develop techniques to approximate the *region of attraction* for a local minimum point. The goal of many stochastic methods is to develop good approximations for the regions of attraction for local minima so that the search for that local minimum is performed only once.

Some stochastic methods try to determine all local minima for the function. Then, the best local minimum is claimed as the global minimum point. One difficulty is that the number of local minima for the problem is not known a priori. Therefore it is difficult to determine when to end the search for local minima. Usually a statistical estimate for the number of local minima is used in practice. The methods usually have two phases: a global phase and a local phase. In the *global phase*, the function is evaluated at a number of randomly sampled points. In the *local phase*, local searches are performed from the sample points to yield candidate global minima. The global phase is necessary because just a local strategy cannot give a global minimum. There are many stochastic methods for global optimization, such as multistart, clustering, controlled random search, simulated annealing, acceptance-rejection, stochastic integration, and genetic algorithms. We shall describe only the basic ideas of some of the methods. More details can be found in Refs. 7 and 17 and works cited therein. It is important to note that since some stochastic methods use random processes, an algorithm run at different times can generate different iteration histories and local minima. Therefore, a particular problem needs to be run several times before the solution is accepted as the global optimum.

Multistart Method. The basic idea of multistart methods is to perform search for a local minimum from each sample point. The best local minimum point found is taken as the global minimum. The stopping criterion for the method is based on a statistical estimate of the number of local minima for the problem. The method is reliable but it is not efficient since many sample points lead to the same local minimum. Therefore, strategies to eliminate this inefficiency in the algorithm have been developed.

Clustering Methods. The basic idea of clustering methods is to remove inefficiency of the multistart method by trying to use the local search procedure only once for each local minimum point.[51] The random sample points are linked into groups to form clusters. Each cluster is considered to represent one region of attraction such that a search initiated from any point in the region converges to the same local minimum point. Four clustering methods have been used for development of the regions of attraction: density clustering, single linkage, mode analysis, and vector quantization multistart.

Controlled Random Search. The controlled random search has both global and local phases in its algorithm. It uses the idea of a *simplex* which is a geometric figure formed by a set of $n+1$ points in the n-dimensional space (n is the number of variables). In two dimensions, the simplex is just a triangle and in three dimensions, it is a tetrahedron. The method does not use gradients of the cost function and so continuity of the functions is not required. In the global phase, one starts with $n+1$ sample points. The worst point (having the largest value for the cost function) is replaced by a trial point evaluated using the centroid for the n sample points including the worst point. If the trial point is feasible and has better cost function value, then it replaces the worst point of the selected set. Otherwise, the process is repeated until a better point is found. In the local phase, the worst point among the current $n+1$ sample points is reflected about the centroid of the simplex. The point is then expanded or contracted to obtain a better point. The worst point is replaced by this point. The two phases are repeated until a stopping criterion is satisfied.

Acceptance-Rejection Methods. The acceptance-rejection methods use ideas from statistical mechanics to improve efficiency of the multistart algorithm.[49] The strategy is to start the local minimization procedure only when the randomly generated point has smaller cost function value than that of the local minimum previously obtained. This forces the algorithm to tunnel below the local minima in search for a global minimum. This modification, however, has been shown to be inefficient, and therefore the tunneling process has been pursued only by means of deterministic algorithms, as explained earlier. The acceptance-rejection based methods modify this tunneling procedure which is sometimes called *random tunneling*. The idea of acceptance phase is to some times start local minimization from a randomly generated point even if it has a higher cost function value than that at a previously obtained local minimum. This involves calculation of certain probabilities. If the local minimization procedure started from an accepted point produces a local minimum that has higher cost function value than a previously obtained minimum, then the new minimum point is rejected. This is called the rejection phase.

Stochastic Integration. In these methods, a stochastic perturbation of the system of differential equations for the trajectory methods is introduced in order to force the trajectory to a global minimum point. This is achieved by monitoring the cost function value along the trajectories. By changing some coefficients in the differential equations we get different solution processes starting from the same initial point. This idea is similar to simulated annealing but here a parameter in the differential equation is decreased continuously.

6. Concluding Remarks

Basic concepts and terminology used for optimization of structural and mechanical systems are described. Various types of optimization models are presented and discussed. Optimality conditions for continuous variable optimization problems are presented. Concept related to algorithms for continuous variable optimization problems are presented and discussed. Basic concepts of methods for discrete variable, multiobjective and global optimization problems are described. The material is introductory in nature, and so, several references are cited for readers interested in more in-depth study of various topics.

References

1. A.V. Fiacco and G.P. McCormick, *Nonlinear Programming: Sequential Unconstrained Minimization Techniques* (Society for Industrial and Applied Mathematics, Philadelphia, 1968).
2. P.E. Gill., W. Murray and M.H. Wright, *Practical Optimization* (Academic Press, New York, 1981).
3. D.G. Luenberger, *Linear and Nonlinear Programming* (Addison-Wesley, MA, 1984).
4. B.N. Pshenichny and Y.M. Danilin, *Numerical Methods in Extremal Problems* (Mir Publishers, Moscow, 1982).
5. A.D. Belegundu and T.R. Chandrupatla, *Optimization Concepts and Applications in Engineering* (Prentice Hall, New Jersey, 1999).
6. E.J. Haug and J.S. Arora, *Applied Optimal Design* (John Wiley, New York, 1979).
7. J.S. Arora, *Introduction to Optimum Design,* (Elsevier Academic Press, Boston, 2004).
8. J.S. Arora, in *Advances in Structural Optimization*, Ed. J. Herskovits (Kluwer Academic Publishers, Boston, 1995), p. 47.
9. J.S. Arora, Ed, *Guide to Structural Optimization* (American Society of Civil Engineering, Reston 1997).
10. J.S. Arora, in *Structural Dynamic Systems: Computational Techniques and Optimization*, Ed. C.T. Leondes (Gordon and Breech Publishers, Newark, 1999), p. 1.
11. J.S. Arora, in *Proceedings of the 14th Analysis and Computation Conference* (American Society of Civil Engineers, Reston, 2000).
12. J.S. Arora, in *Recent Advances in Optimal Structural Design*, Ed. S. Burns (American Society of Civil Engineers, Reston, 2002), p. 1.
13. J.S. Arora and E.J. Haug, *AIAA J.,* 17 (1979).

14. J.S. Arora and M.W. Huang, *Struc. Mech. Earth. Eng.*, 113 (1996).
15. J.S. Arora, A.I. Chahande, and J.K. Paeng, *Num. Meth. in Eng.*, 32 (1991).
16. J.S. Arora, M.W. Huang and C.C. Hsieh, *Str. Mult. Opt.*, 8 (1994).
17. J.S. Arora, O.A. Elwakeil, A.I. Chahande and C.C. Hsieh, *Str. Mult. Opt.*, 9 (1995).
18. A.D. Belegundu and J.S. Arora, *Num. Meth. in Eng.*, 20 (1984).
19. A.D. Belegundu and J.S. Arora, *AIAA J.*, 22 (1984).
20. A.D. Belegundu and J.S. Arora, *Num. Meth. in Eng.*, 21 (1985).
21. O.A. Elwakeil and J.S. Arora, *AIAA J.*, 33 (1995).
22. O.A. Elwakeil and J.S. Arora, *Num. Meth. in Eng.*, 39 (1996).
23. C.C. Hsieh and J.S. Arora, *Comp. Meth. Appl. Mech. Eng.*, 43 (1984).
24. M.W. Huang and J.S. Arora, *Num. Meth. in Eng.*, 39 (1996).
25. M.W. Huang and J.S. Arora, *Num. Meth. in Eng.*, 40 (1997).
26. M.W. Huang and J.S. Arora, *Str. Mult. Opt.*, 14 (1997).
27. M.W. Huang, C.C. Hsieh and J.S. Arora, *Num. Meth. in Eng.*, 40 (1997).
28. F.Y. Kocer and J.S. Arora, *J. of Str. Eng.*, 122 (1996), p. 804.
29. F.Y. Kocer and J.S. Arora, *J. of Str. Eng.*, 122 (1996), p. 1347.
30. F.Y. Kocer and J.S. Arora, *J. of Str. Eng.*, 123 (1997).
31. O.K. Lim and J.S. Arora, *Comp. Meth. Appl. Mech. Eng.*, 57 (1986).
32. T.R. Marler and J.S. Arora, *Str. Mult. Opt.*, 26 (2004).
33. P.B. Thanedar, J.S. Arora, C.H. Tseng, O.K. Lim and G.J. Park, *Num. Meth. in Eng.*, 23 (1987).
34. C.H. Tseng and J.S. Arora, *Num. Meth. in Eng.*, 26 (1988).
35. C.A. Coello-Coello, D.A. Van Veldhuizen and G.B. Lamont, *Evolutionary Algorithms for Solving Multi-Objective Problems* (Kluwer Academic Publishers, Boston, 2002).
36. K. Deb, *Multi-Objective Optimization using Evolutionary Algorithms* (John Wiley, New York, 2001).
37. A. Osyczka, *Evolutionary Algorithms for Single and Multicriteria Design Optimization* (Physica-Verlag, Germany, 2002).
38. E.K.P. Chong and S.H. Żak, *An Introduction to Optimization* (John Wiley, New York, 2001).
39. C.H. Tseng, L.W. Wang and S.F. Ling, *J. of Str. Eng.*, 121 (1995).
40. J.H. Holland, *Adaptation in Natural and Artificial System* (University of Michigan Press, Ann Arbor, 1975).
41. D.E. Goldberg, *Genetic Algorithms in Search, Optimization and Machine Learning* (Addison-Wesley, Reading, 1989).
42. M. Mitchell, *An Introduction to Genetic Algorithms* (MIT Press, Cambridge, 1996).
43. S. Pezeshk, and C.V. Camp, in *Recent Advances in Optimal Structural Design*, Ed. S. Burns (American Society of Civil Engineers, Reston, 2002).
44. M. Ehrgott and X. Gandibleux, Eds., *Multiple Criteria Optimization: State of the Art Annotated Bibliographic Surveys* (Kluwer Academic Publishers, Boston, 2002).
45. A. Schrijver, *Theory of Linear and Integer Programming* (John Wiley, New York, 1986).
46. Y.G. Evtushenko, *Numerical Optimization Techniques* (Optimization Software, New York, 1985).
47. C.A. Floudas, P.M. Pardalos, et al, *Handbook of Test Problems in Local and Global Optimization* (Kluwer Academic Publishers, Boston, 1999).

48. A.V. Levy and S. Gomez, in *Numerical Optimization 1984*, Eds. P.T. Boggs, R.H Byrd and R.B. Schnabel (Society for Industrial and Applied Mathematics, Philadelphia, 1985).

49. S. Lucidi and M. Piccioni, *Opt. Th. Appl.*, 62 (1989).

50. P.M. Pardalos, A. Migdalas and R. Burkard, *Combinatorial and Global Optimization* (World Scientific Publishing, New Jersey, 2002).

51. A.H.G. Rinnooy Kan and G.T. Timmer, *Math. Prog.*, 39 (1987).

52. A. Törn and A. Žilinskas, *Global Optimization* (Springer-Verlag, Germany, 1989).

CHAPTER 2

OPTIMIZATION OF LARGE SCALE SYSTEMS

José Herskovits, Evandro Goulart and Miguel Aroztegui

Mechanical Engineering Program
COPPE - Federal University of Rio de Janeiro
Caixa Postal 68503, 21945 970 Rio de Janeiro, Brazil
E-mail: jose@optimize.ufrj.br

Numerical algorithms for real life engineering optimization must be strong and capable of solving very large problems with a small number of simulations and sensitivity analysis. In this chapter we describe some numerical techniques to solve engineering problems with the Feasible Arc Interior Point Algorithm (FAIPA) for nonlinear constrained optimization. These techniques include quasi- Newton formulations that avoid the storage of the approximation matrix. They include also numerical algorithms to solve in an efficient manner the internal linear systems of FAIPA. Numerical results with large size test problems and with a structural optimization example shows that FAIPA is strong an efficient for large size optimization.

1. Introduction

The engineering optimization task consists in finding the design variables $x_1, x_2, ..., x_n$ that

$$\text{minimize } f(x)$$
$$\text{subject to } g(x) \leq 0 \tag{1}$$
$$\text{and } h(x) = 0,$$

where $x \equiv [x_1, x_2, ..., x_n]^t$, the scalar function $f(x)$ is the objective function and $g(x) \equiv [g_1(x), g_2(x), ..., g_m(x)]^t$ and $h(x) \equiv [h_1(x), h_2(x), ..., h_p(x)]^t$ represent inequality and equality constraints. We assume that $f(x)$, $g(x)$ and $h(x)$ are continuous in $\Re^n$ as well as their first derivatives. In engineering applications most of these functions are nonlinear. Then, (1) is a smooth nonlinear constrained mathematical programming problem.

Real life engineering systems involve a very large number of design variables and constraints. Evaluation of functions and of derivatives coming from engineering models is very expensive in terms of computer time. In practical applications, calculation and storage of second derivatives are impossible to be carried out. Then, numerical techniques for engineering optimization must be capable of solving very large problems with a reasonable number of function evaluations and without need-ing second derivatives. Robustness is also a crucial point for industrial applications.

Quasi-Newton method creates an approximation matrix of second deriva-tives.[13,25,43,48,50] With this method large problems can be solved in a reasonable number of iterations. Employing rank two updating rules, like BFGS or DFP, it is possible to obtain positive definite approximation matrices. This is a requirement of optimization algorithms that include a line search procedure[24,25,26,43,48,50] to ensure global convergence. However, the classic quasi-Newton method cannot be applied for large problems since it requires the calculus and storage of approximation matrices, which are always full.

Limited memory quasi-Newton method avoids the storage of the approximation matrix.[11,42,47,48] Positive definite matrices can also be obtained with this technique. It was first developed for unconstrained optimization and then extended to prob-lems with side constraints. Employing the Feasible Arc Interior Point Algorithm (FAIPA), the limited memory method can also be applied for constrained optimiza-tion problems.[19,31,32,44]

Another approach to solve large problems with a quasi-Newton technique con-sists in obtaining sparse approximation matrices. This idea was first exploited by Toint in the 70th[56,57,58] and by Fletcher et al. in the 90th.[17,18] In both cases sparse matrices were obtained in a very efficient way. However, those methods cannot be applied for optimization algorithms with a line search, since it is not guaranteed that the approximation matrices are positive definite. In booth cases, the authors worked with a trust region algorithm, but the numerical results were poor.

The numerical techniques described in this chapter are based on the Feasible Arc Interior Point Algorithm (FAIPA)[34] for nonlinear constrained optimization. FAIPA, that is an extension of the Feasible Directions Interior Point Algorithm,[22,23,24,26,49] integrates ideas coming from the modern Interior Point Algorithms for Linear Pro-gramming with Feasible Direction Methods. At each point, FAIPA defines a "Fea-sible Descent Arc". Then, it finds a new interior point on the arc, with a lower objective. Newton, quasi - Newton and first order versions of FAIPA can be ob-tained.

FAIPA is supported by strong theoretical results. Global convergence to a lo-cal minimum of the problem is proved with relatively weak assumptions. The search along an arc ensures superlinear convergence for the quasi - Newton version, even when there are highly nonlinear constraints, avoiding the so called "Maratos' effect"[45]. FAIPA, that is simple to code, does not require the solution of quadratic programs and it is not a penalty or a barrier method. It merely requires the solution of three linear systems with the same matrix per iteration. This one includes the second derivative of the Lagrangian, or a quasi - Newton approximation. Several practical applications of the present and previous versions of FAIPA, as well as several numerical results, show that FAIPA constitutes a very strong and efficient technique for engineering design optimization,[1,2,4,5,6,7,8,27,28,29,30,38,39,40,41,54] and also for structural analysis problems with variational inequalities.[3,60,62]

The main difficulty to solve large problems with FAIPA comes from the size and sparsity of the internal linear systems of equations. Since the quasi-Newton matrix is included in the systems, limited memory and sparse quasi - Newton techniques can produce important reductions of computer calculus and memory requirements.

In this chapter we present a new sparse quasi - Newton method that works with diagonal positive definite matrices and employ this technique for constrained optimization with FAIPA. This approach can be employed also in the well known sequential quadratic programming algorithm (SQP)[25,51,52] or in the interior point methods for nonlinear programming, as primal-dual or path following algorithms.[20,48,61] We also describe numerical techniques, to solve large problems with FAIPA, employing exact or iterative linear solvers and sparse or limited memory quasi-Newton formulations.

Quasi - Newton method is described in the next section, including limited memory formulation and our proposal for sparse quasi - Newton matrices. FAIPA is described in Sec. 3 and the structure of the internal systems and some numerical techniques to solve them are discussed in Sec. 4. Numerical experiments with a set of test problems are reported in Sec. 5, followed with some results in structural optimization. Finally, we present our conclusions in the last section.

2. Quasi-Newton Method for Nonlinear Optimization

We consider now the unconstrained optimization problem

$$\text{minimize } f(x); x \in \Re^n \tag{2}$$

Modern iterative algorithms define, at each point, a descent direction of $f(x)$ and make a line search looking for a better solution. The quasi - Newton method works with a matrix that approximates the Hessian of the objective function or its inverse. The basic idea is to build the quasi - Newton matrix with information gathered while the iterations progress.

Let the symmetric matrix $B^k \in \Re^{n \times n}$ be the current approximation of $\nabla^2 f(x^k)$. An improved approximation B^{k+1} is obtained from

$$B^{k+1} = B^k + \Delta B^k. \tag{3}$$

Since

$$\nabla f(x^{k+1}) - \nabla f(x^k) \approx [\nabla^2 f(x^k)](x^{k+1} - x^k),$$

the basic idea of quasi - Newton method consist in taking ΔB^k in such way that

$$\nabla f(x^{k+1}) - \nabla f(x^k) = [B^{k+1}](x^{k+1} - x^k), \tag{4}$$

called *"secant condition"*, is true.

Let

$$\delta = x^{k+1} - x^k \text{ and } \gamma = \nabla f(x^{k+1}) - \nabla f(x^k).$$

Then, (4) is equivalent to

$$\gamma = B^{k+1}\delta. \tag{5}$$

The substitution of (3) into (5) gives us n conditions to be satisfied by ΔB^k. Since $\Delta B^k \in \Re^{n \times n}$, the secant condition is not enough to determine B^{k+1}. Several updating rules for B^{k+1} were proposed.[13,43,52] The most successful is the BFGS (Broyden, Fletcher, Shanno, Goldfarb) formula

$$B^{k+1} = B^k + \frac{\gamma\gamma^t}{\delta^t\gamma} - \frac{B^k\delta\delta^t B^k}{\delta^t B^k \delta}. \tag{6}$$

If B^k is positive definite, it can be proved that

$$\delta^t\gamma > 0 \tag{7}$$

is a sufficient condition to have B^{k+1} positive definite. Under certain assumptions about $f(x)$, (7) is satisfied if an appropriate line search procedure is employed.[25]

A quasi-Newton algorithm can then be stated as follows:

ALGORITHM 1.

Data. Initial $x^0 \in \Re^n$ and $B^0 \in \Re^{n \times n}$ symmetric and positive definite. Set $k = 0$.

Step 1. Computation of the search direction $d^k \in \Re^n$, by solving the linear system

$$B^k d^k = -\nabla f(x^k) \tag{8}$$

Step 2. Line search

Find a step length t^k that reduces $f(x)$, according to a given line search criterium.

Step 3. Updates

Take

$$x^{k+1} := x^k + t^k d^k$$

$$B^{k+1} := B^k + \Delta B^k$$

$$k := k + 1$$

Step 4. Go back to Step 1.   □

Working with an approximation of the inverse, $H^k \approx [\nabla^2 f(x)]^{-1}$, is advantageous since it allows the search direction d^k to be calculated with a simple matrix-vector multiplication.

We have that the secant condition (5) is equivalent to $\delta = H^{k+1}\gamma$. Thus, an updating rule for H can be easily obtained by interchanging B and H as well as δ and γ in (6). We have

$$H^{k+1} = H^k + \frac{\delta\delta^t}{\delta^t\gamma} - \frac{H^k\gamma\gamma^t H^k}{\gamma^t H^k\gamma}, \tag{9}$$

called DFP (Davidon, Fletcher, Powell) updating rule. In general, the approximation matrix H^{k+1} that is obtained with this rule is not the inverse of B^{k+1} given by BFGS rule. An expression for H^{k+1} corresponding to the BFGS rule can be obtained from (6) by computing the inverse of B^{k+1} employing the Sherman - Morrison - Woodbury formula,[13]

$$H^{k+1} = H^k + \left(1 + \frac{\gamma^t H^k\gamma}{\gamma^t\delta}\right)\frac{\delta\delta^t}{\delta^t\gamma} - \frac{\delta\gamma^t H^k + H^k\gamma\delta^t}{\gamma^t\delta}. \tag{10}$$

2.1. *Limited Memory Quasi-Newton Method*

With the limited memory formulation, the product of the quasi-Newton Matrix H^{k+1} times a vector $v \in \Re^n$, or a matrix, can be efficiently computed without the explicit assembly and storage of H^{k+1}. It is only required the storage of the q last pairs of vectors δ and γ. In particular, this technique can be employed for the computation of the search direction in a quasi - Newton algorithm for unconstrained optimization.

The updating rule (10) for H can be expressed as follows:

$$H^{k+1} = H^{k-q} + [\Delta \quad H^{k-q}\Gamma]E[\Delta \quad H^{k-q} \tag{11}$$

where

$$\Delta = [\delta^{k-q}, \delta^{k-q+1}, \delta^{k-q+2}, ..., \delta^{k-1}]; \; \Delta \in \Re^{n \times q}$$

$$\Gamma = [\gamma^{k-q}, \gamma^{k-q+1}, \gamma^{k-q+2}, ..., \gamma^{k-1}]; \; \Gamma \in \Re^{n \times q}$$

$$E = \begin{bmatrix} R^{-t}(D + \Gamma^t H^{k-q}\Gamma)R^{-1} & -R^{-t} \\ -R^{-1} & 0 \end{bmatrix}; \; E \in \Re^{2q \times 2q}$$

$$R = \text{upper}(\Delta^t\Gamma); \; R \in \Re^{q \times q}$$

$$D = \text{diag}(R)$$

We write $A = upper(B)$ when $A_{ij} = B_{ij}$ for $j \geq i$ and $A_{ij} = 0$ for $j < i$.

The limited memory method takes $H^{k-q} = I$. Then, the following expression for $H^{k+1}v$ is obtained:

$$H^{k+1}v = v + [\Delta \quad \Gamma]E[\Delta \quad \Gamma]^t v. \tag{12}$$

This formulation is very strong and efficient for unconstrained optimization. Even in very large problems, taking $q \approx 10$, the number of iterations employing the Limited Memory method is quite similar to the original quasi - Newton Algorithm.

2.2. *Sparse Quasi-Newton Matrices*

The technique proposed by Toint, works with quasi-Newton matrices having the same sparsity as the real second derivative of the function. The new matrix $\boldsymbol{B}^{k+1}$ is the symmetric matrix, closest to $\boldsymbol{B}^k$ and with the prescribed sparsity, that satisfies the secant condition (4). Thus, $\boldsymbol{B}^{k+1}$ is the solution of the following constrained optimization problem

$$
\begin{aligned}
&\text{minimize } \|\boldsymbol{B}^{k+1} - \boldsymbol{B}^k\|_F^2 \\
&\text{subject to } \boldsymbol{B}^{k+1}\delta = \gamma, \\
&\qquad\qquad \boldsymbol{B}^{k+1} = [\boldsymbol{B}^{k+1}]^t \\
&\text{and } \boldsymbol{B}_{ij}^{k+1} = 0 \text{ for } (i,j) \in \boldsymbol{I}^s,
\end{aligned}
\tag{13}
$$

$\|\boldsymbol{M}\|_F = \sqrt{\displaystyle\sum_{i=1}^{n}\sum_{j=1}^{n} M_{ij}^2}$ is the Frobenius norm of $\boldsymbol{M} \in \Re^{n \times n}$ and the set $\boldsymbol{I}^s$ defines the required structure of the quasi-Newton matrix.

Toint obtained $\boldsymbol{B}^{k+1}$ by solving a linear system with the same structure $\boldsymbol{I}^s$. However, $\boldsymbol{B}^{k+1}$ is not guaranteed to be positive definite.

The method proposed by Fletcher et al. relaxes the secant condition and works with the set of the q previous pairs of vectors $\{\delta^i, \gamma^i\}$; for $i = k, k-1, k-2, ..., k-q+1$, as well as in the limited memory method.

Let be

$$
\Delta^k = [\delta^k, \delta^{k-1}, \delta^{k-2}, ..., \delta^{k-q+1}]
$$

and

$$
\Gamma^k = [\gamma^k, \gamma^{k-1}, \gamma^{k-2}, ..., \gamma^{k-q+1}],
$$

where $\Delta^k, \Gamma^k \in \Re^{n \times q}$. The following optimization problem defines $\boldsymbol{B}^{k+1}$

$$
\begin{aligned}
&\text{minimize } \|\boldsymbol{B}^{k+1}\Delta^k - \Gamma^k\|_F^2 \\
&\text{subject to } \boldsymbol{B}^{k+1} = [\boldsymbol{B}^{k+1}]^t \\
&\text{and } \boldsymbol{B}_{ij}^{k+1} = 0 \text{ for } (i,j) \in \boldsymbol{I}^s,
\end{aligned}
\tag{14}
$$

Since the secant condition is relaxed, this problem has a solution whatever it is the structure of B^{k+1}. Then, any sparse structure can be chosen, even in the case when the second derivative matrix is full. This approach does not require the storage of B^k. However B^{k+1} is not ensured to be positive definite, as in Toint's method.

2.3. *Diagonal Quasi-Newton Matrices*

We present a new approach[21] based on the previous formulation, but employing a structure for the quasi- Newton matrix such that checking if this one is positive definite becomes easy. This check is then included as a constraint of the optimization problem. The most simple case is that one in which the approximation matrices are diagonal.

We define the following problem

$$\begin{aligned} \text{minimize} \quad & \|\boldsymbol{B}^{k+1}\Delta^k - \Gamma^k\|_F^2 \\ \text{subject to} \quad & \boldsymbol{B}_{ii}^{k+1} \geq \varepsilon \text{ for } i = 1, 2, ..., n \\ \text{and} \quad & \boldsymbol{B}_{ij}^{k+1} = 0 \text{ for } i \neq j, \end{aligned} \tag{15}$$

where $\varepsilon > 0$ is given.

Let us call $\beta_i \equiv B_{ii}^{k+1}$. It can be shown [21] that (15) is equivalent to the quadratic programming problem in β

$$\begin{aligned} \text{minimize} \quad & \tfrac{1}{2}\beta^t Q\beta - \beta^t \boldsymbol{b} + c \\ \text{subject to} \quad & \beta_i \geq \varepsilon \text{ for } i = 1, 2, ..., n, \end{aligned} \tag{16}$$

where $Q = 2 \times \sum_{i=k-q}^{k} [Diag(\delta^i)]^2$, $b = 2 \times \sum_{i=k-q}^{k} Diag(\delta^i)\gamma^i$ and $c = \sum_{i=k-q}^{k} (\gamma^i)^t \gamma^i$.
$Diag(\boldsymbol{v})$, for $\boldsymbol{v} \in \Re^n$, is a diagonal matrix such that $Diag(\boldsymbol{v})_{ii} = \boldsymbol{v}_i$.

In Ref. 21 it is proved that β, solution of problem (16), can be easily computed as follows:

ALGORITHM 2.

For $i = 1, 2, ..., n$,

- If $\frac{b_i}{Q_{ii}} > \varepsilon$, then set $\beta_i = \frac{b_i}{Q_{ii}}$.

- Else, set $\beta_i = \varepsilon$. □

The above formulation is very simple. The required computational effort is negligible, in terms of calculus and memory. However, limited memory technique has a stronger theoretical support and seems to be more appropriate for unconstrained optimization. We shall employ sparse updating for constrained optimization, in those situations that the limited memory update is not appropriate.

3. The Feasible Arc Interior Point Algorithm

In this section we describe a quasi-Newton version of FAIPA and present some procedures to solve large optimization problems employing the sparse and the limited memory quasi - Newton methods. The best procedure in each case depends on the structure of the problem. In particular, of the sparsity of the matrix of constraints derivatives and of number of variables and constraints.

FAIPA requires an initial point at the interior of the inequality constraints and generates a sequence of interior points. When the problem has only inequality constraints, the objective function is reduced at each iteration. An auxiliary potential function is employed when there are also equality constraints.

ALGORITHM 3. FAIPA - Feasible Arc Interior Point Algorithm

Parameters. $\alpha, \nu \in (0, 1)$ and $\varphi > 0$.
Data. Initial values for $x \in \Re^n$, such that $g(x) < 0$, and for $\lambda \in \Re^m$, $\lambda > 0$, $B \in \Re^{n \times n}$ symmetric and positive definite and $c \in \Re^p, c \geq 0$.

Step 1. Computation of a feasible descent direction

(i) Solve the linear systems:

$$\begin{bmatrix} B & \nabla g(x) & \nabla h(x) \\ \Lambda \nabla g^t(x) & G(x) & 0 \\ \nabla h^t(x) & 0 & 0 \end{bmatrix} \begin{bmatrix} d_0 \\ \lambda_0 \\ \mu_0 \end{bmatrix} = - \begin{bmatrix} \nabla f(x) \\ 0 \\ h(x) \end{bmatrix} \tag{17}$$

and

$$\begin{bmatrix} B & \nabla g(x) & \nabla h(x) \\ \Lambda \nabla g^t(x) & G(x) & 0 \\ \nabla h^t(x) & 0 & 0 \end{bmatrix} \begin{bmatrix} d_1 \\ \lambda_1 \\ \mu_1 \end{bmatrix} = - \begin{bmatrix} 0 \\ \lambda \\ 0 \end{bmatrix}, \tag{18}$$

where $G(x) = \text{Diag}[g(x)]$, $\Lambda = \text{Diag}(\lambda)$ and $\nabla g(x) \in \Re^{n \times m}$ and $\nabla h(x) \in \Re^{n \times p}$ are respectively the matrices of derivatives of the inequality and the equality constraints.
    Let

$$\phi_c(x) = f(x) + \sum_{i=1}^{p} c_i |h_i(x)| \tag{19}$$

be the auxiliary potential function.

(ii) If $c_i < -1.2\mu_0(i)$, then set $c_i = -2\mu_0(i)$; $i = 1, ..., p$.

(iii) If $d_1^t \nabla \phi_c(x) > 0$, set

$$\rho = \inf \left[\varphi \parallel d_0 \parallel_2^2 ; (\alpha - 1) \frac{d_0^t \nabla \phi_c(x)}{d_1^t \nabla \phi_c(x)} \right]. \tag{20}$$

    Otherwise, set

$$\rho = \varphi \parallel d_0 \parallel_2^2 . \tag{21}$$

(iv) Compute the feasible descent direction $d = d_0 + \rho d_1$

Step 2. Computation of the "restoring direction" $\tilde{d}$

Compute:

$$\tilde{\omega}_i^I = g_i(x + d) - g_i(x) - \nabla g_i^t(x)d; \ i = 1, ..., m$$

$$\tilde{\omega}_i^E = h_i(x + d) - h_i(x) - \nabla h_i^t(x)d; \ i = 1, ..., p$$

Solve:

$$\begin{bmatrix} B & \nabla g(x) & \nabla h(x) \\ \Lambda \nabla g^t(x) & G(x) & 0 \\ \nabla h^t(x) & 0 & 0 \end{bmatrix} \begin{bmatrix} \tilde{d} \\ \tilde{\lambda} \\ \tilde{\mu} \end{bmatrix} = - \begin{bmatrix} 0 \\ \Lambda \tilde{\omega}^I \\ \tilde{\omega}^E \end{bmatrix} \tag{22}$$

Step 3. Line search along the feasible descent arc $x(t) = x + td + t^2 \tilde{d}$

Find $t = inf \{1, \nu, \nu^2, \nu^3, ...\}$, such that:

$$\phi(x + td + t^2 \tilde{d}) < \phi(x) + t\eta \nabla \phi^t(x)d \tag{23}$$
$$g(x + td + t^2 \tilde{d}) < 0$$

Step 4. Updates.

(i) Set the new point

$$x := x + td + t^2 \tilde{d}$$

(ii) Define new values for $\lambda > 0$ and B symmetric and positive definite.
(iii) Go back to Step 1.   □

The present algorithm converges to a Karush - Kuhn - Tucker point of the problem for any initial interior point. This is true no matter how $\lambda > 0$ and B, positive definite, are updated. We employ the following updating rule for λ:

Updating Rule for λ

Set, for $i = 1, ..., m$,

$$\lambda_i := \max [\lambda_0; \epsilon \parallel d_0 \parallel_2^2]. \tag{24}$$

If $g_i(x) \geq -\bar{g}$ and $\lambda_i < \lambda^I$, set $\lambda_i = \lambda^I$.

□

The parameters $\epsilon, \bar{g}$ and λ^I are taken positive. In this rule, λ_i is a second order perturbation of λ_0, given by Newton iteration. If $\bar{g}$ and λ^I are taken small enough, then after a finite number of iterations, λ_i becomes equal to λ_0 for the active constraints.

The linear system (17) in (d_0, λ_0, μ_0) is derived from a Newton's iteration to solve Karush - Kuhn - Tucker optimality conditions. Solving (17), we obtain d_1, that improves feasibility. In the calculus of $\tilde{d}$ it is involved an estimate of the second derivatives of the constraints. The feasible descent direction $d = d_0 + \rho d_1$ and the feasible descent arc $x(t) = x + td + t^2 \tilde{d}$ are represented in Fig. 1, for the case when an inequality constraint is active.

The line search described in Step 3 is an extension of Armijo's scheme for unconstrained optimization. More efficient inexact line search algorithms, based on Wolfe's or Goldfarb's criteria, can also be employed.[00,00]

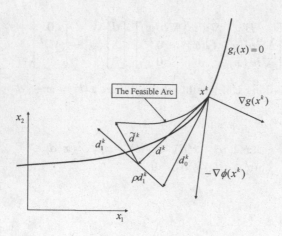

Fig. 1.   The feasible arc

3.1. *BFGS updating rule for constrained optimization*

Let $l(\boldsymbol{x}, \lambda, \mu) = f(\boldsymbol{x}) + \lambda^t \boldsymbol{g}(\boldsymbol{x}) + \mu^t \boldsymbol{h}(\boldsymbol{x})$ be the Lagrangian of Problem (1). Quasi - Newton method for constrained optimization works with an approximation of the Hessian of the Lagrangian:

$$\boldsymbol{B} \approx \boldsymbol{L}(\boldsymbol{x}, \lambda, \mu) = \nabla^2 f(\boldsymbol{x}) + \sum_{i=1}^{m} \lambda_i \nabla^2 g_i(\boldsymbol{x}) + \sum_{i=1}^{p} \mu_i \nabla^2 h_i(\boldsymbol{x}).$$

The same updating rules used for unconstrained optimization can be employed, but taking

$$\gamma = \nabla_x l(\boldsymbol{x}^{k+1}, \lambda_0^k, \mu_0^k) - \nabla_x l(\boldsymbol{x}^k, \lambda_0^k, \mu_0^k).$$

However, since $L(\boldsymbol{x}, \lambda, \mu)$ is not necessarily positive definite at a local minimum, it is not always possible to get positive definite quasi-Newton matrices. When employing BFGS updating rule, γ can be modified in such way to have $\delta^t \gamma > 0$, forcing $\boldsymbol{B}^{k+1}$ to be positive definite. The following rule was proposed by Powell,[51,52]
    If

$$\delta^t \gamma < 0.2 \delta^t \boldsymbol{B}^k \delta,$$

then compute

$$\phi = \frac{0.8 \delta^t \boldsymbol{B}^k \delta}{\delta^t \boldsymbol{B}^k \delta - \delta^t \gamma}$$

and take

$$\gamma := \phi \gamma + (1 - \phi) \boldsymbol{B}^k \delta.$$

$\square$

In Ref. 26 it was proved that the convergence of the quasi-Newton version of FAIPA is two-step superlinear. The search along an arc ensures that the step length can be taken equal to one after a finite number of iterations. This is a requirement to prove superlinear convergence.

4. The Internal Linear Systems in FAIPA

The linear systems (17), (18) and (22) are called *"Primal-Dual"* systems since the unknowns are related to the primal and the dual variables of the problem. These systems can be reformulated in some equivalent ways that we shall describe here. The most favorable formulation can be chosen in each case, depending on the structure of the optimization problem and the numerical technique employed to solve the systems.

The primal-dual systems have a unique solution[49] if the optimization problem satisfies the following assumption:

Regularity Condition - For all x such that $g(x) \leq 0$ and $h(x) = 0$, the vectors $\nabla g_i(x)$, for $i = 1, 2, ..., m$ such that $g_i(x) = 0$ and $\nabla h_i(x)$ for $i = 1, 2, ..., p$ are linearly independent.

However, the primal-dual matrix is not symmetric neither positive definite. Equivalent symmetric primal-dual systems can be obtained with the following coefficient matrix:

$$\begin{bmatrix} B & \nabla g(x) & \nabla h(x) \\ \nabla g^t(x) & \Lambda^{-1}G(x) & 0 \\ \nabla h^t(x) & 0 & 0 \end{bmatrix} \tag{25}$$

When there are inactive constraints at the solution, the corresponding Lagrange multipliers go to zero. In consequence, the symmetric primal-dual matrix becomes ill-conditioned. However, it is not difficulty to obtain preconditioners to overcome this kind of ill-conditioning.

It follows from (17) that

$$d_0 = -B^{-1}[\nabla f(x) + \nabla g(x)\lambda_0 + \nabla h(x)\mu_0]. \tag{26}$$

and that

$$\begin{bmatrix} [\Lambda\nabla g^t(x)B^{-1}\nabla g(x) - G(x)] & \nabla g^t(x)B^{-1}\nabla h(x) \\ \nabla h^t(x)B^{-1}\nabla g(x) & \nabla h^t(x)B^{-1}\nabla h(x) \end{bmatrix} \begin{bmatrix} \lambda_0 \\ \mu_0 \end{bmatrix} =$$

$$- \begin{bmatrix} \Lambda\nabla g^t(x)B^{-1}\nabla f(x) \\ \nabla h^t(x)B^{-1}\nabla f(x) - h(x) \end{bmatrix} \tag{27}$$

Then, (26) and (27) is an alternative formulation to compute λ_0, μ_0 and d_0. Similar expressions to get (d_1, λ_1, μ_1) and $(\tilde{d}, \tilde{\lambda}, \tilde{\mu})$ can be deduced. The system (27) is called *"Dual System"*. Equivalent expressions can be obtained, involving *"Symmetric Dual Systems"*, with the following coefficient matrix:

$$\begin{bmatrix} [\nabla g^t(x) B^{-1} \nabla g(x) - \Lambda^{-1} G(x)] & \nabla g^t(x) B^{-1} \nabla h(x) \\ \nabla h^t(x) B^{-1} \nabla g(x) & \nabla h^t(x) B^{-1} \nabla h(x) \end{bmatrix}. \tag{28}$$

The comments concerning the conditioning of the symmetric primal-dual matrix (25) are also valid for (28).

From (17), we have

$$\lambda_0 = -G(x)^{-1} \Lambda \nabla g^t(x) d_0 \tag{29}$$

and

$$\begin{bmatrix} [B - \nabla g(x) G^{-1}(x) \Lambda \nabla g^t(x)] & \nabla h(x) \\ \nabla h^t(x) & 0 \end{bmatrix} \begin{bmatrix} d_0 \\ \mu_0 \end{bmatrix} = - \begin{bmatrix} \nabla f(x) \\ h(x) \end{bmatrix}. \tag{30}$$

The coefficient matrix of the system (30) is symmetric and positive definite, as well as the *Primal Matrix*

$$\left[B - \nabla g(x) G^{-1}(x) \Lambda \nabla g^t(x) \right]. \tag{31}$$

This one is ill-conditioned when the inequality constraints are small.

The dual formulation is generally most favorable when the number of design variables is much smaller than the number of constraints while, in the opposite situation, the primal formulation is preferable. The primal-dual formulation involves a larger system of equations but it is more advantageous when sparse matrix techniques are employed.

Three linear system with the same matrix are solved at each iteration of FAIPA. In general, the coefficient matrices and right sides of the systems have small changes from one iteration to the following one. Then, it should be possible to take advantage of this fact when solving them.

4.1. *Solving the primal-dual systems*

The primal-dual system is particulary advantageous when the constraints derivative matrix is sparse and a sparse quasi-Newton matrix is employed. We solve the linear systems employing the Harwell Subroutine Library[35], code MA27. This is a set of FORTRAN subroutines for solving sparse symmetric linear systems by Gaussian elimination that includes some procedures to take advantage of the matrix structure[15,16]. The solution process is divided into three stages:

i) An *analysis phase* that examines the structure of the matrix in order to produce a suitable ordering and data structures for an efficient factorization.

ii) A *factorization phase* that performs the actual numerical factorization.

iii) A *solution phase* which performs the forward and backward substitutions.

We can assume that the zero elements are the same for all iterations. Thus, the analysis phase must be only carried out for the first system in the first iteration. Sice at each iteration the linear systems have the same coefficient matrix, the factorization is done once per iteration only.

4.2. *Solving the dual systems*

The dual and the symmetric dual matrices can be computed employing the limited memory formulation to determine the products $B^{-1}\nabla g(x)$ and $B^{-1}\nabla h(x)$ without needing the computation nor storage of B.

Iterative methods for linear systems compute at each iteration the product of the coefficient matrix by a vector. When employing limited memory formulation, this product can be done without storing the coefficient matrix. This can be extremely efficient when the constraints derivative matrix is not sparse.

A new technique based on a preconditioned conjugate gradient algorithm to solve the symmetric dual system is now described, see Ref. 14. Let us consider:

$$Ax = b, \tag{32}$$

where $A \in \Re^{N \times N}$ and $x, b \in \Re^N$. The present algorithm is based on the equivalent systems

$$\tilde{H}L^{-1}AL^{-t}y = \tilde{H}L^{-1}b, \tag{33}$$

$$L^{-t}y = x. \tag{34}$$

The lower triangular matrix $L \in \Re^{N \times N}$ is an incomplete Choleski factorization preconditioner and $\tilde{H} \in \Re^{N \times N}$ is a preconditioner based on quasi-Newton limited memory method to minimize the quadratic function

$$Q(y) \equiv \frac{1}{2}y^t L^{-1}AL^{-t}y - y^t L^{-1}b.$$

That is, $\tilde{H}$ is a quasi - Newton approximation of $(L^{-1}AL^{-t})^{-1}$.

Limited memory preconditioners were proposed by Morales et al.[46]. Dubeux[14] proposed a criterium to select the "best" set of pairs (δ, γ) to construct $\tilde{H}$. The set of pairs (δ, γ), that were obtained in the solution of one of this systems, can be employed as initial set of pairs for the solution of next one since three systems with the same matrix are solved at each iteration of FAIPA.

The following preconditioned conjugate gradient algorithm is based on Algorithm 9.1 in Ref. 55.

ALGORITHM 4. Preconditioned Conjugate Gradient

Data. Initial values: x^0, $r^0 = L^{-1}(b - Ax^0)$, $z^0 = \tilde{H}r^0$ and $p^0 = L^{-t}z^0$.
Set $k = 0$.

Step 1. Compute

$$\alpha^k = \frac{(r^t)^k z^k}{(Ap^k)^t p^k}$$

$$x^{k+1} = x^k + \alpha^k p^k$$

$$r^{k+1} = r^k - \alpha^k L^{-1} Ap^k$$

$$z^{k+1} = \tilde{H} r^{k+1}$$

$$\beta^k = \frac{(r^t)^{k+1} z^{k+1}}{(r^t)^k z^k}$$

$$p^{k+1} = L^{-t} z^{k+1} + \beta^k p^k$$

Step 2. Set $k = k + 1$. Return to Step 1.   □

The product $\tilde{H} r$ is computed employing (12). The products Ax and Ap can be computed without storing A.

5. Numerical Experiments

We present in this section some numerical results obtained with:

i) The classical full-matrix quasi-Newton version of FAIPA, FAIPA_qN.
ii) The limited memory quasi-Newton version, FAIPA_LqN. The internal systems are solved iteratively.
iii) Faipa with a diagonal quasi-Newton matrix, FAIPA_DqN.

Wolfe's line search, described in Ref. 33, is employed in all the cases . The tables with the results employ the same notation as in the paper. The number of box constraints is called nb.

All the problems were solved with the same set of parameters: $\alpha = 0.7$, $\phi = 1$, $\eta_1 = 0.1$, $\eta_2 = 0.7$, $\gamma = 0.5$ and $\epsilon = 0.1$. The initial values were $\lambda_i = 1$; for $i = 1, 2, ..., m$, $S = I$ and $c = 0$.

If the inequality constraints are not verified by the initial points, a feasible initial point was found with the help of the auxiliary mathematical program

$$\begin{cases} \min_{(x,z)} z \\ \text{s. t. } g(x) \leq z, \end{cases} \tag{35}$$

where z is a scalar auxiliary variable. Making iteration with FAIPA to solve (35), a feasible point is obtained once z becomes negative.

5.1. *Results with a collection of test problems*

We report first our results with some test problems largely employed in Mathematical Programming literature. Our experience with 107 problems, compiled by Hock and Schittkowski,[37] are presented in Table 1, where *"nprob"* is the number of the problem.

For all the problems, the optimal function value according to Ref. 37 was obtained. The number of iterations required to have the optimal value of the function with a relative error less than 10^{-5} is reported. All the iterates satisfy the inequality constraints. A tolerance of 10^{-6} was established in the stopping criterium for the equalities. The number of iteration in the line search is very small, since Wolfe's criterion is quite wide. In general, the line search requires only one or two evaluations of the objective function and the constraints.

Table 1 also presents the numerical results obtained with the Sequential Quadratic Programming algorithm included in MATLAB Library[10], with the same stopping criteria. Since the inequality constraints are not always satisfied, a tolerance of 10^{-6} was also imposed.

The numerical results for a set of large problems is reported in Table 2. The are described in Ref. 12. We propose here the parametric problem HS43_ nf, based on Problem 43 in Ref. 37, stated as follows:

$$f(x) = \sum_{t=1}^{nf} x_{4t-3}^2 + x_{4t-2}^2 + 2x_{4t-1}^2 + x_{4t}^2 - 5x_{4t-2} - 21x_{4t-1} + 7x_{4t}$$
$$g_{3j-2}(x) = -(8 - x_{4j-3}^2 - x_{4j-2}^2 - x_{4j-1}^2 - x_{4j}^2 - x_{4j-3} + x_{4j-2} - x_{4j-1} + x_{4j})$$
$$g_{3j-1}(x) = -(10 - x_{4j-3}^2 - 2x_{4j-2}^2 - x_{4j-1}^2 - 2x_{4j}^2 + x_{4j-3} + x^{4j})$$
$$g_{3j}(x) = -(5 - 2x_{4j-3}^2 - x_{4j-2}^2 - x_{4j-1}^2 - 2x_{4j-3} + x_{4j-2} + x_{4j}),$$

for $j = 1, 2, ..., nf$.

Our results are compared in terms of the required number of iterations with the code Knitro described in Ref. 9

5.2. *Experiments with a structural optimization problem*

We study the numerical behavior of the limited memory quasi-Newton version of FAIPA when applied to two examples whose objective is the volume minimization under Von-Misses stress constraints of rectangular plates submitted to in-plane distributed loadings. The supports, loads and the design domains are shown in the Figs. 2 and 5 respectively. In Problem 1 the domain is dicretized in 300, 1200 and 4800 elements and, 3200 elements for Problem 2. Quadrilateral bilinear plane stress elements are employed. Young modulus is assumed to be $E = 210$ GPa and Poisson's ratio $\nu = 0.3$ for all elements.

The thickness is constrained to be larger than 0.1 cm and smaller than 1.0 cm. Von-Misses stresses, computed at the center of each element, must be lower than 250 MPa. The optimal structures are shown in Figs. 3 and 6, when the elements with thickness equal to the lower bound were removed. The iterations histories are represented in Figs. 4 and 7.

Table 1.　Numerical Results, Test problems in Ref. 37

prob	n	m	p	box	SQP	FAIPA qN	LqN	DqN	prob	n	m	p	box	SQP	FAIPA qN	LqN	DqN
1	2	1	0	1	33	36	37		55	6	14	6	8	2†	5	5	12
2	2	1	0	1	17	16	15		56	7	4	4	0	9	9	9	10
3	2	1	0	1	5	15	14	208	57	2	3	0	2	16	28	23	5
4	2	2	0	2	1	4	4	4	59	2	7	0	4	20	21	20	13
5	2	4	0	4	14	4	4	6	60	3	7	1	6	8	9	9	14
6	2	1	1	0	10	9	10	12	61	3	2	2	0	1†	10	10	9
7	2	1	1	0	8	11	11	28	62	3	7	1	6	7	4	4	4
8	2	2	2	0	4	9	9	9	63	3	5	2	3	7	9	9	10
9	2	1	1	0	5	5	5	6	64	3	4	0	3	23	23	22	15
10	2	1	0	0	11	7	7	12	65	3	7	0	6	7	13	13	19
11	2	1	0	0	7	6	6	6	66	3	8	0	6	5	10	10	9
12	2	1	0	0	16	4	4	12	68	4	10	2	6	13	19	29	21
13	2	3	0	2	31	25	40	9	69	4	10	2	8	15	11	11	19
14	2	2	1	0	5	14	14	6	70	4	9	0	8	36	64	31	53
15	2	3	0	1	2	6	6	6	71	4	10	1	8	8	15	15	16
16	2	5	0	3	6	18	18	36	72*	4	10	0	8	14	17	34	15
17	2	5	0	3	8	20	19	14	73*	4	7	1	4	4	16	16	19
18	2	6	0	4	8	12	12	12	74	4	13	3	8	8	27	71	67
19	2	6	0	4	5	74	66	46	75	4	13	3	8	6	50	39	44
20	2	5	0	2	5	9	10	9	76	4	7	0	4	4	8	8	9
21	2	5	0	4	2	4	4	4	77	5	2	2	0	20	18	18	21
22	2	2	0	0	4	10	10	9	78	5	3	3	0	8	7	7	9
23	2	9	0	4	6	9	9	9	79	5	3	3	0	10	10	10	14
24	2	5	0	2	4	4	4	4	80	5	13	3	10	6	8	8	10
25	3	6	0	6	1†	65	11	38	81	5	13	3	10	9	10	10	12
26	3	1	1	0	37	24	32	25	83	5	16	0	10	3	12	12	12
27	3	1	1	0	116	19	22	22	84	5	16	0	10	13	4	4	4
28	3	1	1	0	31	4	4	28	86	5	15	0	5	8	12	12	29
29	3	1	0	0	11	11	11	15	93	6	8	0	6	20	7	7	11
30	3	7	0	6	9	6	6	6	95	6	16	0	12	1	4	4	4
31	3	7	0	6	8	9	9	9	96	6	16	0	12	1	7	7	9
32	3	5	1	3	3	11	11	14	97	6	16	0	12	8	8	8	7
33	3	6	0	3	6†	14	14	9	98	6	16	0	12	8	42	31	6
34	3	8	0	4	7	18	18	9	99	7	16	2	14	27	12	21	6
35	3	4	0	6	5	6	6	11	100	7	4	0	0	13	11	17	9
36	3	7	0	3	1	12	13	10	101	7	20	0	14	26	22	39	28
37	3	8	0	6	6	14	16	11	102	7	20	0	14	26	28	27	17
38	4	8	0	6	28	16	15	17	103	7	20	0	14	25	18	22	28
39	4	2	2	8	199‡	13	13	12	104	8	22	0	16	14	18	17	13
40	4	3	3	0	6	6	6	7	105	8	17	0	16	33	54	63	50
41	4	9	1	0	9	12	12	14	106*	8	22	0	16	81	125	54	123
42	4	2	2	8	8	8	8	6	107	9	14	6	8	6	15	15	15
43	4	3	0	0	11	52	40	9	108	9	14	0	1	14	17	23	9
44	4	10	0	0	5	20	20	15	109	9	26	6	16	15	8	8	8
45	5	10	0	4	5	22	22	10	110	10	20	0	20	63‡	4	4	5
46	5	2	2	10	36	10	10	19	111	10	23	3	20	124‡	35	34	17
47	5	3	3	0	14	12	12	20	112	10	13	3	10	39	11	11	8
48	5	2	2	0	9	3	3	9	113	10	8	0	0	49	23	28	32
49	5	2	2	0	31	10	10	42	114*	10	31	3	20	31	116	93	93
50	5	3	3	0	13	11	13	13	116	13	41	0	26	199‡	51	55	18
51	5	3	3	0	5	4	4	4	117	15	20	0	15	23	40	46	29
52	5	3	3	0	4	5	5	7	118	15	59	0	30	13	45	51	29
53	5	13	3	0	5	5	5	7	119	16	40	8	32	9	77	100	89
54	6	13	1	12	1†	54	48	8									

Note: † The algorithm converges to a local minimum.
‡ Convergence is not achieved.
* The initial point is infeasible. Previous iterations were required in searching for a feasible initial point.

Table 2.   Numerical Results, Test problems in Ref. 12

Problem	n	m	p	box	qN	LqN	DqN	Knitro
DTOC5_50	98	0	49	0	16	18	14	22
DTOC5_100	198	0	99	0	18	19	22	23
DTOC5_500	998	0	499	0	24	33	75	29
DTOC5_1000	1998	0	999	0	31	59	210	20
DTOC1L_5998	5994	0	3996	0	14	14	43	21
OPTCTRL6_40	119	0	80	0	67	42	19	116
OPTCTRL6_100	299	0	200	0	51	149	24	*
OPTCTRL6_400	1199	0	800	0	128	377	31	*
ORTHRDM2_100	203	0	100	0	12	13	8	12
ORTHRDM2_2000	4003	0	2000	0	10	16	9	15
ORTHRDM2_4000	8003	0	4000	0	12	*	19	12
ORTHRDS2_50	103	0	50	0	38	144	44	60
ORTHRDS2_100	203	0	100	0	36	96	107	50
ORTHRDS2_250	502	0	250	0	41	49	123	50
ORTHRDS2_500	1003	0	500	0	35	18	36	50
ORTHRDS2_2500	5003	0	2500	0	45	54	396	40
ORTHREGC_50	105	0	50	0	25	33	52	172
ORTHREGC_500	1005	0	500	0	29	31	214	48
ORTHREGC_2500	5000	0	2500	0	43	79	246	43
ORTHREGD_50	103	0	50	0	11	15	16	16
ORTHREGD_250	503	0	250	0	14	16	18	14
ORTHREGD_500	1003	0	500	0	14	17	17	16
ORTHREGD_2500	5003	0	2500	0	15	19	14	19
ORTHREGD_5000	10003	0	5000	0	*	*	12	*
ORTHRGDS_250	503	0	250	0	48	26	174	23
ORTHRGDS_500	1003	0	500	0	46	33	252	16
GILBERT_5000	5000	0	1	1	58	77	112	45
GILBERT_1000	1000	0	1	1	55	77	67	48
GILBERT_10	10	0	1	1	17	19	18	20
SVANBERG_5000	5000	5000	0	10000	*	*	75	420
SVANBERG_1000	1000	1000	0	2000	137	89	63	247
SVANBERG_100	100	100	0	200	48	56	64	76
POLYGON_25	48	324	0	96	43	33	28	*
POLYGON_50	98	1274	0	196	13	18	38	*
POLYGON_75	148	2849	0	296	18	19	37	*
POLYGON_100	198	5049	0	396	19	13	34	*
HS43NF_500	2000	1500	0	0	17	17	24	*
HS43NF_1000	4000	3000	0	0	19	16	29	*
HS43NF_1500	6000	4500	0	0	17	17	25	*
HS43NF_2000	8000	6000	0	0	19	17	29	*
HS43NF_2250	9000	6750	0	0	*	16	27	*
HS43NF_3500	14000	10500	0	0	*	18	26	*

Note: * Not tested.

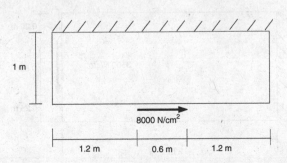

Fig. 2.   Problem 1 - Description

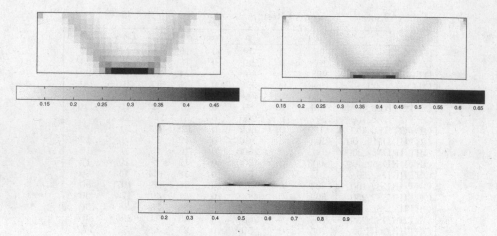

Fig. 3.   Optimal design for 300, 1200 and 4800 elements

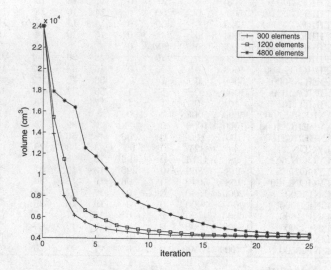

Fig. 4.   Problem 1 — Iterations history

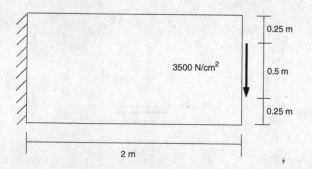

Fig. 5.   Problem 2 — Description

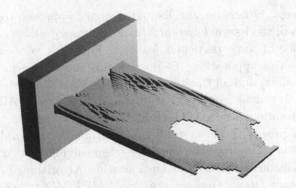

Fig. 6. Problem 2 — Optimal design

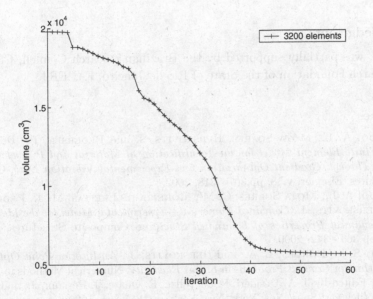

Fig. 7. Problem 2 — Iterations history

6. Conclusions

All the test problems were solved very efficiently with the same set of parameters. The number of iterations remains comparable when the size of the problem is increased.

The numerical results shown here suggest that very large problems can be solved with FAIPA, depending only on the capacity of solving the linear internal systems and storing the required data. The diagonal quasi-Newton Matrix is more efficient for sparse problems. Otherwise, the limited memory approach together with the iterative solution of the internal systems should be employed.

In consequence of some particular features, FAIPA is very advantageous for large scale engineering applications. Engineering optimization usually requires only inequality constraints, and all the iterates given by FAIPA are strictly verified. In consequence the iterations can be stopped at any time. Since FAIPA solves linear systems at each iteration, instead of Quadratic or Linear Programs,[43] a large number of existing techniques for linear systems can be employed. Also algebraical transformations can lead to improve the efficiency when solving particular applications, as in the Simultaneous Analysis and Optimization Algorithm, FAIPA_SAND, or the Multidisciplinary Design Optimization one, FAIPA_MDO, described in Ref. 33. The fact that global convergence is proved for any way of updating B and λ makes FAIPA very strong. In particular, when we substitute the BFGS quasi-Newton matrix by a diagonal approximation.

Acknowledgments

This work was partially supported by the Brazilian research Council, CNPq, and The Research Foundation of the State of Rio de Janeiro, FAPERJ.

References

1. ARAUJO, A. L., MOTA SOARES, HERSKOVITS, J. and PEDERSEN, P., *Development of a Finite Element Model for the Identification of Material and Piezoeletric Properties Through Gradient Optimization and Experimental Vibration Data*, Composite Structures, Elsevier, v.58, pp.307 - 318, 2002.
2. ARAUJO, A. L., MOTA SOARES, C. M., MOREIRA DE FREITAS, M. J., PEDERSEN, P. and HERSKOVITS, J., *Combined Numerical - Experimental Model for the Identification of Mechanical Properties of Laminated Stuctures*, Composite Structures, Elsevier , v.50, pp.363 - 372, 2000.
3. AUATT, S. S., BORGES, L. A., and HERSKOVITS, J., *An Interior Point Optimization Algorithm for Contact Problems in Linear Elasticity*, Numerical Methods in Engineering'96, Edited by J. A. Désidéri, P. Le Tallec, E. Oñate, J. Périaux, J., and E. Stein, John Wiley and Sons, New York, New York, pp. 855-861, 1996.
4. BARON, F. J., *Constrained Shape Optimization of Coupled Problems with Electromagnetic Waves and Fluid Mechanics*, PhD Thesis, University of Malaga, Malaga, Spain, 1994, (in Spanish).
5. BARON, F. J., *Some Shape Optimization Problems in Electromagnetism and Fluid Mechanics*, DSc Thesis, Paris VI University, Paris, France, 1998, (in French).
6. BARON, F. J., DUFFA, G., CARRERE, F., and LE TALLEC, P., *Optimisation de Forme en Aerodynamique*, CHOCS, Revue Scientifique et Technique de la Direction des Applications Militaires du CEA, France, No. 12, pp. 47 - 56, 1994, (in French).
7. BARON, F. J., and PIRONNEAU, O., *Multidisciplinary Optimal Design of a Wing Profile*, Proceedings of Structural Optimization 93, Rio de Janeiro, Brazil, 1993, Edited by J. Herskovits, Vol. 2, pp. 61-68, 1993.
8. BIJAN, M., and PIRONNEAU, O., *New Tools for Optimum Shape Design* , CFD Review, Special Issue, 1995.

9. BYRD, R. H.,HRIBAR, M. E. and NOCEDAL, J., *An Interior Point Algorithm for Large Scale Nonlinear Programming*,SIAM J. Optim., Vol 9(4), pp. 877-900, 1999.

10. BRANCH, M. A. and GRACE, A. *Optimization Toolbox User Guide*, The Mathworks, 24 Prime Park Way, MA,USA, 1996.

11. BYRD, R. H.,NOCEDAL, J. and Schanabel, R. B., *Represetation of Quasi-Newton Matrices and their Use in Limited Memory Methods*, Mathematical Programming, 63, 4, pp. 129-156,1994.

12. BONGARTZ I., CONN A.R., GOULD N., and TOINT PH. L., *CUTE: Constrained and unconstrained testing environment*, ACM Transactions on Mathematical Software, v.21, 123-160, 1995.

13. DENNIS, J. E., and SCHNABEL, R., *Numerical Methods for Constrained Optimization and Nonlinear Equations*, Prentice Hall, Englewood Cliffs, New Jersey, 1983.

14. DUBEUX, V. J. C.*Nonlinear Programming Techniques for Large Size Optimization*, D.Sc. Thesis, COPPE/Federal University of Rio de Janeiro, Mechanical Engineering Program, Caixa Postal 68503, 21945 970 Rio de Janeiro, Brazil, 2005. (in Portuguese)

15. DUFF I. S. and REID J. K. *The multifrontal solution of indefinite sparse symmetric linear equations* ACM Transactions on Mathematical Software, Vol 9. (3) pp. 302-325, 1983.

16. DUFF I. S. and REID J. K. *MA27: A set of Fortran subroutines for solving sparse symmetric sets of linear equations*, Her Majesty's Stationery Office, AERE R10533.

17. FLETCHER, R.*An Optimal Positive Definite Update for Sparse Hessian Matrices*,SIAM Journal on Optimization, v. 5, n. 1, pp. 192-218, 1995.

18. FLETCHER, R., GROTHEY, A. and Leyffer, S.*Computing Sparse Hessian and Jacobian Approximations with Hereditary Properties*,Tech. Report, Department of Mathematics, University of Dundee, 1996.

19. GOLDFELD, P., DUARTE, and A., HERSKOVITS, J., *A Limited Memory Interior Points Technique for Nonlinear Optimization*, Proceedings of ECCOMAS, European Congress on Computational Methods in Applied Science and Engineering, Paris, Sept. 1996.

20. GONZAGA, C. C., *Path following methods for linear programming*, SIAM Review v. 34 No. 2 pp. 167-224, 1992.

21. GOULART, E. *Sparse Quase-Newton Matrices for Large Size Nonlinear Optimization*, D. Sc. Thesis, COPPE/ Federal University of Rio de Janeiro, Mechanical Engineering Program, Caixa Postal 68503, 21945-970, Rio de Janeiro, Brazil, 2005. (in Portuguese)

22. HERSKOVITS, J., *Developement of a Numerical Method for Nonlinear Optimization*, Dr. Ing. Thesis, Paris IX University, INRIA Rocquencort, 1982.

23. HERSKOVITS, J., *A Two-Stage Feasible Directions Algorithm Including Variable Metric Techniques for Nonlinear Optimization* , Research Report 118, INRIA, Le Chesnay, France, 1982.

24. HERSKOVITS, J., *A Two-Stage Feasible Directions Algorithm for Nonlinear Constrained Optimization*, Mathematical Programming, Vol. 36, pp. 19-38, 1986.

25. HERSKOVITS, J., *A View on Nonlinear Optimization*, in Advances in Structural Optimization, Edited by J. Herskovits, Kluwer Academic Publishers, Dordrecht, Holland, pp. 71-117, 1995.

26. HERSKOVITS, J., *A Feasible Directions Interior Point Technique For Nonlinear Optimization*, JOTA, Journal of Optimization Theory and Applications, , Vol.99, N.1, pp.121 - 146, Plenum, London, 1998.

27. HERSKOVITS, J., DIAS, G.,FALCÓN, G. and MOTA SOARES, C. M. *Shape Structural Optimization with an Interior Point Mathematical Programming Algorithm*, Structural And Multidisciplinary Optimization. Springer, v.20, p.107 - 115, 2000.

28. HERSKOVITS, J., DUBEUX, V. MOTA SOARES, C. M. and ARAÚJO, A., *Interior Point Algorithms for Nonlinear Constrained Least Squares Problems*, Inverse Problems in Engineering, Taylor & Francis, USA, v.12, n.2, p.211 - 223, 2004.

29. HERSKOVITS, J., LAPPORTE, E., LE TALLEC P., and SANTOS, G., *A Quasi-Newton Interior Point Algorithm Applied for Constrained Optimum Design in Computational Fluid Dynamics*, European Journal of Finite Elements, Vol. 5, pp. 595-617, 1996.

30. HERSKOVITS, J., LEONTIEV, A., DIAS, G. and FALCON, G. S. *Contact Shape Optimization: a Bilevel Programming Approach*, Structural And Multidisciplinary Optimization, Springer, v.20, n.3, pp.214 - 221, 2000.

31. HERSKOVITS, J., MAPPA, P. and JUILLEN, L. *FAIPA_SAND, An Interior Point Algorithm for Simultaneous ANalysis and Design Optimization*, WCSMO4, Fourth World Congress of Structural and Multidisciplinary Optimization Dalian, China, June 2001.

32. HERSKOVITS, J., MAPPA, P. and MOTA SOARES, C. M., *FAIPA_MDO, A Mathematical Programming Algorithm for Multidisciplinary Design Optimization*, WCSMO5, Fifth World Congress of Structural and Multidisciplinary Optimization, Lido di Jessolo, Italy, May 2003.

33. HERSKOVITS, J., MAPPA, P., GOULART, E. and MOTA SOARES C. M.*Mathematical programming models and algorithms for engineering design optimization* , Computer Methods in Applied Mechanics and Engineering, v. 194/30-33 pp. 3244-3268, 2005.

34. HERSKOVITS, J. and SANTOS, G., *Feasible Arc Interior Point Algorithm for Nonlinear Optimization*, Computational Mechanics, New trends and Applications, Ed. by S. Idelsohn, E. Oñate and E. Dvorkin, CIMNE, Barcelona, 1998.

35. HSL it The Harwell Subroutine Library , http://www.cse.clrc.ac.uk/nag/hsl/

36. HIRIART-URRUTY, J. B., and LEMARÉCHAL, C., *Convex analysis and Minimization Algorithms*, Springer Verlag, Berlin, Germany, 1993.

37. HOCK, W., and SCHITTKOWSKI, K., *Test Examples for Nonlinear Programming Codes*, Lecture Notes in Economics and Mathematical Systems, Springer Verlag, Berlin, Germany, v.187, 1981.

38. LAPORTE, E. and LE TALLEC, P., *Numerical Methods in Sensitivity and Shape Optimization*, Birkhauser, 2002.

39. LEONTIEV, A. and HERSKOVITS, J. *Interior Point Techniques For Optimal Control Of Variational Inequality*, Structural and Multidisciplinary Optimization, Springer , v.14, n.2-3, pp.100 - 107, 1997.

40. LEONTIEV, A., HERSKOVITS, J. and EBOLI, C., *Optimization Theory Application to Slitted Plate Bending Problem* , International Journal of Solids and Structures,United Kingdom: , Vol.35, pp.2679 - 2694, 1997.

41. LEONTIEV, A., HUACASI, W. and HERSKOVITS, J., *An Optimization Technique for the Solution of Signorini Problem Using the Boundary Element Method*, Structural and Multidisciplinar Optimization, Springer - Verlag, v.24, n.1, pp.72 - 77, 2002.

42. D.C.LIU and J. Nocedal,*On the limited memory BFGS method for large scale optimization*, Mathematical Programming,v.45, pp. 503-528,1989.

43. LUENBERGER, D.G., *Linear and Nonlinear Programming* , 2nd. Edition, Addison-Wesley,London, 1984.

44. MAPPA, P., *FAIPA_ SAND: An Algorithm for Simultaneous Analysis and Design Optimization*, DSc Thesis, COPPE, Federal University of Rio de Janeiro, Mechanical Engineering Program, Rio de Janeiro, Brazil, 2004, (in Portuguese).

45. MARATOS, N., *Exact Penalty Function Algorithms for Finite Dimensional Optimization Problems,* PhD Thesis, Imperial College of Science and Technology, London, England, 1978.

46. MORALES, J. L. and NOCEDAL, J., *Automatic Preconditioning by Limited Memory Quasi-Newton Updating*, SIAM Journal on Optimization, v. 10(4),pp. 1079-1096, 2000.

47. NOCEDAL, J.,*Updating quasi-Newton matrices with limited storage*, Mathematics of computation, v.35, pp.773-782, 1980.

48. NOCEDAL, J. and WRIGHT, S. J., *Numerical optimization*, Springer-Verlag, New York, 1999.

49. PANIER, E. R., TITS A. L., and HERSKOVITS J., *A QP-Free, Globally Convergent, Locally Superlinearly Convergent Algorithm for Inequality Constrained Optimization*, SIAM Journal of Control and Optimization, Vol. 26, pp. 788-810, 1988.

50. POLAK, E. *Optimization Algorithm and Consistent Approximations*, Springer - Verlag, 1997.

51. POWELL, M.J.D., *The Convergence of Variable Metric Methods for Nonlinearly Constrained Optimization Calculations*, Nonlinear Programming 3, Edited by O.L. Mangasarian, R.R. Meyer, and S.M. Robinson, Academic Press, London, England, pp. 27-64, 1978.

52. POWELL M. J. D., *Variable Metric Methods for Constrained Optimization*, Mathematical Programming - The State of the Art, Edited by A. Bachem, M. Grotschet, and B. Korte, Springer Verlag, Berlin, Germany, pp. 288-311, 1983.

53. QIAN, Z., XUE, C. and PAN, S., *FEA Agent for Multidisciplinary Optimization*, Structural and Multidisciplinary Optimization, Springer, v.22, pp. 373-383, 2001.

54. ROMERO, J., MAPPA, P. C., HERSKOVITS, J. and MOTA SOARES, C. M., *Optimal Truss Desing Including Plastic Collapse Constraints*, Structural and Multidisciplinary Optimization, Springer, v.27, n.1-2, p.20 - 26, 2004.

55. SAAD, Y., Iterative Methods for Sparse Linear Systems, 2nd Ed., SIAM, 2003.

56. TOINT, PH. L.*On Sparse and Symmetric Matrix Updating Subject to a Linear Equation*, Mathematics of Computation, v. 31, n. 140, pp. 954-961, 1977.

57. TOINT, PH. L.*A Sparse Quasi-Newton Update Derived Variattionally With a Nondiagonally Weighted Frobenius Norm*, Mathematics of Computation, v. 37, n. 156, pp. 425-433, 1981.

58. TOINT, PH. L.*A Note About Sparsity Exploiting Quasi-Newton Updates*, Mathematical Programming, v. 21, pp. 172-181, 1981.

59. VANDERPLAATS, G. N., *Numerical Optimization Techniques for Engineering Design: with applications*, McGraw-Hill, 1984.

60. VAUTIER, I., SALAUN, M. and HERSKOVITS, J., *Application of an Interior Point Algorithm to the Modeling of Unilateral Contact Between Spot-Welded Shells*, Proceedings of Structural Optimization 93, Rio de Janeiro, Brazil, 1993, Edited by J. Herskovits, Vol. 2, pp. 293 - 300, 1993.

61. WRIGHT, S. J.*Primal-Dual Interior-Point Methods*, SIAM, 1997.

62. ZOUAIN, N. A., HERSKOVITS, J., BORGES, L. A., and FEIJOO, R., *An Iterative Algorithm for Limit Analysis with Nonlinear Yield Functions*, International Journal on Solids and Structures, Vol. 30, pp. 1397-1417, 1993.

CHAPTER 3

STRUCTURAL OPTIMIZATION USING EVOLUTIONARY COMPUTATION

Christopher M. Foley

Department of Civil & Environmental Engineering, Marquette University
P.O. Box 1881, Milwaukee, Wisconsin 53201
E-mail: chris.foley@marquette.edu

In the past three decades, evolutionary computation has been shown to be a very powerful tool for structural engineers. Application of evolutionary computation methodologies have been spread far and wide throughout the field of structural engineering ranging from selection of shapes for relatively simple structural systems to designing active control systems to mitigate seismic response to determining the location and extent of damage within structural systems. The present chapter provides an overview of evolutionary computation including a brief history of its development and the types of algorithms that are considered to be forms of evolutionary computation. A basic discussion of the genetic algorithm and evolutionary strategy is provided within the context of application to a very simple structural engineering design problem. The chapter provides a bird's eye view and discussion of many applications of evolutionary computation in the field of structural engineering. A brief synthesis of recent applications of evolutionary computation in the field of structural engineering is provided and recommendations for future work are given.

1. Introduction

It is very easy for the cognitive human being to appreciate the seemingly perfect designs that have been generated through natural processes. The great white shark is but one example of a naturally evolved "machine" that has been optimized within an evolutionary timescale for living in the open ocean. Another interesting example of natural design optimization comes from the field of ethology. Although the ant is essentially without vision, ant colonies survive and thrive as a result of their methods used to locate food. The trial and error methodology used by ants to locate food sources is really nothing more than an optimization (search) algorithm used by these seemingly simplistic animals.

Classical or traditional algorithms for structural optimization are most often driven with deterministic mathematical re-sizing procedures and essentially one design is changed through iteration until a convergence criterion is reached. These methods have been implemented in a wide range of structural optimization applications and several excellent textbooks describing their fundamentals are available[1,2]. These direct and gradient-based solution algorithms are founded upon a uniquely human field describing physical and natural phenomena – mathematics.

While direct and gradient-based algorithms can be considered as very powerful search and optimization tools, they are not without difficulties in application for many practical engineering design optimization problems[3]. Direct mathematical methods are point-to-point search algorithms that use objective function and constraint values to guide the search through feasible decision space. Gradient-based methods use derivatives of objective functions and/or constraint equations to guide the search. As a result, convergence of these algorithms depends upon selection of an initial solution for subsequent modification through iteration and design variable changes. Poor initial design selections can set the algorithms off in unprofitable directions and oftentimes, the algorithms can get stuck in sub-optimal regions of the decision space.

Mathematical algorithms are often problem-specific and the efficiency and capability of the algorithm in finding optimal solutions for general classes of problems varies. Practical engineering problems often utilize discrete decision variables (*e.g.* structural steel cross-section sizes in building design). · This is usually circumvented when mathematical optimization algorithms are employed by re-casting the discrete decision variables into continuous functions that facilitate differentiation. While this is an acceptable work-around, it implies that the algorithm is allowed to consider infeasible locations in decision space and computational time can be wasted determining objective function values for infeasible solutions. In addition, all possible combinations of two values for each decision variable (nearest upper- and lower-neighbor) need to be evaluated for final solution feasibility. This is a significant amount of extra effort and there is no guarantee that these combinations of nearest-neighbor decision variables will lead to the optimal solution. Finally, gradient-based algorithms do not lend themselves to easy implementation in parallel computing environments. Low cost multiple-processor computers and networks of computers have resulted in parallel computing environments becoming widely available and awaiting exploitation.

Taking advantage of the perfect vision of hindsight, one might surmise that it was only a matter of time for scientists and engineers to begin to look towards natural systems for design methodology inspirations. Much of the robustness found in natural systems comes from the ability to adapt to ever-changing environments. The ability of natural systems to change and adapt to their environments has undeniable parallels within the engineering design field. The concept of developing engineering designs by generating populations of potential solutions (rather than single solutions) and mimicking the evolutionary process found in nature to guide the search towards an optimal solution is the highest-level definition of *evolutionary computation* (EC). These characteristics are what distinguish EC from mathematical computation when applied to structural optimization problems. Evolutionary computation includes stochastic components not present in classical mathematically-based optimization algorithms.

It is commonly believed that the field of evolutionary computation has evolved (pun intended) from the activities of three research communities working in widely dispersed geographic locations[3-5]. Taxonomically speaking, three distinct flavors of evolutionary computation emerged in the 1960's. The first is called the *evolution strategy*[6-8]. A second approach, principally related to the evolution strategy, is called *evolutionary programming*[9-11]. A third algorithmic approach that was developed through the desire to simulate the adaptive behavior seen in natural systems has been called the *genetic algorithm*[12-16].

A historical perspective regarding the development of evolutionary computation has recently been provided[4,5]. Essentially four decades (1960's through 1990's) are contained in this perspective. Although conceptualizing evolution as an optimization problem and subsequent development of computer algorithms for optimization using evolution as a metaphor occurred in the 1930's and late 1950's[4]; it wasn't until widespread availability of computers in the 1960's that the tremendous possibility for automating the optimization process using evolutionary-based algorithms was realized. Thorough study of evolutionary algorithms then proceeded as the desire to improve algorithm performance increased. This was termed the "explorative 1970's"[4]. As the theory and behavior of the evolutionary algorithm became better defined and understood, researchers began seeking more widely varied applications of evolutionary computation in a variety of engineering design fields. This period has been called the "exploitative 1980's"[4]. The 1990's has seen the evolutionary computation community begin putting together a unifying view of evolution strategies, evolutionary programming and genetic algorithms and as a result, this decade has been called the "unifying 1990's"[4].

The engineering field has had significant involvement in the development of theoretical foundations to help explain the workings of evolutionary algorithms.

At present, the field of structural engineering appears to be in an exploitative period of its own. In the decades following the 1980's, the field of structural engineering has seen a large number of applications of evolutionary computation in design. These applications have ranged from simple linear elastic analysis-based design, to design of structural control systems, to inelastic analysis-based design, to inelastic time history analysis-based design of structural systems for optimized performance. With these decades of exploiting applications of evolutionary computation in structural design came exploration of evolutionary algorithm parameters, applications of parallel computing, and novel methods to represent systems within the evolutionary algorithm. A bird's eye view of this progress has yet to be developed.

There a several goals of the present chapter. First of all, it is hoped that it can become a very useful starting point for researchers and students in the field of structural engineering in their journey to understand and apply evolutionary computation to the increasingly complex design problems that are being tackled by the modern structural engineer. Secondly, the chapter seeks to provide the reader with a concise, yet complete, summary of recent research efforts in the field of optimal design that utilize evolutionary computation. The chapter will, unfortunately, focus on civil engineering design applications (mainly in the field of structural engineering) as these are most familiar to the author. The application review in the chapter focuses on contributions to the body of knowledge made during the years 2000 through 2006. Excellent resources for structural optimization research prior to 1999 are available[17]. Furthermore, state-of-the-art reviews of evolutionary computation applications in structural design have recently been published[18]. It is the goal of this chapter to provide the reader with an additional review to complement these former efforts. Finally, the author hopes to provide the reader with sources of further information that relate to both the theory and application of evolutionary computation as time marches on.

2. Optimization Problems and Complexity

It is prudent to begin the discussion of evolutionary computation by refreshing one's memory with regard to optimization problem statements that are typically found in engineering design. Structural optimization algorithms are generally formulated to tackle optimization problems whose statements take the following general form;

$$\textit{Maximize:} \qquad f_m(\mathbf{x}),\ \frac{1}{f_n(\mathbf{x})} \qquad m=1,2,\ldots,M \ \text{ and } \ n=1,2,\ldots N$$

Subject To:

$$g_j(\mathbf{x}) \leq 0 \qquad j = 1, 2, \ldots, J$$
$$h_k(\mathbf{x}) = 0 \qquad k = 1, 2, \ldots, K$$
$$x_i^L \leq x_i \leq x_i^U \qquad i = 1, 2, \ldots, I$$

The algorithm employed to solve the multiple objective optimization problem illustrated above will seek to define a vector of design variables, $\mathbf{x}$, within upper- and lower-boundaries that satisfies all inequality, $g(\mathbf{x})$, and equality, $h(\mathbf{x})$, constraints while maximizing the objectives. It should be noted that the present problem illustration assumes that maximizing the inverse of an objective is, in essence, minimization.

The vector of design variables can be used to define multi-dimensional *decision space*, and when the design variable vector is completely defined, there is a mapping of decision space to *objective space* that is unique to the problem being solved[3].

The complexity of the design problem can vary dramatically depending upon the number of objectives, the number of constraints, and the size of the decision space. Furthermore, the mapping of decision space to objective space can result in increased problem complexity. A relatively simple example is characterized by: a single objective being minimized (*e.g.* weight); a small design variable space (*e.g.* 10 discrete cross-sectional areas); relatively few constraints; and instances where constraints and objectives are evaluated using linear elastic structural analysis (*e.g.* elastic analysis of a truss). A relatively complex design problem characteristic of modern structural engineering can be described as having: multiple objectives (*e.g.* minimizing fabrication complexity, minimizing weight, maximizing confidence in meeting a desired performance level during earthquake); very large decision variable space (*e.g.* 250+ discrete steel wide-flange shapes found in buildings); many constraints (*e.g.* buckling, plastic hinge rotation, collapse load limit); and instances where constraints and objectives require advanced analysis methods (*e.g.* inelastic static analysis; inelastic time-history analysis). Optimization algorithms based upon evolutionary computation have been shown to be applicable to wide ranges of problem complexity.

When multiple objectives are considered in the optimal design problem, one must be sure to evaluate the objectives chosen to ensure that they do indeed conflict with one another. In other words, the objectives should have no interdependence. If non-conflicting objectives are chosen when defining the optimization problem, there will be one unique solution. When optimal design problems involve multiple competing objectives are formulated, Pareto optimal fronts in objective space can be defined and a single optimal design is defined by the preference expressed by the designer. When one seeks optimal solutions to these multi-objective problems, diversity in the solutions generated is highly useful. Evolutionary computation has been demonstrated to be very useful for providing diverse solutions along Pareto optimal fronts for multiple-objective optimization[3].

To date, evolutionary computation has been shown to be applicable in the widest range of problem complexities of any algorithm capable of selecting the decision (design) variables in an optimal manner. It is for this reason that evolutionary computation should be included in the arsenal of algorithms used by the modern structural engineer. The next section of the chapter proceeds to discuss the fundamental components of algorithms based upon evolutionary computation principles and elucidates differences between evolution strategies, evolutionary programming, and genetic algorithms when applied to structural optimization problems.

3. Fundamentals of Optimal Design Using Evolutionary Computation

The traditional engineering design problem needs to be cast into a form that is suitable for application of evolutionary computation. The basis of this transformation can be an analogy to Darwinian evolution. The characteristics of a Darwinian evolutionary system are[4]:

- single or multiple populations of individuals competing for resources that are limited;
- birth and death of individuals over time resulting in dynamically changing populations;
- the definition of *fitness* that characterizes the quality of an individual in the given environment thereby reflecting its ability to survive and reproduce;
- the notion of inheritance where parental offspring have characteristics of both parents, but are not identical to the parents.

Given the desire to look at solutions to optimal design problems within the context of a Darwinian evolutionary system, it is useful to select a relatively simple optimal design problem statement and use this problem statement in subsequent discussions. Therefore, let us consider a simple optimization problem based upon design of a simple cantilever beam with defined length and tip loading similar to that shown in Figure 1 adapted from the outstanding example given by Deb[3].

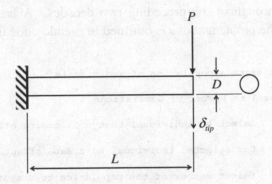

Figure 1. Simple Cantilever for Design Optimization Problem.

An optimal design problem for this simple cantilever can be posed as a single-objective, or multiple-objective problem. The most general case with multiple objectives is stated below:

Minimize: $f_1(\mathbf{x}) = Wt = \rho \cdot A \cdot L = \rho \cdot \dfrac{\pi}{4} \cdot D^2 L$

$$f_2(\mathbf{x}) = \delta_{tip} = \frac{64P}{3E\pi} \cdot \frac{L^3}{D^4}$$

Subject To: $\sigma_{max} = \dfrac{32P}{\pi} \cdot \dfrac{L}{D^3} \le \sigma_{allow}$

$\delta_{tip} \le \delta_{allow}$

$10\ mm \le D \le 40\ mm$ (2 *mm* increments)

$200\ mm \le L \le 830\ mm$ (10 *mm* increments)

Several parameters used in the problem are defined as follows: ρ is the density of the material; E is the elastic modulus of the material; σ_{allow} is the allowable maximum stress in the cross-section; and δ_{allow} is the allowable deflection at the tip of the cantilever. The vector of design variables is $\mathbf{x} = \begin{bmatrix} D & L \end{bmatrix}^T$. Combinations of diameter and length of the cantilever will be sought to minimize one or both objectives depending upon the problem considered.

Evolutionary computation facilitates relatively easy conceptualization of an algorithm to search for values of the design variables that define the optimal design. Initially, we can begin to talk about a simple evolutionary algorithm form that does not contain many nuances that research efforts have sought to

C. M. Foley

define and refine throughout the preceding two decades. A basic evolutionary
algorithm to solve the problem above is outlined in pseudo-code in Figure 2.

```
Generate a population of individuals

for 1 to Number_Of_Generations

        Select an individual that will create offspring

        Use selected individual to create offspring

        Select member of the population to die off

        Evaluate stopping criterion

Report "best" individual found through evolution
```

Figure 2. Pseudo-Code for Simply Evolutionary Algorithm (adapted from De Jong[4]).

Prior to proceeding forward to look at formulations for GA, ES, and EP
approaches to solving the simple cantilever optimization problem, it is prudent to
discuss how populations of individuals may be created. When applying
evolutionary computation, and individual is nothing more than a combination of
design variables. These design variable vectors can be given an evolutionary
reference by generating a slightly different characterization for the individual (*i.e.*
a solution to the optimization problem) as shown below;

$$\mathbf{x} = \langle diameter \quad length \rangle$$

With reference to genetics, individual design variable vectors can be thought of
as the *genotype* or *phenotype* for the individual. The genotype for an individual
within the current optimization problem is a chromosome with two genes. When
these genes are given values, they result in a unique individual with its own
physical traits: diameter and length. The phenotype for the individual in the
design problem is the observable physical traits of the solution: diameter and
length.

The unique features of an individual (*e.g.* diameter and length) can take on a
range of values for the problem being considered. Again, with a genetics
analogy, these values can be referred to as *alleles*. In the current optimization
problem, the alleles making up an individual come from the ranges of the design
variables contained in the problem statement. The alleles describing the
phenotypic trait of diameter can take on values in a range from *10 mm* to *40 mm*

with *2 mm* increments; and the alleles describing cantilever span can take on values in the range of *200 mm* to *830 mm* in *10 mm* increments.

In general, two forms for the alleles have been considered in optimization problems solved using evolutionary computation. These forms loosely correspond to the flavor of evolutionary computation being implemented. Binary alphabets have been used to define alleles within the chromosomal representations of individuals used in genetic algorithms. Real-valued alleles have been used in both genetic algorithms and evolution strategies. Object-oriented representations of individuals have also recently emerged in structural optimization.

One very important aspect to the evolutionary algorithm shown in Figure 2 is the need to make decisions regarding individuals chosen to reproduce and individuals chosen to die off during the evolution. Within the field of biology, the quality of an individual is judged using its ability to survive and produce viable offspring. As a result, quality can only be observed *after* a large portion of the evolutionary process has already taken place. Therefore, quality is judged using fitness based upon hindsight from a vantage point that is somewhere along the evolutionary timescale (we often assume we are looking back from the end of the evolutionary scale). Applications of evolutionary computation as implied in the algorithm shown in Figure 2 demand that the quality of the individual be judged regularly *during* the evolutionary process. Therefore, applications of evolutionary computation generally define the quality of an individual using *objective fitness*, which is an objective measure of its ability to satisfy the constraints and objectives in the optimization problem posed during the evolutionary process. This is often shortened in the optimization literature to *fitness* and technically this use of the term is in conflict with its biological origination[4]. In this chapter, we will use *objective fitness* to describe the quality of an individual during the evolutionary process.

The fundamental differences in the three forms of evolutionary computation: genetic algorithms (GA), evolution strategies (ES), and evolutionary programming (EP): can be easily seen if they are formulated within the context of the simple evolutionary algorithm pseudo-code shown in Figure 2 to tackle the optimization problem described in Figure 1. We can now turn our attention to specific forms of evolutionary algorithms and skeletal applications of how these algorithms would be used to tackle the optimization problem postulated in Figure 1. There are many good resources available describing the details of applying GA's, ES and EP[3-5, 19-24] to solve engineering problems and these details will not be reproduced here. Instead, an overview of how these evolutionary algorithms are used to solve the multiple-objective optimization problem posed in Figure 1 is provided in the following sections.

3.1. *Objective Fitness*

Judging the quality of individuals generated throughout the evolutionary process is essential for selecting individuals for reproduction. The objective fitness of an individual is a function of how well that individual meets the desired objectives and satisfies the constraints put forth in the problem. Assigning this fitness begins with evaluation of the individuals. This evaluation can be quite complex and is dependent upon the complexities of the analysis methods needed to conduct the evaluation. It should be noted that the notion of objective fitness is common to all forms of evolutionary algorithm, and as such, it is treated first as a stand alone concept prior to discussion of specific forms of evolutionary computation.

For the current cantilever problem given in Figure 1, the most basic form of objective fitness comes from simply considering the objective values corresponding to the individual as implied below;

$$f_1(\mathbf{x}_i) = \rho \cdot \frac{\pi}{4} \cdot D_i^2 L_i \tag{2}$$

$$f_2(\mathbf{x}_i) = \frac{64P}{3E\pi} \cdot \frac{L_i^3}{D_i^4} \tag{3}$$

Single-objective fitness can be defined by simply applying weighting factors to each objective whose magnitude is dependent upon user preference as indicated below;

$$F(\mathbf{x}_i) = w_1 \cdot f_1(\mathbf{x}_i) + w_2 \cdot f_2(\mathbf{x}_i) \tag{4}$$

It should be noted that equations (2) and (3) involve weight and deflection, respectively. As a result, the weighting factors might also include normalization so that each component in the fitness definition has appropriate scale when defining a single objective fitness for the individual.

Engineering optimization problems always include one or more constraints to which potential solutions must adhere. The genetic algorithm handles constraint satisfaction through imposition of penalties and therefore, the constrained optimization problem statements must be recast in an unconstrained format.

The most direct way to do this is through imposition of linear or nonlinear penalty functions that depend upon the degree to which a penalty is violated. A penalized objective fitness for the cantilever problem with two constraints is easily formulated using equation (4) as follows;

$$F(\mathbf{x}_i) = \left[w_1 \cdot f_1(\mathbf{x}_i) + w_2 \cdot f_2(\mathbf{x}_i) \right] \cdot \prod_{k=1}^{N_{cons}} (1 + \Phi_i)_k \tag{5}$$

The individual penalty functions corresponding to the deflection and stress constraints in the optimization problem, shown in Figure 1, can take a linearized form and a nonlinear form. The linearized form can have slope variation to enhance the severity of the penalty as the violation increases. The linear penalty form can be taken as[25];

$$\Phi_\delta = \begin{cases} 0 & if \; |\delta_{max}| / \delta_{allow} \leq 1 \\ k_\delta \cdot \dfrac{|\delta_{max}|}{\delta_{allow}} & if \; |\delta_{max}| / \delta_{allow} > 1 \end{cases} \tag{6}$$

$$\Phi_\sigma = \begin{cases} 0 & if \; |\sigma_{max}| / \sigma_{allow} \leq 1 \\ k_\sigma \cdot \dfrac{|\sigma_{max}|}{\sigma_{allow}} & if \; |\sigma_{max}| / \sigma_{allow} > 1 \end{cases} \tag{7}$$

A nonlinear form can be written as[25];

$$\Phi_\delta = \begin{cases} 0 & if \; |\delta_{max}| / \delta_{allow} \leq 1 \\ k_\delta \cdot \left(\dfrac{|\delta_{max}|}{\delta_{allow}} - 1 \right)^{n_\delta} & if \; |\delta_{max}| / \delta_{allow} > 1 \end{cases} \tag{8}$$

$$\Phi_\sigma = \begin{cases} 0 & if \; |\sigma_{max}| / \sigma_{allow} \leq 1 \\ k_\sigma \cdot \left(\dfrac{|\sigma_{max}|}{\sigma_{allow}} - 1 \right)^{n_\sigma} & if \; |\sigma_{max}| / \sigma_{allow} > 1 \end{cases} \tag{9}$$

The scaling multipliers, k_δ and k_σ, are defined by the algorithm user and can be used to enhance the penalty in a manner that is proportional to the violation. The exponents, n_δ and n_σ, can be used to scale the penalty in a manner that is non-proportional to violation.

Identifying the appropriate scaling multipliers and exponents to apply in a particular problem requires some measure of user intuition. Poor selection of these parameters can cause premature convergence to sub-optimal solutions and also epistatic behavior in the algorithm. Automatically adjusting penalties within the evolutionary process has been proposed as a solution to this dilemma[26, 27].

Generating algorithms that can handle multiple objective fitness quantities is an active area of research in the communities of evolutionary algorithm theory and engineering. Genetic algorithm researchers often refer to genetic algorithms designed to handle multiple-objective problems as Multiple-Objective Genetic Algorithms (MOGA's).

The optimization problem currently considered can also be maintained as a multiple objective problem without pre-assigning preference through weighting factors. If one were to maintain the constraint handling process described previously, but preserve the multiple objective nature of the design problem, the objective fitness should now consider two distinct components,

$$F_1(\mathbf{x}_i) = F_1(\mathbf{x}_i) \cdot \prod_{k=1}^{N_{cons}} (1 + \Phi_i)_k \tag{10}$$

$$F_2(\mathbf{x}_i) = F_2(\mathbf{x}_i) \cdot \prod_{k=1}^{N_{cons}} (1 + \Phi_i)_k \tag{11}$$

It should be noted that the form of equations (10) and (11) are not the only ways to handle constraints in the definition of objective fitness. Improvements in constraint handling in MOGA's have been proposed[28].

When multiple objective optimization problems are considered, the use of weighting factors to impose preference as done in equation (5) is omitted. This leaves sets of potential solutions in objective fitness space that the engineer can use to aid in decision making. The Pareto optimal set of solutions is defined as the set of solutions lying along a Pareto optimal front[3]. When a candidate solution is said to be better than another candidate solution in meeting the objectives, it is said to dominate the other solution. The set of non-dominated solutions form the Pareto optimal set of solutions. The landmark resource for multiple objective optimization using evolutionary computation is the text by Deb[3] and a very nice summary of the state-of-the-art in multiobjective evolutionary algorithm developments has been generated[29].

Generating a genetic algorithm or evolution strategy to handle multiple objective optimization problems is not trivial. When iterations in the evolutionary process are completed, a non-dominated surface will form in objective fitness space. Identifying non-dominated Pareto front must consider potential objective fitness surfaces that are both convex and non-convex. Furthermore, if large populations are utilized, maintaining these large populations with complex and time consuming fitness evaluations can result in significant solution time. Procedures for identifying Pareto sets of solutions have been proposed[30-35].

Research has found that non-dominated surfaces of solutions generated for multi-objective optimization problems can include clustering of solutions within objective fitness space. Ensuring that an evolutionary algorithm can generate solutions spread out along the Pareto optimal surface has also been addressed in research activity[30, 32, 36, 37].

Selecting individuals for reproduction[‡] can make for rather interesting difficulties when implementing evolutionary algorithms. The non-dominated set of solutions should be given some measure of preference in the selection process, but if too much emphasis is placed on these solutions, genetic diversity may be lost. To combat this, local selection algorithms[38] and improved sampling methods have been proposed[39].

It may be useful to maintain individuals with high objective fitness throughout the evolutionary process. When the objective fitness is defined as a single individual, this individual is the *elite* individual. When a Pareto set of solutions is considered *elite*, which solution or solutions are to be maintained? Researchers have addressed this issue and made recommendations on how to incorporate elitism[§] within the evolution strategy[40].

The flurry of developments in multiple objective evolutionary algorithms prior to 2000 demanded that an assessment be made to objectively evaluate the performance of these algorithms on test problems. Detailed comparisons of multiple objective evolutionary algorithm performance are available[3, 41].

When multiple objectives are incorporated into an optimal design problem, the engineer will be required to eventually decide on the solution from the Pareto set or front that is most appropriate. This suggests that input from the engineer *during the evolutionary process* may be of benefit to the algorithm. Such an environment has been proposed for multiple objective problems[42] and the interaction of user with algorithm can result in dynamic redefinition of objective fitness space during execution of the algorithm.

3.2. *Genetic Algorithm*

The genetic algorithm (GA) maintains the closest link to genetics of any of the three main types of evolutionary algorithms. The phenotypes and/or genotypes of the individual are most-often described using the genetic concept of a *chromosome*. Furthermore, the genetic algorithm also includes more formalized "laws of motion"[4] to simulate the evolutionary system. A simple flowchart illustrating implementation of a genetic algorithm is given in Figure 3. One iteration through the genetic algorithm is often called a *generation*. The termination criterion is often a user-defined maximum number of generations or lack of significant improvement in solutions.

[‡] Reproduction will be the term used later in the chapter to describe recombination and mutation.
[§] Elitism will be discussed in greater detail later in the chapter.

The genetic algorithm contains many more steps than those implied in the simple evolutionary algorithm pseudo-code given in Figure 3. First of all, each individual in the population needs to be evaluated. This evaluation includes computing the objective function value(s) for the individuals as well as determining if the constraints are satisfied and these computations are often based upon the results of structural analysis in structural engineering systems.

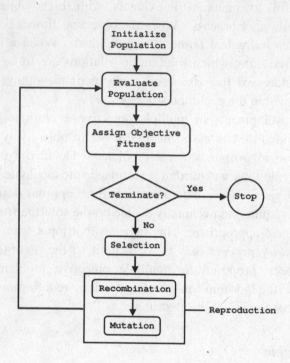

Figure 3. Flowchart of Typical Genetic Algorithm.

The quality of the individual in a GA is often based upon the objective fitness of the individual, which is evaluated using the value of the objective function(s) as well as satisfaction of the constraints. Constraint handling is also a very rich area of research in the GA community, but most often the objective fitness of the individual is scaled using penalties that are a function of the degree to which the constraints are violated.

The selection of individuals to participate in reproduction can vary with GA implementation. The reproduction phase of the algorithm (*i.e.* recombination and mutation phases) depends upon the phenotypic/genotypic representation for the individuals that is used in the GA.

The following sections of the chapter will outline typically employed methods for representing individuals, evaluating objective fitness, conducting selection of individuals for reproduction, methods of recombination, and mutation techniques.

3.2.1. *Individual Representation*

The phenotypic or genotypic representation of the individuals within a genetic algorithm is most often a chromosome of binary digits. In keeping with the biologic analogy, each binary digit can be thought of as a *gene*. There have been theorems (*e.g.* the schema theorem) proposed that suggest binary alphabet representation facilitate very efficient searching[21].

If we consider the two allele individual of the cantilever beam problem, a binary string chromosome and its decoded design variable vector are,

$$0110\,110101 \quad \Rightarrow \quad \mathbf{x} = \lfloor D \quad L \rfloor^T = \lfloor 22 \quad 730 \rfloor^T$$

The decoding of the alleles in each segment of the chromosome can be done very simply using the following expression[3, 21];

$$x = x_{min} + \left[\frac{x^{max} - x_{min}}{2^{L_{bs}} - 1} \right] \cdot DV(bs) \tag{1}$$

where: L_{bs} is the length of the binary string used to encode the design variable; x_{min} is the lower-bound for the design variable; x^{max} is the upper-bound for the design variable; and $DV(bs)$ is the decoded value of the binary string.

Individuals need not be represented in the form of binary string chromosomes. In fact, binary string chromosomes can make search difficult because neighboring solutions can have very significant differences in their binary string representations resulting in Hamming cliffs[3]. Obtaining high-precision solutions can also result in very long binary strings. Long binary strings also imply larger population sizes. Real-value GA's have been proposed to address these issues[43]. Sometimes referred to as decimal-coding, these representations of candidate solutions have been shown to be effective in structural optimization problems[44]. Additional treatment of real parameter GA's can be found elsewhere[21].

Recently, hierarchical representations of individuals similar to representations used in genetic programming have also been proposed[45]. Representing individuals within rational hierarchies alleviates the need for long chromosomal representations for problems with large numbers of design variables. This concept has been extended to object-oriented representations for individuals in complex design problems involving significant numbers of design variables[45, 46]. The use of network random keys in conjunction with tree representations of individuals has been shown to facilitate very efficient solution algorithms[47].

3.2.2. *Selection*

The selection mechanism chosen for implementation in the evolutionary algorithm is very important. The primary goal of selection is to find those candidate designs that are *good* and those that are *not so good* within the population. It is hoped that the algorithm employed will then create additional copies (not exact copies though) of good solutions and allow the not so good solutions to simply die off.

Difficulty often arises in applying selection operators as part of evolutionary algorithms. Some candidate designs that are not rated highly in objective fitness may have redeeming qualities that other solutions may benefit from. To only select the *premier* solutions for subsequent reproduction is analogous to "throwing the baby out with the bathwater". Within the context of search, this implies that the algorithm is honing in on specific regions of the search space without adequate exploration. On the other hand, if the algorithm allows too much disparity in solution quality of individuals selected for reproduction, the algorithm can simply search too large a space thus making convergence of the algorithm difficult. The exploration of the search space within the context of the genetic algorithm is sometimes evaluated using the concept of *selection pressure*. High selection pressure implies that only the elite solutions will be selected for reproduction and low selection pressure is more egalitarian and allows many more individuals to participate in reproduction.

The challenge in developing selection mechanisms or selection operators for evolutionary algorithms is to facilitate exploration, while maintaining exploitation of good solutions. Various selection mechanisms have been developed to meet this challenge.

The *tournament* selection mechanism is a procedure whereby contests or tournaments are held between two members of the population at a stage in the evolution. The winners of these tournaments (those with better objective fitness) are then placed in a *mating pool* until it is filled.

The *fitness proportionate* selection mechanism assumes that copies of individuals are generated for the mating pool. The number of copies is proportional to that individual's objective fitness relative to the average objective fitness of the population. Therefore, better individuals will have greater number of copies. When the proportionate selection mechanism is executed, it is often easiest to implement by simulating a *roulette wheel*. In other words, the slice of the wheel for an individual with better objective fitness will be larger than a lower quality individual. Therefore, the probability of the better individual being chosen for the mating pool is higher. There are a number of modifications to roulette-wheel selection that have been proposed[3].

When proportionate selection is used, scaling issues can arise when there is large disparity among the objective fitness magnitudes for individuals in the population. If individuals within the population begin to have objective fitness magnitudes that greatly overshadow others, their slice of the roulette wheel can become exceedingly large to the point where certain individuals *take over* the subsequent population. Measuring this tendency is done using the concept of *take over time*[21]. Scaling objective fitness is one way to inhibit or slow take over by dominant individuals.

A selection method that avoids scaling issues is *rank-based* selection. Implementing this mechanism simply involves ranking individuals from worst objective fitness to best. The rank-based fitness is then the individual's rank in the population. Selection pressure and take over time have been studied within populations with ranked fitness and selection mechanisms capable of dynamically controlling selection pressure and take over time through user-defined parameters have been proposed[48].

3.2.3. *Recombination*

Recombination is a process by which two candidate solutions to the optimization problem are *combined* to create one or more new potential solutions. These new solutions contain aspects of both *parent* solutions and assuming the parents have "high quality" genetic material, the offspring will as well.

Creation of new genetic material in the population, or new candidate designs, is most often accomplished within the realm of genetic algorithms using the *crossover* operator. When individuals are represented using binary chromosomes, three crossover operations are most often conducted: single-point crossover; multi-point crossover; and uniform crossover. The selection mechanism carries out the task of identifying solutions to become members of the mating pool. Two candidate designs are then selected for mating (recombination) and the crossover mechanism is applied to create offspring that have "genetic" material from both parents. The most common crossover mechanisms are illustrated in Figure 4.

Single point crossover begins with identification of the parent strings. A point along the string is chosen and the segment of the parent chromosome to the right or left of the crossover point is exchanged to create new candidate designs. For the present cantilever design problem, the two parent designs at the left of the figure result in offspring that maintain their diameter gene, but have altered genes defining the span. It should be emphasized that new genes will only be introduced into the offspring when the crossover point lies within the gene. When the crossing site corresponds to the end of one gene and the beginning of the next gene, crossover will simply result in a gene exchange with no new genetic material being introduced. Thus, one can think of single point crossover as minimally disruptive to the genetic material and exploration of the design variable space remains relatively close to the original solutions.

$\lfloor 22 \quad 730 \rfloor^T$ 0 1 1 0 1 1 $\vert$ 0 1 0 1 $\qquad$ 0 1 1 0 1 1 $\vert$ 1 0 1 1 $\lfloor 22 \quad 790 \rfloor^T$

$\lfloor 34 \quad 630 \rfloor^T$ 1 1 0 0 1 0 $\vert$ 1 0 1 1 $\implies$ 1 1 0 0 1 0 $\vert$ 0 1 0 1 $\lfloor 34 \quad 570 \rfloor^T$

Single Point Crossover

$\lfloor 22 \quad 730 \rfloor^T$ 0 $\vert$ 1 1 0 1 1 0 $\vert$ 1 0 1 $\qquad$ 0 $\vert$ 1 0 0 1 0 1 $\vert$ 1 0 1 $\lfloor 18 \quad 650 \rfloor^T$

$\lfloor 34 \quad 630 \rfloor^T$ 1 $\vert$ 1 0 0 1 0 1 $\vert$ 0 1 1 $\implies$ 1 $\vert$ 1 1 0 1 1 0 $\vert$ 0 1 1 $\lfloor 38 \quad 710 \rfloor^T$

Multi-Point Crossover

$\lfloor 22 \quad 730 \rfloor^T$ 0 1 1 0 1 1 0 1 0 1 $\qquad$ 0 1 0 0 1 1 1 0 0 1 $\lfloor 18 \quad 770 \rfloor^T$

$\lfloor 34 \quad 630 \rfloor^T$ 1 1 0 0 1 0 1 0 1 1 $\implies$ 1 1 1 0 1 0 0 1 1 1 $\lfloor 38 \quad 590 \rfloor^T$

Uniform Crossover $\left(p_{cross} = 0.50 \right)$

Figure 4. Crossover Operations Commonly used in Binary String Genetic Algorithms.

The second crossover mechanism in Figure 4 is *multi-point crossover*. In this recombination, two crossing sites are chosen at random and the central segment of the parent chromosomes are exchanged. As indicated in Figure 4, this crossover operator has a higher probability of creating offspring that have do not have common genes with the parent strings. It should be noted that this may only be true for chromosomes with few genes. As the design variable numbers increase, more than two crossing sites may be required to carry out recombination that adequately searches the design space.

Uniform crossover is the final mechanism commonly used. This crossover operator is applied by simply moving bit-by-bit along the chromosomes and flipping a fair (two-sided) coin. If the toss results in heads, the bits (alleles) are exchanged. The results in Figure 4 were obtained using a random number generator with bit exchange occurring when the random number exceeded 0.50.

The "optimal" crossover scenario for all problems remains to be defined. In fact, recombination needs to be approached in significantly different ways when binary chromosomes are not used to represent individuals. When the hierarchical representation of individuals is used[46], recombination as envisioned using crossover cannot create new genetic material. As a result, different crossover mechanisms that have similarities with genetic programming have been developed. These recombination mechanisms, termed homologous and non-homologous crossover[45, 46, 49], help to facilitate recombination similar to that seen in more traditional crossover operations on binary strings. These hierarchical representations have a significant correlation to genetic programming representations of candidate solutions.

Representing candidate solutions in tree structures has great similarity to genetic programming representations for solutions. The schema theorem for binary string genetic algorithms[21] has been shown to be the driving force behind the success of these search algorithms. A schema theorem for sub-tree swapping crossover recombination has recently been proposed[50, 51]. This theorem is very useful for understanding how building blocks in hierarchical solutions can be generated and exploited in evolutionary algorithms.

When genetic algorithms are applied to real-parameter optimization problems, recombination is a tricky issue. Recombination operators using probability distributions around parents have been proposed. A parent-centric operator[43] along with other algorithmic advancements have been demonstrated to show very good performance on test problems and also demonstrate scalability.

It should be noted that there is no guarantee that the crossover mechanism employed will result in a better candidate solution. The evolutionary computation literature is populated with reporting of many research efforts seeking to determine the best recombination operators. It is left to the reader to review the evolutionary computation literature using the sources referenced in the chapter.

3.2.4. *Mutation*

The next component in reproduction for genetic algorithms is *mutation*. Mutation is a process by which an individual in the offspring population is selected at random and is mutated through exchanging alleles in the individual chromosome. Mutation seeks to generate diversity in the population and therefore, explore the design space. The location of the "bit-flipping" is usually randomly chosen and the severity of mutation will be dependent upon the position of the mutated bit. Mutation can create better solutions or worse solutions – there is no guarantee to improve the candidate with mutation.

As stated earlier, the objective of mutation is to create diversity in the population. This diversity can help the genetic algorithm avoid being trapped in local minimums in objective space. As a result, mutation is sometimes thought of as a global search mechanism, while crossover is sometimes thought to be a local search mechanism.

More complicated mutation operations have been proposed. Intelligent mutation operators designed to simulate local and global search characteristics of crossover and mutation in problems that involve representations of individuals that do not involve bit-string chromosomes have been thoroughly evaluated[45, 46].

3.2.5. Elitism

Because crossover and mutation can result in candidate solutions that are worse than the best solution resident in the mating pool, genetic algorithms often maintain the best solution from a previous population through a mechanism termed *elitism*. In simplest terms, elitism is simply taking the best solution (or solutions) from one generation and simply carrying them over to the next generation.

3.3. Evolution Strategy

The evolution strategy (ES) was developed to solve problems that involved significant effort in evaluating objective fitness and focused on problems with real parameter design variable representation. The evolution strategy was formulated to involve small population sizes and utilize reproductive operations that are mutation-based. The *strength* of the mutation can be varied. The simple, two-member $(1+1)ES$ is flowcharted in Figure 5.

As implied in the flowchart, the $(1+1)ES$ involves one individual in the parent population that develops one offspring through mutation. There is no recombination in the two-member ES. The mutation operator in the ES is often based upon a zero-mean normal probability density function. This is symbolically denoted $PDF_{norm}(0, \sigma)$ in the flowchart. The mutation *strength*,

σ, is user defined and it is nothing more than the standard deviation for this distribution. If the mutation strength is increased, the new design variable vector for the next iteration will vary significantly from the parent. If the mutation strength is reduced, this variation will be smaller as exhibited by the smaller magnitude for the standard deviation chosen.

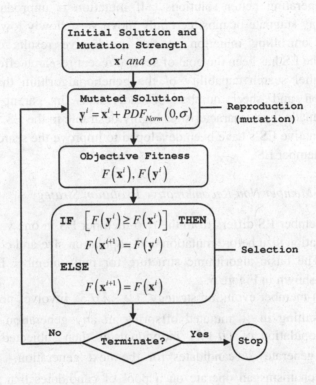

Figure 5. Flowchart for Two-Member $(1+1)ES$.

One can certainly surmise that when the mutation strength is large, the algorithm will have significant global search capability. However, if the mutation strength is small, the algorithm will search locally around the current parent solution. This also implies that the ES is naturally well suited to dynamically changing the mutation strength with advancing generation with the goal being to balance exploration of the search space and then exploitation of good solutions when profitable regions of the search space is found.

A typical ES implements only mutation in the generation of new solutions. Therefore, the typical ES is often called *non-recombinative*. Researchers soon realized that the genetic algorithm operator of crossover is useful within the context of the evolutionary strategy and *recombinative* ES's have been developed.

The use of a single solution to start the two-member ES should give significant concern with regard to the ability of an algorithm started off in this manner to find optimal solutions. The two-member ES is highly dependent upon mutation generating better solutions. If mutation is unproductive, then the algorithm may stagnate or most certainly move very slowly toward the optimal solution. Controlling mutation so that productive results can maintained throughout the ES has been the goal of several recent research efforts[52].

The parallel search capability of the genetic algorithm that results from recombination and sheer number of individuals in parent and offspring populations made these characteristics highly desirable in the ES. Multi-member and recombinative ES's have been developed to improve the searching capability of the two-member ES.

3.3.1. *Multi-Member Non-Recombinative Evolution Strategy*

The multi-member ES differs from the two-member ES in one very fundamental way – population size before mutation and population size and constituents after mutation. The basic algorithmic structure for multi-member ES remains the same as that shown in Figure 5.

The multi-member evolution strategy, $(\mu + \lambda)ES$, involves populations of μ solutions resulting in λ mutated offspring in any generation (iteration). A temporary population of $(\mu + \lambda)$ individuals is then subjected to a selection operator to generate μ candidates for the next generation. Therefore, the selection mechanism can operate on a pool of candidates that is significantly larger than the initial population. Furthermore, if mutation does not yield significant improvements in the solutions, this version of the ES will tend to stagnate because the same solutions will be used over and over again. The $(\mu + \lambda)ES$ is also an *elitist algorithm*. The reason is that since the candidate pool prior to selection involves μ parent solutions, the selection operator may result in these same solutions propagating to the next generation.

The second form of multi-member evolution strategy is the $(\mu, \lambda)ES$. In this "flavor", μ parent solutions are used to generate λ new candidate designs with $\lambda \geq \mu$. The pool of candidate designs then advanced to the next generation through the selection operator does not include the original μ candidates. In this manner, the $(\mu, \lambda)ES$ is *non-elitist*.

3.3.2. *Recombinative Evolution Strategy*

A parallel with genetic algorithms can be maintained through further classifying evolution strategies as being recombinative or non-recombinative[3]. A non-recombinative ES does not include crossover-type operators (only mutation is used), while a recombinative ES incorporates crossover-type mechanisms. The original development of the ES involved real-parameter design variables although this really isn't a restriction on application.

The recombinative ES is implemented with parent designs being recombined prior to mutation. If binary string chromosomes are convenient for this recombination, then they most certainly can be used. If hierarchical (object-based) representations are best, then these can be used as well. The number of parents chosen for recombination is often denoted as ρ [3]. Therefore, the typical notation for a multi-member recombinative evolution strategy is $(\mu/\rho+\lambda)ES$, or $(\mu/\rho,\lambda)ES$.

3.3.3. *Self-Adaptive Evolution Strategy*

The use of the mutation operator alone and mutation strength and its representation as a normal probability density function with strength defined using the standard deviation in this distribution, provides the engineer with a parameter than can be optimized by the algorithm itself and even attached to specific design variables.

Self-adaptation within the context of the ES applied in a solution to a design problem has been classified into three types: isotropic, non-isotropic, and correlated[3]. In all three, the design variable vector for the problem is augmented to include the mutation strength as a design variable. There are multi-level or meta-ES methods of adaptation as well[3].

Isotropic self-adaptation assumes that the mutation strength is a single value applied to all design variables during the evolution. This strength is included as a single design variable in the candidate solution vector. Non-isotropic self-adaptation assumes that a different mutation strategy parameter is applied to each design variable and the design variable vector is augmented to include a number of mutation strength parameters equal to the number of design variables in the problem. This self-adaptation is designed to all the ES to learn and adapt to scenarios where the design variables contribute unequally to the objective fitness. Correlated self-adaptation includes covariance in the decision variable vector and is intended to tackle problems where design variables may be correlated to one another.

4. Applications in Structural Engineering

The field of civil-structural engineering can be thought of as being in the throws of an "exploitation" period with regard to application of evolutionary computation in optimized design. There have been many examples in the literature of successful application of evolutionary computation in structural engineering around the world. As of the date of publication of this chapter, application of evolutionary computation remains strong and the types of optimal design problems being tackled using evolutionary algorithms are increasing in complexity.

Generating a review is always subject to the possibility of omission. The present review will focus contributions during the period 2000 – 2006. The interested reader can consult several other resources for literature reviews covering the time period prior to the year 2000[18, 53].

It should be understood that the list of fields where evolutionary computation has been successfully applied grows each year and the contents of this section should not be used to construe limitations in the application of evolutionary computation. The reality is that application of evolutionary computation is only limited by the imagination.

4.1. *Bridge Maintenance and Management*

Bridge management systems have been developed to aid structural engineers with the arduous task of deciding when and what type of maintenance interventions to interject into transportation infrastructure (e.g. bridges) to extend service live and minimize cost. Using these systems to actively manage bridge maintenance cycles and minimize facility life-cycle costs has been shown to be possible using evolutionary computation.

Genetic algorithms have been applied to develop maintenance plans for existing bridges[54]. The proposed bridge management system employed a fairly traditional genetic algorithm to solve a multiple objective optimization problem that sought to minimize the cost of maintenance measures (*e.g.* repair, strengthening) and maximize typical rating measures of load carrying capability and durability during an "analysis period"[54]. The multiple objective optimization problem considered was recast as essentially a single-objective optimization problem using a prioritization scheme. A binary string chromosomal representation for maintenance plans was used to facilitate 10 scenarios ranging from epoxy-injection repair, to FRP strengthening, to no repair and no strengthening.

Diagnoses of damage within bridge systems include a fair amount of subjectivity. True experts with the ability to make accurate diagnoses of damage within the bridge structural system are not widely available and transportation agencies are often required to make decisions regarding damage using trained individuals making visual inspections of the bridge super- and substructure. Decision algorithms have been derived using a combination of genetic algorithms and data mining to reduce the computation time required to evaluate all possible combinations of condition attributes seen at a bridge structure to generate minimal decision algorithms describing damage[55]. Conditional attributes of damage in the bridge structures considered ranged from longitudinal cracking, to grid-like cracks, to reinforcement corrosion. Causes considered ranged from excessive wheel loads, to insufficient distribution of flexural reinforcement, to inappropriate curing. A hybrid rough-set data mining and genetic algorithm was employed to generate minimal decision tables that related a minimal number of conditional attributes of damage to its underlying cause[55]. It was shown that the methodology used could lead the way to developing minimal decision tables/algorithms for individuals with training in bridge inspection to arrive at damage condition assessments consistent with expert diagnoses.

It is recognized that life-cycle cost minimization through rational and targeted maintenance interventions is the most economical way to allow fixed funding sources to be rationally distributed throughout the life of a bridge structure. A modified genetic algorithm, coined a virus-evolutionary (VE) genetic algorithm[56] was proposed to generate optimal maintenance scenarios to minimize repair costs, strengthening costs and scaffolding costs. A comparison of the maintenance scenarios found using a traditional genetic algorithm and the VE-GA was provided. Examining the trade-offs between life-cycle maintenance costs, lifetime condition, and lifetime safety of bridges and bridge networks is a perfect application for the genetic algorithm. The GA's ability to generate suites of maintenance scenarios can facilitate the MOGA becoming a decision making tool for the bridge manager. Genetic algorithms have been applied to study these tradeoffs for individual bridges[57] and networks of bridges[58].

There is uncertainty associated with the deterioration process in bridge structures and their components. Including this uncertainty in the development of optimal maintenance scenarios for bridges and groups of bridges within a given transportation network is a logical step forward in improving the effectiveness of bridge management systems. Multiple objective optimization problems solved using genetic algorithms and Monte Carlo simulation and Latin-Hypercube sampling have been able to provide optimal bridge maintenance scenarios to minimize cumulative life-cycle maintenance cost, maximum the bridge condition and safety indices[59-61].

4.2. *Structural System and Component Optimization*

Evolutionary computation approaches to optimal design tend to be favored when discrete design variables are considered in the optimization problem tackled. These optimization problems are ubiquitous in civil-structural engineering and therefore, evolutionary computation procedures have seen wide-ranging application. The fact that evolutionary computation approaches do not utilize gradients of objective functions or constraints makes them ideal to handle optimal design problems where evaluation of objective and constraint satisfaction involves complex analysis procedures.

The goal of this section is to synthesize past research efforts related to discrete variable optimization applications of evolutionary computation for civil engineering structural design. There has been a large variety of problems attacked using evolutionary computation. However, these efforts can be classified in one of three categories: (a) deterministic structural optimization – DSO; (b) performance-based structural optimization - PBSO; and (c) reliability-based structural optimization – RBSO.

For purposes of this review, DSO can be classified as a Level 1 reliability format and is characterized by optimal selection of member/component sizes using partial representation of the uncertainty in loads applied to the structural system and the uncertainty in the resistance of the structural system as a whole and its components. This is the design format used if optimization is conducted using current load and resistance factor design codes[62, 63]. For the purpose of this synthesis, DSO procedures will include formulations where constraints are formulated using allowable stresses even though these types of constraints do not make an attempt at considering uncertainty. RBSO can be considered to follow a Level 2 reliability format and it includes full characterization of uncertainty in load and resistance through use of random variable representations. PBSO lies in between and it considers uncertainty in loading and response, but material properties of the components within the system are defined at the median (or other) values in lieu of a random variable cumulative distribution function.

The three characterizations above facilitate taxonomy of the approaches to structure and component optimization in the field of structural engineering and they will be used as an integral component of the synthesis to follow. The extent to which DSO, PBSO, and RBSO problems have been considered and the types of structural systems and response behavior that have been assumed within the optimization formulations can be inferred from the synthesis. Therefore, the reader will gain an appreciation for gaps in application so that the exploitative phase of evolutionary phase can continue. Furthermore, the reader will gain an appreciation for the increased complexity in the problems being tackled using evolutionary computation.

4.2.1. *Deterministic Structural Optimization (DSO)*

DSO procedures using evolutionary computation have been thoroughly spread out within the field of structural engineering ranging from optimization of reinforced concrete to steel structural systems. The analysis methodologies used to define individual fitness in these applications have ranged from linear static to nonlinear time-history analysis. This section of the chapter will generate a synthesis of DSO applications that have used evolutionary computation.

If one were to tally the number of research efforts that sought to tackle DSO problems using evolutionary computation, the structural steel system, would by far, be the most popular; however, reinforced concrete systems have also received some attention. Both of these systems are naturally populated with discrete design variables ranging from members (*e.g.* beams or columns) to connection strength and stiffness to reinforcing bar designations, etc.

Single-objective DSO applications within the realm of structural steel using evolutionary computation include frames and trusses involving both linear and nonlinear analysis. When single-objective problems are considered, it most often involves weight minimization of the structural system, but more advanced forms of single-objective optimization have been addressed.

DSO for minimum weight design of steel planar and space frame systems using linear elastic analysis has been considered[64,65]. Design variable optimization for spatial and planar steel frame systems using a multi-level genetic algorithm has been proposed[64]. In this effort, it is pointed out that as the number of design variables increases the search space required to seek the optimal solution increases exponentially. Therefore, in order to improve the search for the optimal solution using the GA, the researchers proposed a two-level procedure whereby the design variable set is divided in a systematic manner into subsets as the evolution progresses. This process provides a controlled constriction of the search space. A large variety of structure topologies were considered from space trusses to tapered cantilever beams. DSO for minimum weight was applied to spatial structural steel frames using displacement constraints and member strength constraints derived from U.S. design specifications[65]. Wind loading was considered in this effort and member strength constraints were formulated using two U.S. design specification methodologies: allowable stress design; and load and resistance factor design. Comparisons of designs generated using these two constraint formulations suggested that load and resistance factor design methods generated more economical structural systems (based upon the least weight formulation considered)[65]. Pre-engineered buildings are a perfect vehicle for design optimization as the topologies and loading scenarios are often replicated many times over. The topology of these systems often involves rafter members and column members with additional haunches at the rafter-to-column connection. DSO of these systems using a genetic algorithm using the following design variables: rafter cross-sections; column cross-sections; supplemental T-section haunch length into the span; and supplemental T-section haunch depth at the column; has recently been demonstrated[66]. Member strength constraints were formulated using British design specifications and discrete design variables were designed using U.K. member cross-sections. A typical genetic algorithm was used and constraint handling was implemented through penalty multipliers and the design analysis assumed linear elastic behavior.

Application of evolutionary computation to DSO of large- and small-scale steel truss systems for minimum weight design has also received significant attention. Large-scale truss systems are frequent application targets. Preliminary designs of long-span king-post trusses often used for indoor stadium roofing systems has been generated with genetic algorithms[67]. A simplified (preliminary) design analysis of the king-post truss system was formulated such that 6 critical sizing parameters could be considered in lieu of detailed matrix structural analysis. The king-post system was resolved into a first-degree statically indeterminate system thereby allowing the conventional force method to be employed. Canadian steel design specifications were used to establish member strength constraints and the simplified preliminary design-oriented formulation could allow architectural, fabrication and shipping constraints to be included in the preliminary design. As a result, the preliminary design formulation included practical topology constraints for the completed system. Constraint handling was accomplished without penalty functions through a simple filtering of feasible (non-constraint violation) and infeasible (constraint violation) designs. A thorough study of constraint activity during the evolution is provided as well as the results of multiple GA runs. Several long-span truss systems are designed and comparisons of the designs are provided.

Single-objective non-DSO of lattice dome-type trusses, latticed towers used in electrical transmission and supply, and other three-dimensional structural steel truss systems have been fertile grounds for application of evolutionary computation procedures. Moderate-scale transmission tower structures (*e.g.* 25-m or 83-foot) tall have been considered[68]. This study included a minimum weight objective, 3D elastic time-history analysis, and design constraints formulated using a U.S. design code for transmission structures, which were based upon allowable stresses. Three design variable resizing algorithms were considered. A genetic algorithm was employed along with a discrete-continuous and a continuous design variable resizing algorithm. Algorithm performance comparisons were made. It was found that the genetic algorithm could find designs with material volumes that were lower than the other algorithms compared, but CPU time to generate the GA design was significantly greater. It should be noted that study assumed elastic time-history analysis and more complex design analysis (*e.g.* inelastic time history analysis) was not considered. As a result, the ability of the non-GA solution algorithms discussed to consider more complex analyses as the design basis was not evident.

Two additional non-DSO applications to design of space trusses and lattice towers involve object-oriented (OO) representations[69,70] and a third involves an adaptive penalty scheme[71]. The object-oriented approaches to lattice and space trusses have a very similar theme to an object-oriented evolutionary algorithm applied to inelastic analysis-based design of steel frames[49,72,73]. With regard to space truss optimization[69], the OO methodology was applied to the genetic algorithm components (*e.g.* chromosome class) and operators (*e.g.* crossover, mutation). A very simple minimum-weight optimization problem with allowable stress constraints was considered. The OO representations used in this effort appear to be exclusive to programming implementation of the genetic algorithm applied. Several moderately practical structures of relatively large number of members were considered and algorithm performance is discussed. The lattice structure effort[70] has even closer resemblance to a prior effort related to frame structures[46,49,72,73] in that the phenotypic representation of individuals is formulated in an object-oriented manner. However, classification of this effort as object-oriented is a bit misleading because the authors appear to be simply grouping portions of the lattice structure into "objects" rather than implementing object-oriented programming methodology (*e.g.* utilizing inheritance or classes). The constraints implemented in the effort were allowable stresses and the analysis used was linear static.

An adaptive penalty scheme, adaptive crossover and adaptive mutation have also been proposed as integral components of a GA applied in the design of 3D lattice trusses for allowable stresses established using U.S. design specifications[71]. These adaptive schemes were proposed to alleviate the user of the often difficult *a-priori* decisions that need to be made with regard to constant multipliers and exponent magnitudes associated with penalizing infeasible solutions and defining mutation and crossover rates. A two-phase member grouping to reduce the design variable space was proposed. Rather than having members grouped *a-priori* by the user of the GA, the members are grouped based upon the result of an initial analysis of the structural system (with member sizes defined as the same cross-section). Discussion of displacement constraint activity, stress constraint activity, and algorithm performance using the adaptive strategies and two-phase grouping procedure are given. It was shown that a GA utilizing these strategies was successfully able generate very competitive solutions to allowable stress optimization problems. It should be noted that intelligent GA operators (*e.g.* mutation, crossover) similar to the adaptive strategies used in this work have been introduced many years prior[45, 46, 49, 72, 73].

Genetic algorithm solutions to DSO problems involving braced framing systems using British design specifications and drift constraints have also been proposed[74]. Typical member grouping scenarios were utilized in this work and example applications of the proposed GA formulation included multistory moment resisting frames (MRFs), concentrically braced frames (CBFs), CBFs with hat or outrigger truss, chevron or inverted V-bracing, and a structural bracing topology that involved two-story K-bracing. A very traditional GA was applied to generate optimized designs and comparisons of the normalized material weights among the different frame topologies with grouped design variables were drawn. The results of the GA-driven designs confirmed long-standing knowledge regarding the economy of the topologies considered when one considers weight of material as the only design objective.

Reinforced concrete structural systems have also been considered as targets for DSO using genetic algorithms and their variants. A hybrid genetic algorithm was recently proposed to drive resizing of design variables for large-scale high-rise reinforced concrete wall systems[75]. The optimality criteria optimization algorithm was combined with the a traditional genetic algorithm to enhance the local searching capability of the GA and facilitate more efficient solution to real structural systems involving a large number of design variables through minimizing the number of re-analyses required. Linear elastic analysis was assumed and wall sizing constraints and displacement constraints were used and a single objective of minimum weight of materials was included. Reinforced concrete moment resisting frames designed using U.S. load and resistance factor design codes have also been considered[76,77]. The moment resisting frames considered in these efforts are significantly different, but both attest to the capabilities of the typical GA in solving DSO problems for RC MRF systems. Objective functions in these efforts considered cost of steel reinforcement, concrete material, and formwork. Other research efforts have considered DSO of concrete beams using Indian design standards[78] and Eurocode design specifications[79]. These efforts consider very detailed design problem formulations that involve bar arrangement templates, shear reinforcement, tension steel cut off, etc. Objectives included minimization of materials costs, formwork costs, and steel reinforcement volume. Both of these efforts related to continuous beam DSO contain very useful information that may result in these detailed design considerations to be moved forward toward large-scale R.C. systems involving walls, beam-columns, etc. DSO of flat-slab reinforced concrete building systems using British standards has also recently been considered[80,81]. These efforts considered a whole-building approach to the optimization ranging from floor slab thickness and reinforcement to column footings. A review of the DSO efforts related to reinforced concrete systems indicate that there is significant opportunity to bring these widely varying approaches together into a (more or less) unified approach for concrete structural systems.

Multiple-objective DSO of framing has also recently begun to receive significant attention in the research community. Researchers have defined a *min-max* optimum to implement a genetic algorithm in the solution two spatial truss design optimization problems with multiple objectives[82]. The objectives considered were: (a) minimization of weight; (b) minimization of displacement at a user-defined node; and (c) minimization of the stress that each member must support. Objectives (b) and (c) considered in this study are considered as penalized constraints in many GA applications. Authors provide detailed comparisons of many optimization algorithm implementations and illustrate that linear combination of objectives can lead to undesirable results. Structural steel framing systems have also been considered. It has long been recognized that minimum weight is not the only objective that should be considered in DSO of structural steel systems. To this end, researchers have begun including constructibility constraints in the GA-based optimization problem formulations[83] and it has been shown that consideration of a design complexity measure (*e.g.* the number of different member shapes used in the solution) can significantly affect the minimum weight design. As a result, it is clearly indicated that design complexity (from a fabrication standpoint) and minimum weight are competing objectives in the design of steel systems and that trade-offs among these objective is necessary. It is also clearly shown that the typical GA is a very powerful tool for providing Pareto optimal sets of solutions to these problems. An attempt has also been made to elucidate performance comparisons and the associated construction costs among composite steel-concrete, steel, and reinforced moment resisting framing systems in terms of potential energy[84]. An irregular (set back) framing system was considered and design followed U.S. based seismic codes. The multiple-objective optimization decision making tools developed using a GA and Pareto fronts developed in this study illustrate the power of the genetic algorithm.

One of the powerful attributes of evolutionary computation is that gradients of objective functions and/or constraints need not be evaluated during the application of the algorithm. As a result, evolutionary algorithms are, relatively speaking, immune to difficulties associated with sensitivity analysis of nonlinear systems. Of course, the increased computational time required to execute an EA is present, but the application of evolutionary computation is not limited to a certain class of problems. Researchers have recognized this and have begun to apply evolutionary computation to facilitate DSO of systems involving nonlinear response.

There are two types of nonlinear behavior commonly encountered in either static or dynamic analysis. The first is geometric nonlinearity and involves equations of equilibrium formulated for elements and structural systems on the deformed rather than undeformed structure or element. DSO is often accomplished in these systems through application of design specifications or codes. Nonlinear geometric behavior in codes and specifications is often included through amplification of first order forces and deformations. However, most modern design codes and design specifications do allow the direct use of geometrically nonlinear (second-order) analysis. Optimized design of two dimensional fully-restrained (FR) steel framed structural systems has been accomplished using a genetic algorithm and linear and nonlinear structural analysis[85]. U.S. design specifications for steel structures were utilized and a single objective of member weight minimization was considered. It was shown that drift constraints tend to limit geometrically nonlinear behavior in the typical steel moment-resisting frames and group selection mechanisms and adaptive crossover operators are effective. The impact of partially restrained (PR) connections on steel frame response has been well-known for decades. DSO using genetic algorithms for design of PR and FR steel frames has also been demonstrated[86,87]. Elastic geometrically nonlinear analysis using well-known stability functions was utilized in this effort and design variables considered were limited to wide-flange shapes. Member strength constraints were based upon British design standards. Traditional interstory drift and deflection constraints were implemented in the algorithm. Frame designs were generated with FR connections and a variety of PR connection configurations. Comparisons of structural system weights generated through application of the GA with various connection configurations are provided. It was shown that when gravity loading dominates the loading combinations considered, PR frame designs can be lighter than a corresponding FR frame. The use of nonlinear geometric analysis in the evolution of frame designs sometimes resulted in more material for the columns and beams and sometimes resulted in reduced material volume.

Application of a genetic algorithm in the minimum weight design of planar steel frames subjected to interstory drift and limit loading capacity constraints has using plastic mechanism analysis has also been demonstrated[88]. Plastic mechanism analysis is utilized to evaluate the frame limit load capacity in this effort. Two planar steel frames with fully-restrained connections are designed using the proposed genetic algorithm. The study outlines the need to carefully select the population size as it can affect the computing time significantly and the ability to reach the optimum solution. Recommendations on mutation and cross-over rates are made. It was found that when gravity loading alone is present on the frame, both load factor constraints and deflection (vertical) constraints can be active. However, when lateral and gravity loading combinations were applied to the framework, drift constraints become the dominant consideration in the algorithm. It is also found that when nonlinear analysis is utilized instead of linear elastic analysis as the design basis, greater economy can be achieved..

As computational tools progressed over the last two decades, researchers began to explore design methodologies that could exploit software capabilities. The concept of *advanced analysis* grew from these efforts. In a nutshell, advanced analysis is a design method, where the structural analysis employed to evaluate a structural system candidate design is able to include all pertinent phenomenological behavior upon which design specification equations are based. Thus, if advanced analysis is employed, there is no need to include evaluation of design specification equations in the design process. Combination of advanced analysis methods with genetic algorithms has recently been proposed[46,49,72,73,89]. Single-objective DSO was considered in these effort. Design variables in these efforts included beam and column sizes selected from wide-flange shapes were considered. The design variable sets have also been expanded to include a variety of partially restrained connections[46,49,72,73]. Extensive constraint listings consistent with the needs imposed by inelastic analysis-based design methods were included in these efforts. Distributed (fiber-based, plastic zone)[46,49,72,73] and concentrated plastic hinge with gradual formation[89] models for nonlinear material behavior were also included in the studies. Stability functions[89] and geometric stiffness matrices[46,49,72,73] were used to simulate geometric nonlinearity. These two research efforts can be considered as a final demonstration of the power of EC in handling very complicated DSO problems using static analysis.

4.2.2. *Performance-Based Structural Optimization (PBSO)*

The development of performance-based design specifications and model codes for steel and concrete building structural systems[90-92] has ushered in new applications of evolutionary computation in the optimized structural design[93,94]. PBSO formulations usually involve multiple objective optimization problem statements. The constraints often involve target probabilities and/or confidence levels in meeting performance objectives. Efforts that involve minimum life-cycle cost design also falls into this category of structural optimization efforts.

Life-cycle cost optimization for steel framing systems has been a very fertile area of research that can be classified as PBSO. When the structural system life-cycle is included in the optimization, various ground motion levels need to be considered. U.S. design specifications and nonlinear pushover analysis was used in conjunction with a genetic algorithm to solve a performance-based structural optimization problem that involved objectives related to initial construction expense (material weight), the number of different steel sections used in the design (diversity), and future seismic risk associated with interstory drift resulting from both frequent and infrequent ground motions[95]. Life-time seismic damage and initial material cost have also been considered as objectives in a GA-based optimization algorithm using nonlinear pushover analysis as the analytical basis for determining performance[96]. Nonlinear pushover analysis used as the fitness-evaluation engine for genetic algorithms[97, 98] and evolution strategies[99] has been used to solve PBSO problems involving objectives of minimum structure weight and confidence levels in meeting performance objectives during frequent and infrequent ground motion events.

While static pushover analysis is a useful method for defining performance expectations during ground motion events, it is well known that this analytical method tends to give inaccurate results for irregular structural systems. Inelastic time history analysis (THA) is a better predictor of performance during ground motion events. Inelastic THA has been implemented as the foundation of multiple-objective PBSO for 2D frame structures using genetic algorithms[32,93,100-102]. In order to consider multiple objectives in the GA, a novel radial fitness was defined[32,93]. Objectives in these efforts for the PBSO statement were confidence in meeting performance objectives during frequent and infrequent ground motion events.

The work related to PBSO of building systems is laying the foundation for application of genetic algorithms in RBSO of buildings systems. Approaches to accomplish RBSO with evolutionary computation are described in the following sections.

4.2.3. *Reliability-Based Structural Optimization (RBSO)*

Rather than assuming resistance and loads are established at single values, RBSO assumes that these critical components to structural performance remain cumulative distribution functions thereby maintaining non-deterministic measures of uncertainty. There have been relatively few efforts that have successfully formulated RBSO problems for frame and truss structures. Applications of RBSO to maintenance planning were described earlier in this chapter.

In many instances, objectives of RBSO problems include minimization of material cost (*i.e.* weight) and minimization of the cost of failure of the system. Although uncertainty can be pervasive in both cost and failure, most studies limit consideration of uncertainty to failure likelihood. As found in the review of previous structural optimization efforts and evolutionary algorithms, establishing the parameters for the genetic algorithm (*e.g.* crossover probability, population size, mutation probability, crossover type, penalty multiplier magnitude, and penalty exponent magnitude) can be difficult. When the structural design demands uncertainty is considered, there is a pressing need to choose these parameters wisely as the analysis effort needed to evaluate individuals in the populations becomes very time consuming. Multiple-population GA's have been applied in the RBSO of planar truss structures[103,104] to alleviate the user from defining these parameters. The sub-populations within a meta-GA in this study included various parameters and the meta-GA assigned resources to the most-fit sub-populations. The meta-GA implemented was shown to generate optimized solutions in a manner that is more efficient than a standard single population GA. Robustness of the algorithm with respect to finding the optimal solution to the RBSO problem was also improved.

Single objective RBSO problems for minimum weight and cost of members and connections have been formulated for 2D and 3D truss structures with constraints on acceptable probabilities of failure[105]. This research effort considered selection of member cross-sections and member cross-section with truss topology using genetic and modified branch-and-bound algorithms. It is recommended that the inherent parallelism of the GA be exploited for future work in RBSO.

When large-scale structural systems are considered in RBSO, the effort required to evaluate the fitness of individuals within the population becomes significant. As a result, researchers have proposed methodologies for RBSO that involve evolution strategies and Monte-Carlo simulation[106].

The uncertainty present in RBSO problems can make the design cumbersome and as implied, there is uncertainty in the solution. A recent application of non-dominated sorting genetic algorithm (NSGA) modified to become a non-dominated sorting evolution strategy (NSES) to structural optimization involves Robust Structural Optimization (RSO)[107]. For purposes of this synthesis, this can be classified as a subset of RBSO. In RSO, the objective of the optimal design problem is to attain a solution that is insensitive to variation in uncontrollable parameters. A RSO problem for a 3D steel structure subjected to a series of ground motions was considered and nonlinear pushover analysis was used as the basis for fitness definition[107]. A unique aspect to this study is that evolutionary computation is utilized to validate coefficients used in the Greek seismic design code for steel building systems.

4.2.4. Miscellaneous Applications

Applications of evolutionary computation in the structural engineering field have been numerous and extend to a range much wider than that implied in the previous three sub-sections of this synthesis. Aerospace structural engineering applications often require that structural components maintain stability throughout rather wide variation in temperature. To this end, the researchers utilized the evolutionary strategy to determine ply orientations and stacking sequence in a composite laminate such that the buckling load of an edge-supported (simple supports) and edge-loaded plate is maximized[108]. Two constraint handling procedures are used. The first is a traditional penalty applied to the objective function and the second is simply to reject infeasible designs during the evolution. Three $(\mu + \eta)$ evolution strategies are implemented: 4+4, 10+10, and 20+20. It is recommended that for the problem considered, a 4+4 evolution strategy was able to find good solutions with reasonable computation cost.

Operations research has always been a fertile area for the development of optimization algorithms. In a similar vein, researchers have utilized a genetic algorithm to optimize precast member production scheduling using a flow-shop operations research model[109]. The research indicated that the traditional GA implemented can not only produce single optimized solutions, but it could also easily provide a family of good solutions that can be used as a decision-making tool for the production facility. Optimization of mass concrete construction procedures has also been done using a genetic algorithm[110]. A fairly traditional genetic algorithm was implemented in this effort and the design variables included: type of concrete; placing temperature, height of concrete lifts, and placing frequency. Constraints related to structure cracking were formulated and the objective was one of cost minimization. A finite element analysis of the mass concrete structure (dam in the study considered) was conducted and was used as the basis for constraint evaluation and definition of fitness.

Evolutionary computation has not been limited to selection of framing components (*e.g.* beams or columns). Genetic algorithms have been used to develop optimal seismic zoning scenarios[111] and guide selection and scaling of ground motion records[112,113]. As RBSO becomes an integral component of the structural engineering design world, selecting ground motion records that satisfy targeted recurrence intervals and intensities will become more important. Furthermore, being able to rationally establish seismic zoning to minimize losses (*e.g.* deaths, dollars, and downtime) removes the appearance of arbitrarily assigned boundaries for seismic zones.

Applications of evolutionary computation have also been used in the selection of light-gauge steel deck, wide-flange purlin shape, and purlin spacing for typical wide-flange structural steel roofing systems when subjected to unevenly distributed loading caused by snow[114]. Genetic algorithms have also been used develop design charts to aid structural engineers in selecting steel floor framing systems composed of wide-flange shapes and light-gauge deck to minimize cost and satisfy all pertinent strength, deflection and vibration constraints[115]. Both of these studies illustrate that the genetic algorithms implemented could generate either single optimal solutions or families of candidate solutions of common cost, which indicates that the GA is suitable for use as a decision-making tool for the structural engineer.

Design of cold-formed members can be a fairly tedious process due to the significant number of instabilities that can arise within the member being designed. It is well-known that the population size required for successful implementation of a GA can vary with the problem considered. In order to use small population sizes and still tackle a structural engineering optimization problem involving cold-formed steel member design with design variables of cross-section depth, width, thickness and bend radius; researchers proposed a micro-GA application[116].

It is obviously important to consider performance of the structural system when conducting a seismic design of an industrial facility. However, one must not overlook the importance of ensuring that the process components housed within that facility perform as intended during ground motion events. One example of these process components is the networks of piping systems within nuclear or other facilities. The dynamic structural behavior of these relatively complex systems makes selecting the number and location of piping supports very difficult. To this end, researchers have proposed a GA-based decision support system for optimizing the response of piping systems when subjected to seismic accelerations[117]. The implementation of the GA as part of a "...joint-cognitive system..."[117] seeks to take advantage of the strengths present in the experienced engineer as well as the novel-solution generating ability of the genetic algorithm.

The design of steel plate girders for a bridge superstructure using modern U.S. design specifications for highway structures can be a daunting task. Researchers have sought to take advantage of the genetic algorithm to aid in the design of superstructure components[118]. Design variables used in this effort include: presence of a longitudinal stiffener, spacing of girders, depth of the steel girder web plate, thickness of the web plate, width of top and bottom flange, and the thickness of the top and bottom flange. Constraints are formulated using U.S. allowable stress design specifications. A parameter study using the GA developed allowed the researchers to demonstrate optimized bridge superstructure parameters for a wide range of girder spans for the two-span configuration considered. As a result, the GA is again shown to be a very useful decision making tool for the structural engineer.

4.3. *Topology Optimization of Truss-Type Structures*

Topology optimization of structural systems involves selection of topology, geometry and sizes of the components within the system. A related area of topology optimization is that of *Evolutionary Structural Optimization (ESO)*. In this approach, system topology is "evolved" through removal of material or components from a based (ground) topology that are very lightly utilized in terms of stress or force. ESO is not covered in this review as the method does not fit the prototypical evolutionary computation method in which a population of individuals is manipulated through mutation and/or recombination. However, ESO is similar to a $(1+1)ES$ without random mutation. The interested reader can find a plethora of information related to ESO through simply searches with this keyword in the literature listings at the end of this chapter.

System topology optimization is most often associated with conceptual design of structural systems found in the early stages of the design process. It is at these stages when design creativity can be at its peak and it is natural to explore the implementation of stochastic design procedures to generate alternative possibilities. When genetic algorithms are used to solve optimization problems that involve selection of the system topology or geometry in unstructured and structured domains (*e.g.* the perimeter boundary of a frame is undefined or defined, respectively), representation of individuals using binary string chromosomes can generate difficulties because as design variables are added or removed from the system, the string length for the individuals no longer remain consistent. Furthermore, if a very dense base structure topology (or ground structure) is assumed, chromosome lengths can be very long initially and search efficiency via GA can degrade. These difficulties with standard GA formulations as applied to topology and geometry optimization of truss-type structures has led to the development of Implicit Redundant Representation (IRR) GA's[119,120] and other formulations capable of handling variable string lengths[121]. Topology optimization algorithms utilizing GA's that do not use binary string chromosomal representations for individuals have also been proposed[122]. The concern for control of overlapping members, unstable configurations, and zero-force members in truss topologies developing during evolution using a GA has also led to alternate definitions of topology for the system. Rather than define topology and geometry using nodes and members, researchers have examined defining topology using series of inherently stable triangular substructures and have implemented GA formulations to generate optimized solutions[123].

The search for optimum topologies for truss structures considered competing deflection and weight objectives has been done through multiple objective genetic algorithms (MOGA's)[124]. Several unique GA operators were introduced in this effort: active unit-based crossover; and unit-based mutation. A hierarchical chromosomal representation was also proposed and discussion of "total Pareto optimal sets"[124] used to define optimum topologies within the context of objective space is given. Truss structures were considered in this research effort and optimal designs were generated considering topology (bar/member presence in the system); geometry (node or connection locations in 2D space); and sizing (cross-sectional areas). Topology and geometry have also been considered in optimization of a cantilevered truss structure considering objectives of mass minimization, deflection minimization, and/or stress minimization[125].

Comparatively speaking, evolution strategies have not seen as wide spread use in topology optimization of truss structures. Various evolution strategy implementations applied to truss topology optimization for a fixed loading scenario and fixed initial node layout have been compared to simulated annealing and a newly proposed random cost methodology[126]. Various initial (ground) structure topologies and geometries were considered. $(1 + \lambda)ES$ with $\lambda = 1$ and $\lambda = 1024$ were shown to be relatively ill-behaved and $(1, \lambda)ES$ with $\lambda = 64$ was shown to be able to come very close to finding optimal solutions when compared to simulated annealing. The number of fitness (function) evaluations was considered in the comparisons.

4.4. *Structural Control and Supplemental Damping*

Minimizing the impact of seismic excitation on building and truss structures through use of passive supplemental damping mechanisms and active control devices has been a fertile area of research in structural engineering in the past two decades. Examples of supplemental damping mechanisms often considered are: friction dampers; tuned mass dampers (TMD's); added damping and stiffness (ADAS) devices; triangular-plate added damping and stiffness (TADAS) devices; and passive constrained layer damping (PCLD) patches. Examples of active control devices are diagonal brace actuators; piezoelectric patches; magnetorheological dampers; and TMD with active mass drivers (AMD's).

Genetic algorithms have been shown to be very effective tools for determining the positioning and characteristics of passive damping devices within structural engineering systems (*e.g.* buildings and trusses). The torsional response of buildings subject to earthquake and wind excitations can be effectively controlled through positioning TMD's. Genetic algorithms have been used to guide the design of TMD parameters and their locations for multiple story buildings for the following performance criteria: drift; acceleration; drift-based second norm; and acceleration-based second norm[127]. Design variables considered for the tuned mass dampers were: mass ratio; frequency tuning ratio; damping ratio; and damper position. Performance functions appear to have been combined to define single-objective fitness for individuals and the GA implementation was shown to be very effective in providing TMD parameters and location to reduce response. The suitability of genetic algorithms in addressing multiple objective optimization problems involving characteristics and placement of TMD's within a building structure has also been demonstrated[128]. The objectives to be simultaneously minimized using Pareto-optimal fronts in this effort were: maximum nondimensional peak displacement; maximum nondimensional peak acceleration; and maximum nondimensional peak rotation when the system was subjected to a suite of ground motion records. Design variables for the TMD's included: mass, stiffness, damping; mass moment of inertia; rotational stiffness of the coupled dampers; rotational damping characteristics of coupled dampers; and eccentricity of the TMD's from the center of mass in orthogonal directions.

Optimal number and placement of classical viscous fluid dampers, solid viscoelastic dampers, and fluid viscoelastic dampers within a torsionally excited structural system has been accomplished using a genetic algorithm[129]. The objective fitness in this study was defined using drift-based performance and acceleration performance. The design variables considered in the GA implementation was the total number of dampers at locations to be placed in the structure. The GA design tool was used to compare optimal distributions of viscous devices and viscoelastic devices throughout the building for base shear or floor acceleration performance measures in a single-objective optimization format. Optimal placement of passive fluid dampers within a 20-story building was defined using a genetic algorithm[130]. Four single-objective optimization problems using 2-norm and ∞-norm measures of RMS response and frequency-shaped transfer functions were formulated and solved using the GA implementation. The GA generated solutions for damper distribution were shown to include a configuration that significantly reduced inelastic response measured through ductility demand.

Yielding and friction damper devices are also very useful in structural systems that utilized diagonal bracing to resist seismic loading (*e.g.* chevron braces). Optimal parameters of TADAS and ADAS devices placed at the stories within a 10-story planar building system have been determined using a genetic algorithm to meet the following objectives: equal-weight combination of interstory drift and floor acceleration; and an objective of reducing floor accelerations[131]. Optimal parameters for metallic-yielding and friction dampers within the 10-story building considered were defined using a fairly traditional genetic algorithm.

Consideration of optimal placement and optimal type of passive damping device within multi-story buildings is also very important because as building systems become taller, the best type of passive damping device to be used to minimize damage resulting from acceleration or drift may not be uniform over the building height. A genetic algorithm with mortality constraint has recently been proposed to efficiently generate optimal designs for multistory structural systems that include triangular plate energy absorbers (TPEA), linear viscous dampers (LVD), and viscoelastic dampers (VED) distributed throughout the stories[132]. This study illustrates that the best solutions obtained using the genetic algorithm formulation proposed did not include uniform distribution of dampers throughout the height of the structural system and that the GA could generate interesting combinations of TPEA, LVD, and VED devices within a building system. Furthermore, it is suggested that the GA formulation can be used to "…clarify advantages and disadvantages of the various device types as design circumstances change"[132], which is a perfect example of the power of evolutionary computation as an automated and optimized design generation tool.

Passive viscous damper placement within truss systems typically found in aircraft structures has also been guided using a genetic algorithm[133]. The designs generated using the GA formulated were found to match intuition and therefore, it is another example illustrating that the GA can be used as a design tool to generate alternatives. Definition of optimal parameters for passive constrained layer damping (PCLD) treatments for controlling displacement response of simply supported beams has also been accomplished using genetic algorithms[134]. The GA-generated solutions to a design problem involving minimization of beam displacement at mid-span when subjected to wide-frequency-range force excitation illustrated that the shear modulus of the viscoelastic layer and the location and length of the layer are the most important parameters to consider in design.

Design of active and hybrid control systems for building structures has also been a fertile area of research application of genetic algorithms. One very interesting side problem often encountered in control system optimization is that feedback from the system being controlled is often based upon limited information (*e.g.* accelerometer data at limited story locations within the structure). As a result, design of the feedback-control systems for large-scale civil engineering structures is a very fertile area of active research as well.

Use of a genetic algorithm to guide placement of actuators within regular multi-story buildings is the most basic form of optimal control problem[135]. The placement of predetermined actuator types and their corresponding control algorithms using a genetic algorithm was determined using binary string representations for the actuator position. The objective considered was drift at the upper floor of a 16-story regularly framed building. GA operators used were very straightforward. Optimal placement of magnetorheological dampers within building systems using a GA has also been demonstrated[136]. The GA design variables for position were formulated with emphasis on practical installation. Pre-defined controller strategies were considered and the objectives for the GA optimization were norms of RMS of absolute acceleration and interstory drift (considered independently).

The design of a control algorithm/system is not trivial when one considers the environment likely to be present. For example, electronic measurements are being taken at limited locations. This generates two important issues/concerns:

(a) the electronic signals will contain noise; and (b) one must be able to predict and control response throughout the structure using the limited measurements as feedback. As a result, designing the control algorithms for fixed position and type of devices is a challenging problem. Design of controller gains to control response of building systems using binary string GA's has been demonstrated[137,138]. A wind-excited 76-story shear building structure with fixed-positioned ATMD devices at the roof level with fixed sensor locations was used as the basis for an optimization problem that involved generation of optimal controller gains in the presence of sensor noise[138]. A single-objective fitness function involving peak accelerations, peak displacements, RMS accelerations, and RMS displacements at selected floors and actuators was utilized. Robustness criteria for the controller design were also applied in the GA formulation. Hybrid system control gains for 2D buildings have also been optimized using binary string genetic algorithms[137]. Comparison of response with active (including controller optimization), passive, and hybrid control systems were also provided. Real-coded GA's have also been used design optimized controller gains for ATMD devices with limited sensor arrangements[139]. Comparisons of optimal designs of passive and active control devices generated using genetic algorithms for 3D structural systems have also been made[140].

The robustness of a controller can be assessed through consideration of its ability to remain stable and control of system performance in a realistic environment[138]. Sensor output noise and the inability to measure exact structural parameters (*e.g.* mass, stiffness) are two examples of uncertainty in real structures that a robust controller system must address. Optimal controller design in the presence of these uncertainties has been enhanced through use of fuzzy-logic principles. As a result, fuzzy-logic controller (FLC) design algorithms have emerged. Design of an FLC for first-floor actuators in regular buildings using genetic algorithms with an single objective of minimizing roof-level displacement response of a 3-story building has been demontrated[141]. Design of FLC using genetic algorithms within the context of multiple objective optimization using Pareto optimal sets has also been demonstrated. GA-based designs for two-dimensional systems have been generated by simultaneously considering peak interstory drift and peak floor accelerations[142]. Three-dimensional systems have also been considered[143]. Optimal controller design using genetic algorithms and fuzzy logic concepts for smart base isolation systems has also been illustrated. Single objective[144] and multiple objective optimization problems[145] have been considered.

Optimal structural design problems that consider the number, placement, controller gain and type of control system can be considered the most challenging. Selection of the optimal number of actuators, the position of said actuators, and the control algorithm type (linear quadratic regulator, or acceleration feedback control) using a multi-level genetic algorithm has been illustrated and discussed[146]. Consideration of nonlinear structural response has also been included in generation of optimal design algorithms incorporating fuzzy logic methodologies to solve multiple-objective problems using a MOGA where the position, number, and control algorithm are considered as design variables[147]. Optimal placement and controller design have been considered as design variables in a single objective optimization problem solved using genetic algorithms in problems that involve a regular 40-story shear building and a 9-story irregular building[148]. Placement, sizing, and feedback control gains of a novel piezoelectric sensor-actuator to minimize the vibration of shell structures has also been demonstrated[149].

Placement of sensors and actuators within the structural system as well as design of the control algorithm (*i.e.* gains) on a high-rise structural system has been demonstrated through application of a genetic algorithm for actuator/sensor placement and gradient-based optimization methods for sensor/actuator gains[150]. The objective function used in this study was a combination of minimizing building response and control effort.

4.5. *Damage Detection*

There are instances where structural systems have been instrumented to measure response during loading events. An example of this is accelerometers placed within building systems to measure response during seismic events. These systems generally contain very few instruments placed in locations felt to give useful data for extrapolation post-event. It has long been desired to use the data from relatively sparse instrumentation arrays to determine the existence, the extent, and the location of damage in structural systems. This is not a trivial endeavor and genetic algorithms have been used to detect damage in structural systems using measured data.

The location of damaged members within truss structures has been identified using simple genetic algorithms and the implicit redundant representation GA[151]. The proposed GA implementations were shown to be capable of identifying damage (defined as a reduction in axial stiffness of a member) to truss members in statically determinate and indeterminate truss structures when subjected to moving fixed-axle vehicles. A small number of measured static displacements were used to successfully locate damaged members in these truss structures.

In lieu of static loading conditions, changes in the vibrational characteristics of structural systems (*e.g.* frequencies of mode shapes) have also been used to detect damage in structural systems. An optimization problem is formulated in this instance to minimize the differences between measured vibration data and analytically generated data. Binary string GA's[152] and real-coded GA's[153] have been shown to be successful tools in this endeavor. Cantilever beam structures, planar truss structures and plane frame structures have all been considered for application of the damage detection algorithms proposed.

When faults (*e.g.* damage, flaws) are sought within an existing structural system, the engineer must rely on a limited number of sensors distributed in some manner throughout the systems to be his eyes and ears. Genetic algorithms have been used to generate the number and location of sensors to give information regarding the probable position of damage within a cantilever plate[154]. Fitness of candidate solutions were defined using "observability" measures and resistance to clustering of sensors within the system.

4.6. *Parameter, Model, or Structure Identification*

Engineers are well aware that the models for structural analysis that are developed to simulate behavior are just that – models. Parameter identification in the area of structural engineering is the process of generating more realistic structural analysis models through examination of response of a real system to known input (loading). This is sometime called solving an inverse problem in that one is generating an analytical model from measured output data. Parameter, Model, or Structure Identification can be used in structure health monitoring and damage detection.

Locating superimposed mass (either moving or static) on a structural system has been an area where genetic algorithms have been shown to be useful. A computational procedure founded on a genetic algorithm for determining the size and location of a concentrated mass within the boundary of an isotropic plate has been proposed[155]. The objective fitness for the GA is based upon minimization of an output error criterion defined using changes in natural frequencies. Identification of moving masses along continuous (multiple-span) beams typical of bridge superstructures has also been accomplished using a GA[156]. Minimization of the error between measured and reconstructed accelerations is used as the objective in the optimization problem.

One of the few applications of evolutionary programming (EP) in civil engineering design has been in the area of solving inverse problems[157]. Identifying the elastic modulus of an isotropic plate loaded uniformly at its edges and conducting pavement quality inspection were two example problems considered in this effort.

When a large number of parameters needs to be identified in a large structural system, convergence of a numerical algorithm for parameter identification can be unreliable. As a result, researchers have proposed methodologies using genetic algorithms to conduct parameter identification in these large systems. A Modal GA has been proposed to reduce the search space required for parameter identification of large dynamically loaded systems[158]. The single objective optimization problem proposed in this effort involves minimization of the difference in the norms of measured and predicted response in the modal domain. Substructuring and a staged application of a genetic algorithm for large system parameter identification has also been proposed[159]. Using a genetic algorithm to identify prosperous locations in objective space along with a compatible local search methodology (one without the need for gradient information) has also been proposed to conducted parameter identification in large structural systems[160].

Pareto optimal theory and evolution strategies have also been used to identify structural engineering parameters in multiple-objective optimization formulations[161]. The two objectives considered in this effort were norm of the difference in measured and predicted natural frequencies of a selected number of mode shapes and norm of the difference between measured and predicted mode shapes. The Strength Pareto Evolutionary Algorithm (SPEA) proposed in this effort is also used to make predictions regarding reliability of the structural system using the Pareto fronts generated. An extension of the SPEA procedure to nonlinear structural systems is also discussed.

As outlined earlier, when GA's are asked to search for a large number of design variables, the search space can become quite large and the effectiveness of the typical genetic algorithm can suffer. A real-coded genetic algorithm has been proposed to minimize mode shape and mode frequency differences between measured and predicted results[162]. The GA is employed first in the parameter identification, with subsequent implementation of a localized hill climbing algorithm based upon eigen-sensitivity. A search space reduction method (SSRM) has also been proposed to enhance the accuracy and reliability of a genetic algorithm employed in structural parameter identification[163]. Integral to the SSRM is a modified GA that adaptively reduces the search space using individual parameter convergence rates.

Pedestrian bridge structures have been found to vibrate significantly while in service. Some rather famous instances of severe vibration of pedestrian bridges have been reported (e.g. Millennium Bridge in London) and the dynamic loading and response of these systems has been the subject of much research. There are many components that partake in defining a pedestrian bridge's dynamic characteristics (*e.g.* handrails) and defining the loading functions resulting from pedestrians is not trivial. A genetic algorithm has been used to help define forcing functions that are capable of simulating human walking forces acting on slender pedestrian bridge type structures[164]. The identification parameters considered were: heel contact duration; heel period; heel impact coefficient, tip of toe period, tip of toe impact coefficient, and impact of heel to tip of toe. These parameters were defined using a Genetic Algorithm and with the single objective function being a simple normalized summation of the differences between measured and predicted natural frequencies for the first two modes of simply supported pedestrian bridges.

4.7. *Conceptual Design of Building Systems*

Evolutionary computations has the capability to serve as an artificial intelligence mechanism whereby literally thousands of candidate designs can be generated and evaluated automatically using user-defined criteria. As a result, structural engineers have sought to exploit EC to aid in generating conceptual designs of mid- and high-rise structural systems to understand the impact of multiple competing objectives in defining the best compromise solution for building systems.

At early stages of design of a building system, many factors need to be considered. Economy of design demands that initial construction cost, maintenance costs, operating costs, and anticipated income or loss over time be considered. When one considers the myriad of factors that define these costs, he/she may find it impossible to rationally determine relationships among the relative importance in each of these factors (either individually or collectively) in defining the most economical solution with which to proceed to detailed design. A genetic algorithm has been applied to the conceptual design of building systems to simultaneously minimize capital cost, minimize operating cost, and maximize income revenue[165-167]. Various structural systems were considered as well as HVAC costs, land, lease rates, mortgage rates, inflation rates, and many other practical parameters. A novel coloring algorithm is proposed to understand the impact of structure type, number of stories, bay area, window ratio, and design profitability within 3D objective space.

Conceptual design of the structural system has also been the target of application of EC. A structured genetic algorithm (SGA) has been proposed to allow alternative structural systems (*e.g.* precast concrete construction, composite construction) to be represented in a hierarchical binary chromosome structure[168]. The SGA implemented in this study was intended to be a decision making tool for the structural engineer. A parameter study was also undertaken after which recommendations regarding the structural system were made through consideration of the variation in land cost.

In the wake of September 11, 2001, the use of high-rise structural systems was placed under increased scrutiny. The genetic algorithm and multiple-objective coloring/filtering algorithm previously discussed[165-167], was applied to a multiple-objective optimization problem that examined the trade-off between life-cycle profitability and their robustness (their load path safety against progressive collapse)[169]. Robustness was evaluated using a measure of force redundancy in the structural system.

4.8. *Parallel Processing Applications*

The emergence of readily available clusters of networked computers and multiple processor personal desktop computers facilitated exploitation of the inherent parallelism of evolutionary computation procedures and opportunity to reduce computation times in the solution to structural engineering optimization problems.

Large structural steel systems have been optimized for minimum weight with strength constraints defined by U.S. steel design specifications and conventional drift constraints using a multi-level GA and fuzzy GA with MPI and OpenMP parallel programming methods[170]. Comparisons of parallel algorithm performance for the bilevel GA with MPI and OpenMP implementations are provided.

A very large design problem (1080 design variables) was used as a numerical experiment to evaluate the extent to which a typical GA would benefit from having multiple processors[171]. Numerical experiments using as many as 128 processors demonstrated "... radical elapsed time reductions..." approaching linear speed up with appropriate algorithm modification[171].

Determination of the type of support and the support location for piping systems subjected to seismic excitation has been done using a parallel genetic algorithm with the goal being to generate solutions that are near the optimal design in objective space, but as different as possible in decision space[172]. A network of 10 Solaris Ultra-10 workstations was utilized in the parallel GA solution and a single objective optimization problem was considered.

A network of workstations and MPI protocol have also been used to reduce computation time necessary for a simple genetic algorithm[173]. It is shown that if proper load balancing among processors is considered in the parallel algorithm formulation, near linear speed up can be attained on a "…homogeneous hardware cluster…"[173].

5. Other Sources of Information

As with any review, there will always be a need to frame the period for the review an in so doing, useful references will slip through. It is therefore, prudent to guide the reader to additional sources of information in the form of journals and conferences whereby he/she can obtain additional literature related to evolutionary computation and its application in structural engineering.

There are many journals where applications of evolutionary algorithms can be found and where their efficiency in generating solutions to structural engineering optimization problems is evaluated. A listing of some of the more popular journals available in the archival literature is:

Evolutionary Computation
Journal of Computing in Civil Engineering
Journal of Structural Engineering
Journal of Structural and Multidisciplinary Optimization
Computers & Structures
Engineering Structures
Journal of Constructional Steel Research
Journal of Bridge Engineering
Journal of Computer-Aided Civil and Infrastructure Engineering

Conferences that contain written proceedings are also sources for reviewing the evolving state-of-the-art in evolutionary computation. Some of the most pertinent and popular conferences related to EC are:

Genetic and Evolutionary Computation Conference (GECCO)
Foundations of Genetic Algorithms (FOGA)
International Conference on Genetic Algorithms (ICGA)
Congress on Evolutionary Computation (CEC)
SCE-SEI Structures Congress

6. Concluding Remarks

The review contained in this chapter demonstrates the truly staggering range of applicability for evolutionary computation. EC is a very powerful tool for automated and optimized design and its application in structural engineering appears limitless. It also appears that EC methodologies can form the basis of unified automated and optimized design algorithms for structural engineering. It is prudent at this point to provide some concluding remarks that can serve as a brief synthesis for the information reviewed in this chapter. It is hoped that this short summary of observations made at the end of the chapter can stimulate new research directions dealing with application of EC in the field of structural engineering.

The vast majority of applications of evolutionary strategies has been in Europe. Although incredibly useful, they have not seen as wide spread application as genetic algorithms in the field of structural engineering. Evolutionary (genetic) programming concepts have seen limited application as well. Further exploitation and comparison of all EC methodologies in structural engineering (as applicable of course) needs to occur.

There is opportunity for researchers to begin to tackle far more complex structural engineering design problems using EC and future research efforts may need to examine and further exploit alternate methods for expressing phenotypes[45,46,49,72,73] as problems become more an more complex.

Design of supplemental damping and control mechanisms, performance-based engineering design, and reliability-based engineering design appear to be fertile areas for continued exploitation of multiple objective optimization using evolutionary computation.

As design specifications and codes become more and more complicated and computer software becomes more necessity than tool, EC has the opportunity to become an integral part of artificially intelligent design algorithms in the future.

Researchers need to continue exploitation of parallel processing environments (*e.g.* parallel computer clusters) on difficult structural engineering problems beyond those of numerical experiments. Real problems need to be tackled in this newly evolving computational environment.

Acknowledgments

The author would like to acknowledge the support and understanding of Prof. Jasbir Arora of the University of Iowa. He would also like to acknowledge Prof. Donald Grierson of the University of Waterloo, and Prof. Shahram Pezeshk of The University of Memphis for support in his research endeavors. The author would also like to acknowledge former students, Arzhang Alimoradi, Benjamin Shock, Christopher Erwin, Daniel Schinler, and Mark Voss for the many fruitful discussions that have contributed to his understanding of evolutionary computation and its usefulness in structural engineering optimization.

References

1. J. S. Arora, *Introduction to Optimum Design, 2nd Edition* (Elsevier Academic Press, San Diego, CA, USA, 2004).
2. S. Rao, *An Introduction to Structural Optimization* (John Wiley & Sons, Inc., New York, NY, 1980).
3. K. Deb, *Multi-Objective Optimization Using Evolutionary Algorithms* (John Wiley & Sons, Ltd., Chichester, U.K., 2002).
4. K. A. De Jong, *Evolutionary Computation - A Unified Approach* (The MIT Press, Cambridge, MA, 2006).
5. K. A. De Jong, D. B. Fogel and H.-P. Schwefel, A History of Evolutionary Computation, *Evolutionary Computation 1 - Basic Algorithms and Operators*, Ed. T. Back, D. B. Fogel and T. Michalewicz, (Institute of Physics Publishing, Bristol, U.K., 2000). pp. 6-58.
6. H. J. Lichtfuss, *Evolution eines Rohrkrummers*, Diplomarbeit (Technical University of Berlin, Berlin, Germany, 1965).
7. I. Rechenberg, *Cybernatic Solution Path of an Experimental Problem*, Library Translation 1122, (Royal Aircraft Establishment, Farnborough, U.K., 1965).
8. H.-P. Schwefel, *Projekt MHD-Stausstrahlrohr: Experimentelle Optimierung einer Zweiphasenduse, Teil I*, Technical Report 11.034/68, 35, (AEG Forshungsinstitut, Berlin, Germany, 1968).
9. D. B. Fogel, An Evolutionary Approach to the Traveling Salesman Problem, *Biological Cybernetics*, **60**, 2, 1988) pp. 139-144.
10. L. J. Fogel, Autonomous Automata, *Industrial Research*, **4**, 1, 1962) pp. 14 - 19.
11. L. J. Fogel, A. J. Owens and M. J. Walsh, *Artificial Intelligence Through Simulated Evolution* (John Wiley & Sons, New York, NY, 1966).
12. J. H. Holland, Outline for a Logical Theory of Adaptive Systems, *Journal of the Association for Computing Machinery*, **9**, 1962) pp. 297-314.
13. J. H. Holland, Nonlinear Environments Permitting Efficient Adaptation, *Computer and Information Sciences II*, Ed. (Academic Press, 1967). pp.
14. J. H. Holland, *Adaptation in Natural and Artificial Systems* (MIT Press, Ann Arbor, MI, 1975).
15. J. H. Holland, *Hidden Order: How Adaptation Builds Complexity* (Helix Books - An Addison-Wesley Publishing Company, Reading, MA, 1995).
16. J. H. Holland, *Emergence: From Chaos to Order* (Perseus Books, Reading, MA, 1998).
17. S. A. Burns, Eds., *Recent Advances in Optimal Structural Design*, (American Society of Civil Engineers, Reston, VA, 2002).
18. R. Kicinger, T. Arciszewski and K. A. De Jong, Evolutionary Computation and Structural Design: A Survey of the State-of-the-Art, *Computers & Structures*, **83**, (Elsevier Ltd. (Science Direct), 2005) pp. 1943-1978.
19. T. Back, An Introduction to Evolutionary Algorithms, *Evolutionary Computation 1 - Basic Algorithms and Operators*, Ed. T. Back, D. B. Fogel and T. Michalewicz, (Institute of Physics Publishing, Bristol, U.K., 2000). pp. 6-58.
20. D. A. Coley, *An Introduction to Genetic Algorithms for Scientists and Engineers* (World Scientific Publishing Co., Pte., Ltd., Singapore, 2001).
21. D. E. Goldberg, *Genetic Algorithms in Search, Optimization, and Machine Learning* (Addison-Wesley, Boston, MA, 1989).
22. Z. Michalewicz, *Genetic Algorithms + Data Structures = Evolution Programs; Third Revised and Extended Edition* (Springer, Berlin, Germany, 1996).
23. V. W. Porto, Evolutionary Programming, *Evolutionary Computation 1 - Basic Algorithms and Operators*, Ed. T. Back, D. B. Fogel and T. Michalewicz, (Institute of Physics Publishing, Bristol, U.K., 2000). pp. 6-58.

24. G. Rudolph, Evolution Strategies, *Evolutionary Computation 1 - Basic Algorithms and Operators*, Ed. T. Back, D. B. Fogel and T. Michalewicz, (Institute of Physics Publishing, Bristol, U.K., 2000). pp. 6-58.

25. S. Pezeshk and C. V. Camp, State of the Art on the Use of Genetic Algorithms in Design of Steel Structures, *Recent Advances in Optimal Structural Design*, Ed. S. A. Burns, (American Society of Civil Engineers, Reston, VA, 2002). pp. 55-80.

26. C.-Y. Lin and W.-H. Wu, Self-Organizing Adaptive Penalty Strategy in Constrained Genetic Search, *Structural and Multidisciplinary Optimization*, **26**, 4, (Springer - Berlin, Heidelberg, 2004) pp. 417-428.

27. P. Nanakorn and K. Meesomklin, An Adaptive Penalty Function in Genetic Algorithm for Structural Design Optimization, *Computers & Structures*, **79**, (Pergamon, Elsevier Science, Ltd., Oxford, U.K., 2001) pp. 2527-2539.

28. A. Kurpati, S. Azarm and J. Wu, Constraint Handling Improvements for Multiple Objective Genetic Algorithms, *Structural and Multidisciplinary Optimization*, **23**, 3, (Springer, Berlin/Heidelberg, 2002) pp. 204-213.

29. D. A. Van Veldhuizen and G. B. Lamont, Multiobjective Evolutionary Algorithms: Analyzing the State-of-the-Art, *Evolutionary Computation*, **8**, 2, (Massachusetts Institute of Technology, Cambridge, MA, 2000) pp. 125-147.

30. R. J. Balling and S. Wilson, The Maxi-Min Fitness Function for Multi-Objective Evolutionary Computation: Application to City Planning, *Genetic and Evolutionary Computation Conference (GECCO 2001)*, Ed. San Francisco, CA, (Morgan Kaufmann, 2001). pp. 1079-1084.

31. C. A. Coello Coello and G. T. Pulido, Multiobjective Structural Optimization Using a Genetic Algorithm, *Structural and Multidisciplinary Optimization*, **30**, 5, (Springer - Berlin, Heidelberg, 2005) pp. 388-403.

32. C. M. Foley, A. Alimoradi and S. Pezeshk, Probabilistic Performance-Based Optimal Design of Steel Moment-Resisting Frames - Part I, *Journal of Structural Engineering*, at press, (American Society of Civil Engineers, Reston, VA, 2006) pp.

33. P. Hajela and J. Yoo, GA Based Fuzzy Optimization for Non-Convex Pareto Surfaces, *Genetic and Evolutionary Computation Conference (GECCO) Workshop*, Ed. S. A. Burns, San Francisco, CA, (Morgan Kaufmann, 2001). pp. 85-90.

34. J. D. Knowles and D. W. Corne, Approximating the Nondominated Front Using the Pareto Archived Evolution Strategy, *Evolutionary Computation*, **8**, 2, (Massachusetts Institute of Technology, Cambridge, MA, 2000) pp. 149-172.

35. S. Y. Zeng, L. S. Kang and L. X. Ding, An Orthogonal Multi-Objective Evolutionary Algorithm for Multi-Objective Optimization Problems with Constraints, *Evolutionary Computation*, **12**, 1, (Massachusetts Institute of Technology, Cambridge, MA, 2004) pp. 77-98.

36. A. Farhang-Mehr and S. Azarm, Entropy-Based Multi-Objective Genetic Algorithm for Design Optimization, *Structural and Multidisciplinary Optimization*, **24**, 5, (Springer - Berlin, Heidelberg, 2002) pp. 351-361.

37. M. Laumanns, L. Thiele, K. Deb and E. Zitzler, Combining Convergence and Diversity in Evolutionary Multiobjective Optimization, *Evolutionary Computation*, **10**, 3, (Massachusetts Institute of Technology, Cambridge, MA, 2002) pp. 263-282.

38. F. Menczer, M. Degeratu and W. N. Street, Efficient and Scalable Pareto Optimization by Evolutionary Local Selection Algorithms, *Evolutionary Computation*, **8**, 2, (Massachusetts Institute of Technology, Cambridge, MA, 2000) pp. 223-247.

39. R. Kumar and P. Rockett, Improved Sampling of the Pareto-Front in Multiple Objective Genetic Optimization by Steady-State Evolution: A Pareto Converging Genetic Algorithm, *Evolutionary Computation*, **10**, 3, (Massachusetts Institute of Technology, Cambridge, MA, 2002) pp. 283-314.

40. L. Costa and P. Oliveira, An Adaptive Sharing Elitist Evolution Strategy for Multiobjective Optimization, *Evolutionary Computation*, **11**, 4, (Massachusetts Institute of Technology, Cambridge, MA, 2003) pp. 417-438.

41. E. Zitzler, K. Deb and L. Thiele, Comparison of Multiobjective Evolutionary Algorithms: Empirical Results, *Evolutionary Computation*, **8**, 2, (Massachusetts Institute of Technology, Cambridge, MA, 2000) pp. 173-195.

42. I. C. Parmee, D. Cvetkovic, A. H. Watson and C. R. Bonham, Multiobjective Satisfaction within and Interactive Evolutionary Design Environment, *Evolutionary Computation*, **8**, 2, (Massachusetts Institute of Technology, Cambridge, MA, 2000) pp. 197-222.

43. K. Deb, A. Anand and D. Joshi, A Computationally Efficient Evolutionary Algorithm for Real-Parameter Optimization, *Evolutionary Computation*, **10**, 4, (Massachusetts Institute of Technology, Cambridge, MA, 2002) pp. 371-395.

44. W. M. Jenkins, A Decimal-Coded Evolutionary Algorithm for Constrained Optimization, *Computers & Structures*, **80**, (Pergamon, Elsevier Science, Ltd., Oxford, U.K., 2002) pp. 471-480.

45. M. S. Voss and C. M. Foley, Evolutionary Algorithm for Structural Optimization, *Proceedings of the Genetic and Evolutionary Computation Conference (GECCO)*, Ed. W. Banzhaf, J. Daida, A. E. Eiben, M. H. Garzon, V. Honavar, M. Jakiela and R. E. Smith, Orlando, FL, (Morgan Kaufman, 1999). pp. 678-685.

46. C. M. Foley, D. Schinler and M. S. Voss, *Optimized Design of Fully and Partially Restrained Steel Frames Using Advanced Analysis and Object-Oriented Evolutionary Computation*, (Technical Report Submitted to the National Science Foundation, Award Number CMS-9813216, 2002).

47. F. Rothlauf, D. E. Goldberg and A. Heinzl, Network Random Keys - A Tree Representation Scheme for Genetic and Evolutionary Algorithms, *Evolutionary Computation*, **10**, 1, (Massachusetts Institute of Technology, Cambridge, MA, 2002) pp. 75-97.

48. M. S. Voss and C. M. Foley, The (mu, lambda, alpha, beta) Distribution: A Selection Scheme for Ranked Populations, *Proceedings of the Genetic and Evolutionary Computation Conference (GECCO) - Late Breaking Papers*, Ed. W. Banzhaf, J. Daida, A. E. Eiben, M. H. Garzon, V. Honavar, M. Jakiela and R. E. Smith, Orlando, FL, (Morgan Kaufman, 1999). pp. 284-291.

49. D. Schinler and C. M. Foley, An Object-Oriented Evolutionary Algorithm for Advanced Analysis-Based Design, *Genetic and Evolutionary Computation Conference (GECCO) Workshop*, Ed. S. A. Burns, San Francisco, CA, (Morgan Kaufmann, 2001). pp. 73-78.

50. R. Poli and N. F. McPhee, General Schema Theory for Genetic Programming with Subtree-Swapping Crossover: Part I, *Evolutionary Computation*, **11**, 1, (Massachusetts Institute of Technology, Cambridge, MA, 2003) pp. 53-66.

51. R. Poli and N. F. McPhee, General Schema Theory for Genetic Programming with Subtree-Swapping Crossover: Part II, *Evolutionary Computation*, **11**, 2, (Massachusetts Institute of Technology, Cambridge, MA, 2003) pp. 169-206.

52. W. Gutkowski, Z. Iwanow and J. Bauer, Controlled Mutation in Evolutionary Structural Optimization, *Structural and Multidisciplinary Optimization*, **21**, 5, (Springer - Berlin, Heidelberg, 2001) pp. 355-360.

53. N. D. Lagaros, M. Papadrakakis and G. Kokossalakis, Structural Optimization Using Evolutionary Algorithms, *Computers & Structures*, **80**, (Pergamon, Elsevier Science Ltd., 2002) pp. 571-589.

54. A. Miyamoto, K. Kawamura and H. Nakamura, Bridge Management System and Maintenance Optimization for Existing Bridges, *Comptur-Aided Civil and Infrastructure Engineering*, **15**, (Blackwell Publishers, Oxford, U.K., 2000) pp. 45-55.

55. H. Furuta, M. Hirokane and K. Harakawa, Application of Genetic Algorithms and Rough Sets to Data Mining for Integrity Assessment of Bridge Structures, *2001 Genetic and Evolutionary*

Computation Conference (GECCO) Workshop Program, Ed. San Francisco, CA, (Morgan Kaufmann, 2001). pp. 91-96.

56. H. Furuta, D. M. Frangopol, M. Hirokane and M. Saito, Integrated Bridge Management System Based on Life-Cycle Cost, *2001: A Structural Engineering Odyssey, Proceedings of the 2001 Structures Congress and Exposition*, Ed. P. C. Chang, Washington, D.C., (American Society of Civil Engineers, 2001). pp. CD-ROM.

57. M. Liu and D. M. Frangopol, Multiobjective Maintenance Planning Optimization for Deteriorating Bridges Considering Condition, Safety, and Life-Cycle Cost, *Journal of Structural Engineering*, **131**, 5, (American Society of Civil Engineers, Reston, VA, 2005) pp. 833-842.

58. M. Liu and D. M. Frangopol, Balancing Connectivity of Deteriorating Bridge Networks and Long-Term Maintenance Cost Through Optimization, *Journal of Bridge Engineering*, **10**, 4, (American Society of Civil Engineers, Reston, VA, 2005) pp. 468-481.

59. M. Liu and D. M. Frangopol, Optimal Bridge Maintenance Planning Based On Probabilistic Performance Prediction, *Engineering Structures*, **26**, (Elsevier Ltd., U.K., 2004) pp. 991-1002.

60. M. Liu and D. M. Frangopol, Bridge Annual Maintenance Prioritization Under Uncertainty by Multiobjective Combinatorial Optimization, *Comptur-Aided Civil and Infrastructure Engineering*, **20**, (Blackwell Publishing, Malden, MA, 2005) pp. 343-353.

61. L. A. C. Neves, D. M. Frangopol and P. J. S. Cruz, Probabilistic LIfetime-Oriented Multiobjective Optimization of Bridge Maintenance: Single Maintenance Type, *Journal of Structural Engineering*, **132**, 6, (American Society of Civil Engineers, Reston, VA, 2006) pp. 991-1005.

62. ACI, *Building Code Requirements for Reinforced Concrete - ACI 318-05* (American Concrete Institute, Farmington Hills, MI, 2005).

63. AISC, *Specification for the Design of Steel Buildings - ASCE 358.05* (American Institute of Steel Construction, Chicago, IL, 2005).

64. F. Erbatur, O. Hasancebi, I. Tutuncu and H. Kilic, Optimal Design of Planar and Space Structures with Genetic Algorithms, *Computers and Structures*, **75**, (Elsevier Science, Ltd., Oxford, U.K., 2000) pp. 209-224.

65. M. S. Hayalioglu, Optimum Load and Resistance Factor Design of Steel Space Frames using Genetic Algorithm, *Structural and Multidisciplinary Optimization*, **21**, 4, (Springer, Berlin / Heidelberg, 2001) pp. 292-299.

66. M. P. Saka, Optimum Design of Pitched Roof Steel Frames with Haunched Rafters by Genetic Algorithm, *Computers & Structures*, **81**, (Pergammon, Elsevier Science, Ltd., Oxford, U.K., 2003) pp. 1967-1978.

67. L. Xu and Y. Gong, Preliminary Design of Long-Span King-Post Truss Systems with a Genetic Algorithm, *Computer-Aided Civil and Infrastructure Engineering*, **16**, (Blackwell Publishers, Malden, MA, 2001) pp. 94-105.

68. F. Y. Kocer and J. S. Arora, Optimal Design of Latticed Towers Subjected to Earthquake Loading, *Journal of Structural Engineering*, **128**, 2, (American Society of Civil Engineers, Reston, VA, 2002) pp. 197-2004.

69. C. S. Krishnamoorthy, P. Prasanna Venkatesh and R. Sudarshan, Object-Oriented Framework for Genetic Algorithms with Applications to Space Truss Optimization, *Journal of Computing in Civil Engineering*, **16**, 1, (American Society of Civil Engineers, Reston, VA, 2002) pp. 66-75.

70. P. Sivakumar, A. Rajaraman, G. M. Samuel Knight and D. S. Ramachandramurthy, Object-Oriented Optimization Approach Using Genetic Algorithms for Lattice Towers, *Journal of Computing in Civil Engineering*, **18**, 2, (American Society of Civil Engineers, Reston, VA, 2004) pp. 162-171.

71. V. Togan and A. T. Daloglu, Optimization of 3D Trusses with Adaptive Approach in Genetic Algorithms, *Engineering Structures*, **28**, (Elsevier, U.K., 2006) pp. 1019-1027.

72. C. M. Foley and D. Schinler, Optimized Design of Partially and Fully-Restrained Steel Frames Using Distributed Plasticity, *2001: A Structural Engineering Odyssey, Proceedings of the 2001 Structures Congress & Exposition*, Ed. P. C. Chang, Washington, D.C., (American Society of Civil Engineers, 2001). pp. CD-ROM.

73. C. M. Foley and D. Schinler, Automated Design of Steel Frames Using Advanced Analysis and Object-Oriented Evolutionary Computation, *Journal of Structural Engineering*, **129**, 5, (American Society of Civil Engineers, Reston, VA, 2003) pp. 648-660.

74. E. S. Kameshki and M. P. Saka, Genetic Algorithm Based Optimum Bracing Design of Non-Swaying Tall Plane Frames, *Journal of Constructional Steel Research*, **27**, (Elsevier Science, Ltd., U.K., 2001) pp. 1081-1097.

75. C.-M. Chan and P. Liu, Structural Optimization Using Hybrid Genetic Algorithms, *Genetic and Evolutionary Computation Conference (GECCO) Workshop*, Ed. S. A. Burns, San Francisco, CA, (Morgan Kaufmann, 2001). pp. 108-113.

76. C. V. Camp, S. Pezeshk and H. Hansson, Flexural Design of Reinforced Concrete Frames Using a Genetic Algorithm, *Journal of Structural Engineering*, **129**, 1, (American Society of Civil Engineers, Reston, VA, 2003) pp. 105-115.

77. C. Lee and J. Ahn, Flexural Design of Reinforced Concrete Frames by Genetic Algorithm, *Journal of Structural Engineering*, **129**, 6, (American Society of Civil Engineers, Reston, VA., 2003) pp. 762-774.

78. V. Govindaraj and J. v. Ramaswamy, Optimum Detailed Design of Reinforced Concrete Beams Using Genetic Algorithms, *Computers and Structures*, **84**, (Elsevier, Ltd., U.K., 2005) pp. 34-48.

79. M. Leps and M. Sejnoha, New Approach to Optimization of Reinforced Concrete Beams, *Computers and Structures*, **81**, (Elsevier, Ltd., U.K., 2003) pp. 1957-1966.

80. M. G. Sahab, A. F. Ashour and V. V. Toropov, Cost Optimisation of Reinforced Concrete Flat Slab Buildings, *Engineering Structures*, **27**, (Elsevier, Ltd., Oxford, U.K., 2005) pp. 313-322.

81. M. G. Sahab, A. F. Ashour and V. V. Toropov, A Hybrid Genetic Algorithm for Reinforced Concrete Flat Slab Buildings, *Computers and Structures*, **83**, (Elsevier, Ltd., U.K., 2005) pp. 551-559.

82. C. A. Coello and A. D. Christiansen, Multiobjective Optimization of Trusses Using Genetic Algorithms, *Computers and Structures*, **75**, (Elsevier Science, Ltd., Oxford, U.K., 2000) pp. 647-660.

83. M. Liu, S. A. Burns and Y. K. Wen, Genetic Algorithm Based Construction-Conscious Minimum Weight Design of Seismic Moment-Resisting Frames, *Journal of Structural Engineering*, **132**, 1, (American Society of Civil Engineers, Reston, VA., 2006) pp. 50-58.

84. F. Y. Cheng, D. Li and J. Ger, Multiobjective Optimization of Seismic Structures, *Advanced Technologies in Structural Engineering: Proceedings of Structures Congress 2000*, Ed. Philadelphia, PA, (American Society of Civil Engineers, 2000). pp. CD-ROM.

85. S. Pezeshk, C. V. Camp and D. Chen, Design of Nonlinear Framed Structures Using Genetic Optimization, *Journal of Structural Engineering*, **126**, 3, (American Society of Civil Engineers, Reston, VA, 2000) pp. 382-388.

86. E. S. Kameshki and M. P. Saka, Optimum Design of Nonlinear Steel Frames with Semi-Rigid Connections Using a Genetic Algorithm, *Computer and Structures*, **79**, (Elsevier Science, Ltd., U.K., 2001) pp. 1593-1604.

87. E. S. Kameshki and M. P. Saka, Genetic Algorithm Based Optimum Design of Nonlinear Planar Steel Frames with Various Semi-Rigid Connections, *Journal of Constructional Steel Research*, **59**, (Elsevier Science, Ltd., U.K., 2003) pp. 109-134.

88. M. S. Hayalioglu, Optimum Design of Geometrically Non-Linear Elastic-Plastic Steel Frames via Genetic Algorithm, *Computers and Structures*, **77**, (Elsevier Science, Litd., Oxford, U.K., 2000) pp. 527-538.

89. M. K. Yun and B. H. Kim, Optimum Design of Plane Steel Frame Structures Using Second-Order Inelastic Analysis and a Genetic Algorithm, *Journal of Structural Engineering*, **131**, 12, (American Society of Civil Engineers, Reston, VA, 2005) pp. 1820-1831.

90. ATC, *Seismic Evaluation and Retrofit of Existing Concrete Buildings; Volumes 1 and 2* (Applied Technology Council, Redwood City, CA., 1996).

91. FEMA, *FEMA-350; Recommended Seismic Design Criteria for New Steel Moment-Frame Buildings* (Federal Emergency Management Agency, Washington, D.C., 2000).

92. FEMA, *FEMA 356; Prestandard Commentary for the Seismic Rehabilitation of Buildings* (Federal Emergency Management Agency, Washington, D.C., 2000).

93. A. Alimoradi, *Probabilistic Performance-Based Seismic Design Automation of Nonlinear Steel Structures Using Genetic Algorithms*, Ph.D. Dissertation (Civil Engineering, The University of Memphis, Memphis, TN, 2004).

94. M. Liu, *Development of Multiobjective Optimization Procedures for Seismic Design of Steel Moment Frame Structures*, Ph.D. Dissertation (Civil & Environmental Engineering, University of Illinois, Urbana-Champaign, 2004).

95. M. Liu, S. A. Burns and Y. K. Wen, Multiobjective Optimization for Performance-Based Seismic Design of Steel Moment Frame Structures, *Earthquake Engineering and Structural Dynamics*, **34**, (John Wiley & Sons, Inc., New York, NY, 2005) pp. 289-306.

96. M. Liu, S. A. Burns and Y. K. Wen, Optimal Seismic Design of Steel Frame Buildings Based on Life-Cycle Cost Considerations, *Earthquake Engineering and Structural Dynamics*, **32**, (John Wiley & Sons, Inc., New York, NY, 2003) pp. 1313-1332.

97. M. Liu, S. A. Burns and Y. K. Wen, Multiobjective Optimization for Life Cycle Cost Oriented Seismic Design of Steel Moment Frame Structures, *Building the Past: Securing the Future - Proceedings of the 2004 Structures Congress*, Ed. G. E. Blandford, Nashville, TN, (American Society of Civil Engineers, 2004). pp. CD-ROM.

98. M. Liu, Y. K. Wen and S. A. Burns, Life Cycle Cost Oriented Seismic Design Optimization of Steel Moment Frame Structures with Risk-Taking Preference, *Engineering Structures*, **26**, (Elsevier, Ltd., U.K., 2004) pp. 1407-1421.

99. M. Fragiadakis, N. D. Lagaros and M. Papadrakakis, Performance-Based Multiobjective Optimum Design of Steel Structures Considering Life-Cycle Cost, *Structural and Multidisciplinary Optimization*, **32**, 1, (Springer-Verlag, Berlin, 2006) pp. 1-11.

100. A. Alimoradi, S. Pezeshk and C. M. Foley, Automated Performance-Based Design of Steel Frames, *Building on the Past: Securing the Future*, Ed. G. E. Blandford, Nashville, TN, (American Society of Civil Engineers, 2004). pp. CD-ROM.

101. A. Alimoradi, S. Pezeshk and C. M. Foley, Probabilistic Performance-Based Optimal Design of Steel Moment-Resisting Frames - Part II, *Journal of Structural Engineering*, at press, (American Society of Civil Engineers, Reston, VA, 2006) pp.

102. A. Alimoradi, S. Pezeshk and C. M. Foley, Evolutionary Seismic Design for Optimal Performance, *Intelligent Computational Paradigms in Earthquake Engineering*, Ed. N. D. Lagaros, Tsompanakis, Y., (Idea Group Publishing, Hershey, PA, 2006). pp. (at press).

103. C. K. Dimou and V. K. Koumousis, Genetic Algorithms in Competitive Environments, *Journal of Computing in Civil Engineering*, **17**, 3, (American Society of Civil Engineers, Reston, VA, 2003) pp. 142-149.

104. V. K. Koumousis and C. K. Dimou, Genetic Algorithms in a Competitive Environment with Application to Reliability Optimal Design, *2001 Genetic and Evolutionary Computation Conference Workshop Program*, Ed. S. A. Burns, San Francisco, CA, (Morgan Kaufmann, 2001). pp. 79-84.

105. C. K. Prasad Varma Thampan and C. S. Krishnamoorthy, System Reliability-Based Configuration Optimization of Trusses, *Journal of Structural Engineering*, **127**, 8, (American Society of Civil Engineers, Reston, VA, 2001) pp. 947-956.

106. Y. Tsompanakis and M. Papadrakakis, Large-Scale Reliability-Based Structural Optimization, *Structural and Multidisciplinary Optimization*, **26**, 4, (Springer, Berlin-Heidelberg, 2004) pp. 429-440.
107. N. D. Lagaros and M. Papadrakakis, Robust Seismic Design Optimization of Steel Structures, *Structural and Multidisciplinary Optimization*, (Springer-Verlag, Berlin/Heidelberg, 2006) pp. (On-Line First).
108. R. Spallino and G. Theirauf, Thermal Buckling Optimization of Composite Laminates by Evolution Strategies, *Computers and Structures*, **78**, (Elsevier Science, Ltd., Oxford, U.K., 2000) pp. 691-697.
109. W.-T. Chan and H. Hu, An Application of Genetic Algorithms to Precast Production Scheduling, *Computer and Structures*, **79**, (Civil-Comp, Ltd. and Elsevier Science, Ltd., Oxford, U.K., 2001) pp. 1605-1616.
110. E. M. R. Fairbairn, M. M. Silvoso, R. D. Toledo Filho, J. L. D. Alves and N. F. F. Ebecken, Optimization of Mass Concrete Construction using Genetic Algorithms, *Computers & Structures*, **82**, (Elsevier Ltd., Oxford, U.K., 2004) pp. 281-299.
111. J. Garcia-Perez, F. Castellanos and O. Diaz, Optimal Seismic Zoning for Multiple Types of Structures, *Earthquake Engineering and Structural Dynamics*, **32**, (John Wiley & Sons, Ltd., New York, NY, 2003) pp. 711-730.
112. A. Alimoradi, S. Pezeshk and F. Naeim, Identification of Input Ground Motion Records for Seismic Design Using Neuro-Fuzzy Pattern Recognition and Genetic Algorithms, *Building on the Past: Securing the Future*, Ed. G. E. Blandford, Nashville, TN, (American Society of Civil Engineers, 2004). pp. (CD-ROM).
113. F. Naeim, A. Alimoradi and S. Pezeshk, Selection and Scaling of Ground Motion Time Histories for Structural Design Using Genetic Algorithms, *Earthquake Spectra*, **20**, 2, (Earthquake Engineering Research Institute, Oakland, CA, 2004) pp. 413-426.
114. S. S. S. Sakla and E. Elbeltagi, Design of Steel Roofs Subjected to Drifting Snow Using Genetic Algorithms, *Computers & Structures*, **81**, (Elsevier Science, Ltd., Oxford, U.K., 2003) pp. 339-348.
115. C. M. Foley and W. K. Lucas, Optimal Selection and Design of Composite Steel Floor Systems Considering Vibration, *Building on the Past: Securing the Future*, Ed. G. E. Blandford, Nashville, TN, (American Society of Civil Engineers, 2004). pp. (CD-ROM).
116. J. Lee, S.-M. Kim, H.-S. Park and B.-H. Woo, Optimum Design of Cold-Formed Steel Channel Beams Using Micro Genetic Algorithm, *Engineering Structures*, **27**, (Elsevier, Ltd., U.K., 2005) pp. 17-24.
117. A. Gupta, P. Kripakaran, G. Mahinthakumar and J. W. Baugh, Genetic Algorithm-Based Decision Support for Optimizing Seismic Response of Piping Systems, *Journal of Structural Engineering*, **131**, 3, (American Society of Civil Engineers, Reston, VA, 2005) pp. 389-398.
118. K. C. Fu, Y. Zhai and S. Zhou, Optimum Design of Welded Steel Plate Girder Bridges Using a Genetic Algorithm with Elitism, *Journal of Bridge Engineering*, **10**, 3, (American Society of Civil Engineers, Reston, VA, 2005) pp. 291-301.
119. A. M. Raich, Evolving Structural Design Solutions for Unstructured Problem Domains, *Genetic and Evolutionary Computation Conference (GECCO) Workshop*, Ed. S. A. Burns, San Francisco, CA, (Morgan Kaufmann, 2001). pp. 68-72.
120. A. M. Raich and J. Ghaboussi, Evolving Structural Design Solutions Using an Implicit Redundant Genetic Algorithm, *Structural and Multidisciplinary Optimization*, **20**, 3, (Springer-Verlag, Berlin/Heidelberg, 2000) pp. 222-231.
121. J. Ryoo and P. Hajela, Handling Variable String Lengths in GA-Based Structural Topology Optimization, *Structural and Multidisciplinary Optimization*, **26**, 5, (Springer-Verlag, Berlin/Heidelberg, 2004) pp. 318-325.
122. I. A. Azid, A. S. K. Kwan and K. N. Seetharamu, An Evolutionary Approach for Layout Optimization of a Three-Dimensional Truss, *Structural and Multidisciplinary Optimization*, **24**, 4, (Springer-Verlag, Berlin / Heidelberg, 2002) pp. 333-337.

123. K. Kawamura, H. Ohmori and N. Kito, Truss Topology Optimization by Modified Genetic Algorithm, *Structural and Multidisciplinary Optimization*, **23**, 6, (Springer-Verlag, Berlin / Heidelberg, 2002) pp. 467-473.

124. W.-S. Ruy, Y.-S. Yang and G.-H. Kim, Topology Design of Truss Structures in a Multicriteria Environment, *Computer-Aided Civil and Infrastructure Engineering*, **16**, (Blackwell Publishers, Malden, MA, 2001) pp. 246-258.

125. P. V. Hull, M. L. Tinker and G. Dozier, Evolutionary Optimization of a Geometrically Refined Truss, *Structural and Multidisciplinary Optimization*, **31**, 4, (Springer-Verlag, Berlin / Heidelberg, 2006) pp. 311-319.

126. B. Baumann and B. Kost, Structure Assembling by Stochastic Topology Optimization, *Computers and Structures*, **83**, (Elsevier, Ltd., Oxford, U.K., 2005) pp. 2175-2184.

127. M. P. Singh, S. Singh and L. M. Moreschi, Tuned Mass Dampers for Response Control of Torsional Buildings, *Earthquake Engineering and Structural Dynamics*, **31**, (John Wiley & Sons, Ltd., New York, NY, 2002) pp. 749-769.

128. A. S. Ahlawat and A. Ramaswamy, Multiobjective Optimal Absorber System for Torsionally Coupled Seismically Excited Structures, *Engineering Structures*, **25**, (Elsevier Science, Ltd., Oxford, U.K., 2003) pp. 941-850.

129. M. P. Singh and L. M. Moreschi, Optimal Placement of Dampers for Passive Response Control, *Earthquake Engineering and Structural Dynamics*, **31**, (John Wiley & Sons, Ltd., New York, NY, 2002) pp. 955-976.

130. N. Wongprasert and M. D. Symans, Application of a Genetic Algorithm for Optimal Damper Distribution within the Nonlinear Seismic Benchmark Building, *Journal of Engineering Mechanics*, **130**, 4, (American Society of Civil Engineers, Reston, VA, 2004) pp. 401-406.

131. L. M. Moreschi and M. P. Singh, Design of Yielding Metallic and Friction Dampers for Optimal Seismic Performance, *Earthquake Engineering and Structural Dynamics*, **32**, (John Wiley & Sons, Ltd., New York, NY, 2003) pp. 1291-1311.

132. G. F. Dargush and R. S. Sant, Evolutionary Aseismic Design and Retrofit of Structures with Passive Energy Dissipation, *Earthquake Engineering and Structural Dynamics*, **34**, (John Wiley & Sons, Ltd., New York, NY, 2005) pp. 1601-1626.

133. J. A. Bishop and A. G. Striz, On Using Genetic Algorithms for Optimum Damper Placement in Space Trusses, *Structural and Multidisciplinary Optimization*, **28**, 2-3, (Springer-Verlag, Berlin, 2004) pp. 136-145.

134. H. Zheng, C. Cai and X. M. Tan, Optimization of Partial Constrained Layer Damping Treatment for Vibrational Energy Minimization of Vibrating Beams, *Computers and Structures*, **82**, (Elsevier, Ltd., Oxford, U.K., 2004) pp. 2493-2507.

135. D. K. Liu, Y. L. Yang and Q. S. Li, Optimum Positioning of Actuators in Tall Buildings Using Genetic Algorithm, *Computers and Structures*, **81**, (Elsevier, Ltd., Oxford, U.K., 2003) pp. 2823-2827.

136. O. Yoshida and S. J. Dyke, Response Control of Full-Scale Irregular Buildings Using Magnetorheological Dampers, *Journal of Structural Engineering*, **131**, 5, (American Society of Civil Engineers, Reston, VA, 2005) pp. 734-742.

137. A. Alimoradi, Performance Study of a GA-Based Active/Hybrid Control System Under Near Source Strong Ground Motion, *2001: A Structural Engineering Odyssey, Proceedings of the 2001 Structures Congress & Exposition*, Ed. P. C. Chang, Washington, D.C., (American Society of Civil Engineers, 2001). pp. CD-ROM.

138. Y.-J. Kim and J. Ghaboussi, Direct Use of Design Criteria in Genetic Algorithm-Based Controller Optimization, *Earthquake Engineering and Structural Dynamics*, **30**, (John Wiley & Sons, Inc., New York, NY, 2001) pp. 1261-1278.

139. Y. Arfiadi and M. N. S. Hadi, Optimal (Direct) Static Output Feedback Controller Using Real-Coded Genetic Algorithms, *Computers and Structures*, **79**, (Civil-Comp, Ltd and Elsevier Science, Ltd., Oxford, U.K., 2001) pp. 1625-1634.

140. Y. Arfiadi and M. N. S. Hadi, Passive and Active Control of Three-Dimensional Buildings, *Earthquake Engineering and Structural Dynamics*, **29**, (John Wiley & Sons, Inc., New York, NY, 2000) pp. 377-396.
141. A.-P. Wang and C.-D. Lee, Fuzzy Sliding Model Control for a Building Based On Genetic Algorithms, *Earthquake Engineering and Structural Dynamics*, **31**, (John Wiley & Sons, New York, NY, 2002) pp. 881-895.
142. A. S. Ahlawat and J. v. Ramaswamy, Multi-Objective Optimal Design of FLC Driven Hybrid Mass Damper for Seismically Excited Structures, *Earthquake Engineering and Structural Dynamics*, **31**, (John Wiley & Sons, Inc., New York, NY, 2002) pp. 1459-1479.
143. A. S. Ahlawat and A. Ramaswamy, Multiobjective Optimal FLC Driven Hybrid Mass Damper System for Torsionally Coupled, Seismically Excited Structures, *Earthquake Engineering and Structural Dynamics*, **31**, (John Wiley & Sons, Inc., New York, NY, 2002) pp. 2121-2139.
144. H.-S. Kim and P. N. Roschke, Design of Fuzzy Logic Controller for Smart Base Isolation System Using Genetic Algorithm, *Engineering Structures*, **28**, (Elsevier, Ltd., Oxford, U.K., 2006) pp. 84-96.
145. H.-S. Kim and P. N. Roschke, Fuzzy Control of Base-Isolation System Using Mult-Objective Genetic Algorithm, *Comptur-Aided Civil and Infrastructure Engineering*, **21**, (Blackwell Publishing, Malden, MA, 2006) pp. 436-449.
146. Q. S. Li, P. Liu, N. Zhang, C. M. Tam and L. F. Yang, Multi-Level Design Model and Genetic Algorithm for Structural Control System Optimization, *Earthquake Engineering and Structural Dynamics*, **30**, (John Wiley & Sons, Inc., New York, NY, 2001) pp. 927-942.
147. A. S. Ahlawat and J. v. Ramaswamy, Multiobjective Optimal Fuzzy Logic Controller Driven Active and Hybrid Control Systems for Seismically Excited Structures, *Journal of Engineering Mechanics*, **130**, 4, (American Society of Civil Engineers, Reston, VA, 2004) pp. 416-423.
148. P. Tan, S. J. Dyke, A. Richardson and M. Abdullah, Integrated Device Placement and Control Design in Civil Structures Using Genetic Algorithms, *Journal of Structural Engineering*, **131**, 10, (American Society of Civil Engineers, Reston, VA, 2005) pp. 1489-1496.
149. Y. Yang, Z. Jin and C. K. Soh, Integrated Optimization of Control System for Smart Cylindrical Shells Using Modified GA, *Journal of Aerospace Engineering*, **19**, 2, (American Society of Civil Engineers, Reston, VA, 2006) pp. 68-79.
150. M. M. Abdullah, A. Richardson and J. Hanif, Placement of Sensors/Actuators on Civil Structures Using Genetic Algorithms, *Earthquake Engineering and Structural Dynamics*, **30**, (John Wiley & Sons, Inc., New York, NY., 2001) pp. 1167-1184.
151. J.-H. Chou and J. Ghaboussi, Genetic Algortithm in Structural Damage Detection, *Computers and Structures*, **79**, (Elsevier Science, Ltd., Oxford, U.K., 2001) pp. 1335-1353.
152. M. A. Rao, J. Srinivas and B. S. N. Murthy, Damage Detection in Vibrating Bodies Using Genetic Algorithms, *Computers and Structures*, **82**, (Elsevier Science, Ltd., Oxford, U.K., 2004) pp. 963-968.
153. H. Hao and Y. Xia, Vibration-Based Damage Detection of Structures by Genetic Algorithm, *Journal of Computing in Civil Engineering*, **16**, 3, (American Society of Civil Engineers, Reston, VA, 2002) pp. 222-229.
154. K. Worden, Optimal Sensor Placement for Fault Detection, *Engineering Structures*, **23**, (Elsevier Science, Ltd., Oxford, U.K., 2001) pp. 885-901.
155. W. Ostachowicz, M. Krawczuk and M. Cartmell, The Location of a Concentrated Mass on Rectangular Plates from Measurements of Natural Vibrations, *Computers and Structures*, **80**, (Elsevier Science, Ltd., Oxford, U.K., 2002) pp. 1419-1428.
156. R. J. Jiang, F. T. K. Au and Y. K. Cheung, Identification of Masses Moving on Multi-Span Beams Based on a Genetic Algorithm, *Computers and Structures*, **81**, (Elsevier, Ltd., Oxford, U.K., 2003) pp. 2137-2148.

157. C. K. Soh and Y. X. Dong, Evolutionary Programming for Inverse Problems in Civil Engineering, *Journal of Computing in Civil Engineering*, **15**, 2, (American Society of Civil Engineers, Reston, VA, 2001) pp. 144-150.

158. C. G. Koh, B. Hong and C.-Y. Liaw, Parameter Identification of Large Structural Systems in Time Domain, *Journal of Structural Engineering*, **126**, 8, (American Society of Civil Engineers, Reston, VA, 2000) pp. 957-963.

159. C. G. Koh, B. Hong and C.-Y. Liaw, Substructural and Progressive Structural Identification Methods, *Engineering Structures*, **25**, (Elsevier, Ltd., Oxford, U.K., 2003) pp. 1551-1563.

160. C. G. Koh, Y. F. Chen and C.-Y. Liaw, A Hybrid Computational Strategy for Identification of Structural Parameters, *Computers and Structures*, **81**, (Elsevier Science, Ltd., Oxford, U.K., 2003) pp. 107-117.

161. Y. Haralampidis, C. Papadimitriou and M. Pavlidou, Multi-Objective Framework for Structural Model Identification, *Earthquake Engineering and Structural Dynamics*, **34**, (John Wiley & Sons, Inc., New York, NY, 2005) pp. 665-685.

162. Y. Lu and Z. Tu, Dynamic Model Updating Using a Combined Genetic-Eigensensitivity Algorithm and Application in Seismic Response Prediction, *Earthquake Engineering and Structural Dynamics*, **34**, (John Wiley & Sons, Inc., New York, NY, 2005) pp. 1149-1170.

163. M. J. Perry, C. G. Koh and Y. S. Choo, Modified Genetic Algorithm Strategy for Structural Identification, *Computers and Structures*, **84**, (Elsevier, Ltd., Oxford, U.K., 2006) pp. 529-540.

164. T. Obata and Y. Miyamori, Identification of a Human Walking Force Model Based on Dynamic Monitoring Data from Pedestrian Bridges, *Computers and Structures*, **84**, (Elsevier, Ltd., Oxford, U.K., 2006) pp. 541-548.

165. S. Khajehpour and D. E. Grierson, Conceptual Design Using Adaptive Computing, *Genetic and Evolutionary Computation Conference (GECCO) Workshop*, Ed. S. A. Burns, San Francisco, CA, (Morgan Kaufmann, 2001). pp. 62-67.

166. S. Khajehpour and D. E. Grierson, Conceptual Design Using Adaptive Search, *2001: A Structural Engineering Odyssey, Proceedings of the 2001 Structures Congress & Exhibition*, Ed. P. C. Chang, Washington, D.C., (American Society of Civil Engineers, 2001). pp. CD-ROM.

167. S. Khajehpour and D. E. Grierson, Conceptual Design of Structures Using Multi-Criteria Adaptive Search, *Performance of Structures - From Research to Practice, Proceedings of the 2002 Structures Congress*, Ed. Denver, CO, (American Society of Civil Engineers, 2002). pp. CD-ROM.

168. M. Y. Rafiq, J. D. Mathews and G. N. Bullock, Conceptual Building Design - Evolutionary Approach, *Journal of Computing in Civil Engineering*, **17**, 3, (American Society of Civil Engineers, Reston, VA, 2003) pp. 150-158.

169. S. Khajehpour and D. E. Grierson, Profitability Versus Safety of High-Rise Office Buildings, *Structural and Multidisciplinary Optimization*, **25**, 4, (Springer-Verlag, Berlin / Heidelberg, 2003) pp. 279-293.

170. K. C. Sarma and H. Adeli, Bilevel Parallel Genetic Algorithms for Optimization of Large Steel Structures, *Comptur-Aided Civil and Infrastructure Engineering*, **16**, (Blackwell Publishers, Malden, MA, 2001) pp. 295-304.

171. G. E. Plassman, Experience with a Genetic Algorithm Implemented on a Multiprocessor Computer, *Structural and Multidisciplinary Optimization*, **22**, 2, (Springer - Verlag, Berlin / Heidelberg, 2001) pp. 102-115.

172. A. Gupta, P. Kripakaran, S. Kumar and J. W. Baugh, Optimizing Seismic Response of Secondary Systems on Workstation Clusters, *Performance of Structures - From Research to Practice, Proceedings of the 2002 Structures Congress*, Ed. Denver, CO, (American Society of Civil Engineers, 2002). pp. CD-ROM.

173. S. D. Rajan and D. T. Nguyen, Design Optimization of Discrete Structural Systems using MPI-Enabled Genetic Algorithm, *Structural and Multidisciplinary Optimization*, **28**, 5, (Springer - Verlag, Berlin / Heidelberg, 2004) pp. 340-348.

CHAPTER 4

MULTIOBJECTIVE OPTIMIZATION: CONCEPTS AND METHODS

Achille Messac

Professor, Department of Mechanical, Aerospace and Nuclear Engineering
Rensselaer Polytechnic University, Troy, NY 12180, USA
E-mail: messac@rpi.edu

Anoop A. Mullur

Post-doctoral Research Associate, Department of Mechanical, Aerospace and
Nuclear Engineering, Rensselaer Polytechnic University, Troy, NY 12180, USA
E-mail: mullua@rpi.edu

Decision-making is critical to the success of any product or system design. Multi-objective optimization can provide effective and efficient tools for decision-making under conflicting design criteria. The concept of tradeoff is integral to multiobjective optimization; and several approaches have been developed to resolve this tradeoff – yielding the so-called Pareto optimal solutions. These approaches can be broadly classified as those that require the specification of the designer preferences, and those that generate a set of Pareto optimal solutions from which the designer can choose. These methods and their relative merits and shortcomings are the focus of this chapter. A discussion regarding implementing these methods for practical problems is presented, followed by a discussion on industrial and academic applications.

1. Introduction to Multiobjective Optimization

Ever-increasing demands of system performance and economic competitiveness have necessitated the development and use of formal design methods for each phase of the engineering design process. Engineering design is a decision making process, requiring critical decisions at every stage during the design of a product or a system – from the initial conceptual design stage to the final detailed design stage. Decision making is generally challenging because of the existence of conflicting design requirements. In the presence of only a single design objective, it is a nearly trivial task to identify the optimal design configuration. However, as soon as one introduces a second conflicting design objective, the design process becomes more interesting and challenging. Multiobjective optimization techniques offer a formal methodology for effective design and decision making under multiple conflicting

121

design requirements and objectives. As such, multiobjective optimization can be a critical component of the modern design process.

For example, in a simple beam design problem, the design requirements could include the simultaneous minimization of beam mass and stress. Indeed, these two objectives are in mutual conflict: reducing the beam cross-section size would reduce its mass, but increase the stress at critical failure points in the beam; while increasing the cross-section size would reduce the stress, but would increase the mass. Another practical example would be the tradeoff between fuel efficiency and cargo capacity of cars. Large-sized vehicles can carry more load, but at the cost of low fuel efficiency; while small-sized cars yield higher fuel efficiency, at the cost of lower cargo capacity. One can find many such multiobjective examples in different engineering and non-engineering fields. Many believe that all design problems can (and should) be formulated as multiobjective problems.

In the above beam design example, if the mass objective were more important to a designer than stress, he/she would prefer design configurations that yield lower values of mass. A natural question arises: how does one generate optimal design alternatives that reflect a designer's preferences regarding conflicting design requirements? On the other hand, is it possible to provide a *set of optimal solutions* to the designer from which he/she can choose the most desirable one? In this chapter, we address these and many other pertinent questions relating to multiobjective design and optimization.

1.1. *Why Multiobjective Optimization?*

A question often asked in the design community is: Why not simply minimize one of the design objectives, and include the others as part of the constraints? While, in theory, this approach can lead to the desired solution, it is unfortunately fraught with significant pitfalls. For example, when one moves an objective from being part of the objective function to being a constraint, one simultaneously changes the nature of the preference pertinent to that objective – from a *soft* realistic preference, to a *hard* constraint. The latter is in general not truly reflective of the designer's intent; and s/he may unknowingly settle for an inadequate solution. Furthermore, the final solution typically depends heavily on the chosen value for the constraint, which is largely uncertain. As one includes more and more constraints, the choice of these constraint boundaries becomes a formidable task of it own. Fortunately, the application of effective multiobjective methods can obviate these difficulties.

1.2. *Scope of the Chapter*

In this chapter, we present an overview of some of the basic concepts and solution techniques in multiobjective optimization. Section 2 explains the concept of Pareto optimality and provides relevant definitions. Section 3 discusses popular multiobjective methods, categorized as (i) methods that require articulation of preferences, and (ii) Pareto set generation methods. In Sec. 4, we discuss some of the practical issues in multiobjective optimization, while recent applications and recent advances in the field of multiobjective optimization are discussed in Sec. 5. Summary and concluding remarks are given in Sec. 6.

2. Concept of Pareto Optimality

In this section, we introduce the critical concept of Pareto optimality, and present the relevant terminology and definitions.

2.1. *Multiobjective Optimization Problem Statement*

Multiobjective optimization involves the simultaneous optimization of two or more design objectives that are conflicting in nature. A typical optimization problem statement involving multiple (n_f) objectives can be written as

$$\min_{\mathbf{x}} \left[f_1(\mathbf{x}) \quad f_2(\mathbf{x}) \quad \cdots \quad f_{n_f}(\mathbf{x}) \right]^T \tag{1}$$

subject to

$$g(\mathbf{x}) \leq 0 \tag{2}$$
$$h(\mathbf{x}) = 0 \tag{3}$$
$$\mathbf{x}_l \leq \mathbf{x} \leq \mathbf{x}_u \tag{4}$$

where $\mathbf{x}_l$ and $\mathbf{x}_u$ are the lower and upper bounds on the design variables $\mathbf{x}$, respectively; g is the vector of inequality constraints, and h is the vector of equality constraints.

2.2. *Pareto Optimal Solutions*

One may ask two very pertinent questions regarding Eq. 1: (i) How does one define a "solution" to this problem? (ii) How does one solve a vector optimization problem involving conflicting design criteria? To answer the first question, we must look beyond the field of engineering. The concept of optimality – when trying to optimize two or more objective functions simultaneously – was formalized in the 1900's in the field of economics[1]. This concept has come to be known as Pareto optimality after its developer. The second of the above two questions is answered in Sec. 3.

Definition 1: A Pareto optimal solution is one for which any improvement in one objective will result in the worsening of at least one other objective[2]. That is, a tradeoff will take place.

Mathematically, a point f^* (which is a vector of length n_f) is called Pareto optimal if there *does not* exist a point f^p in the feasible design objective space, such that $f_j^* \geq f_j^p$ for all $j = 1, .., n_f$, and $f_j^* > f_j^p$, for at least one j. This mathematical definition assumes that smaller values of the design objectives are more desirable (minimization in Eq. 1).

The concept of tradeoff is central to multiobjective optimization. It signifies that an optimal solution to the multiobjective problem is one that results in an optimal tradeoff between the conflicting design objectives.

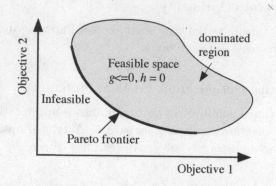

Fig. 1.   Design objective space for a bi-objective problem

In general, there exists an infinite number of Pareto optimal solutions to Eq. 1. Each of these solutions satisfies the definition of Pareto optimality (Def. 1). Most of the current methodologies for multiobjective problems revolve around identifying a representative set of Pareto optimal solutions or a single Pareto optimal design. Figure 1 graphically represents the design objective space of a bi-objective problem. The shaded portion is the feasible space (that is, where all points satisfy the constraints, Eqs. 2 and 3). The points on the edge of this feasible space (thick line) comprise the Pareto optimal set, and satisfy Def. 1. The remainder of the feasible region (shaded) is termed the "dominated region."

Definition 2: A dominated point is a point in the design objective space, for which there exists a point in the feasible space that is better (lower, in the case of minimization) in all objectives.

2.3. Local and Global Pareto Optimality

For design points that are globally Pareto optimal, the definition of Pareto optimality (Def. 1) holds true *with respect to all points in the feasible objective space.* Certain points, however, may be Pareto optimal only in a small region of the feasible design objective space. Such points are called locally Pareto optimal. Multiobjective optimization techniques aim for obtaining globally Pareto optimal points. Gradient-based optimizers typically tend to produce locally Pareto optimal points if the objective functions are multimodal (that is, possessing many local optima).

2.4. *The Pareto Frontier*

Each Pareto solution of a multiobjective problem can be identified using either the design objective values (as in Fig. 1) or the design variable values (**x**). The former results in a critical concept of multiobjective optimization – the Pareto frontier.

Definition 3: The Pareto frontier is the set of all Pareto optimal solutions represented in the design objective (f) space.

The Pareto frontier is a highly useful tool for multiobjective decision making in the design process. For two-objective problems, the Pareto frontier provides a graphical environment (Fig. 1) for making effective tradeoff decisions. In Fig. 1, all design alternatives that lie on the thick line constitute the Pareto frontier. Mathematically, no one Pareto solution is objectively better than any other solution, but to a designer, each Pareto solution represents a different level of desirability.

2.4.1. *Usefulness of the Pareto Frontier*

The Pareto frontier provides the designer a clear picture of the tradeoff characteristics of the different design objectives involved.

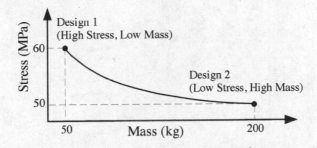

Fig. 2.   Tradeoff characteristics using a Pareto frontier

Figure 2 depicts a typical Pareto frontier for a two-objective optimization problem, which involves simultaneously minimizing the mass of a component and the stress. The design labeled "Design 1" is one that provides the least possible weight (50 kg), subject to design constraints. This design configuration can be obtained by simply ignoring the stress objective, and minimizing only the mass. As a consequence, the stress is adversely affected, and is high (60 MPa) for "Design 1." At the other end of the frontier is "Design 2", which provides the lowest possible stress (50 MPa) – at the cost of worsening the mass objective.

All other points on the Pareto frontier represent varying levels of minimization with respect to each design objective. The selection of the most desirable alternative is dependent on which design objective a designer prefers over the others, and by how much. For example, if he/she prefers the stress objective, then his/her region of interest would be the region of the Pareto frontier that offers low values of stress (in the vicinity of "Design 2"). Thus, the Pareto frontier offers visual guidance to a designer in making a decision regarding the most preferred design in a multiobjective sense.

2.5. *Pareto Frontier in Multiple Dimensions*

So far in the chapter, we have depicted Pareto frontiers only for two-objective problems. For more than two design objectives, visualization of the Pareto frontier can be a challenging task. For example, Fig. 3 shows the Pareto frontier for a three objective problem. Notice that the frontier is no longer a line (as in the case of two dimensions), but is a surface. Beyond three dimensions, we cannot possibly show all the objectives on the same plot. Even in three dimensions, understanding the tradeoffs between the different objectives can be challenging and often impractical. Pareto frontier visualization and tradeoff characterization in multiple dimensions is an open research topic[3].

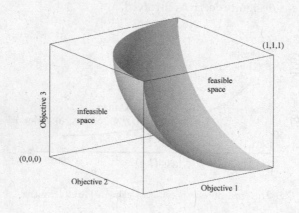

Fig. 3.   Pareto frontier for three objectives

3. Multiobjective Optimization Solution Techniques

In the previous section, we introduced the basic multiobjective optimization terminology and concepts. In this section, we describe some of the popular solution techniques for multiobjective optimization. Multiobjective optimization solution techniques can be broadly classified into two types: (i) Methods requiring designer preferences, and (ii) Methods that yield discrete representations of the Pareto frontier, or Pareto set generating methods.

3.1. *Methods Requiring Designer Preferences*

These methods require and incorporate designer preferences into the multiobjective problem formulation.

Definition 4: Preferences are the wishes or requirements of the designer regarding the different design objectives of a multiobjective problem.

The primary aim of these methods is to provide the designer with a single Pareto optimal design (instead of several solutions – which requires subsequent selection). Consequently, these methods are also called Integrated Generating and Choosing (IGC) methods[2], because they integrate the two aspects of multiobjective optimization and decision making – those of generating a candidate set of Pareto optimal solutions, and subsequently selecting the most desirable solution, based on the designer's preferences.

One of the most challenging tasks for this category of methods is to effectively and unambiguously model the designer's preferences, so as to ensure that the resulting optimal solution is the most desirable one, at least from a practical perspective. To model the designer's preferences, most of the methods belonging to this category adopt a utility function approach. A utility function is a combination (scalarization) of mathematical expressions for each design objective under consideration[4]. Also termed aggregate objective function (AOF) or preference function, this combined performance measure is treated as the objective function that is minimized or maximized. Next, we describe some of the popular approaches that belong to this category.

3.1.1. *Weighted Sum*

The weighted sum (WS) approach to multiobjective optimization is arguably the most popular approach in industry. The weighted sum method uses a linear combination of the objective functions with the help of weights that signify the relative importance between the objectives. Mathematically, the AOF (which is minimized) is defined as

$$J = \sum_{i=1}^{n_f} w_i f_i \tag{5}$$

The designer can prescribe scalar values for the individual weights w_i, depending on his/her preference with respect to each design objective. Often, the weights are chosen so that $w_i \geq 0$ and $\sum_i^{n_f} w_i = 1$. One would prescribe a higher weight corresponding to the more preferred design objective. For instance, if stress and deflection are the two objectives involved, and a designer prefers minimizing stress more than deflection, then he/she could intuitively choose an adequately high weight for stress, compared to that for deflection.

Figure 4 shows that the constant value contours of J in Eq. 5 are straight lines in the objective space. The weights w_i define the slope of the contour of J. The solution to the multiobjective problem is the point where the minimum value AOF contour becomes tangent to the Pareto frontier – for a specified set of weights.

Although quick and easy to implement, the weighted sum approach suffers from some well-known drawbacks, which restrict its use in practical design.
1) It is not always easy to choose a physically meaningful set of weights for the design objectives. This is especially true if the units of the different design objectives involved are disparate. For example, typical stress values are measured in MPa, and those of deflection, in inches. Although it is possible to *scale* the design objectives using a normalization technique, the scalar weights fail to quantify the relative importance between the objectives. Furthermore, a weight of, say, 0.3, provides little information regarding its influence on the stress objective. That is, if a designer needs an improvement of 10 MPa in stress, should the corresponding weight be changed to 0.35, 0.5. or 0.8?

2) The linear AOF (Eq. 5) can miss potentially desirable regions of the Pareto frontier – non-convex regions are unreachable using a linear combination of design objectives. For example, in Fig. 4, point C, although Pareto, cannot be captured using the WS approach. This shortcoming could misrepresent certain regions of the Pareto frontier as non-Pareto (non-optimal), and could lead to undesirable results[5].

The second drawback discussed above can be overcome by a judicious modification of the AOF of the WS approach.

3.1.2. *Compromise Programming*

The compromise programming AOF is a modification of the weighted sum AOF. In this case, the AOF (which is minimized) is a weighted *exponential sum* of the objectives, defined as

$$J = \sum_{i=1}^{n_f} w_i f_i^r \tag{6}$$

where the exponent $r \geq 2$. Typically the exponent is chosen to be an even number. This approach is also known as the weighted exponential sum approach.

The constant value contours of J are curves in the case of compromise programming, which grow increasingly sharper as r increases. Figure 4 shows that this mathematical construct works well for non-convex Pareto frontiers, because the sharp contours of the AOF curves can "reach into" the non-convex regions[5].

However, a critical aspect of the compromise programming approach is the need to specify weights as preferences for the design objectives. Thus, one of the limitations of the weighted sum method is retained in compromise programming as well. In addition, now the designer also must specify the value of r, which is not always an obvious choice.

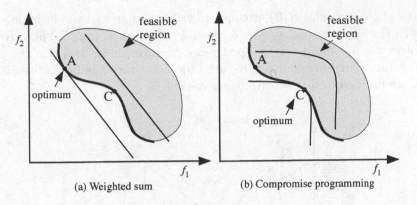

(a) Weighted sum   (b) Compromise programming

Fig. 4.   Non-convex Pareto frontier

A popular variation of the compromise programming AOF includes designer-specified targets as

$$J = \sum_{i=1}^{n_f} w_i \left(f_i - t_i^* \right)^r \tag{7}$$

where t_i^* is the most desirable value (or target) for the i-th design objective. The weights w_i can be manipulated to specify which objective is more (or less) important than the other objectives.

3.1.3. *Weighted Min-Max Method*

The weighted min-max method is another weighted criterion approach. The AOF under this formulation (which is minimized) is given as

$$J = \max_i \left\{ w_i \left(f_i(\mathbf{x}) - f_i^{i*} \right) \right\} \tag{8}$$

where f_i^{i*} represents the i-th objective function value obtained by minimizing only the i-th design objective, subject to constraints[4]. By varying the weights, a designer can control the extent to which each objective can potentially meet its target. However, the selection of weights may yet be an inefficient task. Furthermore, the objective function in Eq. 8 is non-differentiable, and may lead to potential difficulties with gradient-based optimizers. This problem can be overcome by re-formulating the optimization problem with the help of a dummy design variable[4].

3.1.4. *Goal Programming*

Goal programming[4] is one of the early methods that attempted to model the designer preferences in a more physically meaningful manner than weight-based approaches. Goal programming, as the name suggests, requires a designer to specify goals, or targets, for each design objective – values that the designer prefers the most. Goal programming (GP) attempts to yield a design that results in objectives values as close to each target value as possible. To achieve this, GP linearly maps each objective function value to a preference function value, such that the preference function value corresponding to each objective's target is zero, while all other values are mapped to real positive numbers (Fig. 5).

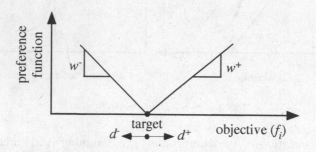

Fig. 5. Goal programming preference function

For each design objective, the designer specifies, in addition to the target value, t_i, two weights, w_i^+ and w_i^-, which represent the slopes of the preference function on either side of the target value. Any deviation to the left or to the right of the target value results in a non-zero preference function value. During the multiobjective optimization, the sum of all preference functions is minimized, which potentially results in a design that minimizes the deviation of each design objective from the specified target.

Mathematically, the goal programming problem formulation can be represented as a linear program (optimization problem) as

$$\min_{\mathbf{x}, d_i^+, d_i^-} \quad J = \sum_{i=1}^{n_f} \left\{ w_i^+ d_i^+ + w_i^- d_i^- \right\} \tag{9}$$

subject to

$$f_i(\mathbf{x}) - d_i^+ \leq t_i \tag{10}$$

$$f_i(\mathbf{x}) + d_i^- \geq t_i \tag{11}$$

$$d_i^+, d_i^- > 0 \tag{12}$$

where d_i^+ and d_i^- are the deviational variables on the positive and negative sides of t_i, respectively (Fig. 5).

Goal programming is an improvement over the typical weighted criteria methods, because the specification of target values is more physically meaningful than the specification of weights. However, goal programming leaves a critical aspect of the AOF formulation to the designer – selection of the slopes of the preference functions. Arbitrary selection of w^+ and w^- in Eq. 9 could lead to numerical scaling issues related to disparate scaling between the objectives.

. Also, a possible source of difficulties is the fact that goal programming penalizes *all* values of the design objective on one side of t_i — equally. This approach is thus incapable of modeling preferences where certain objective values may be significantly more desirable than others. For example a value of 200 kg for mass may be the most desirable, yet a value of 250 kg may be acceptable, and a value between 250 and 300 kg may be undesirable. In such situations (which are common in design problems), identically penalizing all mass values greater than 200 kg is not practical.

3.1.5. *Physical Programming*

Physical programming (PP), developed by Messac[6] (see also Messac et al.[7]), is an approach for multiobjective optimization that is capable of effectively modeling a wide range of complex designer preferences. Furthermore, the designer does not need to specify any scalar weights to reflect his/her preferences – a major drawback of most of the AOF formulation techniques described above.

The PP approach categorizes design objectives as belonging to one of the following sub-classes:

(1) Soft Classes:

    (a) Class 1S: Smaller-is-better (minimization)
    (b) Class 2S: Larger-is-better (maximization)
    (c) Class 3S: Value-is-better
    (d) Class 4S: Range-is-better

(2) Hard Classes:

    (a) Class 1H: Must be smaller
    (b) Class 2H: Must be larger
    (c) Class 3H: Must be equal
    (d) Class 4H: Must be in range

The hard classes become part of the constraints of the optimization problem, while the soft classes are part of the AOF. For each soft class, the designer specifies target values (five each for classes 1S and 2S, nine for class 3S, and ten for class 4S). Figure 6 shows a representative piecewise linear preference function for class 4S. The ten target values on the horizontal (objective function) axis divide the objective space into several regions of desirability, such as ideal, desirable, tolerable, undesirable, highly undesirable, and unacceptable. The target values are specified by the designer, and they can accommodate a wider array of preferences than with the goal programming approach. Compare and contrast this situation with Fig. 5. Observe that as we travel from the central region (the most desirable) to the undesirable regions in Fig. 6, the slope of the preference function increases in magnitude.

A novel aspect of PP is that the vertical axis, which represents the preference function value for each objective, has the same range for all of the design objectives.

This feature avoids potential numerical issues due of disparate scaling between design objectives. Also, PP automatically calculates the slopes of the preference functions using a simple algorithm that also ensures convexity. More details regarding the algorithm can be found in Messac[6] and Messac et al.[7].

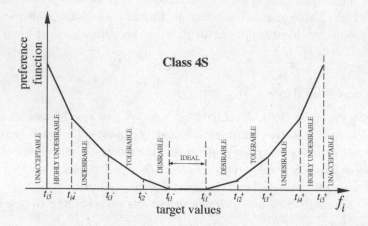

Fig. 6.   Physical programming preference function

The nonlinear version of the PP (Messac[6]), which defines smooth and piecewise nonlinear preference functions. Nonlinear PP uses splines to define the piecewise nonlinear components. Its use is recommended for nonlinear optimization problems.

The linear version of PP (Messac et al.[7]) can be easily formulated as a linear programming problem (if the objectives are linear functions of the design variables) by defining deviational variables on the positive and negative sides of each of the target values for a particular class. The AOF (which is minimized) is the sum of the preference function values of all objectives.

3.2. *Pareto Set Generation Methods*

In the discussion so far, we have focussed on multiobjective methods that attempt to model the designer's preferences and yield a single Pareto optimal solution. There is another distinct, yet important, class of multiobjective approaches that focuses on obtaining a discrete representation of the Pareto frontier. The designer then selects the most desirable design alternative from these representative Pareto optimal points. Such methods are also referred to as Generate First Choose Later (GFCL) methods[2]. The scope of this chapter is limited to methods that yield a discrete representation of the Pareto frontier. In most design problems, a discrete representation is often sufficient for effective decision making.

Desirable features

Before describing some of the popular approaches, we discuss some of the desirable qualities of any Pareto set generation method:

(1) Exploring the *entire* Pareto frontier: The Pareto frontier generation method should be capable of exploring all regions of the Pareto frontier. Some popular methods (such as weighted sum) fail to explore certain regions because of geometrical limitations. In such cases, the decision of the designer could be biased against the regions that are not represented.

(2) Even distribution of Pareto points: It is critical that all regions of the Pareto frontier are represented evenly – no one region should be over- or under-represented. Even representation will ensure that the designer's decision is not biased towards or against any particular region of the Pareto frontier.

(3) Global vs. local Pareto points: It is essential that only global Pareto optimal points are generated. Some of the methods tend to generate locally Pareto or even non-Pareto points. Non-Pareto and locally Pareto points can be removed by using appropriate numeric Pareto filters[2]. In some cases, local Pareto optimum points can be avoided by using global optimizers, such as genetic algorithms.

(4) Computational efficiency: The number of function evaluations required to generate Pareto optimal points is an important issue for any Pareto frontier generation technique, especially in simulation-based design, where each function evaluation could entail a significant computational expense. The recent thrust in the field of multiobjective optimization has been towards developing more computationally efficient methods.

3.2.1. *Weighted Criteria Methods*

As in the case of methods requiring preferences, weighted criteria methods are arguably the most popular Pareto frontier generation techniques. The basic concept is similar to that discussed in Sec. 3.1.1. We first form an aggregate objective function, for example,

$$J = \sum_{i=1}^{n_f} w_i f_i^r \tag{13}$$

where $r = 1$ represents the weighted sum AOF, and $r \geq 2$ represents the compromise programming AOF. In the approach requiring designer preferences, the designer specifies a particular set of weights to obtain a single Pareto optimal solution. However, to generate the Pareto frontier using the weighted criteria approach, we *vary the weights* in a specified range. Each unique combination of weights defines a single objective optimization problem, which when solved yields a Pareto solution. By sequentially varying the weights, we can explore different regions of the Pareto frontier.

A critical issue regarding the inability of the weighted sum approach to explore non-convex Pareto frontier regions was discussed in Sec. 3.1.1. In the case of Pareto frontier generation, we need to address yet another critical issue – that of evenness of the obtained Pareto points. In general, a uniform change in the weights in Eq. 13 does not guarantee an even distribution of points on the Pareto frontier (Fig. 7).

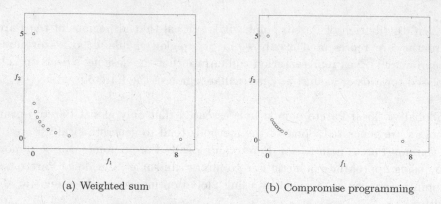

(a) Weighted sum   (b) Compromise programming

Fig. 7.   Uneven distribution of Pareto points - Example 1

Example 1

$$\min_{\mathbf{x}} \{w_1 f_1^r + w_2 f_2^r\} \tag{14}$$

subject to

$$f_1 = \mathbf{x}_1; f_2 = \mathbf{x}_2 \tag{15}$$

$$\left(\frac{f_1 - 10}{10}\right)^8 + \left(\frac{f_2 - 5}{5}\right)^8 - 1 \leq 0 \tag{16}$$

$$-10 \leq \mathbf{x} \leq 10 \tag{17}$$

The above two-objective example is solved using the weighted sum ($r = 1$) and compromise programming ($r = 2$) approaches by uniformly varying the weights between 0 and 1 ($w_i \geq 0, \sum w_i = 1$). Figure 7 shows that the weighted criteria methods do not result in an even distribution for this example. In general, steep and shallow sections of the Pareto frontier are often under-represented. Furthermore, increasing r tends to concentrate solutions in the central region (also termed the "knee") of the Pareto frontier.

3.2.2. ϵ-Constraint Method

The ϵ-constraint method is a precursor to several recently developed Pareto point generation methods. Instead of forming an AOF of all of the objectives, the ϵ-constraint method minimizes only a single objective, while constraining the remaining objectives. By changing the parameters of these additional constraints, we can obtain distinct Pareto points.

For example, for a two-objective problem, the ϵ-constraint method is formulated as

$$\min_{\mathbf{x}} f_1(\mathbf{x}) \tag{18}$$

subject to

$$f_2(\mathbf{x}) \leq \delta \tag{19}$$
$$g(\mathbf{x}) \leq 0 \tag{20}$$
$$h(\mathbf{x}) = 0 \tag{21}$$
$$\mathbf{x}_l \leq \mathbf{x} \leq \mathbf{x}_u \tag{22}$$

The constraint $f_2(\mathbf{x}) \leq \delta$ reduces the effective feasible objective space, while Eq. 18 minimizes the other objective, that is f_1. The Pareto frontier can be explored by sequentially changing δ from the minimum to the maximum possible value of f_2.

Remarks:

1) The ϵ-constraint approach is easy to understand and implement for two-objective problems, because we would only need one additional constraint (Eq. 19). However, for multiple objectives, handling too many additional constraints and their parameters (δ) can be cumbersome.

2) The ϵ-constraint approach may prove ineffective if the Pareto frontier is steep or shallow. In such cases, specifying a reasonable increment (or decrement) for δ can be a challenging task. As a result, this method could miss steep or shallow sections of the Pareto frontier. Also, an even distribution of points is not possible in most practical cases.

3.2.3. *Normal Boundary Intersection Method*

The normal boundary intersection (NBI)[8] method is an improvement over the ϵ-constraint method in that it is effective for multiobjective problems also. Similarly to the ϵ-constraint method, NBI involves solving a series of single objective optimization problems, subject to a shifting constraint. The NBI method is one of the earlier methods to emphasize the importance of obtaining an even distribution of Pareto points.

Mathematically, the NBI method is formulated as

$$\min_{\mathbf{x}, \lambda} \lambda \tag{23}$$

subject to

$$\Phi w + \lambda u = f(\mathbf{x}) - f^u \tag{24}$$

$$g(\mathbf{x}) \leq 0 \tag{25}$$

$$h(\mathbf{x}) = 0 \tag{26}$$

$$\mathbf{x}_l \leq \mathbf{x} \leq \mathbf{x}_u \tag{27}$$

where λ is a dummy design variable, f^u is the utopia point, defined as $f^u = \begin{bmatrix} f_1^{1*} & f_2^{2*} & \cdots & f_{n_f}^{n_f*} \end{bmatrix}^T$ – the coordinates representing the minimum values of each objective. Φ is an $n_f \times n_f$ matrix where the i-th column contains the vector $f(\mathbf{x}_i^*) - f^u$, where $f(\mathbf{x}_i^*)$ is the objective function vector evaluated at the design that minimizes the i-th objective (also termed "anchor point"). w is a vector of positive weights that sum to one, and $u = -\Phi e$, e being a vector of ones.

Geometrically, NBI constrains (through Eq. 24) the solution of Eq. 23 to lie on the normal to the line (hyperplane in multiple dimensions) joining the anchor points. As w is systematically varied, we can obtain a uniform distribution of points on the Pareto frontier. It is possible, however, for the NBI method to generate locally Pareto optimal solutions, or even non-Pareto optimal solutions, regardless of the type of optimizer used (gradient or non-gradient based).

3.2.4. Normal Constraint Method

The normal constraint (NC) method was developed by Messac et al.[9,5] to overcome some of the limitations of the NBI method. The NC method (i) generally reduces the number of non-Pareto optimal points generated, (ii) can be easily extended to guarantee the generation of the entire Pareto frontier[9], and (iii) uses inequality constraints instead of equality, which results in favorable numerical conditioning properties.

The NC method is conceptually similar to the NBI method. The NC method uses anchor points for each objective to define a utopia hyperplane, or a utopia line in two dimensions (as shown in Fig. 8). It then forms a grid of points X_{pk} on this utopia hyperplane (number of points is specified by designer). For each X_{pk}, a single objective optimization problem is defined, which imposes an additional constraint that reduces the feasible region, as shown.

The geometrical details of the NC method are shown in Fig. 8 for a generic utopia hyperplane point X_{pk}. The anchor points for the two objectives are shown as f^{1*} and f^{2*}. The k-th optimization problem statement using the NC method is given as

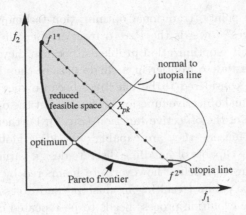

Fig. 8. Normal constraint method

$$\min_{\mathbf{x}} f_2(\mathbf{x}) \tag{28}$$

subject to

$$\left(f^{2*} - f^{1*}\right)^T \left(f(\mathbf{x}) - X_{pk}\right) \leq 0 \tag{29}$$

$$g(\mathbf{x}) \leq 0 \tag{30}$$

$$h(\mathbf{x}) = 0 \tag{31}$$

$$\mathbf{x}_l \leq \mathbf{x} \leq \mathbf{x}_u \tag{32}$$

where $f(\mathbf{x}) = [f_1(\mathbf{x}) \quad f_2(\mathbf{x})]^T$. Solving the above single objective optimization problem projects the point X_{pk} normally onto the Pareto surface, and yields the Pareto optimum, as shown in Fig. 8. By allowing the normal line to intersect at different locations on the utopia line, we can obtain a uniformly distributed set of Pareto optimal points.

The NC method offers important advantages over NBI (as discussed above). However, disparate scales of the objective functions can still pose a problem, which can be avoided using a simple linear normalization scheme (discussed later in Sec. 4.2).

3.2.5. *Multiobjective Genetic Algorithms*

Genetic algorithms (GA) belong to an entirely different class of optimization techniques. Genetic algorithms attempt to mimic the natural evolution process – in order to solve optimization problems. While a detailed discussion regarding their working is beyond the scope of this chapter, we mention some of the notable developments in this field, especially in multiobjective optimization.

Recently, GAs have been effectively used to generate Pareto solutions for multiobjective optimization problems. Genetic algorithms simultaneously process a *pop-*

ulation of candidate points (traditional optimization techniques consider a single point), which "evolves" towards the Pareto frontier. Genetic algorithms are well suited for multiobjective optimization problems, because they can yield a discrete representation of the Pareto frontier in a single pass of the algorithm. Also, multiobjective GAs are considered to be effective, because they do not attempt to aggregate the individual objective functions. Throughout the optimization, they retain the vector form of the objective function (Eq. 1). Furthermore, because GAs are non-gradient optimizers, they are capable of yielding globally Pareto optimal points. Also, they are capable of handling discrete design variables.

Genetic algorithm approaches, however, are hampered by issues such as poor distribution of Pareto points (resulting in clusters) and excessive number of function evaluations. The evolution process needs to be repeated over several hundred "generations," or cycles, which may make multiobjective GAs computationally inefficient. Another criticism often directed towards GAs is the excessive parameter tweaking often needed to obtain useful results.

One of the earliest uses of multiobjective GAs was by Schaffer[10], when he proposed the Vector Evaluated Genetic Algorithm (VEGA) approach. Deb et al.[11] developed the Non-dominated Sorting Genetic Algorithm (NSGA) for obtaining Pareto optimal points. Some of its features include a systematic approach for avoiding under- or over-representation of certain Pareto regions and a fast sorting capability. The Strength Pareto Evolutionary Approach (SPEA) by Zitzler and Thiele[12] is another notable development. Multiobjective GAs have been studied extensively and tested on numerous multiobjective optimization problems in the above-mentioned publications.

Example 2

$$\min_{\mathbf{x}} \{f_1(\mathbf{x}) \quad f_2(\mathbf{x})\} \tag{33}$$

subject to

$$f_1 = \mathbf{x}_1; f_2 = \mathbf{x}_2 \tag{34}$$

$$5e^{-\mathbf{x}_1} + 2e^{-0.5(\mathbf{x}_1-3)^2} \leq \mathbf{x}_2 \tag{35}$$

$$0 \leq \mathbf{x}_1, \mathbf{x}_2 \leq 5 \tag{36}$$

We solve the above bi-objective problem using three representative Pareto set generation approaches, which will highlight some of their advantages and limitations: (i) Weighted Sum, (ii) Normal Constraint, and (iii) Non-dominated Sorting Genetic Algorithm. The MATLAB function 'fmincon' is used to solve each single-objective optimization in the case of the WS and NC approaches. Note that the Pareto frontier for this problem is disconnected (Fig. 9(a)).

Figures 9(b-d) depict the Pareto solutions obtained by the three approaches, in comparison to the actual Pareto frontier. As expected, the weighted sum approach

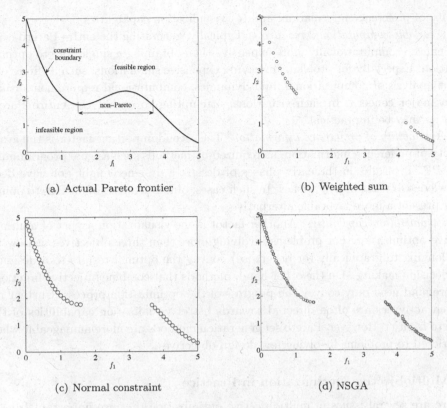

(a) Actual Pareto frontier (b) Weighted sum

(c) Normal constraint (d) NSGA

Fig. 9. Comparison between different Pareto frontier generators

(Fig. 9(b)) does not result in an even distribution of Pareto points, and also misses certain sections of frontier. The NC approach (Fig. 9(c)) results in the best distribution of the three approaches, after easily filtering the non-Pareto solutions. Finally, NSGA (population size 100, evolved over 150 generations) results in an accurate representation of the global Pareto frontier (Fig. 9(d)) – at the cost of too many function evaluations.

3.3. *Choosing an Appropriate Solution Approach*

An important aspect of multiobjective optimization is the choice of an appropriate solution technique for a given design problem. As in most areas of science and engineering, there is no single multiobjective optimization method that works best under all circumstances. However, below we provide some guidelines to help select an appropriate technique.

Preference-based approaches vs. Pareto set generation approaches

The decision regarding whether to choose an approach that requires articulation of designer preferences or one that generates the entire Pareto frontier is generally guided by three factors: (a) available computational resources (and time), (b) availability of preference information, and (c) visualization capabilities.

(a) *Available computational resources*: Typically, generating the entire Pareto frontier can be computationally more expensive than obtaining a single Pareto optimal solution. Especially for problems involving expensive simulations, such as finite element analysis or computational fluid dynamics, computational requirements could be a major concern. In such situations, attempting to obtain the entire Pareto frontier may be impractical.

(b) *Availability of preference information*: The second important factor is the availability of preference information, as required by methods such as goal programming and PP. Typically, in the early phases of design, a designer might not have deep knowledge of what he/she wishes. In such cases, obtaining a set of all Pareto points may present a more favorable alternative.

(c) *Visualization capabilities*: Another factor is the visualization aspect of multiobjective optimization. For problems involving more than three objectives, it may be challenging to graphically (or otherwise) convey the optimal results to a designer for decision making . On the other hand, methods that combine objective functions beforehand need only convey the progress of the optimization process. Current research activities are often directed towards better visualization capabilities of the Pareto frontier. However, Pareto set generation methods are more manageable when restricted to problems involving few design objectives.

4. Multiobjective Optimization in Practice

There are several issues in multiobjective optimization that are important from a practical perspective. Specifically, in this section, we describe some important decision making tools that are beneficial in the context of Pareto set generating methods. We also discuss objective function normalization as an approach to overcome scaling issues. Finally, we discuss the use of Pareto set metrics.

4.1. *Decision Making Tools*

The effectiveness of Pareto frontier based methods can be greatly enhanced by several decision making tools. These are numerical tools that help obtain a more meaningful representation of the Pareto frontier. It is important to note that these tools do not improve the quality of the existing Pareto set; they simply ease the process of selecting desirable Pareto solutions.

4.1.1. *Pareto Filtering Techniques*

In the previous section, we described several Pareto set generation methods. We observed that some methods, such as NBI and NC, can generate locally Pareto or non-Pareto optimal points. However, the generation of such points may not be a serious limitation of the method, because such unwanted points can be identified and removed after the optimization process is complete. Numeric filters have been developed by Mattson et al.[2], which overcome the limitation of the methods that generate non-optimal points. Specifically, a "global Pareto filter" systematically searches through all obtained solutions, and removes the non-globally Pareto optimal designs – leaving a reduced set of points that are globally Pareto. This reduced Pareto set containing only the globally optimal designs can then be presented to the designer for decision making. For mathematical details of the global Pareto filter, refer to Mattson et al.[2].

4.1.2. *Smart Pareto Representation*

One of the desirable qualities of Pareto set generation methods is the ability to yield an even representation in the design objective space. However, recently, there has been increased interest in developing more meaningful Pareto frontier representations. The basic premise is that in practical design problems, certain regions of the Pareto frontier may be of more interest to a designer than other regions.

Particularly, regions that entail practically insignificant tradeoff may not be highly important to a designer. To understand insignificant tradeoff, we refer to Fig. 2, where the Pareto frontier is particularly shallow in the regions of high mass. As a result, if we are willing to give up a small amount in the stress objective, we will be able to significantly improve on the mass objective. Such regions may not need to have the same density of Pareto points as another region with significant stress-mass tradeoff.

Moreover, such a non-uniform distribution could result in a reduction in the total number of Pareto points, and is termed "smart representation" of the Pareto frontier in Mattson et al.[2]. From a decision making perspective, a smaller representative set of Pareto points is more manageable, and could ease the task of decision making for the designer.

4.2. *Objective Function Normalization*

A major source of difficulties in multiobjective optimization can be traced to uneven scaling between the objective functions. Most of the Pareto set generation methods described in this chapter may fail to yield useful results if the objective functions have widely varying scales. However, scaling issues can be overcome by performing a simple linear transformation of the design objectives.

Under this transformation (normalization) scheme, each objective function is first minimized separately to obtain the anchor points $f^{i*} = \begin{bmatrix} f_1^{i*} & f_2^{i*} \cdots f_{n_f}^{i*} \end{bmatrix}^T ; i = 1, .., n_f$. Using the coordinates of these anchor points, the objective function space (that includes the Pareto frontier) is linearly transformed to lie within a unit hypercube (Fig. 10). Mathematically, the transformed domain ($\bar{f}$) is given by

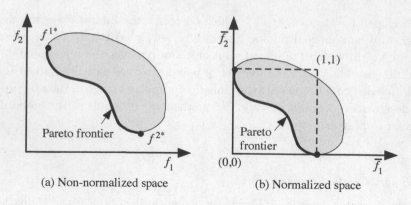

(a) Non-normalized space   (b) Normalized space

Fig. 10.   Objective function normalization

$$\bar{f}_i(\mathbf{x}) = \frac{f_i(\mathbf{x}) - \min\limits_{j} f_i^{j*}}{\max\limits_{j} f_i^{j*} - \min\limits_{j} f_i^{j*}} \tag{37}$$

Pareto set generation can be performed in the normalized space, and the solutions can be mapped back to the original domain using the above transformation in reverse. Although effective, this normalization scheme may be computationally expensive – if obtaining the coordinates of all of the anchor points is expensive. However, a similar normalization may be performed – without obtaining the anchor points – if one has approximate knowledge of the range of each objective function. For a more detailed discussion on normalization in multiobjective optimization, refer to Marler and Arora[13].

4.3. *Pareto Set Accuracy Metrics*

When we have more than three objectives, visualization of the Pareto frontier is no longer a simple matter. Visualization in n-dimension continues to be a subject of considerable interest. Below, we describe a metric that can be used to quantify the quality of Pareto sets obtained using a Pareto set generator.

The Pareto set accuracy metric can be used for testing the effectiveness of a particular Pareto set generation method. It is useful for estimating how an obtained Pareto set compares to the true Pareto set of the test problem. The error metric is defined as follows,[14]

$$P_{\text{eval}}^{f} = \frac{\sum\limits_{i=1}^{n_{\text{obt}}} \left\{ \min\limits_{j} \| f^{(\text{obt})i} - f^{(\text{act})j} \| \right\}}{n_{\text{obt}}} \tag{38}$$

where $f^{(\text{obt})}$ is the set of n_{obt} Pareto points obtained by a given Pareto set generator, and $f^{(\text{act})}$ is the set of true Pareto optimal points. Thus, P_{eval}^{f} is the average of the minimum euclidian distances of each obtained Pareto point from the actual Pareto frontier. A smaller value of P_{eval}^f represents a more accurate set of Pareto points. Specifically, $P_{eval}^f = 0$ indicates that the two Pareto sets coincide.

The above accuracy (quality) metric can also be calculated in the design variable ($\mathbf{x}$) space by simply replacing f with $\mathbf{x}$ in Eq. 38. However, the above metric should be used with caution: it may be highly sensitive to objective function or design variable scaling. One way to avoid scaling problems is to normalize the concerned values using the technique described in Eq. 37.

Other Pareto set evaluation metrics exist that do not require information about the true Pareto frontier. For example, the *hypervolume* metric[15,12] is a measure of the "size of the Pareto space." It is equal to the sum of the hypervolumes defined using each Pareto point and each of the objective axis. However, this measure works well only for convex Pareto frontiers, and could be misleading for non-convex ones. The interested reader may refer to Zitzler and Thiele[12], Wu and Azarm[15], and Velduizen[14] for more information regarding Pareto set metrics. On a similar note, Messac and Mattson[9] propose a measure for evenness of a Pareto set, based on euclidian distances in the objective space.

5. Applications and Recent Advances in Multiobjective Optimization

Multiobjective optimization techniques have been widely used both in industry and in academic research. We briefly mention some of the engineering applications.

Physical programming has been applied to a wide variety of engineering problems. Messac and Hattis[16] have used PP for the design of high speed civil transport. Messac and Wilson[17] apply PP for the multiobjective optimal design of a controller, while Messac et al.[18] use PP for a nine-objective structural optimization problem involving the design of rigidified thin-wall membranes for housing . The NC method has been demonstrated on several test problems involving truss optimization[9,5].

Industrial applications of multiobjective optimization include multiobjective crashworthiness optimization[19,20] and the multiobjective design of a bimorph actuator[21]. Multidisciplinary applications of multiobjective optimization are also common. Tappeta and Renaud[22] present a multiobjective collaborative optimization framework, while McAllister et al.[23] apply multiobjective optimization to a race-car design problem.

5.1. *Recent Advances*

Much recent progress in multiobjective optimization has been towards effectively incorporating the developed techniques into a multidisciplinary simulation-based design environment. Multidisciplinary design optimization (MDO) is characterized by the presence of numerically intensive analysis modules. Thus, each design objective and constraint function of the multiobjective problem could be a computationally expensive simulation. In such situations, it may be difficult to generate even a single Pareto optimal point. Consequently, there have been recent developments in using approximate models of the simulations in the multiobjective problem statement. Effectively constructing these approximations (metamodels) by performing the expensive simulations at judiciously selected design points is an active area of research[24]. Approximation-based design provides a computationally efficient environment for multiobjective decision making.

Another critical area of research in multiobjective optimization is that of visualization. Visualization does not mean simply presenting data in the form of tables and charts. New visualization capabilities, both in terms of optimization progress visualization and Pareto frontier visualization, which are physically meaningful to the designer, are being developed. Physical programming based visualization was developed by Messac and Chen[25], which allows a designer to visualize the relative tradeoffs between design objectives in a physically meaningful manner during the optimization. Mattson and Messac[3] discuss a Pareto frontier visualization approach in multiple dimensions.

6. Summary and Concluding Remarks

Multiobjective optimization can play a critical role in most modern design methodologies. Engineering design is a decision making process, and multiobjective techniques facilitate this decision making by providing the designer a Pareto optimal set of designs, or a single Pareto point that reflects the designer's preferences. Two categories of multiobjective optimization techniques were described in this chapter: methods requiring designer preferences, and methods that generate a representative set of Pareto points. Particularly, we noted some of the deficiencies of the weight-based multiobjective methods, such as the weighted sum and compromise programming approaches – in the context of Pareto frontier representation. Finally, we discussed some practical issues in implementing these multiobjective approaches. Importantly, note that the discussion in this chapter is limited in scope, and the reader is encouraged to consult the references for more details in this ever growing and exciting field.

7. Problems

Problem 1: For a hypothetical bi-objective optimization problem with f_1 and f_2 as two generic design objectives, formulate an appropriate AOF (J) for the following designer preferences.

(a) minimize f_1 and minimize f_2, and both objectives are equally important. (b) minimize f_1 and maximize f_2, and f_1 is twice as important as f_2. (c) Get f_1 as close to 5 as possible, and f_2 as close to 10 as possible. (d) maximize f_2 and get f_2 as close to 10 as possible.

Explain any potential problems with each of your above AOF formulations.

Problem 2: Consider the design of a pinned-pinned beam of length $L = 10$ m, width $b = 0.05$ m, and height h, with a load $P = 400$ N applied at mid-span. The material density is denoted by $\rho = 10^4$ kg/m^3, and the Young's modulus is $E = 10^{11}$ Pa. The quantities of interest to a designer are the mid-span displacement and beam mass. The designer would like both of these quantities to be as low as possible, by controlling the height h.

(a) Express the displacement and mass as a function of the design parameters. (b) Identify the design objective(s) and design variable(s) in this problem. (c) Formulate a multiobjective problem, and state it in the format of Eq. 1.

Problem 3: In Problem 2 above, obtain an optimal design configuration in each of the following cases of designer preferences. Use any multiobjective technique discussed in this chapter. The designer:

(a) is more inclined towards a design that provides a low mass value. (b) would much rather prefer the displacement objective over the mass objective. (c) prefers both the displacement and mass objectives *equally*. (d) would like a design that provides as low a displacement as possible, and mass as close to 295 kg as possible. After obtaining the optimal designs, plot each of them (on the same figure) in the *design objective space*. Using this plot, comment on the performance of the chosen optimization technique. Comment also on the non-uniqueness of your answer, and its practical implication.

Problem 4: Using the weighted sum approach, obtain a discrete representation (30-40 points) of the Pareto frontier for Problem 2 above. Consider two cases: (i) non-normalized design objectives with mass in kg, and displacement in m, and (ii) normalized objectives using the normalization scheme in Eq. 37. How did normalization help (or did not help) in this case, and why?

Problem 5: Consider the following bi-objective optimization problem.

$$\min_{\mathbf{x}} [f_1(\mathbf{x}) \quad f_2(\mathbf{x})]^T$$
$$f_1 = \mathbf{x}^2; f_2 = (\mathbf{x} - 1)^2$$
$$-4 \leq \mathbf{x} \leq 4$$

(a) Obtain several optimal points on the Pareto frontier using the compromise programming method ($r = 2$). Use Matlab's fmincon function for optimization. Plot the Pareto points in the objective space.

(b) Obtain and plot Pareto frontiers for $r = 1, 2, 4, 6, 8$. Comment on the general trend of Pareto point distribution as r increases.

Problem 6: Plot the Pareto frontier (25-30 points) for the following bi-objective problem using: (a) Weighted sum, (b) Compromise programming using an appropriate exponent, and (c) Normal constraint. Explain the advantages and limitations of each method by examining the plots.

$$\min_{\mathbf{x}} \left[f_1(\mathbf{x}) \quad f_2(\mathbf{x}) \right]^T$$

$$f_1 = \mathbf{x}_1; f_2 = \mathbf{x}_2$$

$$\mathbf{x}_1^2 + \mathbf{x}_2^2/9 \geq 1; \mathbf{x}_1^4 + \mathbf{x}_2^4 \geq 16$$

$$\mathbf{x}_1^3/27 + \mathbf{x}_2^3 \geq 1$$

$$0 \leq \mathbf{x} \leq 2.9$$

References

1. V. Pareto. *Manuale di Economica Politica*. Societa Editrice Libraria, Milano, Italy, 1906. Schwier, A. S., Manual of Political Economy, Macmillan, New York, 1971 (English translation).
2. C. A. Mattson, A. A. Mullur, and A. Messac. Smart Pareto filter: Obtaining a minimal representation of multiobjective design space. *Engineering Optimization*, 36(6):721–740, 2004.
3. C. A. Mattson and A. Messac. Pareto frontier based concept selection under uncertainty, with visualization. *Optimization and Engineering*, 6(1):85–115, 2005.
4. R. T. Marler and J. S. Arora. Survey of multi-objective methods for engineering. *Structural and Multidisciplinary Optimization*, 26(6):369–395, 2004.
5. A. Messac, A. Ismail-Yahaya, and C. A. Mattson. The normalized normal constraint method for generating the Pareto frontier. *Structural and Multidisciplinary Optimization*, 25(2):86–98, 2003.
6. A. Messac. Physical programming: Effective optimization for computational design. *AIAA Journal*, 34(1):149–158, 1996.
7. A. Messac, S. Gupta, and B. Akbulut. Linear Physical Programming: A new approach to multiple objective optimization. *Transactions on Operational Research*, 8:39–59, 1996.
8. I. Das and J. E. Dennis. Normal-boundary intersection: a new method for generating the Pareto surface in nonlinear multicriteria optimization problems. *SIAM Journal on Optimization*, 8:631–657, 1998.
9. A. Messac and C. A. Mattson. Normal constraint method with guarantee of even representation of complete Pareto frontier. *AIAA Journal*, 42(10):2101–2111, 2004.
10. J. D. Schaffer. Multiple objective optimization with vector evaluated genetic algorithms. In *Genetic Algorithms and their Applications: Proceedings of the First International Conference on Genetic Algorithms*, pages 93–100. Lawrence Erlbaum Associates, 1985.
11. K. Deb, A. Pratap, S. Agarwal, and T. Meyarivan. A fast and elitist multiobjective genetic algorithm: NSGA-II. *IEEE Transactions on Evolutionary Computation*, 6(2):182–197, 2002.
12. E. Zitzler and L. Thiele. Multiobjective evolutionary algorithms: A comparative case study and the stength Pareto approach. *IEEE Transactions on Evolutionary Computation*, 3(4):257–271, 1999.
13. R. T. Marler and J. S. Arora. Function-transformation methods for multi-objective optimization. *Engineering Optimization*, 2005. In Press.
14. D. A. Van Veldhuizen. *Multiobjective Evolutionary Algorithms: Classifications, Analyses, and New Innovations*. PhD thesis, Air Force Institute of Technology, May 1999.
15. J. Wu and S. Azarm. Metrics for quality assessment of a multiobjective design optimization solution set. *ASME Journal of Mechanical Design*, 123:18–25, 2001.

16. A. Messac and P. Hattis. Physical programming design optimization for high speed civil transport (HSCT). *AIAA Journal of Aircraft*, 33(2):446–449, 1996.

17. A. Messac and B. Wilson. Physical programming for computational control. *AIAA Journal*, 36(2):219–226, 1998.

18. A. Messac, S. Van Dessel, A. A. Mullur, and A. Maria. Optimization of rigidified inflatable structures for housing using physical programming. *Structural and Multidisciplinary Optimization*, 26(1-2):139–151, 2004.

19. H. Fang, M. Rais-Rohani, and M. F. Horstemeyer. Multiobjective crashworthiness optimization with radial basis functions. 10th AIAA/ISSMO Multidisciplinary Analysis and Optimization Conference AIAA 2004-4487, Albany, NY, August-September 2004.

20. J. Andersson and M. Redhe. Response surface methods and Pareto optimization in crashworthiness design. Proceedings of DETC'03, ASME 2003 Design Engineering Technical Conferences and Computers and Information in Engineering Conference DETC2003/DAC - 48752, Chicago, IL, September 2003.

21. D. J. Cappalleri, M. I. Frecker, and T. W. Simpson. Design of a PZT bimorph actuator using a metamodel-based approach. *ASME Journal of Mechanical Design*, 124(2):354–357, 2002.

22. R. V. Tappeta and J. E. Renaud. Multiobjective collaborative optimization. *ASME Journal of Mechanical Design*, 119:403–411, 1997.

23. C. D. McAllister, T. W. Simpson, K. Hacker, K. Lewis, and A. Messac. Integrating linear physical programming within collaborative optimization for multiobjective multidisciplinary design optimization. *Structural and Multidisciplinary Optimization*, 29(3):178–189, 2005.

24. A. Messac and A. A. Mullur. Pseudo response surface (PRS) methodology: Collapsing computational requirements. 10th AIAA/ISSMO Multidisciplinary Analysis and Optimization Conference AIAA-2004-4376, Albany, NY, August-September 2004.

25. A. Messac and X. Chen. Visualizing the optimization process in real-time using physical programming. *Engineering Optimization*, 32(5):721–747, 2000.

CHAPTER 5

SHAPE OPTIMIZATION

Tae Hee Lee

School of Mechanical Engineering
Hanyang University, Seoul 133-791, Korea
E-mail: thlee@hanyang.ac.kr

Recently, shape optimization has been implemented into several commercial finite element programs to meet industrial need to lower cost and to improve performance. This chapter provides a comparatively easy method of shape optimization to implement into a commercial finite element code by using geometric boundary method. The geometric boundary method defines design variables as CAD based curves. Surfaces and solids are consecutively created and meshes are generated within finite element analysis. Then shape optimization is performed outside of finite element program.

1. Introduction

Shape optimization can lead to minimization of mass by changing or determining boundary shape while satisfying all design requirements. Shape optimization has received increasing interest for about 3,000 years to achieve best results within limited sources, beginning with the isoperimetric problem. The history of isoperimetric problem, the determination of the shape of a closed curve of given length and enclosing the maximum area on a plane, begins with the legendary origins in the "Problem of Queen Dido" about 900BC.

Virgil told a story of Queen Dido in his famous epic 'The Aeneid.'[1] Dido, also called Elissa, was princess of Tyre in Phoenicia. Escaping tyranny in her country, Dido came to the coast of Libya and sought to purchase a land from the natives. However, they asserted that they would sell only as much territory as could be enclosed with a bull's hide. Therefore, she had his people cut a bull's hide into thin strips and sew them together to make a single and very long string.

Then Dido took the seashore as one edge for the piece of land and laid the skin into a half-circle. By her institution, she inferred that the maximum area bounded by a line could be enclosed by a semicircle as shown in Fig. 1. In this way, she could purchase the largest site called "Bull's Hide." In that site, Dido and her friends founded Carthage, a city in today's Tunisia.

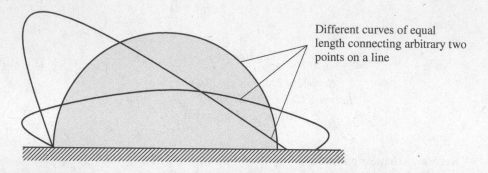

Different curves of equal length connecting arbitrary two points on a line

Fig. 1 Geometric description of isoperimetric problem which originated by Queen Dido in about 900BC

In the engineering field, the first shape optimization problem was defined by Galileo in 1638 at his famous book titled 'Dialogues Concerning Two New Sciences,' where he presented a logical definition and solution for the shape of a cantilever beam for uniform strength as given in Fig. 2.[2]

(From the *Discorsi*, Leiden 1638.)

Fig. 2 Shape optimization problem defined by Galileo in 1638

The subject of shape optimization has been a topic of in-depth research for over three decades since Zienkiewicz and Campbell[3] presented a basic formulation for the problem. Structural optimization methodologies are now noticeably matured and these methods have been implemented into commercial finite element programs. Morphing technology has been developed to treat large shape changes without mesh distortion during shape design process. However, shape optimization within finite element program needs considerable manual efforts to define design variable and added constraints, and to integrate with CAD system and optimizer. In this chapter, we summarize definition of standard shape optimization problem and its solution schemes. Then some limits of methods are described and an emerging technique, the so-called geometric boundary technique, is introduced. The geometric boundary method defines design variables as CAD based curves. Then surfaces and solids are consecutively created and meshes are generated within finite element analysis. An engineering application of geometric boundary technique is presented to give insights for practical design applications.

2. Definition of Shape Optimization Problem

Structural problem can be governed by means of the principle of virtual work for a deformable continuum body in static equilibrium under the action of specified body force f_i and surface traction t_i^0 as follows:

$$\int_V f_i \delta u_i dV + \int_{\Gamma_t} t_i^0 \delta u_i d\Gamma_t = \int_V \sigma_{ij} \delta u_{i,j} dV$$

$$u_i = u_i^0 \text{ on } \Gamma_u \text{ and } t_i = t_i^0 \text{ on } \Gamma_t$$

(1)

for kinematically admissible virtual displacement δu_i. V denotes the known domain in analysis phase, Γ_u and Γ_t represent displacement and traction specified boundaries, respectively, as given in Fig. 3. Note that the summation convention is applied for repeated index.

A shape optimization problem can be defined in general as follows:

Find the boundary of $V(\Gamma)$

to minimize a cost function $m(V, u_i)$

subject to $g_j(V, u_i) \le g_j^0$ and $h_k(V, u_i) \le h_k^0$

while u_i satisfy the governing equations

(2)

where $g_j(V,u_i)$ and $h_k(V,u_i)$ denote inequality and equality constraints respectively. Each constraint describes a design requirement.

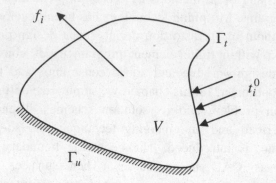

Fig. 3 Definition of a deformable body and applied forces

Definition of design variables characterizes the types of optimization such as sizing, shape and topology as illustrated in Fig. 4. In sizing optimization, a typical size of structure such as thickness of a beam and shell elements and material properties such as density, elastic modulus, thermal conductivity etc are optimized without changing meshes. In shape optimization, the shape of a structure, i.e., boundary of design domain such as length of a beam and boundary of shell is optimized so that meshes are varied as design changes. In topology optimization, topology of a structure is optimized so that shape and connectivity of design domain are altered.

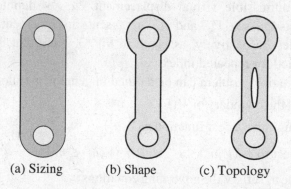

(a) Sizing   (b) Shape   (c) Topology

Fig. 4 Definition of types of design variables

Suppose that a torque arm given in Fig. 4 is modeled by using shell elements for finite element analysis. Then, thickness of each shell element and material properties as shown in Fig 4 (a) can be candidates of sizing design variables. Geometric boundary such as size of holes and outer shape of torque arm as illustrated in Fig. 4 (b) can be candidates of shape design variables. In addition to shape design variables, change of connectivity within design domain such as new hole as shown in Fig. 4 (c) can be candidate of topology design variables.

It is important to note that shape optimization problem may have multiple solutions so unique solution is not guaranteed because it is an ill-posed problem like most design problems. The reason is that the domain in which to look for the final design domain, $V(\Gamma)$, is not determined yet. Moreover, influence of $V(\Gamma)$ on the governing equation Eq. (1) or on the design constraints $g_j(V, u_i)$ and $h_k(V, u_i)$ is not explicit. However, our goal is not to obtain absolute optimum design, but to get a better design or at least best design within the neighborhood of small design changes. Therefore, we don't prove that our solution is global optimum design.

3. Shape Optimization Methods

Shape optimization based on finite element analysis has received increasing interest in the practical design because the finite element analysis can replace physical experiments in many engineering fields. However, it is difficult to provide continuous shape changes during shape optimization without the mesh distortion of finite element analysis. In addition to the formulation for shape optimization given in Eq. (2), mathematical representation of the geometric boundary, mesh generation and manipulation play an important role in shape optimization when finite element program is employed to predict performances.

The design boundary can be properly parameterized by using a parametric language that is often available in preprocessor of finite element programs. Moreover, finite element program is widely integrated with a commercial CAD system under design frameworks for industrial design. Thus, shape optimization with high-fidelity finite element program can be performed with minimum manual efforts.

In this section, techniques for representation of geometric boundary are reviewed and geometric boundary method that defines boundary by using CAD based curves is suggested. Then funnel shape of cathode ray tube for display device is optimized to reduce the depth.

3.1. *Element Nodal Coordinate Method*

Element nodal coordinate method is an early method for shape optimization using finite element nodal coordinates as design variables.[4] However, relocation of boundary nodal points often deteriorates the mesh quality or an unacceptable design as shown in Fig. 5. To avoid the possibility of mesh distortion or unacceptable designs, new constraints must be added to control the movement of each nodal coordinate by trial and error. Therefore, it is natural to integrate with a CAD system to define suitable design boundary and with a good mesh generator to update the finite element model while changing design variables.

The boundary shape can also be obtained as a linear combination of several basis shapes represented by boundary element nodal coordinates such as a mapped mesh generator,[5] prescribed displacements, or fictitious loads.[6] In order to characterize the continuous shape changes with a finite number of design variables, the reduced-basis method where a few of design vectors are usually used to sufficiently describe the shape changes in finite element analysis has been implemented.[7]

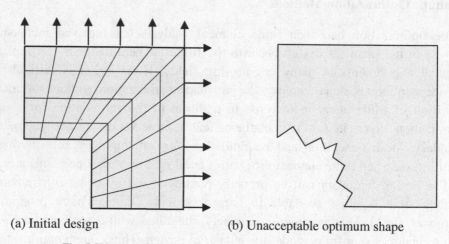

(a) Initial design   (b) Unacceptable optimum shape

Fig. 5 Shape optimization of square plate with square hall

3.2. *Geometric Boundary Method*

Geometric boundary for shape optimization can be defined by CAD-based curves, which are referred to as geometric boundary method. For shell type structure, appropriate curves are predefined, surface is generated from the predefined curves, and automatic mesh generator generates meshes on the surface. Once the design is changed, CAD-based curves are changed. Then surface modification and new mesh generations are sequentially followed during shape optimization procedure. However, note that geometric boundary method is not recommended to 3-dimensional finite elements such as hexahedral element because of limitation in automatic mesh generation.

In this chapter, we performs shape optimization for cathode ray tube of flat panel display device to minimize its depth for space saving.[8] To reduce the depth of cathode ray tubes, the most important component is a funnel that should mechanically withstand the vacuum pressure between the inner and the outer pressures. An ideal arch-like shape of funnel geometry is initially modeled to distribute vacuum stress from the shape of the original product. However, in order to reduce the depth of cathode ray tubes without failure, the arch-like shape of the funnel is optimized to achieve the goal systematically.

3.2.1. *Definition of Shape Design Variable*

In order to generate 3-dimensional funnel geometry of the cathode ray tubes, three axes of funnel geometry are defined as short, long and diagonal axes as illustrated in Fig. 6. Because funnel geometry is symmetric, the axes on the first quarter are generated and expanded to full domain. Curve along each axis can be precisely described by a rational Bezier function of degree 5 with 6 control points that represents accurately the full 3-dimensional funnel geometry as given in Figs. 6 and 7. Rational Bezier curve of degree n is given by a weighted summation of Bernstein polynomials, $B_{i,n}(t)$, as follows:

$$\mathbf{C}(t) = \frac{\sum_{i=0}^{n} B_{i,n}(t) w_i \mathbf{P}_i}{\sum_{i=1}^{n} B_{i,n}(t) w_i} \tag{3}$$

$$B_{i,n}(t) = \frac{n!}{i!\,(n-i)!} t^i (1-t)^{n-i} \tag{4}$$

where the points $\mathbf{P}_i$ are control points and w_i the nonnegative weights, respectively.

To meet the design goal for cathode ray tube, the shortest depth of funnel must be achieved. However, the depth is specified as a target value in this research. Note that we select a control point as design variable along each axis as marked in Fig. 7. And we have examined that 3 design variables can provide smooth shape of funnel.

3 curves are generated from 6 control points of each axis and translated into the smooth surface passing through 3 curves for finite element mesh.[8] Then finite element meshes on the surface are automatically generated by auto-mesh command. If mapped meshes are used, we can control the quality of finite element meshes. Generating curves, surface and meshes are executed in the preprocessor of finite element program. In this study, APDL (ANSYS Parametric Design Language) is used to generate Bezier curves, surface by ASKIN and meshes by AMESH during shape optimization.

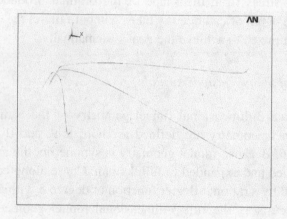

Fig. 6 Definition of shape variables for funnel geometry using 3 curves along each axis

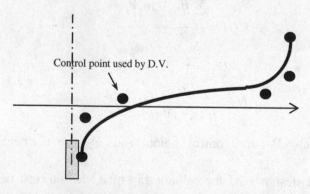

Fig. 7 Rational Bezier curve of degree 5 and its control point on an axis

3.2.2. *Shape Optimization*

Since glass is brittle, maximum principal stress becomes the failure criterion. The first principal stress of the original model is obtained as 3.64 kgf/mm^2 that is beyond the yield strength. Therefore, the maximum principal stress should be minimized below the specified yield strength without adding mass of the funnel.

For simplicity, we optimize the shape of funnel while maintaining its depth that is given as a target value by designer. In this study, one design variable from each axis is selected, but you can add more design variables to obtain better shape. Now we define shape optimization problem with three design variables as follows:

$$\text{Minimize } \sigma_{\max}(w_i)$$
$$\text{Subject to } m(w_i) \le m_0 \quad (5)$$

Table 1 shows optimum result of funnel shape in cathode ray tube. Fig. 8 shows the stress distribution for the optimum shape of the cathode ray tube. The maximum stress is reduced by 12.4 % without increasing mass of the funnel.

Table 1 Optimum result of cathode ray tube

	Initial	Optimum
w_1	1	0.997
w_2	1	0.956
w_3	1	1.092
Maximum stress (MPa)	35.67	31.65
Mass (kg)	24.77	24.77

For highly time-consuming simulation, approximation models can be replaced high-fidelity simulation models to predict performances efficiently during shape optimization.[9] To generate efficient approximation models of performance over design domain, good metamodels, appropriated sampling strategy and validation method of metamodel accuracy are considered.

4. Concluding Remarks

Shape optimization can be employed for daily computer aided design tool because the manual efforts for seamless integration of CAD system, finite element program and optimizer at high fidelity levels are considerably reduced. This chapter addresses representation of design variables for shape optimization problem and its implementation methods.

A comparatively easy method for shape optimization to implement into a commercial finite element code is illustrated by using geometric boundary method. Curve generation, surface generation, and mesh generation are performed in the finite element program by using the so-called parametric language. Shape optimization of the funnel for a cathode ray tube is performed by using a commercial finite element program and a reasonable shape of funnel is obtained.

For highly time-consuming simulation, approximation models need to be employed to predict performances efficiently. Moreover, shape optimization program must be integrated with topology optimization program in order to convert optimum topology into an initial shape for shape optimization automatically.

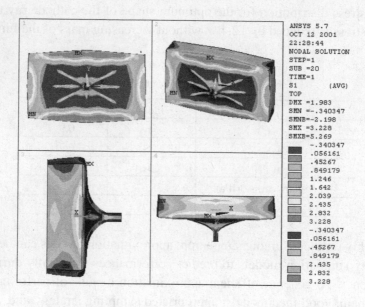

Fig. 8 Distribution of maximum stress on optimum shape of funnel

References

1. Virgil, The Aeneid, translated by W.F. Jackson Knight, Penguin (1958).
2. Galileo Galilei, Dialogues Concerning Two New Sciences, translated by Henry Crew and Alfondo de Salvio, William Andrew Publishing (1991).

3. Zienkiewicz, O.C. and Campbell, J.S., Shape optimization and sequential linear programming, Optimum Structural Design: Chapter 7, edited by Gallagher and Zienkiewicz, John Wiley and Son (1973).
4. Francavilla, A., Ramakrishnan, C.V. and Zienkiewicz, O.C., Optimization of shape to minimize stress concentration, Journal of Strain Analysis, Vol. 10, pp. 63-70 (1975).
5. Kodiyalam, S., Vanderplaats, G.N. and Miura, H., Structural shape optimization with MSC/Nastran, Computers & Structures, Vol. 40, pp. 821-829 (1991).
6. Rajan, S.D. and Belegundu, S.D., Shape optimal design using fictitious loads, AIAA Journal, Vol. 27, pp. 102-107 (1989).
7. Yang, R.J., Lee, A. and McGreen, D.T., Application of basis function concept to practical optimization problem, Structural Optimization, Vol. 5, pp. 55-63 (1992).
8. Lee, T.H. and Lee, K., Multi-criteria shape optimization of a funnel in cathode ray tubes by using response surface model, Structural Multidisciplinary Optimization, Vol. 29, pp. 374-381 (2005).
9. Lee, T.H. and Jung, J.J., Metamodel-based shape optimization of connecting rod considering failure life, Key Engineering Materials, Vols. 306-308, pp. 211-216 (2006).

CHAPTER 6

TOPOLOGY OPTIMIZATION

Martin P. Bendsøe

Department of Mathematics,
Technical University of Denmark,
DK-2800 Lyngby, Denmark
E-mail: M.P.Bendsoe@mat.dtu.dk

Ole Sigmund

Department of Mechanical Engineering,
Technical University of Denmark,
DK-2800 Lyngby, Denmark
E-mail: sigmund@mek.dtu.dk

Taking as a starting point a design case for a compliant mechanism (a force inverter), the fundamental elements of topology optimization are described. The basis for the developments is a FEM format for this design problem and emphasis is given to the parameterization of design as a raster image and the techniques associated with solving this class of problems by computational means.

1. Introduction

Topology optimization is a phrase used to characterize design optimization formulations that allow for the prediction of the lay-out of a structural and mechanical system. That is, the topology or "landscape" of the structure should be an outcome of the procedure. In principle the result of a topology optimization procedure is also optimal with respect to size and shape, but it is here essential to note that fundamental differences in the design parameterization means that direct comparisons are difficult in practise. Moreover, topology optimization is often restricted to design situations with a moderate number of constraints. One should always consider topology optimization as a companion discipline that provides the user with new types of designs that may be processed directly or which may be further refined using size and shape optimization.

The classical concept of lay-out design was created in the early 1900s by Michell[1] and was concerned with design of "thin" frame structures in a setting of plastic limit design. This work was based on analytical methods. Much later, the methods of mathematical programming, more specifically linear programming techniques and the simplex method, were employed for stress constrained minimum weight design of truss structures.[2,3] By allowing for variable cross-sectional areas with a lower bound equal to zero, the single load, minimum compliance optimal topology of a truss can be determined by solving a linear programming problem. The topology here signifies which nodes that are connected, starting from a so-called *ground-structure* consisting of a given, fixed set of nodal points and an associated set of potential bars.

The idea of working from a given reference domain – the ground-structure – was later carried over to the case of continuum structures. Some of the fundamental ideas were first clarified in theoretical studies related to existence of solutions and the application of homogenization techniques for obtaining well-posed problem formulations. In turn, this work constituted the foundation for the computational methods that now typically are called *material distribution* techniques and which work with a design parameterization that allows for the prediction of the optimal material distribution in a given reference domain. While the first computational work[4] relied on·optimality criteria methods for the optimization, today such methods are typically based on mathematical programming together with FEM for analysis. This means that many of the fundamental solution techniques of the material distribution methods are very similar to methods developed for sizing optimization, but with a range of intricacies that relate to the special form of the design parameterization for topology design. Also, the large-scale setting required for topology optimization requires special attention, calling for careful attention when formulating the design problems to be solved.

A recent development in the field is the application of level-set methods for the description of design. This involves an implicit description of design through the level-set curves obtained from a level-set function. This means that such methods rely on sensitivity analysis results from shape design, but in contrast to standard shape design techniques the level-set idea allows for changes in topology.

In the following we concentrate the developments on the material distribution method for structural problems and show examples of the use of the methodology in an industrial setting. Also, current research issues related to multi-physics design problems and to various developments in level-set methods and in new mathematical programming approaches are outlined.

For a thorough historical overview of the field we refer to the detailed review article by Eschenauer and Olhoff[5] and the reader is also referred to the various monographs in the area for further references and overview of the area. This includes works on topology design methods in-the-large, on the so-called homogenization method in particular and on aspects of variational methods.[6,7,8] We note that the references in the following thus emphasize recent works as we try to avoid a lengthy bibliography with a huge overlap with the overview given in these works.

2. Problem Setting

In order to set the scene for the developments of this chapter, we will here consider the problem of topology design of a compliant mechanism, initially in a setting of small

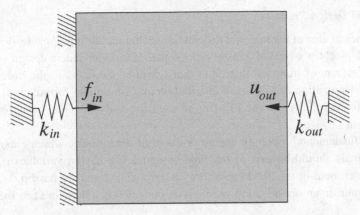

Fig. 1. A basic compliant mechanism analysis problem: the displacement inverter, with spring and load model for the input actuator (left) and workpiece (right).

displacement linear elasticity and using an approach initiated in Ref. 9. An important application of compliant mechanisms lies in MicroElectroMechanical Systems (MEMS) where the small scale makes it difficult to use rigid body mechanisms that attain their mobility from hinges, bearings and sliders.

The problem is formulated as the maximization of displacement at an output port, for a given force at an input port. In order to cater for problem settings where either geometric advantage or mechanical advantage is premium, the workpiece at the output port is modelled by a linear spring with spring constant k_{out}. Choosing a high stiffness results in a small output displacement and a relatively large force, while a small spring stiffness results in a large output displacement and a smaller output force. Also, to simulate the physics of input actuators we here model a linear strain based actuator through a spring with stiffness k_{in} and an input force f_{in}. For example, for a piezoelectric actuator we have a blocking force f_{in} and a free (un-loaded) displacement f_{in}/k_{in}. Alternatively, a non-linear spring model can be applied.

If we consider the analysis problem only for a block of linearly elastic material (filling the domain Ω) undergoing small displacements, see Figure 1, the FEM format for determining the output displacement u_{out} can be written as:

$$u_{\text{out}} = \mathbf{l}^T \mathbf{u}; \qquad \text{where} \qquad \mathbf{K}(\boldsymbol{\rho})\,\mathbf{u} = \mathbf{f} \qquad (1)$$

Here $\mathbf{K}$ is the stiffness matrix, $\mathbf{f}$ is the load vector and $\mathbf{u}$ the displacement vector. Moreover, $\mathbf{l}$ is a vector for which the inner product with $\mathbf{u}$ produces the relevant output displacement u_{out} ($\mathbf{l}$ is interpreted as a (unit) load vector).

In the equilibrium equation of (1) we have already anticipated that the stiffness matrix in a design optimization formulation will depend on a vector of design variables, here denoted by $\boldsymbol{\rho}$.

M. P. Bendsøe and O. Sigmund

2.1. *The 0–1 Design Problem*

The fundamental idea of the material distribution technique for topology design is to associate to each pixel (or voxel) of a raster representation an independent design variable that defines the amount of material utilized in that subset of the domain (the raster representation is a discretization of a material distribution in a continuum formulation). Typically, one would for simplicity use the FEM mesh used for analysis as the basis for the raster representation.

For the fundamental topology design problem of determining which elements of the analysis domain should be part of the final structure, the design variables ρ_e (one real variable per element in the FEM mesh) are *discrete valued* and we have $\rho_e \in \{0,1\}$. We can thus formulate an optimization problem on our reference domain Ω in the following form:

$$\min_{\mathbf{u},\boldsymbol{\rho}} \left\{ u_{\text{out}} = \mathbf{l}^T \mathbf{u} \right\}$$

$$\text{s.t.} : \mathbf{K}(\boldsymbol{\rho})\, \mathbf{u} = \mathbf{f}$$

$$\sum_{e=1}^{N} v_e \rho_e \leq V \tag{2}$$

$$\rho_e \in \{0,1\}, \quad e = 1,\dots,N$$

This is thus a FEM format of the maximum output compliant mechanism problem for a given input load and a prescribed volume V (v_e denotes the volume of element e). Note that when the analysis mesh and the raster mesh coincide, we can write the stiffness matrix in the form:

$$\mathbf{K}_{\text{lin}}(\boldsymbol{\rho}) = \sum_{e=1}^{N} \rho_e \mathbf{K}_e \tag{3}$$

where $\mathbf{K}_e$ is the (global level) element stiffness matrix for an element filled with the basis material used for the structure to be designed (this can be isotropic or not). We have here used an lower index *lin* for the stiffness matrix to signify that it depends linearly on the design variables.

If we wish to view the problem setting in the standard nested format for design optimization problems where the equilibrium conditions are considered as function calls we have to remove the possibility for the stiffness matrix to become singular. This can be done by assigning a low, but non-zero stiffness (given by a value $\rho_{\min} > 0$) to the elements where $\rho = 0$, giving a stiffness matrix in the form

$$\mathbf{K}_{\text{aff}}(\boldsymbol{\rho}) = \sum_{e=1}^{N} \left[\rho_{\min} + (1 - \rho_{\min})\rho_e \right] \mathbf{K}_e \tag{4}$$

We can now write the problem as a problem in the design variables only as:

$$\min_{\boldsymbol{\rho}} \left\{ u_{\text{out}}(\boldsymbol{\rho}) = \mathbf{l}^T \mathbf{K}_{\text{aff}}^{-1}(\boldsymbol{\rho}) \mathbf{f} \right\}$$

$$\text{s.t.} : \sum_{e=1}^{N} v_e \rho_e \leq V \tag{5}$$

$$\rho_e \in \{0, 1\}, \quad e = 1, \dots, N$$

We have now formulated a topology design problem defined on a given and *fixed* reference domain and one could attack this directly by computational means. However, the problem (5) is a non-convex mathematical programming problem with integer variables and with "expensive" function calls that involve FEM analyses. Moreover, in any relevant format the problem is of very large scale as we in this type of topology optimization define the design in terms of a raster representation. Thus, for a suitable resolution one should use a high number of elements – and when addressing 3D design problems the situation is aggravated. This not only means that a high number of design variables have to be dealt with, but it also influences the computational cost of the FEM analyses involved. For high resolution design this makes it computationally very costly to use methods like simulated annealing[10,11] or genetic algorithms[12,13] and experience with deterministic methods is also limited to fairly small scale (bench-mark) examples (see Section 5) or to special design problems such as minimum compliance problems.[14]

The advantage of the topology design formulation above is that the analysis problems to be solved are defined on a fixed domain; this means that if we relax the integer constraint the problem is a standard sizing problem.

In a continuum setting of (2) the problem represents the basic idea of finding the topology of a structure by searching for an optimal indicator function defining the subset of Ω that should be filled with material.[15,16] It is now well understood that this problem is not, in general, well-posed and lacks existence of solutions (see Refs. 7, 8, and references therein). A well-posed problem can be obtained either by restricting the class of subsets considered (see later) or one can extend the set of designs. For compliance it is now known that homogenized multi-scale layered microstructures (composites) constitute an extended design space that provides for existence of solutions and it is interesting to note that this also means that the integer constraint on $\boldsymbol{\rho}$ is relaxed. In essence grey-scale designs are allowed, thus opening up the possibility to apply gradient based optimization techniques. It was the initial mathematical studies of such relaxation schemes that constituted the foundation for the approach[4] which today is referred to as the "homogenization method" for topology design. In this presentation this aspect is not emphasized and we rather see relaxation as a method for obtaining computationally tractable formulations. However, the reader should be aware of this close relation to the theory of composites and to the theory of relaxation of variational problems – much of the litterature refers to these aspects and the use of and the reference to the various concepts are often intertwined.

2.2. Working with a Grey-scale: Interpolation Models

The most straightforward way to obtain a problem formulation that relaxes the integer constraint on the design variables is to consider the problem

$$\min_{\mathbf{u},\boldsymbol{\rho}} u_{\text{out}}$$

$$\text{s.t.}: \sum_{e=1}^{N} [\rho_{\min} + (1-\rho_{\min})\rho_e] \mathbf{K}_e \, \mathbf{u} = \mathbf{f}$$

$$\sum_{e=1}^{N} v_e \rho_e \leq V \tag{6}$$

$$0 \leq \rho_e \leq 1, \quad e = 1,\dots,N$$

where intermediate values of the design variables ρ_e are now allowed.

Unfortunately, this formulation will typically result in large regions (many pixels) with values of ρ_e between zero and one, so one needs to include some additional conditions in order to avoid such "grey" areas. The requirement is that the optimization should result in designs consisting almost entirely of regions of material ($\rho_e = 1$) or no (i.e., weak) material ($\rho_e = 0$) and intermediate values of ρ_e should be penalized in a manner analogous to other continuous optimization approximations of 0-1 problems. One could choose to add directly a penalty function to the objective function in the form

$$\Phi(\boldsymbol{\rho}) = \sum_{e=1}^{N} v_e \rho_e (1-\rho_e) \tag{7}$$

but typically an alternative path is taken.

One very efficient possibility for avoiding grey-scale designs is the so-called penalized, proportional stiffness model with the acronym SIMP for **S**olid **I**sotropic **M**aterial with **P**enalization.[17,18,19] This extremely popular method represents the stiffness matrix as

$$\mathbf{K}_{\text{simp}}(\boldsymbol{\rho}) = \sum_{e=1}^{N} [\rho_{\min} + (1-\rho_{\min})\rho_e^{p}] \mathbf{K}_e \tag{8}$$

where the power p satisfies $p > 1$. This means that the design problem statement becomes

$$\min_{\mathbf{u},\boldsymbol{\rho}} u_{\text{out}}$$

$$\text{s.t.}: \mathbf{K}_{\text{simp}}(\boldsymbol{\rho}) \, \mathbf{u} = \mathbf{f}$$

$$\sum_{e=1}^{N} v_e \rho_e \leq V \tag{9}$$

$$0 \leq \rho_e \leq 1, \quad e = 1,\dots,N$$

In SIMP one chooses $p > 1$ so that intermediate densities are unfavourable. That is, the rigidity/stiffness obtained for intermediate densities is small compared to the volume of the material. For problems where the volume constraint is active the optimization then typically results in black-and-white (0-1) designs, if one chooses p sufficiently big (as a rule of thumb, $p \geq 3$).

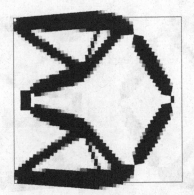

Fig. 2. Topology optimization: A displacement inverter working as a compliant mechanism. Modelled through linear analysis using the formulation of problem (9).

Note that the design optimization problem (9) is now in the format of a standard sizing problem with continuous variables and defined on a fixed domain. This means that techniques described in earlier chapters can now be applied to the problem. An example solution is shown in Fig. 2. Note that there here has been used some additional computational techniques to avoid unstable numerical behaviour and very fine variations of the design, see below for details. We also remark here that since mechanisms intrinsically should provide large deflections one must for all practical purposes actually apply large displacement theory when studying design of such devices; this will also be discussed in a later section.

Several alternatives to the scheme above have been proposed. They are all based on the same principle of being able to interpolate between 0 and 1 – or rather in terms of material properties between weak material ($\rho = \rho_{min}$ representing holes) and strong material ($\rho = 1$).[20,21,22,23] One such model is the RAMP model[22] where one models the stiffness matrix as

$$\mathbf{K}_{ramp}(\boldsymbol{\rho}) = \sum_{e-1}^{N} \left[\rho_{min} + (1 - \rho_{min}) \frac{\rho_e}{1 + q(1 - \rho_e)} \right] \mathbf{K}_e \tag{10}$$

where q has to be chosen reasonably large. This approach was developed in order to have a formulation where the *minimum compliance* design problem becomes concave in the design variables (this requires that $q \geq (1 - \rho_{min})/\rho_{min}$) and thus generates solution that are guaranteed to be integer valued. For an overview of the various possibilities we refer to Ref. 6.

2.3. *Interpreting Grey-scale: Material Models*

If we in problem (9) set $p = 1$ we return to the setting of problem (6). This problem can in 2D be interpreted as a mechanism design problem for a thin disk where the thickness of each element is determined by ρ_e. For the SIMP approach in general one can also ask if this model can be interpreted in physical terms for example such that areas of "grey" can be understood as regions consisting of a composite material constructed by a fine-scale varia-

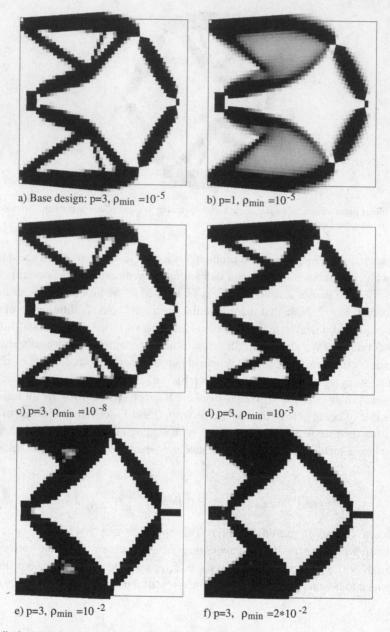

a) Base design: p=3, ρ_{min} =10^{-5} b) p=1, ρ_{min} =10^{-5}

c) p=3, ρ_{min} =10^{-8} d) p=3, ρ_{min} =10^{-3}

e) p=3, ρ_{min} =10^{-2} f) p=3, ρ_{min} =2*10^{-2}

Fig. 3.   A displacement inverter designed with various values of the power p in the SIMP model, and with various values of ρ_{min}. The improvements of the objective functions compared to the base design (a) are respectively: b) 2.6%, c) 0.4%, d) -14.4%, e) -64.7% and f) -98.1%.

tion of the geometric lay-out of the given material. Such a comparison is mainly beneficial in order to – perhaps – understand the behaviour of the computational schemes. However, if a numerical method results in 0-1 designs one can for all practical purposes disregard such considerations. However, the physical realization is important when understanding grey-scale results of a premature termination of the optimization.

If one seeks to construct a material model that mimics the SIMP model one should at least satisfy the well-known Hashin-Shtrikman bounds for the properties of two-phase materials.[24] These express the limits of isotropic material properties of materials with microstructure built from two given, linearly elastic, isotropic materials. In order to satisfy these bounds (in the limit of $\rho_{min} = 0$) the power in the SIMP model p should satisfy some fairly simple condition that depends on the spatial dimension and the Poisson ratio.[25] As an example, in both dimension 2 and 3, the condition is that p should be greater than or equal to 3, when the Poisson's ratio ν is equal to $1/3$. In this case one can also design physical materials that realize the SIMP model.

3. Solution Methods

We will now address a possible computational procedure for solving the topology design problem (9). This not only involves the application of techniques covered in more details in other chapters of this book but also necessitates a course of action specific to the topology design format.

3.1. *Computational Framework*

The approach to solve problem (9) that we propose here is based on using finite elements for the analysis part (as the format of the statement (9) already presupposes) combined with the use of a mathematical programming algorithm as an iterative procedure for treating (9) in its nested format

$$\min_{\rho} \left\{ u_{\text{out}} = \mathbf{l}^T \mathbf{K}_{\text{simp}}^{-1}(\boldsymbol{\rho})\mathbf{f} \right\}$$

$$\text{s.t.} : \sum_{e=1}^{N} v_e\rho_e \leq V \tag{11}$$

$$0 \leq \rho_e \leq 1, \quad e = 1,\ldots,N$$

This means that we treat analysis as a function call and have to perform sensitivity analysis in order to use a derivative based optimization algorithm.

3.1.1. *FEM*

The very nature of the raster representation of the design means that the finite element analysis models involved in the material distribution method become large scale, especially in 3D. However, note that (as mentioned earlier) we are working on a fixed grid and no re-meshing of the design domain is necessary[a]. Moreover, the special format of the stiffness matrix means that all element matrices can be pre-computed; a change of the design variables only affects the relative contribution to the global stiffness matrix. The FEM analysis

---

[a]If adaptive methods[26] are applied a re-meshing will take place, but this is not inherent in the approach.

can be further optimized if rectangular (box-like) domains are used and are discretized with the same element throughout. Then only one element matrix needs to be computed. For large scale computations iterative solvers may be required for storage reasons and parallel implementations[27] are also useful. Typically, solving the equilibrium equations becomes the most time consuming part of topology design problems.

3.1.2. *Sensitivity Analysis*

For topology design problems we work with a huge number of design variables. Also, we typically try to limit the number of constraints in the problem statements. Thus the application of an adjoint method for the computation of derivatives is imperative.

For the functional $u_{\text{out}} = \mathbf{l}^T \mathbf{u}$ we use that $\mathbf{u}$ satisfies the equilibrium equation, i.e., we have that $\mathbf{Ku} - \mathbf{f} = \mathbf{0}$. For any vector $\boldsymbol{\lambda}$ we can thus write

$$u_{\text{out}} = \mathbf{l}^T \mathbf{u} - \boldsymbol{\lambda}^T (\mathbf{Ku} - \mathbf{f}) \tag{12}$$

Differentiating this results in an equation

$$\frac{\partial u_{\text{out}}}{\partial \rho_e} = \mathbf{l}^T \frac{\partial \mathbf{u}}{\partial \rho_e} - \frac{\partial \boldsymbol{\lambda}^T}{\partial \rho_e} (\mathbf{Ku} - \mathbf{f}) - \boldsymbol{\lambda}^T \left(\frac{\partial \mathbf{K}}{\partial \rho_e} \mathbf{u} + \mathbf{K} \frac{\partial \mathbf{u}}{\partial \rho_e} \right) \tag{13}$$

that can be rearranged to the format

$$\frac{\partial u_{\text{out}}}{\partial \rho_e} = \left(\mathbf{l}^T - \boldsymbol{\lambda}^T \mathbf{K} \right) \frac{\partial \mathbf{u}}{\partial \rho_e} - \boldsymbol{\lambda}^T \frac{\partial \mathbf{K}}{\partial \rho_e} \mathbf{u} \tag{14}$$

If the adjoint variable $\boldsymbol{\lambda}$ now satisfies the adjoint equation

$$\mathbf{l}^T - \boldsymbol{\lambda}^T \mathbf{K} = 0 \tag{15}$$

we obtain the following simple expression for the derivative of the output displacement:

$$\frac{\partial u_{\text{out}}}{\partial \rho_e} = -p \, \rho_e^{p-1} \boldsymbol{\lambda}^T \mathbf{K}_e \mathbf{u} \tag{16}$$

where we have used the expression (8) for the matrix $\mathbf{K}$ in terms of the design variables.

3.1.3. *Optimization Algorithm*

A major challenge for the computational implementation of topology design along the lines described above is to cope with the high number of design variables. Optimality criteria methods were first applied[4] but the use of mathematical programming algorithms typically implies greater flexibility and this is crucial for the problems that will be discussed in this chapter. As mentioned earlier, the high number of design variables of topology design is normally combined with a moderate number of constraints, and an algorithm that has proven itself to be versatile and well suited for large scale topology optimization problems is the MMA algorithm, with "MMA" being the acronym for Method of Moving Asymptotes.[28,29,30]

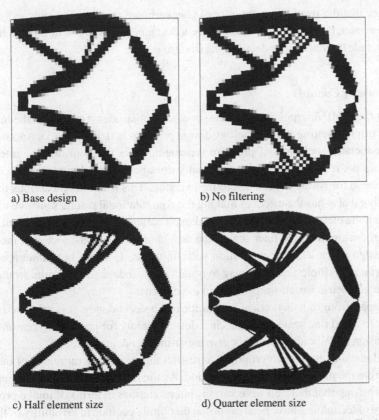

a) Base design   b) No filtering

c) Half element size   d) Quarter element size

Fig. 4. Topology optimization: A displacement inverter working as a compliant mechanism. Checkerboard patterns and mesh-dependent results. The improvements of the objective functions compared to the base design (a) are respectively: b) 4.3%, c) 14.4%, d) 29.9%

The Method of Moving Asymptotes and the related method called CONLIN[b], like Sequential Linear Programming (SLP) and Sequential Quadratic Programming (SQP) methods, work with a sequence of approximate subproblems of given type. The subproblems are constructed from sensitivity information at the current iteration point as well as some iteration history. In the case of MMA and CONLIN these subproblems are separable and convex and the subproblems are solved by for example a dual method or by an interior point primal-dual algorithm.

3.2. *Finer Points*

After implementing a computational scheme for topology design along the lines for standard sizing design problems, as described above, one immediately discovers that additional issues have to be addressed. First, if low order elements are used for the analysis elements

---

[b] See, eg., Refs. 31, 32 for recent papers using this approach.

one will see that the results contain a lot of checkerboard patterns of black and white elements. Moreover, refining the mesh can have a dramatic effect and will change the design so as to include finer and finer detail in the design.

3.2.1. *Geometry Control*

The use of the SIMP interpolation scheme (or other similar ideas) addresses the integer format of the original setting of the topology design problem as defined in (5). Another serious problem associated with the 0-1 problem statement, and a problem that the interpolation scheme does not resolve, is that there normally does not exist solutions to the continuum design problem for which (5) is a discretized version. This is not only a serious theoretical drawback but it also has the effect of making the computational results sensitive to the fineness of the finite element mesh. The physical explanation for the mesh dependent results is that by introducing finer and finer scales, the design space is expanded and in the limit the optimal design is not a classical solution with finite size features, but rather a composite with material at multiple scales. A way to obtain mesh-independent results would thus be to limit the geometric variations possible in the designs.

Quite apart from this theoretical reason for mesh-dependency, it is relevant in its own right to try to avoid fine scale variations in a design, simply for production reasons. Also, a sensible design tool should preferably give mesh-independent results.

Several methods have been proposed to restrict the geometric variations of the designs resulting from the material distribution method. As the method uses a raster representation it is not surprising that these all have various interpretations in terms of image processing.

A range of techniques have been proposed that limits geometric variability of the design field by imposing additional constraints on the problem that restrict the size of the gradient of the density distribution ρ. This can be in terms of a constraint on the *perimeter* or on some L^q-norm of the gradient; in both cases experimentation is needed to find a suitable constraint value.[33,34,35] An alternative is to impose a point-wise limitation on the gradient of the density field. The constraint value here has immediate geometric meaning in terms of the thinnest possible features of a design.[36] Implementation can be problematic, but can be handled by interior point ideas in MMA[37] or via a move limit strategy.[38]

Limiting the geometric variations of the raster representation of the design can also be achieved by alternative means. A popular method is to apply filters as known from image processing. One can thus work with filtered densities in the stiffness matrix so that the equilibrium constraint of problem (9) is modified to the format:

$$\mathbf{K}_{\mathrm{simp}}(H(\boldsymbol{\rho}))\,\mathbf{u} = \mathbf{f} \tag{17}$$

where H denotes a filtering of $\boldsymbol{\rho}$. This can for example be a linear filter with filter radius $r_{\min}$ that gives the modified density $H(\boldsymbol{\rho})_e$ in element e as

$$H(\boldsymbol{\rho})_e = \sum_{k=1}^{N} H_k^e \, \rho_k \tag{18}$$

Here the normalized weight factors H_k^e are defined by

$$H_k^e = \frac{1}{\sum_{k=1}^{N} A_k^e} A_k^e \tag{19}$$

where

$$A_k^e = r_{\min} - \text{dist}(e,k), \quad \{k \in N \mid \text{dist}(e,k) \leq r_{\min}\}, \; e = 1,\ldots,N \tag{20}$$

In this latter expression, $\text{dist}(k,i)$ denotes the distance between the center of the element e and the center of an element k. The convolution weight H_k^e is zero outside the filter area and the weights for the elements k decay linearly with the distance from the element e.

The filtering means that the stiffness in an element e depends on the density ρ in all elements in a neighborhood of e, resulting in a smoothing of the density. The filter radius $r_{\min}$ is fixed in the formulation and implies the enforcement of a fixed length-scale in the designs and convergence with mesh refinement. Generally filtering also results in density fields ρ that are bi-valued, but the stiffness distribution is more "blurred" with grey boundaries.

For implementation, the standard procedure described so far still apply, but the sensitivity information should be modified to cater for the redefined stiffness matrix (this means that the sensitivity of the output displacement with respect to ρ_e will involve information from neighboring elements). Note that the application of a filter does not require any additional constraints to be added to the problem, in contrast to the approaches that work by limiting the gradient or the perimeter.

An alternative to the direct filtering of the densities is a filtering of the sensitivity information of the optimization problem, and computational experience is that this is a highly efficient way to ensure mesh-independency. In scope this is similar to ideas used to ensure mesh-independence for simulations of bone-remodelling and for analysis with plastic-softening materials. The scheme works by modifying the element sensitivities of Eq. (16) as follows

$$\widehat{\frac{\partial u_{\text{out}}}{\partial \rho_e}} = \frac{1}{\rho_e} \sum_{k=1}^{N} H_k^e \, \rho_k \, \frac{\partial u_{\text{out}}}{\partial \rho_k} \tag{21}$$

This heuristic filtering is similar to (18) but it is not the same as applying the filter H to the sensitivities as the densities here also enters as weights. Using these sensitivities in the optimization algorithm produces results very similar to for example those obtained by a local gradient constraint. It requires little extra CPU-time and is simple to implement. The sensitivity (21) converges to the original sensitivity when the filter radius $r_{\min}$ approaches zero and all sensitivities will be equal when $r_{\min}$ approaches infinity. An interesting side-effect of this filtering technique is that it somehow improves the computational behaviour of the topology design procedures and allows for greater design changes before settling on a "good" design; this is due to the inclusion of ρ_k in the filtering expression.

3.2.2. Checkerboards

A description of the material distribution problem is not complete without a mention of the so-called checkerboard problem. In certain cases an implementation of the material distribution method for topology design can generate checkerboard patches where the density of material assigned to contiguous finite elements varies in a periodic fashion similar to a checkerboard consisting of alternating solid and void elements, cf. Fig. 4b. It is now well understood that this is unphysical and arises due to a bad FEM modelling being exploited by the optimization procedure. For example, a checkerboard of material in a uniform grid of square Q4 elements has a stiffness that is greater than any other possible material distribution. Detailed analyses of the problem can be found in Refs. 39, 40.

The occurrence of checkerboards can easily be prevented, and any of the methods for achieving mesh-independency described above will also remove checkerboards (when the mesh becomes fine enough). It is normally recommended always to apply such measures for geometry control and this could then be the end of the story. However, a fixed scale geometric restriction on the designs is counter-productive when using numerical methods to obtain an understanding of the behaviour of optimal topologies at a fairly fine scale, when designing low volume fraction structures, or when composites are used as a basis for the optimization. The most fundamental approach is to use a FEM discretization where checkerboards are not present (their appearance is a FEM phenomenon). This typically involves the use of higher order elements for displacements, with corresponding higher computational cost, so many alternatives have been devised (see, e.g., Ref. 6 for an overview) Here we just mention that the filtering techniques described above can also be used to control only checkerboard formation, without imposing a mesh independent length scale. This just requires that one adjusts the filtering to be mesh-dependent (ie., r_{min} varies with the mesh size).

3.2.3. *Hinges*

When inspecting topology optimization results for compliant mechanism it is noticeable that the resulting mechanisms are often not truly compliant. Instead almost moment-free one-node connected hinges are present, especially for examples with large output displacements (i.e., small transfer of forces). As the structure would break at such hinges, techniques to avoid them are required.

The one-node connected hinges are caused by bad computational modelling that the optimization procedure exploits – just as for the checkerboard problem. In a Q4 model, the hinge is artificially stiff and the stress variations are very badly modelled. However, using higher order elements is only part of the answer and local stress constraint should preferably be added to the formulation; this is computationally problematic and other methods have been devised. Only some of the checkerboard and mesh-independency schemes described above prevent the one-node connected hinges. For example, the filtering of gradients does not prevent hinges if the gain (sensitivity) in building a hinge is big enough, and a local gradient control only partly eliminates the problem and often results in hinges of intermediate density.

We will here elaborate on a special geometry constraint that has been developed with the hinge problem in mind.[41] Being a MOnotonicity based minimum LEngth scale (MOLE) method, it adds one extra constraint to the optimization problem that should have the value

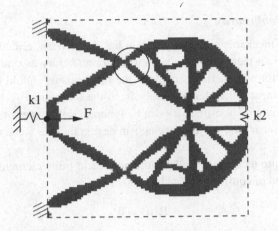

Fig. 5.   The effect of restricting the formation of artificial hinges by the MOLE constraint (from Poulsen[41]).

zero for the minimum length scale restriction to be satisfied. Also, it provides a similar exact control of geometry as when using local gradients, but with just one constraint.

The idea of MOLE is to pass a circular "filter" window over the design and measure if the density ρ along four equally spaced diagonals (horizontal, vertical and at $\pm\pi/4$ from the horizontal) is monotonic or not within this window. The diameter of the window defines the desired minimum length-scale of the design. The monotonicity can be measured by noting that a sequence of real numbers x, y, z is monotonic (increasing, decreasing or constant) if and only if the expression

$$m(x,y,z) = |y-z| + |z-y| - |z-x|$$

is zero and m is strictly positive otherwise. Adding such calculations over all elements and any of the test directions results in a number that should be zero in order for the structure to satisfy the desired length scale control. The computational effort in evaluating the constraint is linear in the number of elements and derivatives can be computed analytically (for a suitably smoothed version of the constraint).

4. Extensions

The challenge of extending the topology optimization method to new areas is to develop sensible combinations of the design description and of objective functions and constraints. Experimentation is the key to working out physically meaningful formulations which can be managed by the optimization algorithms. Here we shall first consider some generalization of the mechanism design problem, both in terms of non-linear analysis modelling and in terms of adding design of the supports to the optimization formulation.

4.1. *Geometrical Nonlinearity*

The analysis of the mechanism design problem has so far been treated in the framework of linear finite element theory. This is somewhat contradictory as one tries to maximize displacements, and for all practical situations the mechanisms should be modelled using geometrically non-linear finite element analysis; in the following, we assume that strains are small and that material non-linearity can be ignored. References on the application of geometrically non-linear analysis for mechanism design include Refs. 42, 43, 44, 45, 46, 47.

Without going into technical details of the non-linear finite element analysis, we here use that the condition of equilibrium can be written as

$$\mathbf{R}(\mathbf{u}) = \mathbf{0} \tag{22}$$

Here $\mathbf{R}$ is the residual in obtaining equilibrium, expressed in terms of Green-Lagrange strains and Piola-Kirchhoff stresses which we assume are related via the standard SIMP relation (as a linear Hooke's law). The finite element equilibrium (the solution of (22)) may be found incrementally or in one load step using a Newton-Raphson method, and both methods require the determination of the tangent stiffness matrix

$$\mathbf{K}_T = \frac{\partial \mathbf{R}}{\partial \mathbf{u}}. \tag{23}$$

With this type of non-linear modelling, a large displacement formulation of our mechanism design problem can be written as

$$\min_{\mathbf{u},\boldsymbol{\rho}} u_{\text{out}}$$

$$\text{s.t.}: \mathbf{R}_{\text{simp}}(\boldsymbol{\rho},\mathbf{u}) = \mathbf{0}$$

$$\sum_{e=1}^{N} v_e \rho_e \leq V \tag{24}$$

$$0 \leq \rho_e \leq 1, \quad e = 1,\dots,N$$

The sensitivity of the output displacement can be found along the lines described in Chapt. 8 of this monograph, and using the adjoint technique we obtain that Eq. (16) is now modified to

$$\frac{\partial u_{out}}{\partial \rho_e} = -\boldsymbol{\lambda}^T \frac{\partial \mathbf{R}}{\partial \rho_e} \tag{25}$$

where $\boldsymbol{\lambda}^T$ is the solution to the adjoint load problem

$$\boldsymbol{\lambda}^T \mathbf{K}_T = \mathbf{l}^T \tag{26}$$

that uses the tangent stiffness matrix at the current design and corresponding displacement. With these developments one can now apply the standard optimization procedure, as outlined for the linear case.

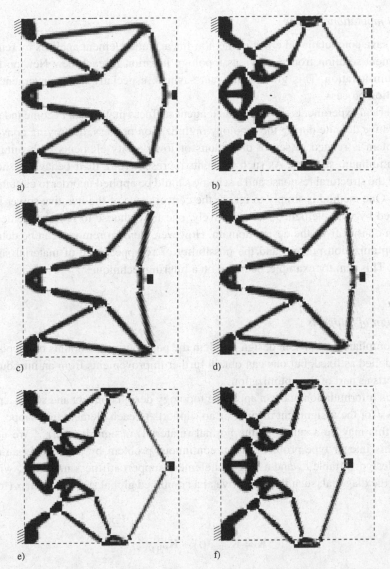

Fig. 6. Taking non-linearity into account. a) Optimized topology using linear modelling, b) optimized topology using non-linear modelling, c) and d) deflection of a) using linear and non-linear modelling, respectively and e) and f) deflection of b) using linear and non-linear modelling, respectively (from Pedersen et al.[42]).

4.1.1. *The Importance of Non-linear Modelling*

We remark here that the use of geometrically non-linear finite element modelling is absolutely essential for mechanism synthesis. If a mechanism is designed using linear analysis one notices that it typically behaves differently when large displacement analysis is applied. In rare situations one merely has inaccurate results but one also risks that the results are useless as large displacement mechanisms, see Fig. 6.

4.1.2. *Computational Issues*

One can save computational effort in the non-linear finite element analysis by reusing the displacement solution from a previous topology iteration in each new Newton-Raphson equilibrium iteration. This gives significant savings, especially near convergence of the optimization process.

Numerical experiments show that the tangent stiffness matrix can become indefinite or even negative definite during the topology optimization process, destroying convergence. The problem is related to severe distortions of low-density elements with minimum or close to minimum stiffness. As such elements represent void, their behaviour should not influence the structural response and a scheme should be applied in order to circumvent the problem. One method is simply to ignore the convergence of the iterative solver in nodes surrounded by void elements.[48] Alternatively, one may choose to remove elements with minimum density from the design domain. However, element removal can be detrimental for the optimization process and the possibility of re-appearance of material should be included. This can, for example, be based on a filtering technique.[49]

4.2. *Design of Supports*

For the compliant mechanism design problem the positions and amounts of supports have been modelled as fixed, but one can obtain further improvements from an introduction of the supports as part of the optimization.

In such a formulation one can apply the topology design concept and set up a problem that allows for the assignment of rigid or no supports to each element in a support design domain (this may be a subset of the normal (material) design domain).[50] We now also convert this integer type problem into a continuous problem by introducing an element support design variable ξ_e and a diagonal element support stiffness matrix $\mathbf{K}_s$ with large values in the diagonal, such that we arrive at a combined global stiffness matrix (for linear analysis):

$$\mathbf{K} = \mathbf{K}_{\text{simp}}(\boldsymbol{\rho}) + \mathbf{K}_{\text{supp}}(\boldsymbol{\xi})$$

where we with a small lower bound ξ_{min} on the support design variables $\boldsymbol{\xi}$ define

$$\mathbf{K}_{\text{supp}}(\boldsymbol{\xi}) = \sum_{e=1}^{N} [\, \xi_{\text{min}} + (1 - \xi_{\text{min}})\, \xi_e{}^q \,] \, \mathbf{K}_{s,e} \,.$$

Here q is a penalization factor corresponding to the power p for the stiffness variables in the SIMP approach.

For the topology design problem a bound on the total support area S is also introduced (for mechanism design this bound is not very important) and we can write a combined

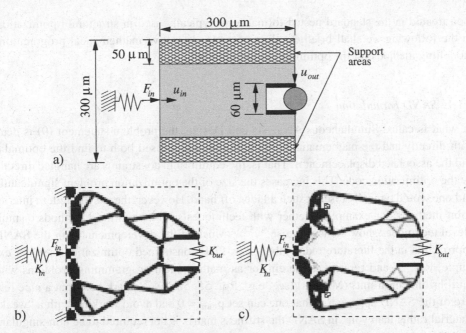

Fig. 7.   Design for supports. a) Design domain, b) Optimized topology without support design ($u_{out} = 10.8\mu m$) and c) Optimized topology including support design ($u_{out} = 19.1\mu m$). The gain in output displacement is 77% (from Buhl[50]).

material distribution and support distribution problem as

$$\min_{\mathbf{u},\boldsymbol{\rho},\boldsymbol{\xi}} u_{out}$$

$$\text{s.t.} : \left[\mathbf{K}_{simp}(\boldsymbol{\rho}) + \mathbf{K}_{supp}(\boldsymbol{\xi})\right] \mathbf{u} = \mathbf{f}$$

$$\sum_{e=1}^{N} v_e \rho_e \leq V, \qquad 0 \leq \rho_e \leq 1, \quad e = 1,\ldots,N \tag{27}$$

$$\sum_{e=1}^{N} v_e \xi_e \leq S, \qquad 0 \leq \xi_e \leq 1, \quad e = 1,\ldots,N$$

Here, the sensitivity of the output displacement with respect to the two sets of variables split into (16) and a similar expression with respect to the support variables (using $\mathbf{K}_{s,e}$).

It turns out that the possible gains in using variable supports for compliant mechanism design is quite dramatic. Thus this slight generalization of the fundamental topology design concept is quite effective even though the extra computational cost is moderate (the analysis problem does not change in complexity).

5. Variations of the Theme

5.1. *Mathematical Programming Issues*

The approach to the material distribution problem for topology design presented so far uses continuous variables to convert the problem to a standard sizing problem. This problem is then treated in the standard nested format most typically used in structural optimization. In the following we shall briefly outline possible alternative mathematical programming modelling methods for the optimization problems at hand.

5.1.1. *SAND Formulation*

In what is called Simultaneous ANalysis and Design, the problem statement (9) is dealt with directly and the mathematical programming code is used both to find the optimal ρ and the associated displacement $\mathbf{u}$. That is, the equilibrium constraints are handled directly by the optimization code. This increases the size of the optimization problem significantly and one would typically discard such an idea off hand. However, the use of modern interior point methods, for example together with techniques such as multigrid methods or multilevel techniques show some promise.[51,52,53] Much of the developments with the SAND approach is in the literature today labelled as PDE-constrained optimization (see, for example, Ref. 54 and references therein) or as mathematical programming problems with equilibrium constraints (MPECs) (see, e.g., Ref. 55, 56). We remark here that a nice feature of the SAND approach is that one can set $\rho_{min} = 0$ and avoid working with a "weak" material rather than void; in SAND the stiffness matrix is not required to be non-singular.

5.1.2. *Mixed-integer Format*

From a modelling point of view the SAND formulation also has some interesting implications for the original 0-1 format of problem (2). One notes here that the only non-linear function that appears in this problem statement is the equilibrium constraint which is bilinear (in ρ and $\mathbf{u}$, cf., Eq. (3)). If we rewrite this problem with an extra set of variables $\mathbf{s}$ that represent element stresses it takes the form

$$\min_{\mathbf{u},\mathbf{s},\rho} \mathbf{l}^T \mathbf{u}$$

$$\text{s.t.} : \mathbf{B}\,\mathbf{s} = \mathbf{f}$$

$$s_e = \rho_e\,\mathbf{E}_e\,\mathbf{B}_e\,\mathbf{u}, \quad e = 1,\dots,N$$

$$u_{min} \leq u_e \leq u_{max}, \quad e = 1,\dots,N \tag{28}$$

$$\sum_{e=1}^{N} v_e \rho_e \leq V$$

$$\rho_e \in \{0,1\}, \quad e = 1,\dots,N$$

where $\mathbf{B}$, $\mathbf{E}_e$, and $\mathbf{B}_e$ are suitably defined matrices and where we have included some displacement constraints (these can be chosen to be redundant). In (28) only the second set of constraints are bilinear, and they are of a form where these, using the combination of the integer constraint on ρ_e and the displacement constraints, can be split into some linear constraints. There exists thus numbers c_e^{max} and c_e^{min} so that problem (28) is equivalent with the problem[57,58]

$$\min_{\mathbf{u},s,\rho} \mathbf{l}^T \mathbf{u}$$

$$\text{s.t.} : \mathbf{B} \, s = \mathbf{f}$$

$$\rho_e \, c_e^{\min} \leq s_e \leq \rho_e c_e^{\max}, \quad e = 1,\dots,N$$

$$\mathbf{E}_e \, \mathbf{B}_e \, \mathbf{u} - s_e \geq (1-\rho_e)c_e^{\min}, \quad e = 1,\dots,N$$

$$\mathbf{E}_e \, \mathbf{B}_e \, \mathbf{u} - s_e \leq (1-\rho_e)c_e^{\max}, \quad e = 1,\dots,N \tag{29}$$

$$u_{\min} \leq u_e \leq u_{\max}, \quad e = 1,\dots,N$$

$$\sum_{e=1}^{N} v_e\rho_e \leq V$$

$$\rho_e \in \{0,1\}, \quad e = 1,\dots,N$$

which has quite a number of additional constraints. However, one notes that this problem is actually now a *linear programming problem*, albeit with mixed integer and continuous variables. This means that one can apply techniques to obtain *globally* optimal designs. Typically, only rather small sized problems can be solved by attacking problem (29) directly with a code like CPLEX, but these can then serve as bench-mark examples for other techniques. Further work on developing variants of (29) with more benign behaviour for computations may alleviate this complication.

5.1.3. *Stress Constraints*

Adding stress constraints to topology optimization problems in the framework of SIMP presents several difficult issues. First, the modelling of the stress constraints is not direct, but comparison with microstructural modelling implies that a reasonable constraint is of the form[59,60]

$$\sigma_{\text{VM}} \leq \rho^p \, \bar{\sigma}, \quad \text{if } \rho > 0 \tag{30}$$

expressed in terms of the von Mises equivalent stress σ_{VM} and a stress limit $\bar{\sigma}$. One here immediately notices one of the problems with stress constraints: they constitute a set of constraints that depend on the value of the design variable. This effect can be removed by writing

$$\rho\sigma_{\text{VM}} \leq \rho^{p+1} \, \bar{\sigma} \tag{31}$$

Unfortunately, it turns out that this is now a type of constraint that generates serious problems for gradient based algorithms if these are applied directly to the nested formulation that is so popular in design optimization. This phenomenon is normally referred to as the stress "singularity" problem and requires special attention as the optimum can be located in degenerated parts of the design space. A constraint relaxation scheme is often used, but also this can be troublesome; note that most of the studies on this mathematical programming issues has mostly been concerned with truss structures (this is not a limitation from a mathematical point of view). We refer the reader to for example Refs. 61, 62, 63, 64, 65

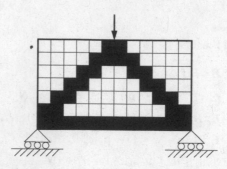

Fig. 8.   A globally optimal solution, with 0-1 variables and stress constraints. By courtesy of M. Stolpe[58].

and references therein for further information on this challenging issue, and to Refs. 66, 67, 68 for further modelling work on and computational experience for stress constrained topology optimization of continuum structures. We finally note that if one applies the original 0-1 formulation, the modelling of the stress constraint is straightforward; moreover, the mixed integer LP format of (29) can be extended to cover stress constraints as well.[57,58]

5.1.4. *Other Algorithmic Approaches*

The approach for topology design described so far works with a raster-based design description and a gradient based mathematical programming technique for the optimization iterations. A whole range of methods have been proposed that maintain the basic raster concept but apply alternative optimization methods, both for the 0-1 format and for the interpolated grey-scale models. These methods typically combine aspects of fully stressed design, OC methods, element removal or structural growth, sometimes applying sensitivity analysis as a basis for the updating schemes. The methods often apply the word "evolutionary", which here should *not* be confused with genetic algorithms (that have also been used[12,13]). We here mention example papers describing some of the more popular methods that are named SKO[69] and ESO[70], see Ref. 5 for an overview of the various approaches. We here also mention the very interesting examples of the paper Refs. 71 that show the limitations of applying heuristic methods.

5.2. *Design of Articulated Mechanisms*

The design problem treated so far has dealt with compliant mechanisms that attain their mobility from flexibility of their constituents, and we have here applied the material distribution technique to find optimized topologies for the given design settings. An alternative is to use a truss representation, as in the *ground-structure* approach mentioned in the Introduction.[9,72] In a sense this representation of all possible designs represents an intermediate class of mechanisms as there is here both flexibility in the truss members and hinges in the truss nodes (joints).

Kinematic diagrams are widely used for conventional mechanism designs and it is both natural and advantageous to represent kinematic diagrams by truss elements and pin joints.

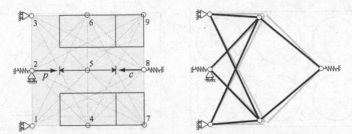

Fig. 9.   An articulated mechanism (a force inverter), designed using a truss representation; the design setting is illustrated to the left and shows the truss ground structure of potential bars, input force etc. Both topology alone (grey design to the left) and topology as well as node locations (solid black design) are optimized, using a mixed-integer formulation that is solved to global optimality by a branch-and-bound algorithm. In case of nodal positions being optimized, these are limited to move in the boxes shown in the left-hand side illustration. By courtesy of M. Stolpe and A. Kawamoto[75].

Also, in a truss ground structure representation it is possible to accommodate large displacement without element distortion problems.

In order to obtain true articulated mechanisms that obtain all of their mobility from their joints the concept of degrees of freedom (DOF) becomes critical for obtaining a proper design. This concept is an insignificant feature for compliant mechanism design as such mechanisms do not have real joints. This means that the design formulation (along the lines of (2)) has to include additional constraints that intrinsically are related to integer-type information. The degrees of freedom constraint to be included can be based on Maxwell's rule.[73] However, this requires that every sub-structure contains no redundant elements and for a ground-structure one then has to count the number of bars in a suitable way, for example by introducing additional constraints or to make proper counting procedures.

For solution of the design problem one can work with various techniques for relaxing the various integer constraints[74] or one can apply branch-and-bound techniques directly on the integer format of the problem.[75] The latter allows for the determination of globally optimal solutions and in this case it turns out that additional constraints on for example the stability of the truss needs to be included in the problem setting as well.

5.3. *Level-set Methods*

A new approach to topology design that has attracted considerable interest in recent years is the application of ideas and methods from level-set methods that have traditionally been used for modelling free and moving surface problems.[76,77] In a level-set method one works with an implicit definition of the boundaries of the design and this allows for a change in topology as the level-set function develops during the iterative optimization procedure. In most implementations (see, e.g., Ref. 78, 79, 80, 81 and references therein) the level-set is not parameterized as such and the updates of the boundaries are based on the so-called Hamilton-Jacobi equation. In turn, the driving term of this equation uses shape sensitivity information, derived as presented in Chapter 6. This makes the optimization scheme similar to steepest descent methods. If a direct parameterization[82] of the level-set is employed general mathematical programming techniques are available, at the cost of a more involved geometry modelling. In the phraseology of image processing, the material distri-

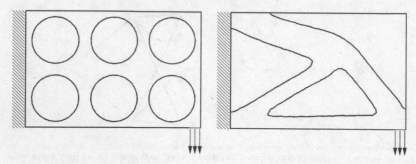

Fig. 10.   A minimum compliance result using level-set method and FEMLAB. Left: Initial design with 6 holes and right: optimized design with two resulting holes.

bution method and the level-set approach are both concerned with segmentation and one can see many analogies between the basic concepts of image segmentation and the field of topology design (this is for example clearly illustrated in Refs. 83, 84).

Another possibility to consider changes in the topology of a design is to apply the bubble method.[85,86] This approach has in recent years been supplemented by the flourishing field of topological derivatives.[87,88,89] The technique is to calculate the sensitivity to infinitesimal topological changes of functionals that depend on the solution to partial differential equations, where a topological change means the appearance or closing of cavities. The approach is closely related to shape sensitivity analysis; for a detailed overview see Ref. 90. This sensitivity analysis cannot be applied in a standard mathematical programming framework as there is no underlying parameter space for the design; instead material removal ideas can be applied[91] or the information can be used in connection with level-set methods.[92]

6. From Theory to Product

6.1. *Industrial Use of Topology Design*

The computer-based topology optimization method was first introduced in 1988 for the minimum weight design of structural components. Since then, the topology optimization method has gained widespread popularity in academia and industry and is now being used to reduce weight and optimize performance of automobiles, aircrafts, space vehicles and many other structures. Today, several commercial software systems provide topology optimization for industry and a main user of topology design in the daily design efforts is the automobile industry, where most major manufacturers and their sub-suppliers now use the methodology.

A recent example of the use of topology design in aeronautics is for the design of integrally stiffened machined ribs for the inboard inner fixed leading edge of the new airliner, the Airbus 380. Two types of software were applied, one which is similar to the method described here, and one that also includes information on the type of (eventually composite) material that is useful for the design. Based on these results and quite a bit of engineering interpretation a new type of structure was devised for the ribs which gave a weight benefit

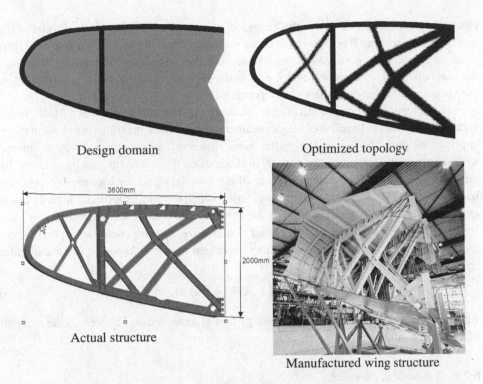

Design domain

Optimized topology

3600mm

2000mm

Actual structure

Manufactured wing structure

Fig. 11.   Mimicking of industrial design process. Top: Design domain and resulting topology for rib structure in frontpart of airplane wing and bottom: actual design and manufactured front part of wing at EADS (courtesy of EADS Military Aircraft).

against traditional (up to 40%) and competitive honeycomb/composite designs.

The process of generating this new type of rib for an aircraft is typical for applications of topology design in many mechanical engineering settings. The technique is not necessarily used for creating the final design, but rather to give inspiration to the experienced engineer that can see new possibilities on the basis of the computational results. In other fields, however, one also sees that the results of the topology design are transferred directly to production, for example when designing and producing new types of photonic crystals, see below.

6.2. *Nano-photonics*

6.2.1. *Wave Propagation Problems*

The governing equation for a number of different wave-propagation problems is the scalar Helmholtz equation

$$\nabla \cdot (A\nabla u) + \omega^2 Bu = 0, \tag{32}$$

where, depending on the physics problem considered, the field u (in 2D or 3D) as well as the material constants will have different physical meanings. For the case of planar transverse electromagnetic polarized waves (the TE-mode), u is the electric field, A is the inverse of the dielectric constant and B is the product of the vacuum permitivity and vacuum permeability, whereas for the other polarization (transverse magnetic waves - the TM-mode), u denotes the magnetic field value, $A = 1$ and B equals the product of the dielectric material value, the vacuum permitivity and the vacuum permeability.

For the topology optimization of, e.g., photonic crystals it turns out that a SIMP model with $p = 1$ suffices to handle the design parameterization, since maximum contrast in many applications gives the best wave-confinement. However, in some cases "grey solutions" with intermediate densities may appear. In those cases $0 - 1$ designs can be obtained by introducing some artificial damping in one of the material phases or we introduce a penalization damping term (called a "pamping" term) which introduces artificial high damping in intermediate density elements.[93] For the objective function of the optimization we can apply the so-called Poynting vector in order to maximize the wave energy transport to specified output domains; the Poynting vector[94] averaged over a time-period is for the scalar case outlined here calculated as

$$I = \frac{\omega}{2} \int_{\Gamma_{\text{out}}} A \Re(i\, u\, \nabla \overline{u})\, d\Gamma, \tag{33}$$

where Γ_{out} is a line through which the energy flow is measured and $\Re$ denotes the real part of a complex number.

6.2.2. A Z-bend in Photonics

The planar photonic crystal is an optical nano-material with periodic modulation of the refractive index. The modulation is designed to forbid propagation of light in certain wavelength ranges, so-called photonic bandgaps. Breaking the crystal symmetry by introducing line defects and other discontinuities allows control of the light on a sub-wavelength scale in the photonic crystals. Therefore, photonic devices based on the bandgap effect may be several length-scales smaller than traditional integrated optical devices.

The idea behind these devises are as follows. Light propagates as waves and if transmitted through a transparent medium like glass, it will propagate essentially without losses. However, if one perforates the glass structures with a periodic arrangement of air holes with hole distances a little less than the wave length of the light (this means that we are talking about length scales smaller than micrometers, i.e. nano-scale), waves at certain frequencies will no more propagate through the glass structure. This effect can be used to produce mirrors in nano-scale or it can be used to guide light in optical chips. The latter can be obtained by filling some of the air holes in channel-like patterns as seen for a Z-bend in Figure 12. Since the light cannot travel through the perforated structure, it will stay within the channel and can be led around sharp corners and may be manipulated in other ways. Such photonic crystal structures will in the future provide the basic building blocks for optical devices and computers.

The idea of loss-less transmission of optical waves through photonic crystals outlined above is a truth with modifications. In reality, the transmission is less than 100% because of leaking waves and reflections at channel corners. It is quite obvious that the efficiency may be optimized by changing the shape, number and position of the holes along the channels. Therefore, the design of photonic crystals is an obviously interesting application for the topology optimization method.

Figure 12 shows the result of the design process for a Z-bend. If we had just removed air holes to obtain a Z-shaped bend, the light transmitted through the bend would have been less than 50% due to losses and reflections. For topology optimization it was chosen to utilize only the outer parts of the two bend regions as design areas. Although one could choose much larger design areas, the numerical experiments showed that relatively small design areas were enough to yield the wanted improvement in efficiency. Had the efficiency been unsatisfactory, the design areas could have been enlarged in order to introduce more freedom in the design. In order to reduce the bend loss, the transmitted energy flow measured by the Poynting vector through the Z-bend waveguide is maximized in the topology optimization procedure (see Fig. 12). The optimization can be performed for any number of frequencies simultaneously, e.g., in a min-max formulation. In the case of the Z-bend it was found that the use of a single frequency in the optimization is sufficient to produce a large bandwidth with low loss.

The result of the topology optimization process resulted in a close to 100% transmission in a wide frequency range. Figure 12 shows the optimized design and the resulting wave propagation through the optimized waveguide. The optimized Z-bend was manufactured by e-beam lithography techniques at the Center for Optical Communication (COM) at DTU (see Figure 12). The manufactured device performs very well with record breaking bandwidth and transmission properties.

Many more examples of topology optimization in wave-propagation problems can be found in the literature.[95,96,97,98,99,93]

7. Challenges in the Field

Topology optimization has become a very popular methodology, both in industrial use and as a research area. It generates very efficient designs for many areas of applications and has had a bigger impact than was envisaged just a decade ago. But there are still many challenges in the area, and we try here to outline a few of central importance.

7.1. *Multiphysics*

The topology optimization method has over the last years been applied to several other design problems. Examples are the design of tailored 'exotic' materials with counter-intuitive properties such as negative Poisson's ratios (materials which expand transversely when pulled) and negative thermal expansion coefficients (material which shrink when heated). Other applications include the design of transducers for underwater sound detection and MicroElectroMechanical Systems for use in hearing aids, air-bag sensors, and micro-robots. Also design of channels flows is now possible.[100,101,102]

These new challenges can be treated within the same basic format of the design parametrization, problem statement and computational procedure, as was also outlined

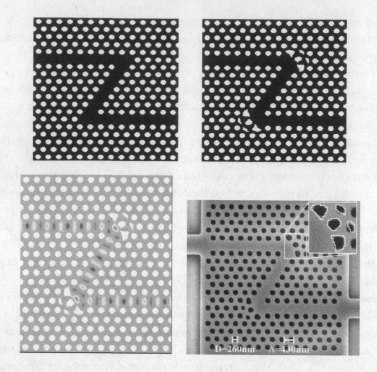

Fig. 12.   Top, left: Standard Z-bend waveguide. Top, right: The optimized design. Bottom, left: TE polarized light propagating through the topology optimized Z-bend. Bottom, right: The manufactured device, produced directly from pixel-representation of the design (from Borel et al.[95]).

above for the photonic crystal design. Thus the design of a acoustic device is closely related to the design of a compliant manipulator and we can again use the topology optimization for the design process. However, several issues need to be addressed. First, and a common feature with most optimal design techniques, is the question of how one formulates objective functions and constraints that result in useful engineering designs. Another central issue – and one particular to topology design – is how to relate gray-scale (density) to physical properties that allow these objective functions and constraints to be evaluated. Finally, a scheme should be imposed to obtain black-and-white designs.

7.2. Algorithms

Structural optimization problems in general constitute a difficult class of optimization problems – notwithstanding the great successes of the field much improvement is probably still possible when it comes to effective mathematical programming methods for large scale problems. Considering topology optimization problems adds to the complications. These are typically large scale in whatever format they are cast. Basically, the integer format is to be preferred, but only small problems can be treated. Thus most work treats the relaxed formats of the problem, using intermediate densities as continuous variables. Never-the-less, the resulting problems can be more tricky than sizing problems, for example through the

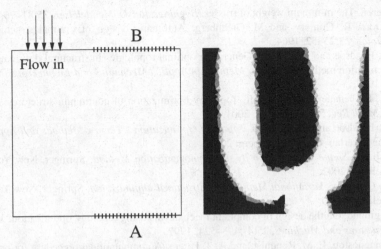

Fig. 13. A non-linear flow example. Design of a fluid transistor. For low Reynolds numbers the in-coming fluid should exit at boundary A and for higher Reynolds numbers the fluid should exit at boundary B (from Gersborg-Hansen et al.[101]).

possibility of design dependent sets of constraints, as is seen for stress constraints and also for buckling constraints. There is thus plenty of work to do for researchers in mathematical programming.

7.3. *Defining the Design*

In the material distribution method one applies a raster representation of the design. This means that the geometry modelling does not have a representation of boundaries where for example a normal vector can be uniquely defined. Also, the interpolation schemes depend on the possibility to define the physical field in every part of the reference domain (in fluids this can be done via a Brinkman model for porous flow). For more general physics situation there may not be any method to do this and one may believe that shape design is the only obvious approach to design optimization in such cases. Level-set methods may provide an answer, but here we need more work on algorithms for general optimization formulations. Thus, there is also still plenty of scope for work in geometry modelling and its relation to modelling of multiple and interacting physics.

Acknowledgments

This work was supported by the Danish Technical Research Council and through a European Young Investigator Award from ESF/Eurohorcs (OS).

References

1. A. G. M. Michell. The limit of economy of material in frame structures. *Philosophical Magazine*, 8(6):589–597, 1904.

2. P. Fleron. The minimum weight of trusses. *Bygningsstatiske Meddelelser*, 35:81–96, 1964.

3. W. Dorn, R. Gomory, and M. Greenberg. Automatic design of optimal structures. *J. de Mecanique*, 3:25–52, 1964.

4. M. P. Bendsøe and N. Kikuchi. Generating optimal topologies in structural design using a homogenization method. *Computer Methods in Applied Mechanics and Engineering*, 71(2):197–224, 1988.

5. H. A. Eschenauer and N. Olhoff. Topology optimization of continuum structures: A review. *Appl Mech Rev*, 54(4):331–390, 2001.

6. M. P. Bendsøe and O. Sigmund. *Topology Optimization - Theory, Methods and Applications*. Springer Verlag, Berlin Heidelberg, 2003.

7. G. Allaire. *Shape Optimization by the Homogenization Method*. Springer, New York Berlin Heidelberg, 2002.

8. A. V. Cherkaev. *Variational Methods for Structural Optimization*. Springer, New York Berlin Heidelberg, 2000.

9. O. Sigmund. On the design of compliant mechanisms using topology optimization. *Mechanics of Structures and Machines*, 25(4):493–524, 1997.

10. G. Anagnostou, E. M. Rønquist, and A. T. Patera. A computational procedure for part design. *Comp. Meth. Appl. Mech. Engng.*, 97:33–48, 1992.

11. P. Y. Shim and S. Manoochehri. Generating optimal configurations in structural design using simulated annealing. *International Journal for Numerical Methods in Engineering*, 40(6):1053–1069, 1997.

12. C. Kane and M. Schoenauer. Genetic operators for two-dimensional shape optimization. *Lecture Notes in Computer Science*, 1063:355–369, 1996.

13. E. Kita and H. Tanie. Topology and shape optimization of continuum structures using GA and BEM. *Structural Optimization*, 17(2-3):130–139, 1999.

14. M. Beckers. Dual methods for discrete structual optimization problems. *Int. J. Numer. Meth. Engng*, 48:1761–1784, 2000.

15. J. Cea, A. Gioan, and J. Michel. Quelques resultat sur l'identification de domaines. *Calcolo III / IV*, 1973.

16. L. Tartar. Estimation de coefficients homogénéisés. In *Computing methods in applied sciences and engineering (Proc. Third Internat. Sympos., Versailles, 1977), I,*, volume 704 of *Lecture Notes in Mathematics*, pages 364–373. Springer-Verlag, Berlin; Heidelberg; London; etc., 1979.

17. M. P. Bendsøe. Optimal shape design as a material distribution problem. *Structural Optimization*, 1:193–202, 1989.

18. G. I. N. Rozvany, M. Zhou, and O. Sigmund. Topology optimization in structural design. In H. Adeli, editor, *Advances in Design Optimization*, chapter 10, pages 340–399. Chapman and Hall, London, 1994.

19. R. J. Yang and C.-H. Chuang. Optimal topology design using linear programming. *Computers and Structures*, 52(2):265–276, 1994.

20. H. P. Mlejnek. Some aspects of the genesis of structures. *Structural Optimization*, 5:64–69, 1992.

21. C. C. Swan and J. S. Arora. Topology design of material layout in structured composites of high stiffness and strength. *Structural Optimization*, 13(1):45–59, 1997.

22. M. Stolpe and K. Svanberg. An alternative interpolation scheme for minimum compliance optimization. *Structural and Multidisciplinary Optimization*, 22:116–124, 2001.

23. N. L. Pedersen. Topology optimization of laminated plates with prestress. *Computers & Structures*, 80(7-8):559–70, 2002.

24. Z. Hashin and S. Shtrikman. A variational approach to the theory of the elastic behaviour of multiphase materials. *Journal of the Mechanics and Physics of Solids*, 11:127–140, March–April 1963.

25. M. P. Bendsøe and O. Sigmund. Material interpolation schemes in topology optimization. *Archives of Applied Mechanics*, 69(9-10):635–654, 1999.

26. K. Maute and E. Ramm. Adaptive topology optimization. *Structural Optimization*, 10(2):100–112, 1998.

27. T. Borrvall and J. Petersson. Large-scale topology optimization in 3D using parallel computing. *Computer Methods in Applied Mechanics and Engineering*, 190(46-47):6201–6229, 2001.

28. K. Svanberg. The method of moving asymptotes - A new method for structural optimization. *International Journal for Numerical Methods in Engineering*, 24:359–373, 1987.

29. K. Svanberg. A class of globally convergent optimization methods based on conservative convex separable approximations. *SIAM Journal on Optimization*, 12(2):555–573, 2002.

30. C. Zillober. SCPIP - an efficient software tool for the solution of structural optimization problems. *Structural and Multidisciplinary Optimization*, 24(5):362–371, 2002.

31. C. S. Jog. Reducing radiated sound power by minimizing the dynamic compliance. In *Proc. IUTAM Conference on Design for Quietness*, Dordrecht, 2002. Kluwer Academic Publishers.

32. W.H. Zhang and P. Duysinx. Dual approach using a variant perimeter constraint and efficient sub-iteration scheme for topology optimization. *Computers and Structures*, 81(22-23):2173–2181, 2003.

33. R. B. Haber, C. S. Jog, and M. P. Bendsøe. A new approach to variable-topology shape design using a constraint on the perimeter. *Structural Optimization*, 11(1):1–11, 1996.

34. J. Petersson. Some convergence results in perimeter-controlled topology optimization. *Computer Methods in Applied Mechanics and Engineering*, 171(1-2):123–140, 1999.

35. T. Borrvall. Topology optimization of elastic continua using restriction. *Archives of Computational Methods in Engineering*, 8(4):351–385, 2001.

36. J. Petersson and O. Sigmund. Slope constrained topology optimization. *International Journal for Numerical Methods in Engineering*, 41(8):1417–1434, 1998.

37. C. Zillober, K. Schittkowski, and K. Moritzen. Very large scale optimization by sequential convex programming. *Optimization Methods and Software*, 19(1):103–120, 2004.

38. M. Zhou, Y. K. Shyy, and H. L. Thomas. Checkerboard and minimum member size control in topology optimization. *Structural and Multidisciplinary Optimization*, 21:152–158, 2001.

39. A. R. Díaz and O. Sigmund. Checkerboard patterns in layout optimization. *Structural Optimization*, 10(1):40–45, 1995.

40. C. S. Jog and R. B. Haber. Stability of finite element models for distributed-parameter optimization and topology design. *Computer Methods in Applied Mechanics and Engineering*, 130(3-4):203–226, 1996.

41. T. A. Poulsen. A new scheme for imposing a minimum length scale in topology optimization. *International Journal of Numerical Methods on Engineering*, 57:741–760, 2003.

42. C. B. W. Pedersen, T. Buhl, and O. Sigmund. Topology synthesis of large-displacement compliant mechanisms. *International Journal for Numerical Methods in Engineering*, 50(12):2683–2705, 2001.

43. T. E. Bruns and D. A. Tortorelli. An element removal and reintroduction strategy for the topology optimization of structures and compliant mechanisms. *International Journal for Numerical Methods in Engineering*, 57(10):1413–1430, 2003.

44. O. Sigmund. Design of multiphysics actuators using topology optimization - Part I: One-material structures. *Computer Methods in Applied Mechanics and Engineering*, 190(49-50):6577–6604, 2001.

45. O. Sigmund. Design of multiphysics actuators using topology optimization - Part II: Two-material structures. *Computer Methods in Applied Mechanics and Engineering*, 190(49-50):6605–6627, 2001.

46. T. Sekimoto and H. Noguchi. Homologous topology optimization in large displacement and buckling problems. *JSME International Journal*, 44(4):616–622, 2001. Series A.

47. A. Saxena and G. K. Ananthasuresh. Topology synthesis of compliant mechanisms for nonlinear force-deflection and curved path specifications. *Transactions of the ASME - R - Journal of Mechanical Design*, 123(1):33–42, 2001.

48. T. Buhl, C. B. W. Pedersen, and O. Sigmund. Stiffness design of geometrically non-linear structures using topology optimization. *Structural and Multidisciplinary Optimization*, 19(2):93–104, 2000.

49. T. E. Bruns and D. A. Tortorelli. Topology optimization of non-linear elastic structures and compliant mechanisms. *Computer Methods in Applied Mechanics and Engineering*, 190(26-27):3443–3459, 2001.

50. T. Buhl. Simultaneous topology optimization of structure and supports. *Structural and Multidisciplinary Optimization*, 23(5):336–346, 2002.

51. B. Maar and V. H. Schulz. Interior point multigrid methods for topology optimization. *Structural and Multidisciplinary Optimization*, 19(3):214–224, 2000.

52. R.H.W. Hoppe, S.I. Petrova, and V. Schulz. Primal-dual newton-type interior-point method for topology optimization. *Journal of Optimization Theory and Applications*, 114(3):545–571, 2002.

53. R. Stainko. An adaptive multilevel approach to minimal compliance topology optimization. *Communications in Numerical Methods in Engineering*, 22:109–118, 2006.

54. L. T. Biegler, O. Ghattas, M. Heinkenschloss, and B. van Bloemen Waanders, editors. *Large Scale PDE-Constrained Optimization*, volume 30 of *Lecture Notes in Computational Science and Engineering*. Springer, Berlin Heidelberg, 2003.

55. M. Kocvara and J. V. Outrata. Optimization problems with equilibrium constraints and their numerical solution. *Mathematical Programming*, 101(1):119–149, 2004.

56. X. Liu and J. Sun. Generalized stationary points and an interior-point method for mathematical programs with equilibrium constraints. *Mathematical Programming*, 101(1):231–261, 2004.

57. M. Stolpe and K. Svanberg. Modelling topology optimization problems as linear mixed 0-1 programs. *International Journal for Numerical Methods in Engineering*, 57(5):723–739, 2003.

58. M. Stolpe. On the reformulation of topology optimization problems as linear or convex quadratic mixed 0-1 programs. *Optimization and Engineering*, to appear, 2006.

59. P. Duysinx and M. P. Bendsøe. Topology optimization of continuum structures with local stress coinstraints. *International Journal for Numerical Methods in Engineering*, 43(8):1453–1478, 1998.

60. R.P. Lipton. Assessment of the local stress state through macroscopic variables. *Philosophical Transactions of the Royal Society London, Series A (Mathematical, Physical and Engineering Sciences)*, 361(1806):921–46, 2003.

61. X. Guo, G. D. Cheng, and K. Yamazaki. A new approach for the solution of singular optima in truss topology optimization with stress and local buckling constraints. *Structural and Multidisciplinary Optimization*, 22(5):364–373, 2001.

62. X. Guo and G. D. Cheng. Extrapolation approach for the solution of singular optima. *Structural and Multidisciplinary Optimization*, 19, 2000.

63. J. Petersson. On continuity of the design-to-state mappings for trusses with variable topology. *International Journal of Engineering Science*, 39(10):1119–1141, 2001.

64. M. Stolpe and K. Svanberg. On the trajectories of the epsilon-relaxation approach for stress-constrained truss topology optimization. *Structural and Multidisciplinary Optimization*, 21:140–151, 2001.

65. A. Evgrafov. On globally stable singular truss topologies. *Structural and Multidisciplinary Optimization*, 29(3):170–177, 2005.

66. R. Lipton. Design of functionally graded composite structures in the presence of stress constraints. *International Journal of Solids and Structures*, 39(9):2575–2586, 2002.

67. G. Allaire, F. Jouve, and H. Maillot. Topology optimization for minimum stress design with the homogenization method. *Structural and Multidisciplinary Optimization*, 28(2-3):87–98, 2004.

68. J. T. Pereira, E. A. Fancello, and C. S. Barcellos. Topology optimization of continuum structures with material failure constraints. *Structural and Multidisciplinary Optimization*, 26(1):50–66, 2004.

69. A. Baumgartner, L. Harzheim, and C. Mattheck. SKO (Soft Kill Option): The biological way to find an optimum structure topology. *International Journal of Fatigue*, 14:387–393, 1992.

70. C. Zhao, G. P. Steven, and Y. M. Xie. Evolutionary optimization of maximizing the difference between two natural frequencies of a vibrating structure. *Structural Optimization*, 13(2-3):148–154, 1997.

71. M. Zhou and G. I. N. Rozvany. On the validity of ESO type methods in topology optimization. *Structural and Multidisciplinary Optimization*, 21:80–83, 2001.

72. M. I. Frecker, G. K. Ananthasuresh, S. Nishiwaki, and S. Kota. Topological synthesis of compliant mechanisms using multi-criteria optimization. *Transactions of the ASME*, 119(2):238–245, June 1997.

73. C. R. Calladine. Buckminster Fuller's 'tensegrity' structures and Clerk Maxwell's rules for the construction of stiff frames. *Int. J. Solid Structures*, 14:161–172, 1978.

74. A. Kawamoto, M. P. Bendsøe, and O. Sigmund. Articulated mechanism design with a degree of freedom constraint. *International Journal for Numerical Methods in Engineering*, 61(9):1520–1545, 2004.

75. M. Stolpe and A. Kawamoto. Design of planar articulated mechanisms using branch and bound. *Mathematical Programming B*, 103(2):357–397, 2005.

76. J. A. Sethian. *Level Set Methods and Fast Marching Methods*. Cambridge University Press, Cambridge, UK, 1999.

77. S. Osher and R. Fedkiw. *Level set methods and dynamic implicit surfaces*. Springer, New York,NY, 2003.

78. X. Wang, M. Y. Wang, and D. Guo. Structural shape and topology optimization in a level-set-based framework of region representation. *Structural and Multidisciplinary Optimization*, 27(1-2):1–19, 2004.

79. G. Allaire, F. Jouve, and A.-M. Toader. Structural optimization using sensitivity analysis and a level-set method. *Journal of Computational Physics*, 194(1):363–393, 2004.

80. M. Y. Wang and X. Wang. "Color" level sets: a multi-phase method for structural topology optimization with multiple materials. *Computer Methods in Applied Mechanics and Engineering*, 193(6-8):469–496, 2004.

81. Z. Liu, J. G. Korvink, and R. Huang. Structure topology optimization: Fully coupled level set method via FEMLAB. *Structural and Multidisciplinary Optimization*, 29(6):407–417, 2005.

82. J. Norato, R. Haber, D. Tortorelli, and M. P. Bendsøe. A geometry projection method for shape optimization. *International Journal for Numerical Methods in Engineering*, 60(14):2289–2312, 2004.

83. B. Bourdin and A. Chambolle. Design-dependent loads in topology optimization. *ESAIM: Control, Optimisation and Calculus of Variations*, 9:19–48, 2003.

84. M. Burger and R. Stainko. Phase-field relaxation of topology optimization with local stress constraints. *SIAM Journal on Control and Optimization*, to appear, 2006.

85. H. A. Eschenauer, V. V. Kobelev, and A. Schumacher. Bubble method for topology and shape optimization of structures. *Structural Optimization*, 8:42–51, 1994.

86. H. A. Eschenauer and A. Schumacher. Topology and shape optimization procedures using hole positioning criteria – theory and applications. In G. I. N. Rozvany, editor, *Topology Optimization in Structural Mechanics, CISM Courses and Lectures*, volume 374, pages 135–196. Springer, Vienna, 1997.

87. J. Sokolowski and A. Zochowski. Energy change due to the appearance of cavities in elastic solids. *International Journal of Solids and Structures*, 40:1765–1803, 2003.

88. A. A. Novotny, R. A. Feijoo, E. Taroco, and C. Padra. Topological sensitivity analysis. *Computer Methods in Applied Mechanics and Engineering*, 192:803–829, 2003.

89. J. Sokolowski and A. Zochowski. Optimality conditions for simultaneous topology and shape optimization. *SIAM Journal on Control and Optimization*, 42:1198–1221, 2003.

90. J. Sokolowski and A. Zochowski. On topological derivatives in shape optimization. In T. Lewinski, O. Sigmund, J. Sokolowski, and A. Zochowski, editors, *Optimal Shape Design and Modeling*, pages 55–144, Warsaw, 2004. Akademicka Oficyna Wydawnicza EXIT.

91. J. Cea, S. Garreau, P. Guillaume, and M. Masmoudi. The shape and topological optimizations connection. *Comput. Methods Appl. Mech. Engrg.*, 188:713–726, 2000.

92. M. Burger, B. Hackl, and W. Ring. Incorporating topological derivatives into level set methods. *Journal of Computational Physics*, 194(1):344 – 362, 2004.

93. J. S. Jensen and O. Sigmund. Topology optimization of photonic crystal structures: A high bandwidth low loss T-junction waveguide. *Journal of the Optical Society of America B*, 22(6):1191–1198, 2005.

94. L. D. Landau and E. M. Lifshitz. *The Classical Theory of Fields, 4th ed.* Pergamon Press, Oxford, 1975.

95. P. I. Borel, A. Harpøth, L. H. Frandsen, M. Kristensen, J. S. Jensen P. Shi, and O. Sigmund. Topology optimization and fabrication of photonic crystal structures. *Optics Express*, 12(9):1996–2001, 2004.

96. J. S. Jensen. Phononic band gaps and vibrations in one- and two-dimensional mass-spring structures. *Journal of Sound and Vibration*, 266(5):1053–1078, 2003.

97. O. Sigmund and J. S. Jensen. Systematic design of phononic band gap materials and structures by topology optimization. *Philosophical Transactions of the Royal Society A: Mathematical, Physical and Engineering Sciences*, 361:1001–1019, 2003.

98. J. S. Jensen and O. Sigmund. Systematic design of acoustic devices by topology optimization. In J. L. Coelho and D. Alarcão, editors, *ICSV12 Proceedings, Twelfth International Congress on Sound and Vibration. Lisbon, Portugal*, page 8, 2005. CD-ROM.

99. J. S. Jensen and O. Sigmund. Systematic design of photonic crystal structures using topology optimization: Low-loss waveguide bends. *Applied Physics Letters*, 84(12):2022–2024, 2004.

100. T. Borrvall and J. Petersson. Topology optimization of fluids in Stokes flow. *International Journal for Numerical Methods in Fluids*, 41:77–107, 2003.

101. A. Gersborg-Hansen, O. Sigmund, and R. B. Haber. Topology optimization of channel flow problems. *Structural and Multidisciplinary Optimization*, 30(3):181–192, 2005.

102. L. H. Olesen, F. Okkels, and H. Bruus. A high-level programming-language implementation of topology optimization applied to steady-state Navier-Stokes flow. *International Journal for Numerical Methods in Engineering*, 65(7):975–1001, 2006.

CHAPTER 7

SHAPE DESIGN SENSITIVITY ANALYSIS
OF NONLINEAR STRUCTURAL SYSTEMS

Nam Ho Kim

Department of Mechanical and Aerospace Engineering
University of Florida, P.O. Box 116250
Gainesville, Florida 32611, USA
E-mail: nkim@ufl.edu

Recent developments in design sensitivity analysis of nonlinear structural systems are presented. Various aspects, such as geometric, material, and boundary nonlinearities are considered. The idea of variation in continuum mechanics is utilized in differentiating the nonlinear equations with respect to design variables. Due to the similarity between variation in design sensitivity analysis and linearization in nonlinear analysis, the same tangent stiffness is used for both sensitivity and structural analyses. It has been shown that the computational cost of sensitivity calculation is a small fraction of the structural analysis cost. Such efficiency is due to the fact that sensitivity analysis does not require convergence iteration and it uses the same tangent stiffness matrix with structural analysis. Two examples are presented to demonstrate the accuracy and efficiency of the proposed sensitivity calculation method in nonlinear problems.

1. Introduction

Engineering design often takes into account the nonlinear behavior of the system, such as the design of a metal forming process and the crashworthiness of a vehicle. Nonlinearities in structural problems include material, geometric, and boundary nonlinearities.[1] Geometric nonlinearity occurs when the structure experiences large deformation and is described using the material or spatial formulation. Material nonlinearity is caused by the nonlinear relationship between stress and strain and includes nonlinear elasticity, hyperelasticity, elastoplasticity, etc. A contact/impact problem is often called a boundary nonlinearity, categorized by flexible-rigid and multibody contact/impact conditions. These nonlinearities are often combined together in many structural applications. In the sheet metal forming process,[2] for example, the blank material will experience contact with the punch and die (boundary nonlinearity), through which the blank material will be deformed to a desired shape (geometric nonlinearity). At the same time, the blank material will experience permanent, plastic deformation (material nonlinearity).

Design sensitivity analysis[3,4] of nonlinear structures concerns the relationship between design variables available to the design engineers and performance measure determined through the nonlinear structural analysis. We use the term "design sensitivity" in order to distinguish it from parameter sensitivity. The performance measures include: the weight, stiffness, and compliance of the structure; the fatigue life of a mechanical component; the noise in the passenger compartment of the automobile; the vibration of a beam or plate; the safety of a vehicle in a crash, etc. Any system parameters that the design engineers can change can serve as design variables, including the cross-sectional geometry of beams, the thickness of plates, the shape of parts, and the material properties.

Design sensitivity analysis can be thought of as a variation of the performance measure with respect to the design variable.[5] Most literature in design sensitivity analysis focuses on the first–order variation, which is similar to the linearization process. In that regard, sensitivity analysis is inherently linear. The recent development of second–order sensitivity analysis also uses a series of linear design sensitivity analyses in order to calculate the second–order variation.[6,7]

Different methods of sensitivity calculation have been developed in the literature, including global finite differences,[8,9] continuum derivatives,[10-12] discrete derivatives,[13-15] and automatic differentiation.[16-18] The global finite difference method is the easiest way to calculate sensitivity information, and repeatedly evaluates the performance measures at different values of the design variables. Engineering problems are often approximated using various numerical techniques, such as the finite element method. The continuum equation is approximated by a discrete system of equations. The discrete derivatives can be obtained by differentiating the discrete system of equations. The continuum derivatives use the idea of variation in continuum mechanics to evaluate the first–order variation of the performance function. After the continuum form of the design sensitivity equation is obtained, a numerical approximation, such as the finite element method, can be used to solve the sensitivity equation. The difference between discrete and continuum derivatives is the order between differentiation and discretization. Finally, automatic differentiation refers to a differentiation of the computer code itself by defining the derivatives of elementary functions, which propagate through complex functions using the chain rule of differentiation. The accuracy, efficiency, and implementation efforts of these methods are discussed by van Keulen *et al.*[19]

In this text, only the continuum derivatives are considered, assuming that the same idea can be implemented to the discrete derivatives. In the finite difference and computational differentiation, there is no need to distinguish linear and nonlinear problems, as these two approaches are identical for both problems.

In spite of the rigorous development of the existence and uniqueness of design sensitivity in linear systems,[20] no literature is available regarding existence and uniqueness of design sensitivity in nonlinear problems. In this text, the relation between design variables and the performance measures is assumed to be continuous and differentiable. However, by no means should this important issue in differentiability be underestimated.

The organization of the text is as follows. In Section 2, the design sensitivity formulation of nonlinear elastic problems is presented. The unique property of the problems in this category is that the sensitivity equation needs to be solved once at the end of the converged configuration. Thus, the sensitivity calculation is extremely inexpensive; basically, it is the same as that of linear problems.

In Section 3, the design sensitivity formulation of elastoplastic problems is presented. Because the constitutive relation is given as a rate form and the problem at hand is history–dependent, the sensitivity equation needs to be solved at each load step. However, the sensitivity calculation is still inexpensive compared with the nonlinear structural analysis, because the convergence iteration is not required in the sensitivity calculation. After the convergence iteration is finished, the linear sensitivity equation is solved using the decomposed coefficient matrix from the structural analysis.

In Section 4, the design sensitivity formulation of contact problems is presented. The penalty–regularized variational equation is differentiated with respect to design variables.

This chapter is by no means comprehensive in terms of deriving sensitivity formulations. The reader interested in detailed derivations is referred to the literature.[21-29]

2. Design Sensitivity Analysis of Nonlinear Elastic Problems

When the deformation of a structure is significant, the initial (undeformed) domain (Ω_X) is distinguished from the deformed domain (Ω_x). A material point $\mathbf{X} \in \Omega_X$ is deformed to a point $\mathbf{x} \in \Omega_x$, such that $\mathbf{x}(\mathbf{X}) = \mathbf{X} + \mathbf{z}(\mathbf{X})$, with $\mathbf{z}(\mathbf{X})$ being the displacement (see Fig. 1).

The weak form of a static problem, whether it is elastic or elastoplastic, can be stated that to find the solution $\mathbf{z} \in V$, such that

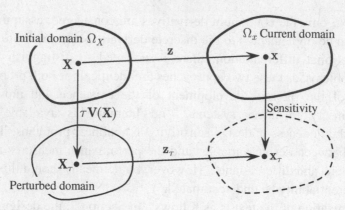

Figure 1. Illustration of shape design perturbation in a nonlinear structural problem. The initial domain is deformed to the current domain. For a given shape design variable, the design velocity field $\mathbf{V}(\mathbf{X})$ is defined in the initial domain. Design sensitivity analysis is then to estimate the deformation of the perturbed domain without actually performing additional nonlinear analysis.

$$a_\Omega(\mathbf{z}, \overline{\mathbf{z}}) = \ell_\Omega(\overline{\mathbf{z}}), \tag{1}$$

for all $\overline{\mathbf{z}} \in \mathbb{Z}$. In Eq. (1), $\mathbb{V}$ is the solution space and $\mathbb{Z}$ is the space of kinematically admissible displacements. $a_\Omega(\mathbf{z}, \overline{\mathbf{z}})$ and $\ell_\Omega(\overline{\mathbf{z}})$ are the energy and load forms, respectively, whose expressions depend on the formulations. In many cases, the load form is simple and often it is independent of the deformation. Thus, emphasis will be given to the energy form.

In nonlinear structural analysis, two approaches have been introduced: the total and the updated Lagrangian formulations.[1] The former refers to Ω_X as a reference, whereas the latter uses Ω_x as a reference. In both formulations, equilibrium equations are obtained using the principle of virtual work. These equations are then linearized to yield the incremental form. As noted by Bathe[1], these two formulations are analytically equivalent.

2.1. *Total Lagrangian Formulation*

2.1.1. *Incremental Solution Procedure*

When Ω_X is the reference, the energy form in Eq. (1) can be written as

$$a_{\Omega_X}(\mathbf{z}, \overline{\mathbf{z}}) = \iint_{\Omega_X} \mathbf{S}(\mathbf{z}) : \mathbf{E}(\mathbf{z}; \overline{\mathbf{z}}) \, d\Omega_X, \tag{2}$$

where $\mathbf{S}(\mathbf{z})$ is the second Piola–Kirchhoff stress tensor, ':' is the double contraction operator, and $\mathbf{E}(\mathbf{z}; \overline{\mathbf{z}})$ is the variation of the Green–Lagrange strain tensor, whose expression is given as

$$\mathbf{E}(\mathbf{z}; \overline{\mathbf{z}}) = sym\left(\nabla_0 \overline{\mathbf{z}}^T \cdot \mathbf{F}\right), \tag{3}$$

where $sym(\mathbf{A}) = \frac{1}{2}(\mathbf{A} + \mathbf{A}^T)$ represents the symmetric part of a tensor, $\mathbf{F} = \nabla_0 \mathbf{x}$ is the deformation gradient, and $\nabla_0 = \partial/\partial \mathbf{X}$ is the gradient operator in the initial domain. Note that $\mathbf{E}(\mathbf{z}; \bullet)$ is linear with respect to its argument, while $\mathbf{S}(\mathbf{z})$ is generally nonlinear.

The load form is independent of the deformation, and is defined as

$$\ell_{\Omega_X}(\overline{\mathbf{z}}) = \iint_{\Omega_X} \overline{\mathbf{z}}^T \cdot \mathbf{f}^B \, d\Omega + \int_{\Gamma_X^S} \overline{\mathbf{z}}^T \cdot \mathbf{f}^S \, d\Gamma, \tag{4}$$

where $\mathbf{f}^B$ is the body force and $\mathbf{f}^S$ the surface traction on the boundary Γ_X^S. The deformation–dependent load form can be found in Schweizerhof.[30]

Since the energy form is nonlinear, an incremental solution procedure, such as the Newton–Raphson method, is often employed through linearization. Let $\mathbb{L}[\bullet]$ denote the linearization operator with respect to incremental displacement $\Delta \mathbf{z}$. Then the energy form in Eq. (2) can be linearized, as

$$\mathbb{L}[a_{\Omega_X}(\mathbf{z}, \overline{\mathbf{z}})] = \iint_{\Omega_X} [\mathbf{E}(\mathbf{z}; \overline{\mathbf{z}}) : \mathbf{C} : \mathbf{E}(\mathbf{z}; \Delta \mathbf{z}) + \mathbf{S}(\mathbf{z}) : \mathbf{H}(\Delta \mathbf{z}, \overline{\mathbf{z}})] \, d\Omega_X$$
$$\equiv a_{\Omega_X}^*(\mathbf{z}; \Delta \mathbf{z}, \overline{\mathbf{z}}), \tag{5}$$

where $\mathbf{C}$ is the material tangent moduli, obtained from $\mathbb{L}[\mathbf{S}(\mathbf{z})] = \mathbf{C} : \mathbf{E}(\mathbf{z}; \Delta \mathbf{z})$, and the increment of $\mathbf{E}(\mathbf{z}; \overline{\mathbf{z}})$ is given as

$$\mathbf{H}(\Delta \mathbf{z}, \overline{\mathbf{z}}) = sym\left(\nabla_0 \overline{\mathbf{z}}^T \cdot \nabla_0 \Delta \mathbf{z}\right). \tag{6}$$

The notation of $a_{\Omega_X}^*(\mathbf{z}; \Delta \mathbf{z}, \overline{\mathbf{z}})$ is selected such that it implicitly depends on the total displacement $\mathbf{z}$, and has two parameters $\Delta \mathbf{z}$ and $\overline{\mathbf{z}}$. Note that $a_{\Omega_X}^*(\mathbf{z}; \bullet, \bullet)$ is bilinear and symmetric with respect to its two arguments.

In the solution procedure of a nonlinear problem, the applied load is divided by N load steps and a convergence iteration is carried out at each load step. Let the left superscript n denote the current load step and the right superscript k the iteration counter. Then, the incremental equation can be written as

$$a_{\Omega_X}^*({}^n \mathbf{z}^k; \Delta \mathbf{z}^k, \overline{\mathbf{z}}) = \ell_{\Omega_X}^n(\overline{\mathbf{z}}) - a_{\Omega_X}({}^n \mathbf{z}^k, \overline{\mathbf{z}}), \tag{7}$$

for all $\overline{\mathbf{z}} \in Z$. $\ell_{\Omega_X}^n(\overline{\mathbf{z}})$ is the load form at the current load step. After solving the incremental displacement $\Delta \mathbf{z}^k$, the total displacement is updated using ${}^n \mathbf{z}^{k+1} = {}^n \mathbf{z}^k + \Delta \mathbf{z}^k$. The iteration in Eq. (7) is repeated until the right–hand side (residual term) vanishes. After the solution is converged, the load step is increased. This procedure is repeated until the last load step N.

Note that Eq. (7) is still in the continuum form, and the discretization is not introduced yet. If the finite element method is used to approximate Eq. (7), the discrete counter part will be

$$[{}^{n}\mathbf{K}^{k}]\{\Delta\mathbf{U}^{k}\} = \{{}^{n}\mathbf{R}^{k}\}, \tag{8}$$

where $[{}^{n}\mathbf{K}^{k}]$ is the tangent stiffness matrix, $\{\Delta\mathbf{U}^{k}\}$ the vector of incremental nodal displacements, and $[{}^{n}\mathbf{R}^{k}]$ the vector of residual forces.

2.1.2. *Shape Sensitivity Formulation*

A shape design variable is defined in order to change the geometry of the structure. The concept of design velocity is often used for this purpose, which represents the direction of design change for a given shape design variable. By introducing a scalar parameter τ that can control the magnitude of the design change, the perturbed design, as shown in Fig. 1, in the direction of the design velocity can be obtained as

$$\mathbf{X}_{\tau} = \mathbf{X} + \tau\mathbf{V}(\mathbf{X}). \tag{9}$$

The perturbation process in Eq. (9) is similar to the dynamic process by considering τ as time. Because of this analogy, the direction $\mathbf{V}(\mathbf{X})$ of the design change is called the design velocity.

For a given design velocity, the sensitivity of a function is defined as a material derivative with respect to the parameter τ. For example, the material derivative of displacement can be written as

$$\dot{\mathbf{z}} \equiv \frac{d}{d\tau}[\mathbf{z}_{\tau}(\mathbf{X}_{\tau})]\bigg|_{\tau=0} = \lim_{\tau\to 0}\frac{\mathbf{z}_{\tau}(\mathbf{X}_{\tau}) - \mathbf{z}(\mathbf{X})}{\tau}. \tag{10}$$

As in continuum mechanics, the above material derivative can be decomposed by the partial derivative and the convective term, as

$$\dot{\mathbf{z}}(\mathbf{X}) = \mathbf{z}'(\mathbf{X}) + \nabla_{0}\mathbf{z} \cdot \mathbf{V}(\mathbf{X}). \tag{11}$$

Even if the partial derivative is interchangeable with the spatial gradient, the material derivative is not.[3] The following relation should be used for the material derivative of the spatial gradient:

$$\frac{d}{d\tau}(\nabla_{0}\mathbf{z})\bigg|_{\tau=0} = \nabla_{0}\dot{\mathbf{z}} - \nabla_{0}\mathbf{z} \cdot \nabla_{0}\mathbf{V}. \tag{12}$$

Since stress and strain include the gradient of the displacement, their material derivative will include the unknown term $\nabla_{0}\dot{\mathbf{z}}$ (implicit dependence) and the known term $\nabla_{0}\mathbf{z} \cdot \nabla_{0}\mathbf{V}$ (explicit dependence). The design sensitivity analysis solves for the first using the second. For example, the material derivative of the strain variation in Eq. (3) can be written as

$$\frac{d}{d\tau}\mathbf{E}(\mathbf{z};\overline{\mathbf{z}})\bigg|_{\tau=0} = \mathbf{H}(\dot{\mathbf{z}},\overline{\mathbf{z}}) + \mathbf{H}_{V}(\mathbf{z};\overline{\mathbf{z}}), \tag{13}$$

with the explicitly dependent term being

$$\mathbf{H}_V(\mathbf{z};\overline{\mathbf{z}}) = -sym\big[(\nabla_0\overline{\mathbf{z}} \cdot \nabla_0\mathbf{V})^T \cdot \mathbf{F}\big] - sym\big[\nabla_0\overline{\mathbf{z}}^T \cdot (\nabla_0\mathbf{z} \cdot \nabla_0\mathbf{V})\big]. \tag{14}$$

Let the incremental equation (7) be converged at the last load step, which means that the nonlinear variational equation (1) is satisfied. Then, Eq. (1) is differentiated with respect to the parameters τ to obtain the following design sensitivity equation:

$$a^*_{\Omega_X}(\mathbf{z};\dot{\mathbf{z}},\overline{\mathbf{z}}) = \ell'_V(\overline{\mathbf{z}}) - a'_V(\mathbf{z},\overline{\mathbf{z}}), \tag{15}$$

for all $\overline{\mathbf{z}} \in \mathbb{Z}$. The first term on the right–hand side is the explicit term from the load form and the second from the energy form. These explicit terms can be obtained after differentiating with respect to parameter τ, as

$$\ell'_V(\overline{\mathbf{z}}) = \iint_{\Omega_X} \overline{\mathbf{z}}^T \cdot \big[(\nabla_0\mathbf{f}^B \cdot \mathbf{V}) + \mathbf{f}^B div\mathbf{V}\big] d\Omega$$
$$+ \int_{\Gamma_X^s} \overline{\mathbf{z}}^T \cdot \big[(\nabla_0\mathbf{f}^S \cdot \mathbf{V}) + \kappa\mathbf{f}^B(\mathbf{V} \cdot \mathbf{N})\big] d\Gamma \tag{16}$$

and

$$a'_V(\mathbf{z},\overline{\mathbf{z}}) = \iint_{\Omega_X} [\mathbf{E}(\mathbf{z};\overline{\mathbf{z}}) : \mathbf{C} : \mathbf{E}_V(\mathbf{z}) + \mathbf{S} : \mathbf{H}_V(\mathbf{z};\overline{\mathbf{z}}) + \mathbf{S} : \mathbf{E}(\mathbf{z};\overline{\mathbf{z}})div\mathbf{V}] d\Omega, \tag{17}$$

where $div\mathbf{V}$ is the divergence of the design velocity, and

$$\mathbf{E}_V(\mathbf{z}) = -sym\big[(\nabla_0\mathbf{z} \cdot \nabla_0\mathbf{V}) \cdot \mathbf{F}\big] \tag{18}$$

is the explicitly dependent term from the Green–Lagrange strain. In Eq. (16), κ is the curvature of the boundary, and $\mathbf{N}$ the unit normal vector to the boundary.

The design sensitivity equation (15) in continuum form can be discretized using the same method with the nonlinear structural analysis. We assume that the nonlinear problem has been solved up to the final load step N and the final iteration K. If the finite element method is used to approximate Eq. (15), the discrete form of the sensitivity equation will be

$$[^N\mathbf{K}^K]\{\dot{\mathbf{U}}\} = \{\mathbf{R}^{fic}\}, \tag{19}$$

where $[^N\mathbf{K}^K]$ is the tangent stiffness matrix at the last analysis, which is already factorized from the structural analysis; $\{\dot{\mathbf{U}}\}$ the vector of nodal displacement sensitivity; and $[\mathbf{R}^{fic}]$ the fictitious load representing the right–hand side of Eq. (15).

If Eq. (7) is compared with Eq. (15), the left–hand sides are identical except that the former solves for $\Delta\mathbf{z}$, while the latter for $\dot{\mathbf{z}}$. The computational advantage of sensitivity analysis comes from the fact that the linear equation (15) is solved once at the last converged load step. In addition, the LU–decomposed tangent stiffness matrix can be used in solving for $\dot{z}$ with a different right–hand side, often called the fictitious load[3] or the pseudo load.[11]

If $a^*_{\Omega_X}(z; \Delta z, \overline{z})$ is a true linearization of $a_{\Omega_X}(z, \overline{z})$, this method provides a quadratic convergence when the initial estimate is close to the solution. Even if the tangent operator is inexact, the structural analysis may still converge after a greater number of iterations are performed. However, in sensitivity analysis the inexact tangent operator produces an error in the sensitivity result because no iteration is involved. Without accurate tangent stiffness, sensitivity iteration is required,[31] which significantly reduces the efficiency of sensitivity calculation.

In shape sensitivity analysis, the total Lagrangian formulation has been more popular than the updated Lagrangian formulation.[32-35] This is partly because the reference configuration Ω_X is the same as the design reference. However, it will be shown in the next section that the sensitivity expressions of the two formulations are identical after appropriate transformation.

2.2. *Updated Lagrangian Formulation*

2.2.1. *Incremental Solution Procedure*

The updated Lagrangian formulation uses Ω_x as a reference. The energy form in the updated Lagrangian formulation can be written as

$$a_{\Omega_x}(z, \overline{z}) = \iint_{\Omega_x} \sigma(z) : \varepsilon(\overline{z}) \, d\Omega, \tag{20}$$

where $\sigma(z)$ is the Cauchy stress tensor, $\varepsilon(\overline{z})$ the variation of the engineering strain tensor, whose expression is given as

$$\varepsilon(\overline{z}) = sym(\nabla_x \overline{z}), \tag{21}$$

and $\nabla_x = \partial / \partial x$ is the spatial gradient operator. The same load form in the total Lagrangian formulation is used.[36]

Even if Eqs. (2) and (20) seem different, it is possible to show that they are identical using the following relations:

$$\varepsilon(\overline{z}) = \mathbf{F}^{-T} \cdot \mathbf{E}(z; \overline{z}) \cdot \mathbf{F}^{-1}, \tag{22}$$

$$\sigma(z) = \frac{1}{J} \mathbf{F} \cdot \mathbf{S} \cdot \mathbf{F}^T. \tag{23}$$

The same transformation as in Eq. (22) can be applied for $\mathbf{E}(z; \Delta z)$. In Eq. (23), J is the Jacobian of the deformation, such that $d\Omega_x = J d\Omega_X$.

The linearization of Eq. (20) is complicated because not only the stress and strain, but also the domain Ω_x depends on the deformation. Thus, instead of directly linearizing Eq. (20), it is first transferred to the undeformed configuration (pull-back). After linearization, the incremental form (the same as Eq. (5)) is transferred to the deformed configuration (push-forward) to obtain

$$a_{\Omega_x}^*(\mathbf{z}; \Delta\mathbf{z}, \overline{\mathbf{z}}) \equiv \iint_{\Omega_x} [\boldsymbol{\varepsilon}(\overline{\mathbf{z}}) : \mathbf{c} : \boldsymbol{\varepsilon}(\Delta\mathbf{z}) + \boldsymbol{\sigma}(\mathbf{z}) : \boldsymbol{\eta}(\Delta\mathbf{z}, \overline{\mathbf{z}})] d\Omega, \tag{24}$$

where $c_{ijkl} = F_{iI} F_{jJ} F_{kK} F_{lL} C_{IJKL}$ is the spatial tangent moduli[37] and

$$\boldsymbol{\eta}(\Delta\mathbf{z}, \overline{\mathbf{z}}) = sym(\nabla_x \overline{\mathbf{z}}^T \cdot \nabla_x \Delta\mathbf{z}) \tag{25}$$

is the transformation of $\mathbf{H}(\Delta\mathbf{z}, \overline{\mathbf{z}})$ in Eq. (6).

The same incremental equation as in Eq. (7) can be used for the Newton–Raphson iterative solution procedure with different definitions of $a_{\Omega_x}(\mathbf{z}, \overline{\mathbf{z}})$ and $a_{\Omega_x}^*(\mathbf{z}; \Delta\mathbf{z}, \overline{\mathbf{z}})$. There is one difficulty in the expression of Eq. (20): the reference Ω_x is unknown. For computational convenience, the domain at the previous iteration is often chosen as a reference domain, assuming that as the solution converges, the difference between the two domains can be ignored.

2.2.2. *Shape Sensitivity Formulation*

From the viewpoint of the shape design, the sensitivity formulation of the updated Lagrangian can be done in two ways: either differentiating the energy form in Eq. (20) directly, or differentiating the total Lagrangian form first and then transforming it to the current configuration. The first is relatively complex because the reference Ω_x depends on both the design and the deformation. Cho and Choi[38] differentiate the energy form in Ω_x. Since the design velocity $\mathbf{V}(\mathbf{X})$ is always defined in Ω_X, they update the design velocity at each load step, which requires additional steps in the sensitivity calculation. In addition, this approach cannot take advantage of the computational efficiency, because the sensitivity equation must be solved at each load step.

From the idea that the total and updated Lagrangian formulations are equivalent, the second approach is taken; i.e., transforming the sensitivity Eq. (15) to the deformed configuration to obtain

$$a_{\Omega_x}^*(\mathbf{z}; \dot{\mathbf{z}}, \overline{\mathbf{z}}) = \ell_V'(\overline{\mathbf{z}}) - a_V'(\mathbf{z}, \overline{\mathbf{z}}), \tag{26}$$

for all $\overline{\mathbf{z}} \in \mathbb{Z}$. In Eq. (26), the same $\ell_V'(\overline{\mathbf{z}})$ in Eq. (16) is used, since the difference between two formulations is in the energy form, not in the load form. The explicitly dependent term from the energy form can be obtained, after transformation, as

$$a_V'(\mathbf{z}, \overline{\mathbf{z}}) = \iint_{\Omega_x} [\boldsymbol{\varepsilon}(\overline{\mathbf{z}}) : \mathbf{c} : \boldsymbol{\varepsilon}_V(\mathbf{z}) + \boldsymbol{\sigma} \cdot \boldsymbol{\eta}_V(\mathbf{z}, \overline{\mathbf{z}}) + \boldsymbol{\sigma} : \boldsymbol{\varepsilon}(\overline{\mathbf{z}}) div\mathbf{V}] d\Omega, \tag{27}$$

where

$$\varepsilon_V(\mathbf{z}) = -sym(\nabla_0 \mathbf{z} \cdot \nabla_x \mathbf{V}),\tag{28}$$

$$\eta_V(\mathbf{z}, \overline{\mathbf{z}}) = -sym\left[\nabla_n \overline{\mathbf{z}}^T \cdot (\nabla_0 \mathbf{z} \cdot \nabla_x \mathbf{V})\right] - sym\left[\nabla_0 \overline{\mathbf{z}} \cdot \nabla_x \mathbf{V}\right].\tag{29}$$

Note that the sensitivity Eq. (26) solves for the sensitivity of the total displacement, not its increment. Thus, the same efficiency as with the total Lagrangian approach can be expected.

3. Design Sensitivity Analysis of Elastoplastic Problems

In addition to the nonlinear elastic material in the previous section, the elastoplastic material is important in engineering applications. The major difference is that the former has a potential function so that the stress can be determined as a function of state, whereas the latter depends on the load history. In that regard, the elastoplastic problem is often called history–dependent. One of the main disadvantages of this type of problem is that the sensitivity analysis must follow the nonlinear analysis procedure closely.[39-43] Two formulations are discussed in this section: the rate form and the total form.

3.1. *Small Deformation Elastoplasticity*

3.1.1. *Incremental Solution Procedure*

When deformation is small (i.e., infinitesimal), the constitutive relation of elastoplasticity can be given in the rate form, and stress can be additively decomposed into elastic and plastic parts. The elastic part is described using the traditional linear elasticity, while the plastic part (permanent deformation) is described by the evolution of internal plastic variables.

Due to the assumption of small deformation, it is unnecessary to distinguish the deformed configuration from the undeformed one. Since the problem depends on the path of the load, it is discretized by N load steps: $[t_0, t_1, \ldots, t_N]$ with the current load step being t_n. In order to simplify the presentation, only isotropic hardening is considered in the following derivations, in which the plastic variable is identical to the effective plastic strain, e_p.

Let the incremental solution procedure converge at load step t_{n-1} and the history–dependent variable $^{n-1}\xi = \{^{n-1}\sigma, \, ^{n-1}e_p\}$ be available. Then, the energy form at t_n can be written as

$$a_\Omega(^{n-1}\xi; \, ^n\mathbf{z}, \overline{\mathbf{z}}) = \iint_\Omega \varepsilon(\overline{\mathbf{z}}) : \, ^n\sigma \, d\Omega.\tag{30}$$

In Eq. (30), the left superscripts n and $n-1$ represent the load steps t_n and t_{n-1}, respectively. However, they will often be omitted whenever there is no confusion. The notation of the energy form is selected such that it implicitly depends on the history–dependent variable at the pervious load step.

The energy form is nonlinear with respect to its arguments. In order to linearize the energy form, it is necessary to consider the update procedure of the stress and the plastic variable. In the displacement–driven procedure, it is assumed that the displacement increment $\Delta \mathbf{z}$ is given from the previous iteration. Mathematically, elastoplasticity can be viewed as a projection of stress onto the elastic domain, which can be accomplished using a trial elastic predictor followed by a plastic corrector. Then, the stress and the plastic variable can be updated according to

$$ {}^{n}\boldsymbol{\sigma} = {}^{n-1}\boldsymbol{\sigma} + \mathbf{C} : \Delta\boldsymbol{\varepsilon} - 2\mu\gamma\mathbf{N}, \tag{31} $$

$$ {}^{n}e_p = {}^{n-1}e_p + \sqrt{\tfrac{2}{3}}\gamma, \tag{32} $$

where $\mathbf{C} = (\lambda + \tfrac{2}{3}\mu)\mathbf{1} \otimes \mathbf{1} + 2\mu\mathbf{I}_{dev}$ is the fourth–order isotropic constitutive tensor; λ and μ are Lame's constants; $\Delta\boldsymbol{\varepsilon} = \boldsymbol{\varepsilon}(\Delta\mathbf{z})$ is the incremental strain; $\mathbf{N}$ is a unit deviatoric tensor, normal to the yield function; and γ is the plastic consistency parameter. In Eq. (31), the first two terms on the right–hand side correspond to the trial stress; i.e., ${}^{tr}\boldsymbol{\sigma} = {}^{n-1}\boldsymbol{\sigma} + \mathbf{C} : \Delta\boldsymbol{\varepsilon}$.

The plastic consistency parameter can be obtained from the relation that the stress stays on the boundary of the yield function during the continuous yielding:

$$ f({}^{n}\mathbf{s}, {}^{n}e_p) = \| {}^{n}\mathbf{s}\| - \sqrt{\tfrac{2}{3}}\kappa({}^{n}e_p) = 0, \tag{33} $$

where ${}^{n}\mathbf{s} = \mathbf{I}_{dev} : {}^{n}\boldsymbol{\sigma}$ is the deviatoric stress tensor, $\mathbf{I}_{dev}$ is the fourth–order unit deviatoric tensor, $\kappa({}^{n}e_p)$ is the radius of the elastic domain in the isotropic hardening plastic model. In general, the above equation is nonlinear, so that the local Newton–Raphson method can be used for the plastic consistency parameter. When there is no plastic deformation, γ is equal to zero.

Using the update procedure described in Eqs. (31)–(33), the energy form can be linearized to obtain

$$ a_{\Omega}^{*}({}^{n-1}\boldsymbol{\xi}; \Delta\mathbf{z}, \bar{\mathbf{z}}) = \iint_{\Omega} \boldsymbol{\varepsilon}(\bar{\mathbf{z}}) : \mathbf{C}^{\text{alg}} : \boldsymbol{\varepsilon}(\Delta\mathbf{z})\, d\Omega, \tag{34} $$

where the algorithmic tangent operator[44] is defined by

$$ \mathbf{C}^{\text{alg}} = \mathbf{C} - \frac{4\mu^2}{A}\mathbf{N} \otimes \mathbf{N} - \frac{4\mu^2\gamma}{\| {}^{tr}\mathbf{s}\|}[\mathbf{I}_{dev} - \mathbf{N} \otimes \mathbf{N}], \tag{35} $$

where $A = 2\mu + \tfrac{2}{3}\partial\kappa/\partial e_p$; $\otimes$ is the tensor product; and ${}^{tr}\mathbf{s}$ is the deviatoric stress at the trial state, which can be obtained by assuming that all incremental displacements are elastic. To guarantee the quadratic convergence of the nonlinear analysis, the algorithmic tangent operator must be consistent with the stress update algorithm.[44]

Once the energy form and the linearized energy form are available, the same linear equation as Eq. (7) can be used to solve for the incremental displacement. After the residual term vanishes, the stress and the plastic variable are updated according to Eqs. (31) and (32), the analysis moves to the next load step, and proceeds until the last load step.

3.1.2. Shape Sensitivity Formulation

In the shape design sensitivity formulation for the elastoplastic material, it is assumed that the structural problem has been solved up to the load step t_n and the sensitivity analysis has been finished up to the load step t_{n-1}. The goal is to solve the sensitivity equation at the load step t_n. This is necessary because the problem at hand is history–dependent. At each load step, the sensitivity of the incremental displacement is solved, and the sensitivity of the stress and the plastic variable is updated for the sensitivity calculation at the next load step.

By differentiating the variational equation (1) with the energy form in Eq. (30), the sensitivity equation can be obtained as

$$a_\Omega^*(^{n-1}\xi; \Delta\dot{z}, \overline{z}) = \ell_V'(\overline{z}) - a_V'(^n z, \overline{z}) - a_p'(^{n-1}\xi, {}^n z, \overline{z}), \tag{36}$$

for all $\overline{z} \in Z$. The linearized energy form $a_\Omega^*(^{n-1}\xi; \bullet, \bullet)$ is identical with that of Eq. (34). Two differences can be observed in the above sensitivity equation compared to the elastic problem: (1) it solves for the sensitivity of incremental displacement $\Delta\dot{z}$, and (2) it depends on the sensitivity results at the previous load step. In Eq. (36), the explicit term from the load form is similar to Eq. (16), and the explicit term from the energy form is defined as

$$a_V'(z, \overline{z}) = \iint_\Omega [\varepsilon_V(\overline{z}) : \sigma + \varepsilon(\overline{z}) : \mathbf{C}^{\text{alg}} : \varepsilon_V(z) + \varepsilon(\overline{z}) : \sigma \, div\mathbf{V}] \, d\Omega, \tag{37}$$

where

$$\varepsilon_V(z) = -sym(\nabla z \cdot \nabla \mathbf{V}) \tag{38}$$

is the explicit term from the material derivative of the strain tensor. The last term in Eq. (36), the history–dependent term, is given as

$$a_p'(z, \overline{z}) = \iint_\Omega [\varepsilon(\overline{z}) : \sigma^{fic}] \, d\Omega, \tag{39}$$

where

$$\sigma^{fic} = {}^{n-1}\dot{\sigma} - \frac{2\mu}{A}\left[\mathbf{N} : {}^{n-1}\dot{\mathbf{s}} - \sqrt{\frac{2}{3}}\frac{\partial\kappa}{\partial e_p}{}^{n-1}\dot{e}_p\right]\mathbf{N} - \frac{2\mu\gamma}{\|{}^{tr}\mathbf{s}\|}(\mathbf{I} - \mathbf{N}\otimes\mathbf{N}) : {}^{n-1}\dot{\mathbf{s}} \tag{40}$$

is the plastic variable from the sensitivity calculation at the previous load step.

After the sensitivity equation is solved, the sensitivity of the total displacement can be updated by

$$^n\dot{\mathbf{z}} = {}^{n-1}\dot{\mathbf{z}} + \Delta\dot{\mathbf{z}}. \tag{41}$$

In addition, the sensitivity of the stress and the plastic variable must be updated for the calculation in the next load step, using the following formulas:

$$^n\dot{\sigma} = {}^{n-1}\dot{\sigma} + \mathbf{C} : [\varepsilon(\Delta\dot{\mathbf{z}}) + \varepsilon_V(\Delta\mathbf{z})], \tag{42}$$

$$^n\dot{e}_p = {}^{n-1}\dot{e}_p + \frac{1}{A}\sqrt{\frac{2}{3}}\left[\mathbf{N} : {}^n\dot{\mathbf{s}} - \sqrt{\frac{2}{3}}\frac{\partial\kappa}{\partial e_p}{}^{n-1}\dot{e}_p\right]. \tag{43}$$

The fact that the sensitivity equation needs to be solved at each load step may decrease the computational advantage. However, the sensitivity calculation is still inexpensive compared to the structural analysis. First, the convergence iteration in the nonlinear problem is avoided and the linear sensitivity equation is solved at the end of each load step. Second, the LU–decomposed stiffness matrix from structural analysis can be used for sensitivity calculation. Considering the fact that most computational cost in the matrix equation is involved in the decomposition, the proposed sensitivity calculation method provides a significant advantage. The major cost in sensitivity calculation is involved in the construction of the fictitious load and updating the history–dependent variables.

3.2. *Finite Deformation Elastoplasticity*

When a structure undergoes a large deformation, the elastoplasticity theory with the infinitesimal deformation needs to be modified. A new method for expressing the kinematics of finite deformation elastoplasticity using the hyperelastic constitutive relation is becoming a desirable approach for isotropic material. This method defines a stress–free intermediate configuration composed of plastic deformation, and obtains the stress from the elastic strain energy density function defined in the intermediate configuration (see Fig. 2).

In this model, the deformation gradient is decomposed by the elastic and plastic parts,[45] as

$$\mathbf{F}(\mathbf{X}) = \mathbf{F}_e(\mathbf{X}) \cdot \mathbf{F}_p(\mathbf{X}), \tag{44}$$

where $\mathbf{F}_p(\mathbf{X})$ is the deformation through the intermediate domain, which is related to the plastic deformation, and $\mathbf{F}_e^{-1}$ is the stress–free, unloaded process.

3.2.1. *Incremental Solution Procedure*

Similar to the previous section, the load is discretized by N load steps and the current load step is t_n. In order to simplify the presentation, only isotropic hardening is considered in the following derivations. In the incremental solution process, it is assumed that the nonlinear analysis has been converged and plastic variables $^{n-1}\boldsymbol{\xi} = \{^{n-1}\mathbf{F}_p, \,^{n-1}e_p\}$ are available from load step t_{n-1}.

The variational equation is similar to that of the updated Lagrangian formulation, and the energy form is defined as

$$a_{\Omega_X}(^{n-1}\boldsymbol{\xi}; \,^{n}\mathbf{z}, \overline{\mathbf{z}}) = \iint_{\Omega_X} \boldsymbol{\varepsilon}(\overline{\mathbf{z}}) : \,^{n}\boldsymbol{\tau} \, d\Omega. \tag{45}$$

Note that the energy form is defined using the integral over domain Ω_X, and the Kirchhoff stress tensor $\boldsymbol{\tau} = J\boldsymbol{\sigma}$ is used so that the effect of Jacobian is included in the constitutive relation.[46]

In order to solve the nonlinear equation (45), the procedure of stress update is presented first. At load step t_n, with given displacement increment, the deformation gradient is calculated by

$$^{n}\mathbf{F} = \mathbf{f} \cdot \,^{n-1}\mathbf{F} = \,^{tr}\mathbf{F}_e \cdot \,^{n-1}\mathbf{F}_p, \tag{46}$$

where $\mathbf{f} = 1 + \nabla_x \Delta \mathbf{z}$ is the relative deformation gradient, and $^{tr}\mathbf{F}_e = \mathbf{f} \cdot \,^{n-1}\mathbf{F}_e$ is

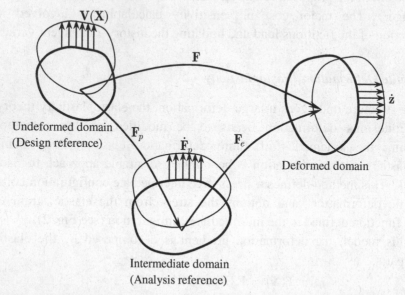

Figure 2. Analysis and design configurations for large deformation elastoplasticity. Plastic deformation is applied to the intermediate domain. The constitutive relation is hyperelasticity between the intermediate and deformed domains. The design velocity is always defined in the undeformed domain.

the trial elastic deformation gradient, which is obtained by assuming that the relative deformation gradient is purely elastic.

Since the trial state assumes that all incremental deformation is elastic, it goes out of the elastic domain when a part of it is plastic deformation. Thus, the trial state needs to return to the elastic domain, which is called the return–mapping. In this model, the return–mapping is achieved in the principal stress space with a fixed principal direction. By using the constitutive relation between the principal stress and logarithmic strain, better accuracy is obtained for a large elastic strain problem than with the classical elastoplasticity.

Let $\mathbf{e} = \{e_1, e_2, e_3\}^T = \{\log(\lambda_1), \log(\lambda_2), \log(\lambda_3)\}^T$ be the logarithmic principal stretch of the elastic left Cauchy–Green deformation tensor, defined by

$$^{tr}\mathbf{b}^e = {}^{tr}\mathbf{F}_e \cdot {}^{tr}\mathbf{F}_e^T = \sum_{i=1}^{3} \lambda_i^2 \mathbf{n}^i \otimes \mathbf{n}^i. \tag{47}$$

Then, the Kirchhoff stress tensor, after plastic deformation, can be calculated by

$$\boldsymbol{\tau} = \sum_{i=1}^{3} \tau_i^p \mathbf{n}^i \otimes \mathbf{n}^i, \tag{48}$$

where $\boldsymbol{\tau}^p = \{\tau_1^p, \tau_2^p, \tau_3^p\}$ is the principal Kirchhoff stress. Note that for the isotropic material, $\boldsymbol{\tau}$ and $^{tr}\mathbf{b}^e$ share the same principal directions. Equation (48) means that the principal direction is fixed during the plastic deformation, and the principal Kirchhoff stress is updated, including plastic deformation, as

$$^n\boldsymbol{\tau}^p = \mathbf{c}^e \cdot \mathbf{e} - 2\mu\gamma\mathbf{N}, \tag{49}$$

where $\mathbf{c}^e = (\lambda + \frac{2}{3}\mu)\tilde{\mathbf{1}} \otimes \tilde{\mathbf{1}} + 2\mu\mathbf{1}_{dev}$ is the 3×3 elastic constitutive tensor for the isotropic material; $\tilde{\mathbf{1}} = \{1, 1, 1\}^T$ is the first–order tensor; $\mathbf{1}_{dev} = 1 - \frac{1}{3}(\tilde{\mathbf{1}} \otimes \tilde{\mathbf{1}})$ is the second–order deviatoric tensor; $\mathbf{N}$ is a unit vector, normal to the yield function; and γ is the plastic consistency parameter. If Eq. (49) is compared with Eq. (31), two formulations yield a very similar return–mapping procedure. The differences are that Eq. (49) is in the principal stress space, and the logarithmic principal stretch is used instead of the engineering strain tensor.

The plastic consistency parameter can be obtained from the relation that the stress stays on the yield function during the continuous yielding:

$$f(^n\mathbf{s}, {}^n e_p) = \|{}^n\mathbf{s}\| - \sqrt{\tfrac{2}{3}}\kappa(^n e_p) = 0, \tag{50}$$

where $^n\mathbf{s} = \mathbf{1}_{dev} : {}^n\boldsymbol{\tau}^p$ is the deviatoric part of $^n\boldsymbol{\tau}^p$, and $\kappa(^n e_p)$ is the radius of the yield surface after plastic deformation.

The linearization of the energy form is similar to that of the updated Lagrangian formulation, except that the integration domain is changed to the undeformed one:

$$a_{\Omega_X}(^{n-1}\boldsymbol{\xi}, \mathbf{z}; \Delta\mathbf{z}, \overline{\mathbf{z}}) = \iint_{\Omega_X} [\boldsymbol{\varepsilon}(\overline{\mathbf{z}}) : \mathbf{c} : \boldsymbol{\varepsilon}(\Delta\mathbf{z}) + \boldsymbol{\tau} : \boldsymbol{\eta}(\Delta\mathbf{z}, \overline{\mathbf{z}})] d\Omega. \tag{51}$$

The tangent stiffness moduli $\mathbf{c}$ in the above equation must be consistent with the stress update procedure that is explained between Eqs. (46) and (50). The explicit form of $\mathbf{c}$ is available in Simo.[46]

Using the energy form in Eq. (45) and its linearization in Eq. (51), the Newton-Raphson method, similar to Eq. (7), can be employed to solve for the incremental displacement. Once the residual term is converged through iteration, the plastic variables are updated and analysis moves to the next load step.

Different from the classical elastoplasticity, it is not necessary to store stress because, as is clear from Eq. (49), stress can be calculated from hyperelasticity. Instead, the intermediate configuration, which is represented by $\mathbf{F}_p$ or counter part $\mathbf{F}_e$, is stored for the calculation in the next load step. For that purpose, first the relative plastic deformation gradient is calculated by

$$\mathbf{f}_p = \sum_{i=1}^{3} \exp(-\gamma N_i) \mathbf{n}^i \otimes \mathbf{n}^i, \tag{52}$$

from which the elastic part of the deformation gradient is updated by $^n\mathbf{F}_e = \mathbf{f}_p \cdot {}^{tr}\mathbf{F}_e$, and the plastic part can be obtained from $^n\mathbf{F}_p = {}^n\mathbf{F}_e^{-1} \cdot {}^n\mathbf{F}$. In addition, the effective plastic strain that determines the radius of the yield surface can be updated by

$$^n e_p = {}^{n-1}e_p + \sqrt{\tfrac{2}{3}}\gamma. \tag{53}$$

After the plastic variables are updated, the sensitivity analysis is performed at each converged load step.

3.2.2. *Shape Sensitivity Formulation*

As mentioned before, the reference for the design is always the undeformed configuration. When the references for the design and analysis are different, transformation is involved in sensitivity differentiation. In the case of finite deformation elastoplasticity, functions in the intermediate configuration are transformed to the undeformed configuration (pull–back). After differentiation, they are transformed to the deformed configuration (push–forward) in order to recover the updated Lagrangian formulation.

By differentiating the nonlinear variational equation (45) with shape design, the following sensitivity equation can be obtained:

$$a_{\Omega_X}^*(^{n-1}\boldsymbol{\xi}, \mathbf{z}; \dot{\mathbf{z}}, \overline{\mathbf{z}}) = \ell_V'(\overline{\mathbf{z}}) - a_V'(\mathbf{z}, \overline{\mathbf{z}}) - a_p'(^{n-1}\boldsymbol{\xi}; \mathbf{z}, \overline{\mathbf{z}}), \tag{54}$$

where the explicit term from the load form is given in Eq. (16), and the explicit term from the energy form is given by

$$a'_V(\mathbf{z}, \overline{\mathbf{z}}) = \iint_{\Omega_X} [\varepsilon(\overline{\mathbf{z}}) : \mathbf{c} : \varepsilon_V(\mathbf{z}) + \boldsymbol{\tau} : \boldsymbol{\eta}_V(\mathbf{z}, \overline{\mathbf{z}}) + \boldsymbol{\tau} : \varepsilon(\overline{\mathbf{z}}) div\mathbf{V}] d\Omega. \tag{55}$$

The expressions of $\varepsilon_V(\mathbf{z})$ and $\boldsymbol{\eta}_V(\mathbf{z}, \overline{\mathbf{z}})$ are identical to those in the updated Lagrangian formulation in Sec. 2.2. The last term on the right–hand side of Eq. (54) is the history–dependent term, which is contributed by the plastic deformation, given as

$$a'_p({}^{n-1}\boldsymbol{\xi}, \mathbf{z}, \overline{\mathbf{z}}) = \iint_{\Omega_X} [\varepsilon(\overline{\mathbf{z}}) : \mathbf{c} : \varepsilon_p(\mathbf{z}) + \boldsymbol{\tau} : \boldsymbol{\eta}_p(\mathbf{z}, \overline{\mathbf{z}}) + \boldsymbol{\tau}^{fic} : \varepsilon(\overline{\mathbf{z}})] d\Omega. \tag{56}$$

The first two integrands are related to the material derivative of the intermediate configuration, and are defined as

$$\varepsilon_p(\mathbf{z}) = -sym(\mathbf{F}_e \cdot \dot{\mathbf{F}}_p \cdot \mathbf{F}^{-1}), \tag{57}$$

$$\boldsymbol{\eta}_p(\mathbf{z}, \overline{\mathbf{z}}) = -sym(\nabla_x \overline{\mathbf{z}}^T \cdot \mathbf{F}_e \cdot \dot{\mathbf{F}}_p \cdot \mathbf{F}^{-1}). \tag{58}$$

In addition, the last term in Eq. (56) is related to the history–dependent plastic variable,

$$\boldsymbol{\tau}^{fic} = \sum_{i=1}^{3} \left[\frac{\partial \tau_i^p}{\partial e_p} {}^{n-1}\dot{e}_p \right] \mathbf{n}^i \otimes \mathbf{n}^i. \tag{59}$$

Note that the sensitivity equation (54) solves for the sensitivity of the total displacement, which is different from the classical elastoplasticity.

After the sensitivity equation is solved for $\dot{\mathbf{z}}$, the sensitivities of history–dependent terms are updated. For that purpose, the sensitivity of the plastic consistency parameter is first obtained as

$$\dot{\gamma} = \frac{1}{A} \left(2\mu \mathbf{N} \cdot \dot{\mathbf{e}} - \frac{\partial \kappa}{\partial e_p} {}^{n-1}\dot{e}_p \right), \tag{60}$$

where $\dot{e}_i = (\mathbf{n}^i \otimes \mathbf{n}^i) : [\varepsilon(\dot{\mathbf{z}}) + \varepsilon_V(\mathbf{z}) + \varepsilon_p(\mathbf{z})]$. Then, the sensitivity of the effective plastic strain is updated by

$$ {}^n\dot{e}_p = {}^{n-1}\dot{e}_p + \sqrt{\tfrac{2}{3}}\dot{\gamma}. \tag{61}$$

The sensitivity of the intermediate domain is also history–dependent, and can be updated by

$$ {}^n\dot{\mathbf{F}}_p = {}^n\mathbf{F}_e^{-1} \cdot {}^n\dot{\mathbf{F}} - {}^n\mathbf{F}_e^{-1} \cdot {}^n\dot{\mathbf{F}}_e \cdot {}^n\mathbf{F}_e^{-1}, \tag{62}$$

where ${}^n\dot{\mathbf{F}} = \nabla_0 \dot{\mathbf{z}} - \nabla_0 \mathbf{z} \cdot \nabla_0 \mathbf{V}$ and ${}^n\dot{\mathbf{F}}_e = \dot{\mathbf{f}}_p \cdot {}^{tr}\mathbf{F}_e + \mathbf{f}_p \cdot {}^{tr}\dot{\mathbf{F}}_e$. In the above equation, the sensitivity of $\dot{\mathbf{f}}_p$ can be obtained by differentiating Eq. (52). After updating the plastic variables, the nonlinear analysis moves to the next load step.

4. Design Sensitivity Analysis of Contact Problems

Contact problems are common and important aspects of mechanical systems. Metal forming, vehicle crashes, projectile penetration, various seal designs, and bushing and gear systems are only a few examples of contact problems. In this section, the contact condition of a 2D flexible body–rigid wall is considered. This problem can easily be extended to 3D flexible–flexible body contact problems, as shown by Kim et al.[28]

4.1. *Contact Problems with the Rigid Surface*

Contact between two bodies can be described using the impenetrability condition, which prevents one body from penetrating into another.[47,48] Figure 3 illustrates a contact condition with a rigid surface in R^2. A natural coordinate ξ is used to represent the location on a rigid surface. For example, the contact point $\mathbf{x}_c$ corresponds to the natural coordinate ξ_c, so that $\mathbf{x}_c = \mathbf{x}_c(\xi_c)$.

The impenetrability condition can be imposed on the structure by measuring the gap $g_n(\mathbf{x})$ between $\mathbf{x} \in \Gamma_c$ and the rigid surface, as shown in Fig. 3:

$$g_n \equiv (\mathbf{x} - \mathbf{x}_c(\xi_c)) \cdot \mathbf{e}_n(\xi_c) \geq 0, \quad \mathbf{x} \in \Gamma_c, \tag{63}$$

where $\mathbf{e}_n(\xi_c)$ is the unit outward normal vector of the rigid surface. The contact point $\mathbf{x}_c$ that corresponds to body point $\mathbf{x} \in \Gamma_c$ is determined by solving the following nonlinear equation:

$$(\mathbf{x} - \mathbf{x}_c(\xi_c)) \cdot \mathbf{e}_t(\xi_c) = 0, \tag{64}$$

where $\mathbf{e}_t(\xi_c)$ is the unit tangential vector. The contact point $\mathbf{x}_c(\xi_c)$ is the closest projection point of $\mathbf{x} \in \Gamma_c$ onto the rigid surface that satisfies Eq. (64).

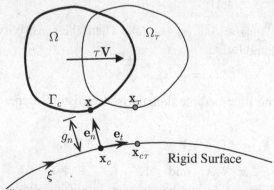

Figure 3. Contact condition between flexible and rigid bodies. The penalty function is established for the region Γ_c where the gap function is less than zero. Shape design change will move the contact point.

The structural problem with the contact condition can be formulated using a variational inequality, which is equivalent to the constrained optimization problem.[49] In practice, this optimization problem is solved using the penalty method. If there is a region Γ_c that violates Eq. (63), then it is penalized using a penalty function. After applying to the structural problem, the variational equation with the contact condition can be written as

$$a_\Omega(\mathbf{z}, \overline{\mathbf{z}}) + b_\Gamma(\mathbf{z}, \overline{\mathbf{z}}) = \ell_\Omega(\overline{\mathbf{z}}), \quad \forall \overline{\mathbf{z}} \in \mathbb{Z}, \tag{65}$$

where the energy and load forms are identical to the previous sections, depending on the constitutive model. The contact form can be defined from the variation of the penalty function, as

$$b_\Gamma(\mathbf{z}, \overline{\mathbf{z}}) = \omega \int_{\Gamma_c} g_n \overline{\mathbf{z}} \cdot \mathbf{e}_n \, d\Gamma, \tag{66}$$

where ω is the penalty parameter. In Eq. (66), ωg_n corresponds to the contact force. The nonlinear contact form in Eq. (66) can be linearized to obtain

$$b_\Gamma^*(\mathbf{z}; \Delta\mathbf{z}, \overline{\mathbf{z}}) = \omega \int_{\Gamma_c} \overline{\mathbf{z}} \cdot (\mathbf{e}_n \otimes \mathbf{e}_n) \cdot \Delta\mathbf{z} \, d\Gamma - \omega \int_{\Gamma_c} \frac{\alpha g_n}{c} \overline{\mathbf{z}} \cdot (\mathbf{e}_t \otimes \mathbf{e}_t) \cdot \Delta\mathbf{z} \, d\Gamma, \tag{67}$$

where

$$\alpha = \mathbf{e}_n \cdot \mathbf{x}_{c,\xi\xi}, \quad c = \|\mathbf{t}\|^2 - g_n\alpha. \tag{68}$$

Note that there is a component in the tangential direction because of curvature effects. If the rigid surface is approximated by a piecewise linear function, then $\alpha = 0$ and $c = \|\mathbf{t}\|^2$.

Suppose the current load step is t_n and the current iteration count is k. Then, the linearized incremental equation of (65) is obtained as

$$a_\Omega^*(^n\mathbf{z}^k; \Delta\mathbf{z}^k, \overline{\mathbf{z}}) + b_\Gamma^*(^n\mathbf{z}^k; \Delta\mathbf{z}^k, \overline{\mathbf{z}}) = \ell_\Omega^n(\overline{\mathbf{z}}) - a_\Omega(^n\mathbf{z}^k, \overline{\mathbf{z}}) - b_\Gamma(^n\mathbf{z}^k, \overline{\mathbf{z}}), \quad \forall \overline{\mathbf{z}} \in \mathbb{Z}. \tag{69}$$

The linearized system of (69) is solved iteratively with respect to incremental displacement until the residual forces on the right–hand side vanish at each load step.

4.2. *Design Sensitivity Analysis for Contact Problems*

The shape design sensitivity formulation of the contact problem has been extensively developed using linear variational inequality.[50,51] The linear operator theory is not applicable to a nonlinear analysis, and the non-convex property of the constraint set makes it difficult to prove the existence of the derivative. Despite such a lack of mathematical theory, the shape design sensitivity formulation for the contact problem is derived in a general continuum setting. As a result of the regularizing property of the penalty method, it is assumed that the solution continuously depends on shape design. As has been well established in the literature, differentiability fails in the region where contact status changes.[50] One good feature of the penalty method is that the contact region is established using a violated region, thus avoiding a non-differentiable region.

It is shown by Kim *et al.*[21] that the design sensitivity analysis of a frictionless contact problem is path–independent, whereas that of a frictional contact problem is path–dependent and requires information from the previous time step to compute sensitivity at the current time.

In order to derive the derivative of the contact form, the gap function in Eq. (63) is first differentiated with respect to the shape design variable, to obtain

$$\dot{g}_n = (\mathbf{V} + \dot{\mathbf{z}}) \cdot \mathbf{e}_n. \tag{70}$$

In the above derivation, the tangential component has been canceled due to the fact that the perturbed contact point also satisfies the consistency condition. Equation (70) implies that, for an arbitrary perturbation of the structure, only the normal component will contribute to the sensitivity of the gap function.

The contact form in Eq. (66) can then be differentiated with respect to the shape design, as

$$\frac{d}{d\tau}[b_{\Gamma_\tau}(\mathbf{z}_\tau, \overline{\mathbf{z}})]\Big|_{\tau=0} = b_\Gamma^*(\mathbf{z}; \dot{\mathbf{z}}, \overline{\mathbf{z}}) + b_V'(\mathbf{z}, \overline{\mathbf{z}}). \tag{71}$$

The first term on the right–hand side represents implicitly dependent terms through $\dot{\mathbf{z}}$, and the second term explicitly depends on $\mathbf{V}$. The implicit term $b_\Gamma^*(\mathbf{z}; \dot{\mathbf{z}}, \overline{\mathbf{z}})$ is available in Eq. (67) by substituting $\dot{\mathbf{z}}$ into $\Delta\mathbf{z}$. The explicit term $b_V'(\mathbf{z}, \overline{\mathbf{z}})$ is defined as the contact fictitious load and can be obtained by collecting all terms that have explicit dependency on the design velocity, as

$$b_V'(\mathbf{z}, \overline{\mathbf{z}}) = b_\Gamma^*(\mathbf{z}; \mathbf{V}, \overline{\mathbf{z}}) + \omega \int_{\Gamma_c} \kappa g_n \overline{\mathbf{z}} \cdot \mathbf{e}_n V_n \, d\Gamma. \tag{72}$$

The design sensitivity equation can then be obtained by differentiating the penalty–regularized variational Eq. (65) with respect to the design variable, as

$$a_\Omega^*(\mathbf{z}; \dot{\mathbf{z}}, \overline{\mathbf{z}}) + b_\Gamma^*(\mathbf{z}; \dot{\mathbf{z}}, \overline{\mathbf{z}}) = \ell_V'(\overline{\mathbf{z}}) - a_V'(\mathbf{z}, \overline{\mathbf{z}}) - b_V'(\mathbf{z}, \overline{\mathbf{z}}), \quad \forall \overline{\mathbf{z}} \in \mathbf{Z}. \tag{73}$$

For the frictionless contact problem, the fictitious load of the contact form in Eq. (72) depends on $\mathbf{z}$ and $\mathbf{V}$. The material derivative formula in Eq. (73) is history–independent. Thus, it is very efficient to compute the design sensitivity of a frictionless contact problem. The design sensitivity equation is solved only once at the last load step with the same tangent stiffness matrix from the structural analysis. As compared with nonlinear response analysis, this property provides great efficiency in the sensitivity computation process.

5. Numerical Examples

5.1. *Shape Design Sensitivity Analysis of the Windshield Wiper Problem*[24]

The continuum forms of the structural equation and the sensitivity equation are approximated using the reproducing kernel particle method (RKPM) , where the structural domain is represented by a set of particles.[52,53] RKPM is an ideal choice since, unlike the traditional finite element method, the solution is much less sensitive to the mesh distortion that causes many difficulties in large deformation analysis as well as in shape optimization.

Figure 4(a) shows the geometry of the windshield blade. The windshield is assumed to be a rigid body. For the convenience of the analysis, a vertical line is added to the windshield for smooth deformation. The upper part of the blade is supported by a steel slab. Hyperelastic material (rubber) is used for the blade, and $\omega = 10^7$ is used for the contact penalty.

As the glass moves to the left, the tip of the blade is in contact with the glass, which is modeled as flexible-rigid body contact. The function of the thin neck is to generate flexibility such that the direction of the blade can be easily turned over when the blade changes its moving direction. The role of the wing is to supply enough contact force at the tip point. Figure 4(b) shows a von Mises stress contour plot with the deformed geometry at the final configuration. The stress concentration is found at the neck and the tip because of the bending effect.

The geometry of the structure is parameterized using nine shape design variables as shown in Fig. 4(a). The design velocity at the boundary is obtained

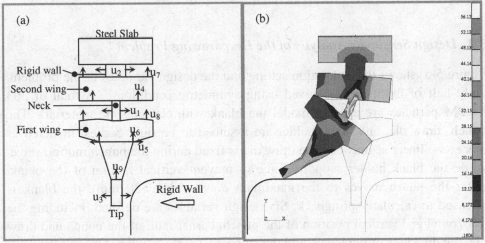

Figure 4. (a) Windshield blade geometry and shape design variables, (b) Contour plot of equivalent stress.

first by perturbing the boundary curve corresponding to the design variable, and the domain design velocity field is computed using an isoparametric mapping method. Four performance measures are chosen: the total area of the structure, two von Mises stresses of the neck region, and the contact force at the tip.

Sensitivity analysis is carried out at each converged load step to compute the material derivative of the displacement. The sensitivities of the performance measures are computed at the final converged load step using $\dot{z}$. The cost of the sensitivity computation is about 4% of that of the response analysis per design variable, which is quite efficient compared to the finite difference method. The accuracy of the sensitivity is compared with the forward finite difference results for the perturbation size of $\tau = 10^{-6}$. Table 1 shows the accuracy of the sensitivity results. In the third column of Table 1, $\Delta\psi$ denotes the finite difference results and the fourth column represents the change of the function from the proposed method. Excellent sensitivity results are obtained.

Table 1. Sensitivity results and comparison with finite difference method

Design	ψ	$\Delta\psi$	$\dot{\psi}$	$(\Delta\psi / \dot{\psi})\%$
1	Area	.28406E–5	.28406E–5	100.00
	$\sigma_{VM}(53)$	.19984E–3	.19984E–3	100.00
	$\sigma_{VM}(54)$	.28588E–3	.28588E–3	100.00
	F_C	.55399E–5	.55399E–5	100.00
3	Area	.68663E–5	.68663E–5	100.00
	$\sigma_{VM}(53)$	.19410E–3	.19410E–3	100.00
	$\sigma_{VM}(54)$	.68832E–4	.68832E–4	100.00
	F_C	.43976E–4	.43976E–4	100.00

5.2. Design Sensitivity Analysis of the Deepdrawing Problem[54]

Figure 5(a) shows the simulation setting and the design variables of the problem. Only half of the model is solved using symmetric conditions. A total of 303 RKPM particles are used to model the blank with elastoplastic material. The punch, draw die, and blank holder are assumed to be rigid bodies, modeled as piecewise linear segments. The draw die is fixed during the punch motion stage, while the blank holder supports force to prevent vertical motion of the blank. After the punch moves to the maximum down–stroke (30 mm), the blank is released to calculate springback. Six design variables are defined, including the horizontal and vertical position of the punch, corner radii of the punch and draw die, the thickness of the blank, and the gap between the blank holder and the die.

Figure 5(b) provides a contour plot of effective plastic strain after springback. A significant amount of sliding is observed between the workpiece and the draw die. High plastic strain distribution is observed in the vertical section. In the optimization, the maximum allowable amount of plastic strain is limited to prevent material failure due to excessive plastic deformation.

Two different types of results are evaluated: the amount of springback and effective plastic strain e_p. The amount of springback is defined as a difference between deformations at the maximum down–stroke and after releasing the blank. Since the sensitivity of effective plastic strain is updated at each load step, no additional computation is required for e_p. The sensitivity of the springback is calculated using the displacement sensitivity.

The accuracy of sensitivity result is compared with the finite difference result by slightly perturbing the design and re-solving the same problem. Table 2 compares the accuracy of the proposed sensitivity $\dot{\psi}$ with the finite difference result $\Delta\psi$. A very good agreement between two methods is observed. A perturbation of $\tau = 10^{-6}$ is used for the finite difference results. In this example, it is hard to find an appropriate perturbation size because the sensitivity

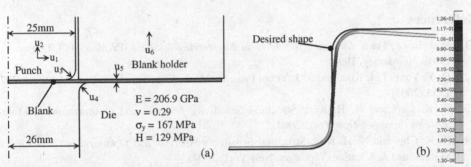

Figure 5. (a) Geometry of the deepdrawing problem and design variables. (b) Effective strain plot after springback. The solid line is the deformed geometry at the maximum down–stroke.

Table 2. Sensitivity results and comparison with finite difference method

Design	ψ	$\Delta\psi$	$\dot{\psi}$	$(\Delta\psi / \dot{\psi})\%$
1	springback	−4.31897E−5	−4.37835E−5	98.64
	$e_p(41)$	1.48092E−8	1.48111E−8	99.99
	$e_p(55)$	2.92573E−8	2.92558E−8	100.01
	$e_p(157)$	−2.08880E−8	−2.08875E−8	100.00
3	springback	1.50596E−5	1.55745E−5	96.69
	$e_p(41)$	−1.81265E−9	−1.81292E−9	99.99
	$e_p(55)$	−1.60858E−8	−1.60891E−8	99.98
	$e_p(157)$	1.14224E−8	1.14229E−8	99.99

magnitudes of the two functions are very different.

The computational cost of the sensitivity analysis is 3.8% of the analysis cost per design variable. Such efficiency is to be expected, since sensitivity analysis uses the decomposed tangent stiffness matrix, and no iteration is required.

6. Conclusions and Outlook

The design sensitivity formulations for various nonlinear problems are presented, including nonlinear elasticity, small and large deformation elastoplasticity, and frictionless contact problems. Even if the structural analysis contains combined nonlinearities, the consistent derivative yields very accurate sensitivity results. One of the most important advantages of the proposed approach is the computational efficiency of calculating sensitivity information, which is critical in the gradient–based optimization. Due to the facts that the proposed approach does not require iteration and uses the decomposed stiffness matrix from the structural analysis, it is shown through numerical examples that the computational cost of the sensitivity calculation is less than 5% of the analysis cost.

References

1. K.-J. Bathe, *Finite Element Procedures in Engineering Analysis* (Prentice-Hall, Englewood Cliffs, New Jersey, 1996).
2. D. Y. Yang, T. J. Kim, and S. J. Yoon, Proc. In. Mech. Eng., Part B: J. Eng. Manufact. **217**, 1553 (2003).
3. K. K. Choi and N. H. Kim, Structural Sensitivity Analysis and Optimization 1: Linear Systems (Springer, New York, 2004).
4. K. K. Choi and N. H. Kim, Structural Sensitivity Analysis and Optimization 2: Nonlinear Systems and Applications (Springer, New York, 2004).
5. J. Cea, in *Optimization of Distributed Parameter Structures*; *Vol. 2*, Ed. E. J. Haug and J. Cea (Sijthoff & Noordhoff, Alphen aan den Rijn, The Netherlands, 1981), p. 1005.
6. R. T. Haftka, AIAA J. **20,** 1765 (1982).
7. C.-J. Chen and K. K. Choi, AIAA J. **32,** 2099 (1994).
8. R. T. Haftka and D. S. Malkus, Int. J. Numer. Methods Eng. **17,** 1811 (1981).
9. C. H. Tseng and J. S. Arora, AIAA J. **27,** 117 (1989).
10. K. K. Choi and E. J. Haug, J. Struct. Mech. **11,** 231 (1983).
11. J. S. Arora, T.-H. Lee, and J. B. Cardoso, AIAA J. **30,** 1638 (1992).
12. D. G. Phelan and R. B. Haber, Comp. Meth. Appl. Mech. Eng. **77,** 31 (1989).
13. G. D. Pollock and A. K. Noor, Comp. Struct. **61,** 251 (1996).
14. S. Kibsgaard, Int. J. Numer. Methods Eng. **34,** 901 (1992).
15. N. Olhoff, J. Rasmussen, and E. Lund, Struct. Multidiscipl. Optim. **21,** 1 (1993).
16. L. B. Rall and G. F. Corliss, in Computational Differentiation: Techniques, Applications, and Tools (1996), p. 1.

17. I. Ozaki and T. Terano, Finite Elem. Anal. Design **14,** 143 (1993).
18. J. Borggaard and A. Verma, SIAM J. Sci. Comp. **22,** 39 (2000).
19. F. van Keulen, R. T. Haftka, and N. H. Kim, Comp. Meth. Appl. Mech. Eng. **194,** 3213 (2005).
20. E. J. Haug, K. K. Choi, and V. Komkov, *Design Sensitivity Analysis of Structural Systems* (Academic Press, London, 1986).
21. N. H. Kim, K. K. Choi, J. S. Chen, and Y. H. Park, Comp. Mech. **25,** 157 (2000).
22. N. H. Kim, K. K. Choi, and J. S. Chen, AIAA J. **38,** 1742 (2000).
23. N. H. Kim, K. K. Choi, and J. S. Chen, Comp. Struct. **79,** 1959 (2001).
24. N. H. Kim, Y. H. Park, and K. K. Choi, Struct. Multidiscipl. Optim. **21,** 196 (2001).
25. N. H. Kim, K. K. Choi, J. S. Chen, and M. E. Botkin, Mech. Struct. Machines **53,** 2087 (2002).
26. N. H. Kim and K. K. Choi, Mech. Struct. Machines **29,** 351 (2002).
27. N. H. Kim, K. K. Choi, and M. Botkin, Struct. Multidiscipl. Optim. **24,** 418 (2003).
28. N. H. Kim, K. Yi, and K. K. Choi, Int. J. Solids and Struct. **39,** 2087 (2002).
29. N. H. Kim, K. K. Choi, and J. S. Chen, Mech. Struct. Machines **51,** 1385 (2001).
30. K. Schweizerhof and E. Ramm, Comp. Struct. **18,** 1099 (1984).
31. S. Badrinarayanan and N. Zabaras, Comp. Meth. Appl. Mech. Eng. **129,** 319 (1996).
32. J. L. T. Santos and K. K. Choi, Struct. Multidiscipl. Optim. **4,** 23 (1992).
33. K. K. Choi and W. Duan, Int. J. Numer. Methods Eng. (2000).
34. I. Grindeanu, K. H. Chang, K. K. Choi, and J. S. Chen, AIAA J. **36,** 618 (1998).
35. Y. H. Park and K. K. Choi, Mech. Struct. Machines **24,** 217 (1996).
36. T. Belytschko, W. K. Liu, and B. Moran, *Nonlinear Finite Elements for Continua and Structures* (Wiley, New York, 2001).
37. J. C. Simo and R. L. Taylor, Comp. Meth. Appl. Mech. Eng. **85,** 273 (1991).
38. S. Cho and K. K. Choi, Int. J. Numer. Methods Eng. **48,** 375 (2000).
39. G. Bugeda, L. Gil, and E. Onate, Struct. Multidiscipl. Optim. **17,** 162 (1999).
40. M. Kleiber, Comp. Meth. Appl. Mech. Eng. **108,** 73 (1993).
41. M. Ohsaki and J. S. Arora, Int. J. Numer. Methods Eng. **37,** 737 (1994).
42. E. Rohan and J. R. Whiteman, Comp. Meth. Appl. Mech. Eng. **187,** 261 (2000).
43. C. A. Vidal and R. B. Haber, Comp. Meth. Appl. Mech. Eng. **107,** 393 (1993).
44. J. C. Simo and R. L. Taylor, Comp. Meth. Appl. Mech. Eng. **48,** 101 (1985).
45. E. H. Lee, J.f Appl Mech **36,** 1 (1969).
46. J. C. Simo, Comp. Meth. Appl. Mech. Eng. **99,** 61 (1992).
47. P. Wriggers, T. V. Van, and E. Stein, Comp. Struct. **37,** 319 (1990).
48. T. A. Laursen and J. C. Simo, Int. J. Numer. Methods Eng. **36,** 2451 (1993).
49. N. Kikuchi and J. T. Oden, Contact Problems in Elasticity: a Study of Variational Inequalities and Finite Element Method (SIAM, Philadelphia, VA, 1988).
50. J. Sokolowski and J. P. Zolesio, *Introduction to Shape Optimization* (Springer-Verlag, Berlin, 1991).
51. A. Haraux, J. Math. Soc. Japan **29,** 615 (1977).
52. W. K. Liu, S. Jun, and Y. F. Zhang, Int. J. Numer. Methods Eng. **20,** 1081 (1995).
53. J. S. Chen, C. T. Wu, S. Yoon, and Y. You, Int. J. Numer. Methods Eng. **50,** 435 (2001).
54. K. K. Choi and N. H. Kim, AIAA J. **40,** 147 (2002).

CHAPTER 8

OPTIMAL CONTROL OF STRUCTURES

Satish Nagarajaiah

Professor, Departments of Civil & Environmental Engineering and Mechanical Engineering & Material Science, Rice University, Houston, Texas, USA
E-mail: nagaraja@rice.edu

Sriram Narasimhan

Assistant Professor, Department of Civil & Environmental Engineering, University of Waterloo, Waterloo, Ontario, Canada
E-mail: snarasim@uwaterloo.ca

Optimal controllers are presented in this chapter for control of structures with emphasis on disturbance modeling. Both time domain and frequency domain methods are presented. Advantages and disadvantages of both the methods are discussed. Techniques for incorporating the excitation characteristics and frequency response information using augmentation techniques are presented. Numerical examples illustrating the control techniques and augmentation procedures for single and multiple degrees of freedom system are presented. The robustness principles in the context of linear optimal control are also discussed briefly.

1 Introduction

Optimal structural control using time domain and frequency domain methods have been proposed and used extensively during the last two decades for mitigating the effects of wind and earthquakes[1,2,3,4,5]. The field of linear optimal control has been the subject of active research for the past few decades. Time domain methods such as LQR (Linear Quadratic Regulator) and its counterpart, LQG (Linear Quadratic Gaussian) have been adopted in various structural control applications for mitigating the effects of wind and earthquakes[1]. The main idea behind the LQR method is the minimization of a performance index under the constraints imposed by the dynamic equations of motion. This minimization is aimed at reducing the structural responses of the system. Typically, two types of weighting matrices are used in the LQR procedure. They correspond to weighting structural responses and control forces. By choosing the appropriate matrices, the desired performance of the system can be achieved. The major limitations in the LQR method are (i) assuming the excitation as a zero-mean white noise to reduce the time varying riccatti matrix equation to an algebraic one, and (ii) ability to measure all the states of the system for full state feedback.

Some of the limitations in the LQR method are addressed in the LQG method. In this method the LQR method is combined with a state estimator (Kalman Bucy Filter); the estimated states (along with the partially measured states) are used in place of unobserved states for full state feedback in LQR using the separation principle[6]. Such a procedure is very useful in structural control where only partial state measurements are available for output feedback.

Though the LQG procedure addresses some of the limitations of the LQR method, there is no assurance of robustness, which can only be ensured by the H_2 and H_∞ frequency domain methods. H_2 and H_∞ derive their name from the norm that is minimized, 2-norm or ∞ norm. The LQG procedure is essentially equivalent to H_2 control as minimizing the LQG cost function is equivalent to minimizing the closed-loop system 2-norm. The control designs in frequency domain provide more physical insights into the problem, especially because the disturbances can be described satisfactorily using their power spectral densities (PSD); and the structural system can be adequately modeled by transfer function. The disturbances (e.g., earthquake) can be modeled by augmenting the state equations (with an appropriate filter excited by white noise). More descriptions of the augmentation procedures are presented in the following sections. H_2 and H_∞ frequency domain methods also incorporate the system uncertainty directly in the problem formulation to provide a robust design. Both the methods, H_2 and H_∞-minimize a prescribed norm of the transfer function from the excitation to the output.

The basics of dynamic systems, time domain methods and frequency domain methods are introduced next.

2 State Space Representation and Transfer Functions

The equations of motion are formulated in state space for the application of control to structural systems subjected to excitations such as earthquakes. State-space representation provides a consistent framework to analyze systems of any degree of complexity. A dynamical system can have multiple realizations, or equivalent forms of system representation. One such realization is the state space representation where a nth order differential equation is converted into n simultaneous first order differential equations cast in matrix form. These equations are in the time domain and are simple to solve using standard methods. In the state space representation, a general linear, time varying structural system excited by an earthquake can be represented as

$$\dot{\mathbf{x}}(t) = \mathbf{A}(t)\mathbf{x}(t) + \mathbf{B}(t)\mathbf{u}(t) + \mathbf{E}(t)\ddot{\mathbf{u}}_g(t)$$
$$\mathbf{y}(t) = \mathbf{C}(t)\mathbf{x}(t) + \mathbf{D}(t)\mathbf{u}(t) + \mathbf{v}(t) \tag{1}$$

where, $\mathbf{x}$ are the states of the system, $\mathbf{A}$, $\mathbf{B}$, $\mathbf{C}$, $\mathbf{D}$ and $\mathbf{E}$ are time-varying system matrices, $\ddot{\mathbf{u}}_g$ is the earthquake excitation, $\mathbf{u}$ is the vector of control forces, $\mathbf{y}$ is the measurement vector, and $\mathbf{v}$ is the measurement noise. For a linear time-invariant system (LTI), where the systems matrices do not change with time, the above equations can be re-written as,

$$
\begin{aligned}
\dot{\mathbf{x}}(t) &= \mathbf{A}\mathbf{x}(t) + \mathbf{B}\mathbf{u}(t) + \mathbf{E}\ddot{u}_g(t) \\
\mathbf{y}(t) &= \mathbf{C}\mathbf{x}(t) + \mathbf{D}\mathbf{u}(t) + \mathbf{v}(t)
\end{aligned} \tag{2}
$$

Two important properties of a system are controllability and observability. A system is said to be controllable if a state can be driven to any specified value from its initial state. A system is said to be observable if a state vector can be determined or estimated from the measured output. The controllability matrix can be formed using the $\mathbf{A}$ and $\mathbf{B}$ matrices as

$$
\mathbf{CO} = \begin{pmatrix} \mathbf{B} \ \mathbf{AB} \ \mathbf{A}^2\mathbf{B} \ ... \ \mathbf{A}^{n-1}\mathbf{B} \end{pmatrix} \tag{3}
$$

and the observability matrix with $\mathbf{C}$ and $\mathbf{A}$ matrices as

$$
\mathbf{OB} = \begin{pmatrix} \mathbf{C} \ \mathbf{CA} \ \mathbf{CA}^2 \ ... \ \mathbf{CA}^{n-1} \end{pmatrix}^T \tag{4}
$$

If the dimension of the state vector is n, then the system is said to be controllable if $rank(\mathbf{CO})=n$ and the system is said to be observable if $rank(\mathbf{OB})=n$.

Let us consider the simplified form of the state space equations in Eq. 2 without the external excitation and the measurement noise.

$$
\begin{aligned}
\dot{\mathbf{x}}(t) &= \mathbf{A}\mathbf{x}(t) + \mathbf{B}\mathbf{u}(t) \\
\mathbf{y}(t) &= \mathbf{C}\mathbf{x}(t) + \mathbf{D}\mathbf{u}(t)
\end{aligned} \tag{5}
$$

Taking the laplace transform on both sides of Eq. 5 (for zero initial conditions), we get

$$
s\mathbf{X}(s) = \mathbf{A}\mathbf{X}(s) + \mathbf{B}\mathbf{U}(s) \Rightarrow \mathbf{X}(s) = (s\mathbf{I} - \mathbf{A})^{-1}\mathbf{B}\mathbf{U}(s) \tag{6}
$$

$$
\mathbf{Y}(s) = \mathbf{C}\mathbf{X}(s) + \mathbf{D}\mathbf{U}(s) = \left[\mathbf{C}(s\mathbf{I} - \mathbf{A})^{-1}\mathbf{B} + \mathbf{D}\right]\mathbf{U}(s) \Rightarrow \mathbf{H}(s)\mathbf{U}(s) \tag{7}
$$

where, $\mathbf{H}(s)$ is the transfer function from the control input, $\mathbf{u}$, to the measurement vector, $\mathbf{y}$. Each term in $\mathbf{H}(s)$ is a proper ratio of polynomials, and for the case $\mathbf{D} = \mathbf{0}$, the ratio of each term in the transfer function matrix is strictly proper. The transfer function is the frequency response function, if the variable s is replaced by the complex variable $j\omega$. As shown in Fig 1, the output is equal to the harmonic excitation input, $e^{j\omega t}$, multiplied by the frequency response function.

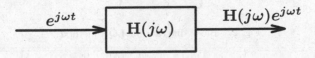

Fig. 1.   Frequency response function for a linear system

3 Time Domain Methods: LQR and LQG

3.1 LQR Method

The problem of finding the optimal control involves the minimization of a specified cost function subject to either static or dynamic constraints. There are special cases when the constraint is dynamic, linear and the cost function is quadratic. Such conditions arise, for example, when the objective is the minimization of strain energy in a structure and whose deformations are predominantly in the linear elastic range. The Linear Quadratic Regulator (LQR) method[6], also known as the quadratic optimal regulator method, provides a systematic way of computing the state feedback control gain matrix under the aforementioned conditions. The LQR method involves computing the feedback gain matrix $\mathbf{K}$ of the optimal control vector, $\mathbf{u} = -\mathbf{K}\mathbf{x}(t)$, given the state equation,

$$\dot{\mathbf{x}} = \mathbf{A}\mathbf{x} + \mathbf{B}\mathbf{u} \tag{8}$$

so as to minimize the quadratic performance index,

$$J = \int_0^\infty \left(\mathbf{x}^T \mathbf{Q} \mathbf{x} + \mathbf{u}^T \mathbf{R} \mathbf{u} \right) dt \tag{9}$$

where, $\mathbf{Q}$ is a positive-definite (or positive-semidefinite) Hermitian or real symmetric matrix, and $\mathbf{R}$ is a positive-definite Hermitian or real symmetric matrix. These weighting matrices determine the relative importance of the responses and the expenditure of control energy. The block diagram representation is shown in Fig. 2. The control gain matrix is obtained by solving the optimization problem and is

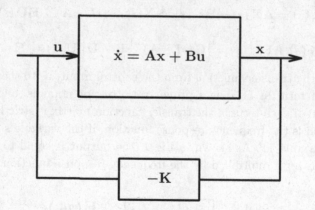

Fig. 2.   Block diagram for LQR method

given as[6],

$$\mathbf{K} = \mathbf{R}^{-1}\mathbf{B}^T\mathbf{P} \tag{10}$$

and the matrix $\mathbf{P}$ is the steady state solution (assuming an excitation of zero mean white noise) of the following simplified algebraic Riccatti matrix equation,

$$\mathbf{A}^T\mathbf{P} + \mathbf{PA} - \mathbf{PBR}^{-1}\mathbf{B}^T\mathbf{P} + \mathbf{Q} = 0 \tag{11}$$

The basic design steps for the LQR method involves the solution of the algebraic Riccatti matrix equation, Eq. 11 for $\mathbf{P}$. If a positive definite matrix $\mathbf{P}$ exists, the system is stable, or the matrix $\mathbf{A} - \mathbf{BK}$ is stable. The matrix $\mathbf{P}$ obtained is substituted in Eq. 10 to obtain the optimal feedback gain matrix. Eqs. 10 and 11 correspond to the continuous case; the discrete counterparts for these equations could be found in other references[6].

- **Example 1**: An idealized two-story building is shown in Fig. 3. The objective is to design a controller based on LQR method. The mass of the two floors are $m_1 = m_2 = 10,000$ kgs and the stiffnesses of the two floors are $k_1 = k_2 = 10,000$ kN/m. Let us assume that the damping coefficients are $c_1 = c_2 = 31.6$ kN-s/m, and the damping matrix is stiffness proportional. Assuming the structural motion is sufficiently small that the nonlin-

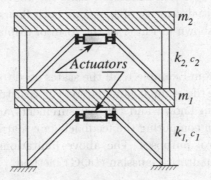

Fig. 3. Idealized two-story building model

ear effects can be neglected, and denoting the displacements relative to the ground by $\bar{\mathbf{x}} = [x_1\ x_2]^T$, the equations of motion for the structural system can be expressed as

$$\mathbf{M}\ddot{\bar{\mathbf{x}}} + \mathbf{C}\dot{\bar{\mathbf{x}}} + \mathbf{K}\bar{\mathbf{x}} = \mathbf{\Gamma u} \tag{12}$$

where $\mathbf{u}$ is a column vector which consists of the forces f_1 and f_2 exerted by the actuators located in the two storys; $\mathbf{\Gamma} = \mathbf{I}_{2\times2}$ is an identity matrix. The mass, damping and stiffness matrices are,

$$\mathbf{M} = \begin{bmatrix} m_1 & 0 \\ 0 & m_2 \end{bmatrix}, \mathbf{C} = \begin{bmatrix} c_1 + c_2 & -c_2 \\ -c_2 & c_2 \end{bmatrix}, \mathbf{K} = \begin{bmatrix} k_1 + k_2 & -k_2 \\ k_2 & k_2 \end{bmatrix} \tag{13}$$

Defining states, $\mathbf{x} = [\bar{\mathbf{x}}\ \dot{\bar{\mathbf{x}}}]^T$, Eq. 12 can be cast in state space form as,

$$\dot{\mathbf{x}}(t) = \mathbf{Ax}(t) + \mathbf{Bu}(t) \tag{14}$$

where,

$$A = \begin{bmatrix} 0 & I \\ -M^{-1}K & -M^{-1}C \end{bmatrix}, B = \begin{bmatrix} 0 \\ M^{-1}\Gamma \end{bmatrix} \tag{15}$$

If the performance function is chosen to be of the form in Eq. 9, with the following weighting matrices Q and R,

$$Q = \begin{bmatrix} 10^{10} & 0 & 0 & 0 \\ 0 & 10^{10} & 0 & 0 \\ 0 & 0 & 10^2 & 0 \\ 0 & 0 & 0 & 10^2 \end{bmatrix}, R = \begin{bmatrix} 10^{-3} & 0 \\ 0 & 10^{-3} \end{bmatrix}$$

The steady-state control that minimizes the cost is obtained as,

$$u(t) = -Kx(t) = -\begin{bmatrix} 4.5255 & 4.2434 & 0.5332 & 0.5317 \\ 4.2434 & 8.7689 & 0.5317 & 1.0649 \end{bmatrix} \times 10^5 x \; x(t) \tag{16}$$

The closed loop system is simulated using the above computed steady-state gain with an initial velocity of 1 m/s at both floors. Sample results of the simulation are shown in Fig. 4 in the form of displacement and force time histories.

As can be readily seen from Example 1, all the states need to be known at each time step in order to calculate the optimal gain matrix. For structural control applications all the states are seldom known and the measurements are often noisy. Lack of excitation information in computing the feedback gain matrix is another limitation for the structural control purposes. The above limitations are addressed, to an extent, in the Linear Quadratic Gaussian (LQG) method.

3.2 Optimal Estimation

Optimal estimation is needed for output feedback wherein unobserved states, or noisy state measurements, are estimated. Estimation of unknown states based on available measurements is accomplished with the aid of the Kalman Bucy filter. Optimal (in the sense of minimum-variance) estimates of the states are obtained for the case when the system is linear, cost function is quadratic, and the inputs and errors are gaussian. The linear optimal estimator minimizes the mean-square estimation error with respect to the choice of a filter gain matrix. The estimate of states is through a linear ordinary differential equation based on a system model with the actual residual measurement errors driving the state propagation through the optimal gain matrix. The covariance estimate is derived from linear ordinary differential equation driven by the statistics of the assumed measurement errors and disturbance inputs[6]. The dynamic system considered in Eq. 2 is repeated here where the excitation is a white, zero-mean gaussian random process, and the matrix,

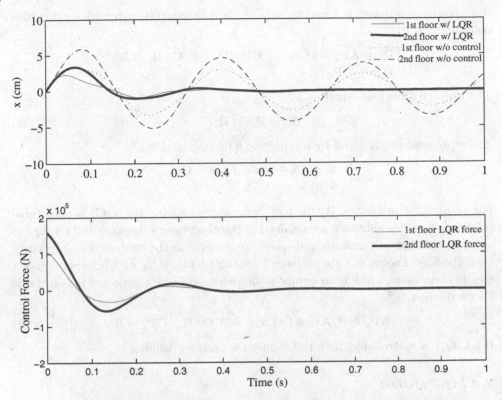

Fig. 4. Simulation results for LQR control - (a) Floor displacements with and without LQR control; (b) LQR force time history for the two floors

$\mathbf{D} = \mathbf{0}$.

$$
\begin{aligned}
\dot{\mathbf{x}}(t) &= \mathbf{A}\mathbf{x}(t) + \mathbf{B}\mathbf{u}(t) + \mathbf{E}\mathbf{w}(t) \\
\mathbf{y}(t) &= \mathbf{C}\mathbf{x}(t) + \mathbf{v}(t)
\end{aligned}
\tag{17}
$$

The known expected values (denoted by $E(\cdot)$) of the initial state and covariance assuming uncorrelated disturbance input and measurement are as follows:

$$
\begin{aligned}
E(\mathbf{x}(0)) &= \hat{\mathbf{x}}_0 \\
E([\mathbf{x}(0) - \hat{\mathbf{x}}_0][\mathbf{x}(0) - \hat{\mathbf{x}}_0]^T) &= \mathbf{P}_0
\end{aligned}
\tag{18}
$$

The disturbance input and measurement error are white, zero-mean gaussian processes with spectral density matrices $\mathbf{Q}_c$ and $\mathbf{R}_c$ defined as follows:

$$
\begin{aligned}
E(\mathbf{w}(t)) &= E(\mathbf{v}(t)) = 0 \\
E(\mathbf{w}(t)\mathbf{w}^T(t)) &= \mathbf{Q}_c(t)\delta(t - \tau) \\
E(\mathbf{v}(t)\mathbf{v}^T(t)) &= \mathbf{R}_c(t)\delta(t - \tau)
\end{aligned}
\tag{19}
$$

The linear estimator is optimal in the sense that the variance of the state estimate error is minimized on the average (expected value). Based on the optimization[6], the covariance estimate is found by solving the following estimator Riccatti matrix equation[6] for $\mathbf{P}_e$:

$$\dot{\mathbf{P}}_e = \mathbf{AP}_e + \mathbf{P}_e\mathbf{A}^T + \mathbf{EQ_cE}^T - \mathbf{P}_e\mathbf{C}^T\mathbf{R}_c^{-1}\mathbf{CP}_e$$
$$\mathbf{P}_e(0) = \mathbf{P}_0 \tag{20}$$

The optimal filter gain equation is[6],

$$\mathbf{K}_c = \mathbf{P}_e\mathbf{C}^T\mathbf{R}_c^{-1} \tag{21}$$

The state estimate is found by solving the following equation[6]:

$$\dot{\hat{\mathbf{x}}} = \mathbf{A}\hat{\mathbf{x}} + \mathbf{Bu} + \mathbf{K_c}[\mathbf{x} - \mathbf{C}\hat{\mathbf{x}}]$$
$$\hat{\mathbf{x}}(0) = \hat{\mathbf{x}}_0 \tag{22}$$

For stable filter estimates, $\mathbf{R}_c^{-1}$ should be positive definite. If $(\mathbf{A}, \mathbf{C})$ is detectable (system's unstable subspace is contained in the observable subspace) and $(\mathbf{A}, \mathbf{Q}_c)$ is stabilizable (system's unstable subspace is contained in the controllable subspace), then the filter known as Kalman Bucy Filter (KBF) is stable and $\mathbf{K}_c$ approaches a steady state value, which is an unique positive semi-definite solution of the algebraic Riccatti equation[6]

$$\mathbf{AP}_e + \mathbf{P}_e\mathbf{A}^T + \mathbf{EQ_cE}^T - \mathbf{P}_e\mathbf{C}^T\mathbf{R}_c^{-1}\mathbf{CP}_e = 0 \tag{23}$$

If $(\mathbf{A}, \mathbf{Q}_c)$ is controllable, then the solution is positive definite.

3.3 LQG Method

From the LQR method, it is seen that the optimal control for a linear system with a quadratic performance index is a linear feedback of the state variables. From the KBF, the estimates of the state variables can be obtained from noisy measurements of linear combinations of the state variables, using a filter that is a model of the system and a feedback signal proportional to the difference between the actual and estimated measurements. The LQG method involves the combination of the KBF optimal estimation filter and the optimal deterministic controller. This optimal feedback controller is combined in the ensemble average sense, for linear-quadratic problems with additive gaussian white noise. Consider the following state equations:

$$\dot{\mathbf{x}}(t) = \mathbf{Ax}(t) + \mathbf{Bu}(t) + \mathbf{Ew}(t)$$
$$\mathbf{y}(t) = \mathbf{Cx}(t) + \mathbf{v}(t) \tag{24}$$

For this system, using the separation principle[6], the controller is designed using the LQR method with the states estimated using the KBF. The estimated states are used for the controller design as though they are the exact states of the system. In other words, the LQG method involves the minimization of the quadratic performance index,

$$J = E[\int_0^\infty (\mathbf{x}^T\mathbf{Qx} + \mathbf{u}^T\mathbf{Ru}) \, dt] \tag{25}$$

under the constraint imposed by the equations of motion. The solution to this problem is[6]:

$$\mathbf{u} = -\mathbf{K}\hat{\mathbf{x}}(t)$$
$$\dot{\hat{\mathbf{x}}} = \mathbf{A}\hat{\mathbf{x}} + \mathbf{B}\mathbf{u} + \mathbf{K}_c[\mathbf{x} - \mathbf{C}\hat{\mathbf{x}}]$$

(26)

If $(\mathbf{A}, \mathbf{B})$ is stabilizable, $(\mathbf{A}, \mathbf{Q})$ is detectable, $(\mathbf{A}, \mathbf{C})$ is detectable, and $(\mathbf{A}, \mathbf{Q}_c)$ is stabilizable, then the closed loop system using the LQG control is stable. The order of the resulting controller is the same as the order of the plant. The control and estimation parts are derived separately and then combined in the LQG implementation as they possess the certainty-equivalence property[6].

- **Example 2**: The structural system considered in this example is the same as in Example 1. The objective in this example is to show the performance of the LQG controller using the measurements of all states with additive Gaussian white noise and with an external excitation being uniformly distributed white noise. In order to design the controller, the estimator gains are computed first as shown in Eqs. 21 and 23. The estimator gains are computed assuming that the input excitation and the measurement error are zero mean Gaussian random processes with variances 1 and 0.01 respectively. The gains of the estimator are given by,

$$\mathbf{K}_c = \begin{bmatrix} 0.0094 & 0.0147 & 0.003 & 0.0172 \\ 0.0147 & 0.0237 & -0.0167 & 0.0005 \\ 0.0003 & -0.0167 & 4.1573 & 5.5093 \\ 0.0172 & 0.0005 & 5.5093 & 8.9965 \end{bmatrix}$$

(27)

The actual and estimated states computed using the estimator gains in Eq. 27 are shown in Fig. 5. The controller gains, $\mathbf{K}$ computed in Eq. (16) are used together with the estimator gains in Eq. 27 to design the LQG controller as shown in Eq. 26. The results of the simulation with and without the LQG controller for the structure under uniform white noise external disturbance and a Gaussian distributed additive noise of variance 0.01 are shown in Fig. 6.

From Fig. 6, it is clear that the LQG control performs well under the broad-band excitations considered in this example. However, the performance of the LQG controller for structural control applications under narrow-band earthquake excitations is only marginal. The performance of the LQG controller can be improved by augmenting the state equations using the frequency information of the earthquakes as shown in the following section.

4 Control in the Frequency Domain

The time-domain control techniques introduced in section 3 involve the minimization of states and control input in the form of a quadratic cost function subject

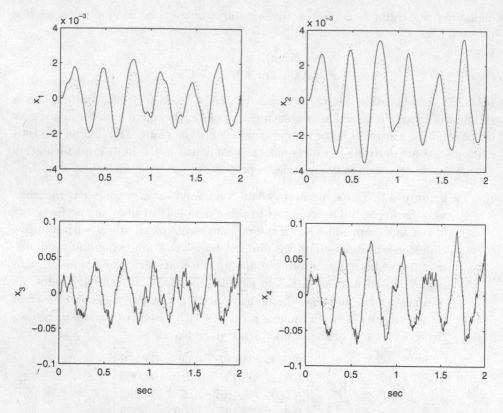

Fig. 5.   Actual states (solid lines) and their respective estimates (dotted lines)

to dynamic constraints. Earlier, we had introduced the concept of a transfer function in Eq. 7 that is obtained by taking a Laplace transform of the time-domain equations. If the variable s in the transfer function is replaced by $j\omega$, where, ω is the circular frequency, the transfer function can be regarded as the frequency response function. Transfer functions and frequency response functions determine the input-output relationships in the frequency domain and the mapping from the time domain to the frequency domain can be achieved through the transformation techniques described earlier. These transformations are applicable to linear systems only and are not valid for nonlinear systems. From the structural control perspective, this restriction is not severe, as most of the structures are designed to operate in the linear range and the localized nonlinearities for many structural systems can be linearized through a variety of techniques available in the literature. Frequency domain control techniques are particulary advantageous for structural control, as the frequency information from the disturbances such as earthquakes and wind can be incorporated in the control design formulation. The main objective of the frequency domain control is a minimization of a norm of a transfer function between the input and output as described in the following sections.

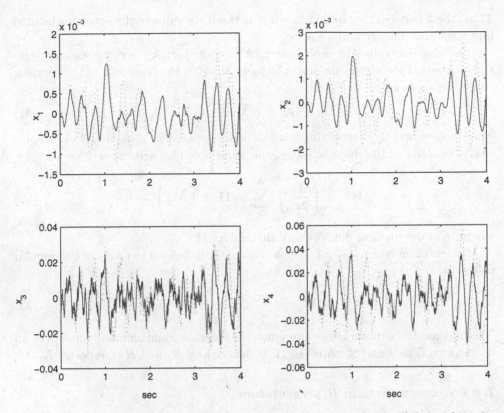

Fig. 6. States with LQG control (solid lines) and their respective uncontrolled time histories (dotted lines)

4.1 H_2 and H_∞ Norms

For a stationary random process, the power spectral density[7] $\mathbf{S}_z$, of the output $\mathbf{z}$, from a frequency response function $\mathbf{H}$, subjected to an input $\mathbf{d}$, of power spectral density, $\mathbf{S}_d(\omega)$, is given by

$$\mathbf{S}_z(\omega) = \mathbf{H}(j\omega)\mathbf{S}_d(\omega)\mathbf{H}^*(j\omega) \tag{28}$$

where, the $*$ indicates the conjugate transpose of $\mathbf{H}$. The root mean square (RMS) value of the output, $\mathbf{z}$ is,

$$\|\mathbf{z}\|_{rms} = \left(\frac{1}{2\pi} \int\limits_{-\infty}^{\infty} trace[\mathbf{H}(j\omega)\,\mathbf{S}_d(\omega)\,\mathbf{H}^*(j\omega)\,d\omega \right)^{1/2} \tag{29}$$

For the case when the input $\mathbf{d}$ is a unit intensity white noise signal, the H_2 norm of the transfer function is defined as,

$$\|\mathbf{z}\|_{rms} = \left(\frac{1}{2\pi} \int\limits_{-\infty}^{\infty} trace[\mathbf{H}(j\omega)\,\mathbf{H}^*(j\omega)\,d\omega \right)^{1/2} \tag{30}$$

Thus, the 2 norm of the transfer function is the RMS value of the output, when the input is a unit intensity white noise.

The singular values of any matrix, $\mathbf{A}$, denoted $\sigma_i[\mathbf{A}]$, are the non-negative square-roots of the eigen values of $\mathbf{A}^*\mathbf{A}$, where, $\mathbf{A}^*$ is the transpose of the complex conjugate of $\mathbf{A}$, given by

$$\sigma_i[\mathbf{A}] = \sqrt{\lambda_i(\mathbf{A}^*\mathbf{A})} \tag{31}$$

The smallest and the largest singular values are denoted by $\underline{\sigma}[\mathbf{A}]$ and $\overline{\sigma}[\mathbf{A}]$ respectively. In terms of the singular values, the 2 norm can be written as

$$\|\mathbf{H}\|_2 = \left(\frac{1}{2\pi}\int_{-\infty}^{\infty}\sum_i^n \sigma_i[\mathbf{H}(j\omega)]^2 d\omega\right)^{1/2} \tag{32}$$

where, n is the smallest dimension of the matrix $\mathbf{H}$.

The ∞ norm of a transfer function matrix, $\mathbf{H}$ is defined in terms of its singular values as,

$$\|\mathbf{H}\|_\infty = \sup_\omega\left(\overline{\sigma}[\mathbf{H}(j\omega)]\right) \tag{33}$$

This means the ∞ norm is the supremum of the maximum singular value over all frequencies. The 2 and ∞ norms of $\mathbf{H}$, is denoted by H_2 and H_∞ respectively.

4.2 Frequency Domain Representation

The basic block diagram used for representing the architecture in the frequency domain control is shown in Fig. 7. The generalized plant is represented by $\mathbf{G}$ and

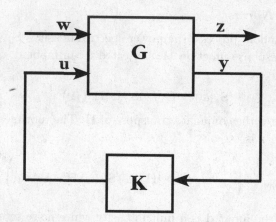

Fig. 7.   Basic block diagram for frequency domain representation

the controller by $\mathbf{K}$. The measurement outputs are represented by $\mathbf{y}$, the outputs to be regulated by $\mathbf{z}$ (may or may not be measured), external disturbance by $\mathbf{w}$, which includes the earthquake excitation and sensor noise, and the control input is represented by $\mathbf{u}$. Frequency domain representation enables the frequency information of the excitation and the regulated variables to be included in the system representation. In order to accomplish this within the framework of the standard block diagram representation, the frequency weighting functions are augmented in the generalized plant. A more detailed description of the procedure is introduced in the following sections. For all the discussions to follow, unless otherwise noted, the system $\mathbf{G}$ is assumed to be Linear Time Invariant (LTI).

In order to explain the main idea of frequency domain control methods, the partitioned form of the transfer function of the plant shown in Fig. 7 is,

$$\begin{pmatrix} \mathbf{G}_{zw} & \mathbf{G}_{zu} \\ \mathbf{G}_{yw} & \mathbf{G}_{yu} \end{pmatrix} \tag{34}$$

The subscripts in the transfer function components of the partitioned matrix denote the input-output pairs. For example, $\mathbf{G}_{zu}$ denotes the transfer function between the control input $\mathbf{u}$ and the regulated output $\mathbf{z}$. By a simple rearrangement of the input-output equations, we obtain the transfer function for the disturbance input, $\mathbf{w}$ to the regulated outputs, $\mathbf{z}$ as:

$$\mathbf{H}_{zw} = \mathbf{G}_{zw} + \mathbf{G}_{zu}\mathbf{K}(\mathbf{I} - \mathbf{G}_{yu}\mathbf{K})^{-1}\mathbf{G}_{yw} \tag{35}$$

The central idea behind the frequency domain control methods is to minimize the norm of $\mathbf{H}_{zw}$. Depending upon whether the 2-norm or the ∞ norm that is minimized, the method is named accordingly as H_2 or H_∞.

For the purposes of structural control, frequency dependent weighting functions are introduced in order to design a controller that is effective for the range of frequencies in the excitation, and the frequency of responses of interest. In order to accomplish this within the frame work of the standard block diagram shown in Fig. 7, the weighting functions are appended to the plant system. The resulting plant is typically of a higher order than its original. However, the larger order is usually not a serious limitation as structural systems may be reduced using model reduction techniques. A schematic representation of the augmentation is shown in Fig. 8. The weighting functions are represented by $\mathbf{W}1$ and $\mathbf{W}2$. $\mathbf{W}1$ is a filter whose output represents the excitation of interest. This filter is designed to simulate the frequency characteristics of the excitation (for example, earthquake) and $\mathbf{W}2$ weights the structural responses at the frequencies of interest to be regulated. The resulting augmented system is represented by $\mathbf{G}_a(s)$ and contains both the weighting functions and the plant, and replaces the plant system, $\mathbf{G}(s)$ in Fig. 7. The weighting procedures are described in detail in Example 3.

4.3 Equivalence of LQG and H_2 Optimal Control

The steady-state LQG control is equivalent to an H_2 optimization problem as it involves finding a feedback controller that internally stabilizes the closed-loop system

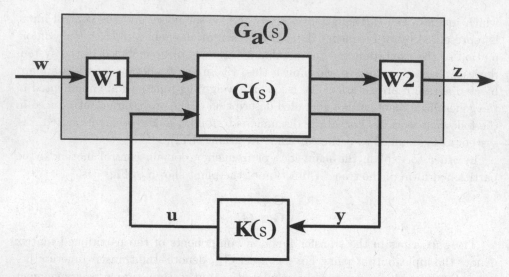

Fig. 8.   Augmented System

and minimizes the 2-norm as shown in Eq. 30. Now, let us consider the quadratic performance function for LQG control given by[8],

$$J = E\left[\mathbf{x}^T\left(\infty\right)\mathbf{Q}\mathbf{x}\left(\infty\right) + \mathbf{u}(\infty)^T\mathbf{R}\mathbf{u}\left(\infty\right)\right] \qquad (36)$$

which can be written as,

$$J = E\left\{\left[\mathbf{Q}^{1/2}\mathbf{x}\left(\infty\right)^T \ \mathbf{R}^{1/2}\mathbf{u}\left(\infty\right)^T\right]\begin{bmatrix}\mathbf{Q}^{1/2}\mathbf{x}\left(\infty\right)\\\mathbf{R}^{1/2}\mathbf{u}\left(\infty\right)\end{bmatrix}\right\} \qquad (37)$$

In the above equation, $\mathbf{x}$ and $\mathbf{u}$ denote the states and control inputs respectively, and ∞ represents the steady-state condition. Minimizing Eq. 37 is equivalent to a 2-norm minimization and hence, LQG and H_2 optimization can be regarded as equivalent under the aforementioned conditions.

5 H_∞ Optimal Control

The discussion so far regarding the H_2 and H_∞ controllers have been primarily in the frequency domain, where transfer function matrices have been presented to describe the system and the controller. However, in order to compute the H_∞ controller, the dynamical system in Eq. 2 is cast in the state space form. The state space equations written in the standard form[9] as follows:

$$\begin{aligned}\dot{\mathbf{x}} &= \mathbf{A}\mathbf{x} + \mathbf{B_1}\mathbf{w} + \mathbf{B_2}\mathbf{u}\\ \mathbf{z} &= \mathbf{C_1}\mathbf{x} + \mathbf{D_{11}}\mathbf{w} + \mathbf{D_{12}}\mathbf{u}\\ \mathbf{y} &= \mathbf{C_2}\mathbf{x} + \mathbf{D_{21}}\mathbf{w} + \mathbf{D_{22}}\mathbf{u}\end{aligned} \qquad (38)$$

In Eq. 38, $\mathbf{D_{11}}, \mathbf{D_{12}}, \mathbf{D_{21}}, \mathbf{D_{22}}, \mathbf{C_1}$ and $\mathbf{C_2}$ are mapping matrices of appropriate dimensions. The basic block diagram[9] is shown in Fig. 7. The generalized plant is represented by $\mathbf{G}$ and the controller by $\mathbf{K}$. The measurement outputs are represented by $\mathbf{y}$, the outputs to be regulated by $\mathbf{z}$, external disturbance by $\mathbf{w}$, which includes the excitation and sensor noise and the control input is represented by $\mathbf{u}$. The purpose of the H_∞ control method is to minimize the ∞-norm of the transfer function from input $\mathbf{w}$ to regulated output $\mathbf{z}$, $\mathbf{G}_{zw}$ and is written as

$$\|\mathbf{G_{zw}}(s)\|_\infty = \sup_\omega \left[\bar{\sigma}\left(\mathbf{G_{zw}}(s)\right)\right] \le \gamma \tag{39}$$

$\bar{\sigma}$ is the largest singular value of the transfer function, *sup* denotes the supremum and γ is a positive bound for the norm. The solution for the controller for the generalized regulator problem[10,11,12] is given by

$$u = -\mathbf{F}_\infty \hat{\mathbf{x}} \tag{40}$$

and the state estimator is given by

$$\dot{\hat{\mathbf{x}}} = \mathbf{A}\hat{\mathbf{x}} + \mathbf{B_2}u + \mathbf{B_1}\hat{\mathbf{w}} + \mathbf{J}_\infty\mathbf{L}_\infty(\mathbf{y} - \hat{\mathbf{y}}) \tag{41}$$

where,

$$\hat{\mathbf{w}} = \gamma^{-2}\mathbf{B_1^T}\mathbf{K}_\infty\hat{\mathbf{x}}$$

and

$$\hat{\mathbf{y}} = \gamma^{-2}\mathbf{D_{21}}\mathbf{B_1}^T\mathbf{K}_\infty\hat{\mathbf{x}} + \mathbf{C_2}\hat{\mathbf{x}}$$

The term, $\hat{\mathbf{w}}$ and $\hat{\mathbf{y}}$ are the estimates of the worst case disturbance and output of the estimator. There exists a stabilizing controller if and only if there exists positive semi-definite solutions to the two Riccatti equations for $\mathbf{K}_\infty$ and $\mathbf{N}_\infty$ and the condition

$$\rho(\mathbf{K}_\infty\mathbf{N}_\infty) < \gamma^2 \tag{42}$$

where $\rho(\mathbf{A})$ is the spectral radius of $\mathbf{A}$ which is defined as the largest singular value of $\mathbf{A}$. The controller written in the packed matrix notation is

$$\mathbf{K}_{sub}(s) = \begin{bmatrix} \hat{\mathbf{A}}_\infty & \mathbf{J}_\infty\mathbf{L}_\infty \\ -\mathbf{F}_\infty & 0 \end{bmatrix} \tag{43}$$

where,

$$\mathbf{F}_\infty = (\mathbf{D_{12}}^T\mathbf{D_{12}})^{-1}(\mathbf{B_2^T}\mathbf{K}_\infty + \mathbf{D_{12}}^T\mathbf{C_1})$$

$$\mathbf{L}_\infty = (\mathbf{N}_\infty\mathbf{C_2^T} + \mathbf{B_1}\mathbf{D_{21}^T})(\mathbf{D_{21}}\mathbf{D_{12}}^T)^{-1}$$

and

$$\mathbf{J}_\infty = (\mathbf{I} - \gamma^{-2}\mathbf{N}_\infty\mathbf{K}_\infty)^{-1}$$

The terms, $\mathbf{K}_\infty$ and $\mathbf{N}_\infty$ are the solutions to the controller and estimator Riccatti equations given by

$$\mathbf{K}_\infty = Ric\begin{pmatrix} \mathbf{A} - \mathbf{B_2}\tilde{\mathbf{D}}_{12}\mathbf{D}_{12}^T\mathbf{C_1} & \gamma^{-2}\mathbf{B_1}\mathbf{B_1^T} - \mathbf{B_2}\tilde{\mathbf{D}}_{12}\mathbf{B_2^T} \\ -\tilde{\mathbf{C}}_1^T\tilde{\mathbf{C}}_1 & -\left(\mathbf{A} - \mathbf{B_2}\tilde{\mathbf{D}}_{12}\mathbf{D}_{12}^T\mathbf{C_1}\right)^T \end{pmatrix} \tag{44}$$

$$N_\infty = Ric \left(\begin{matrix} (A - B_1 D_{21} \tilde{D}_{21} C_2)^T & \gamma^{-2} C_1^T B_1 - C_2^T \tilde{D}_{21} C_2 \\ -\tilde{B}_1 \tilde{B}_1^T & -\left(A - B_1 \tilde{D}_{21}^T \tilde{D}_{21} C_2\right) \end{matrix} \right) \tag{45}$$

where

$$\tilde{C}_1 = \left(I - D_{12} \tilde{D}_{12} D_{12}^T\right) C_1$$

$$\tilde{B}_1 = B_1 \left(I - D_{21}^T \tilde{D}_{21} D_{21}\right)$$

$$\tilde{D}_{12} = \left(D_{12}^T D_{12}\right)^{-1}; \tilde{D}_{21} = \left(D_{21} D_{21}^T\right)^{-1}$$

The computations involving the controller and estimator gains are performed using MATLAB[13] in Example 3.

- **Example 3**: The objective of this example to illustrate the frequency domain augmentation techniques and design of frequency domain controllers for a simple single degree of freedom (SDOF) system subject to earthquake excitation. This SDOF system can be thought of as an idealized base isolated structure[3]. Frequency dependent weighting matrices are chosen for the control design incorporating the outputs and input characterizations. Four types of control designs using both H_2 and H_∞ methods are considered (i) No weighting filters, (ii) Output weighting filter only, (iii) Input excitation filter only, and (iv) both output and input excitation filters. Comparison of the responses for all cases in the frequency domain and some general observations regarding the choice of weighting functions are made. The method of augmenting the system equations with the weighting functions is also presented for each case. The SDOF system chosen for this example has the following system properties:

$$A = \begin{bmatrix} 0 & 1 \\ -6.317 & -0.0503 \end{bmatrix}$$
$$B = \begin{bmatrix} 0 & 1 \end{bmatrix}^T = -E \tag{46}$$

The control objective is to minimize the displacements, which is one of the states of the system and the control energy input. The measurement consists of the noisy measurement of the velocity at all times. The plant with the weighting filters $W1$ and $W2$ is shown in Fig. 9. The vector $\mathbf{w}$ consists of the earthquake excitation vector and the measurement noise,

$$\mathbf{w} = \begin{bmatrix} w & v \end{bmatrix}^T.$$

and the regulated quantities are

$$\mathbf{z} = \begin{bmatrix} z_1 & z_2 \end{bmatrix}^T$$

which are the base displacement and the control input respectively.

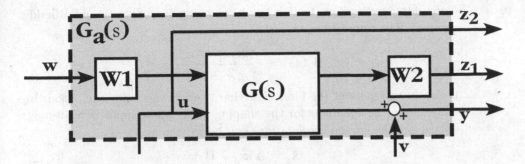

Fig. 9.   General augmented system used in the Example 3

Augmentation with W1

In order to better inform the controller about the frequency content of the ground motion, a input shaping filter is incorporated into the system. Fig. 10 shows the magnitude of the filter as well as the frequency content of

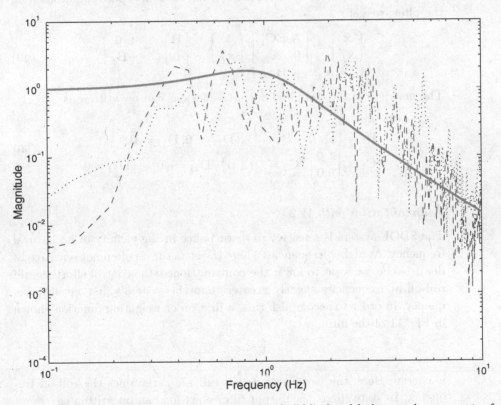

Fig. 10.   Frequency content of input excitation filter (solid line) and fault-normal components of Northridge earthquake (Sylmar-dashed line and Rinaldi-dotted line)

Northridge earthquake (fault-normal components of Sylmar and Rinaldi records). The transfer function of the form,

$$W1(s) = \frac{2\varsigma_g \omega_g s + \omega_g^2}{s^2 + 2\varsigma_g \omega_g s + \omega_g^2}$$

is chosen to represent the filter that characterizes the input excitation. In state space, the equations for the shaping filter, whose input is white and output are the ground accelerations can be written as:

$$\begin{aligned}
\dot{\mathbf{x}}_f &= \mathbf{A}_f \mathbf{x}_f + \mathbf{B}_f w \\
\ddot{u}_g &= \mathbf{C}_f \mathbf{x}_f
\end{aligned} \tag{47}$$

where,

$$\begin{aligned}
\mathbf{A}_f &= \begin{bmatrix} 0 & 1 \\ -\omega_g^2 & -2\varsigma_g \omega_g \end{bmatrix} \\
\mathbf{B}_f &= \begin{bmatrix} 0 & 1 \end{bmatrix}^T \\
\mathbf{C}_f &= \begin{bmatrix} -\omega_g^2 & -2\varsigma_g \omega_g \end{bmatrix}
\end{aligned} \tag{48}$$

Here, $\omega_g = 2\pi$ rad/s and $\varsigma_g = 0.6$. Augmenting the state space equations with the filter, we get

$$\begin{Bmatrix} \dot{\mathbf{x}} \\ \dot{\mathbf{x}}_f \end{Bmatrix} = \begin{bmatrix} \mathbf{A} & \mathbf{EC}_f \\ 0 & \mathbf{A}_f \end{bmatrix} \begin{Bmatrix} \mathbf{x} \\ \mathbf{x}_f \end{Bmatrix} + \begin{bmatrix} \mathbf{B} \\ 0 \end{bmatrix} u + \begin{bmatrix} 0 \\ \mathbf{B}_f \end{bmatrix} w$$
$$\dot{\mathbf{x}}_a = \mathbf{A}_a \mathbf{x}_a + \mathbf{B}_a u + \mathbf{E}_a w \tag{49}$$

The matrices in the state and output equations can be written as

$$\begin{aligned}
\mathbf{A} &= \mathbf{A}_a; \\
\mathbf{B}_1 &= \begin{bmatrix} \mathbf{E}_a & 0 \end{bmatrix}; \mathbf{B}_2 = \mathbf{B}_a; \mathbf{D}_{11} = 0; \mathbf{D}_{12} = \mathbf{B}; \\
\mathbf{C}_1 &= \begin{bmatrix} 1 & 0 \\ 0 & 0 \end{bmatrix}; \mathbf{C}_2 = \begin{bmatrix} 0 & 1 & 0 \end{bmatrix}; \mathbf{D}_{21} = \begin{bmatrix} 0 & 1 \end{bmatrix}; \mathbf{D}_{22} = 0.
\end{aligned} \tag{50}$$

Augmentation with W2

The SDOF system is sensitive to disturbance in the vicinity of its natural frequency. At higher frequencies where the structure is often not sensitive to disturbance, we want to lower the control. Hence, the control effort should roll-off at frequencies slightly greater than the system's first natural frequency. In order to accomplish this, a first order weighting function shown in Fig. 11 of the form,

$$W2 = \frac{a}{s + a}$$

is chosen. Here, the parameter $a = 3.0$ rad/sec determines the roll-off frequency. In state space, the output filter equation can be written as,

$$\dot{\mathbf{x}}_o = \mathbf{A}_o \mathbf{x}_o + \mathbf{B}_o \mathbf{x} \tag{51}$$

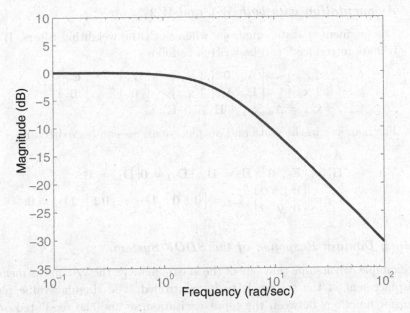

Fig. 11.   Weighting function $W2$

$$y_o = C_o x_o \qquad (52)$$

Where, x_o are the states of the filter, y_o are the outputs, and,

$$A_o = \begin{bmatrix} -3 & 0 \\ 0 & -3 \end{bmatrix}, \ B_o = \begin{bmatrix} 2 & 0 \\ 0 & 2 \end{bmatrix}, \text{ and } C_o = \begin{bmatrix} 1.5 & 0 \end{bmatrix} \qquad (53)$$

As with the input excitation filter, the plant is augmented with the filter as follows:

$$\begin{Bmatrix} \dot{x} \\ \dot{x}_o \end{Bmatrix} = \begin{bmatrix} A & 0 \\ B_o & A_o \end{bmatrix} \begin{Bmatrix} x \\ x_o \end{Bmatrix} + \begin{bmatrix} B \\ 0 \end{bmatrix} u + \begin{bmatrix} E \\ 0 \end{bmatrix} w$$
$$\dot{\tilde{x}}_a = \tilde{A}_a \tilde{x}_a + \tilde{B}_a u + \tilde{E}_a w \qquad (54)$$

The matrices in the state and output equations can be written as

$$
\begin{aligned}
&A = \tilde{A}_a; \\
&B_1 = \begin{bmatrix} \tilde{E}_a & 0 \end{bmatrix}; B_2 = \tilde{B}_a; D_{11} = 0; D_{12} = B; \\
&C_1 = \begin{bmatrix} 0 & 1.5 & 0 \\ 0 & 0 & 0 \end{bmatrix}; C_2 = \begin{bmatrix} 0 & 1 & 0 \end{bmatrix}; D_{21} = \begin{bmatrix} 0 & 1 \end{bmatrix}; D_{22} = 0.
\end{aligned} \qquad (55)
$$

Augmentation with both $W1$ and $W2$

The augmented state equations when both the weighting filters, $W1$ and $W2$ are introduced, can be written as follows:

$$\begin{Bmatrix} \dot{x}_a \\ \dot{x}_o \end{Bmatrix} = \begin{bmatrix} \mathbf{A}_a & \mathbf{0} \\ \mathbf{B}_o & \mathbf{A}_o \end{bmatrix} \begin{Bmatrix} x_a \\ x_o \end{Bmatrix} + \begin{bmatrix} \mathbf{B_a} \\ \mathbf{0} \end{bmatrix} u + \begin{bmatrix} \mathbf{E_a} \\ \mathbf{0} \end{bmatrix} w$$

$$\dot{x}_{aa} = \mathbf{A}_{aa}x_{aa} + \mathbf{B}_{aa}u + \mathbf{E}_{aa}w \tag{56}$$

The matrices in the state and output equations can be written as

$$\begin{aligned} \mathbf{A} &= \mathbf{A}_{aa}; \\ \mathbf{B}_1 &= \begin{bmatrix} \mathbf{E}_{aa} & \mathbf{0} \end{bmatrix}; \mathbf{B}_2 = \mathbf{B}_{aa}; \mathbf{D}_{11} = 0; \mathbf{D}_{12} = \mathbf{B}; \\ \mathbf{C}_1 &= \begin{bmatrix} 0 & 1.5 & 0 \\ 0 & 0 & 0 \end{bmatrix}; \mathbf{C}_2 = \begin{bmatrix} 0 & 1 & 0 \end{bmatrix}; \mathbf{D}_{21} = \begin{bmatrix} 0 & 1 \end{bmatrix}; \mathbf{D}_{22} = 0. \end{aligned} \tag{57}$$

Frequency Domain Response of the SDOF System

In this example, we assume that one of the states, namely, the velocity is measured. The displacement of the system is to be controlled. The singular value plots of the transfer functions between the input excitation, w and the regulated output, namely, the displacement, $z1$ is shown in Fig. 12 and Fig. 13 for both H_2 and H_∞ controls. The closed loop transfer function for both cases, namely, H_2 and H_∞ have been generated using MATLAB[13].

Fig. 12 shows the singular value plots for the transfer functions for the case of H_2 control in all the four cases; namely, without weighting filters, with output weighting filter, $W2$ only, with input excitation filter, $W1$ only, and both $W1$ and $W2$. From Fig. 12, we can see that the case with both input and output filters minimizes the response at higher frequencies. However, the responses corresponding to the peak are not minimized. In comparison, from Fig. 13 we can see that the H_∞ control minimizes the peaks for all cases. As with the H_2 case, the presence of both $W1$ and $W2$ leads to better response reductions; however, in the case of H_∞, the response reductions occur at all frequencies. It is clear from these figures that the H_∞ control is more effective in suppressing the response peaks of the systems compared to the H_2 control. In other words, H_2 control minimizes the responses in an average sense and H_∞ control minimizes the worst case responses. This behavior is very important for the case of structures whose responses are dominated by their fundamental mode, as filters can be designed specifically taking this effect into consideration.

6 A Brief note on Robustness of H_2 and H_∞ Methods

No discussion on the H_2 and H_∞ methods is complete without reference on the robustness of these methods to model uncertainties. A controller that functions adequately for all admissible perturbations is termed robust. Robustness can be defined in terms of stability or performance. A control system is said to be robustly stable if it is stable for all admissible perturbations. A control system is said to

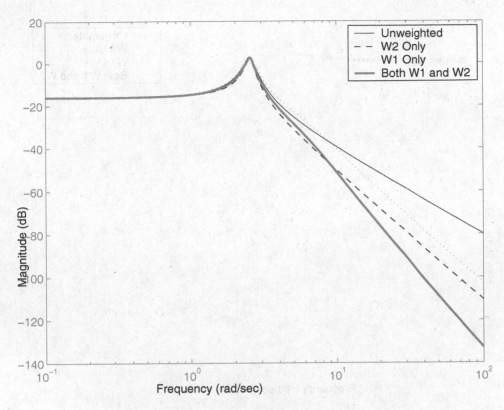

Fig. 12.   Magnitude of the transfer function, H_{z1w} for the case of H_2 control

perform robustly if it satisfies the performance specifications for all admissible perturbations. The stability of feedback systems is determined in terms of gain and phase margins for gain and phase perturbations. In the field of optimal control, two types of uncertainties are considered in the control design: (i) Structured uncertainty where there is information available about the uncertainty, which will restrict the uncertainty to a section of a model process; (ii) unstructured uncertainty where no information about the uncertainty is known except the upper bound of its magnitude. There has been significant research conducted in the areas of structured and unstructured uncertainties[8]. Unstructured uncertainty is modeled by connecting an unknown but bounded perturbation to the plant. The unstructured uncertainty is analyzed by placing them within a common framework discussed in the earlier sections. The system so formed now will have three inputs and three outputs. Combining the nominal plant, $\mathbf{G}(s)$ with the feedback, $\mathbf{K}(s)$, results in a system consisting of a nominal closed loop system, $\mathbf{N}(s)$, with the perturbation, $\Lambda(s)$, in a feedback loop as shown in Fig. 14.

The above feedback system, for the bounded unstructured uncertainty, $\|\Delta\| \leq 1$, is internally stable for all possible perturbations provided the nominal closed loop

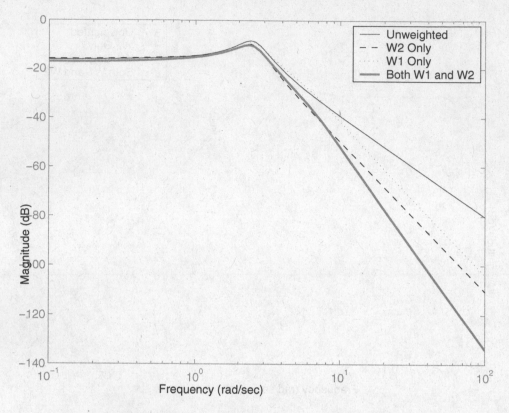

Fig. 13.   Magnitude of the transfer function, H_{z1w} for the case of H_∞ control

system is stable and

$$\|\mathbf{N}_{z_d w_d}\|_\infty = \overset{\text{sup}}{\omega} \{\bar{\sigma}\left[\mathbf{N}_{z_d w_d}\left(j\omega\right)\right]\} < 1 \qquad (58)$$

This is called the small-gain theorem and it used to test for robust stability with respect to bounded perturbations. Eq. 58 is a necessary and sufficient condition for internal stability with respect to unstructured uncertainty.

Structured uncertainty arises when a plant is subjected to multiple uncertainties such as a number of uncertain parameters or multiple unstructured uncertainties. For this case, the structured uncertainty can be written in the block diagonal transfer function form:

$$\mathbf{\Delta}\left(s\right) = \begin{bmatrix} \mathbf{\Delta}_1\left(s\right) & 0 & 0 & 0 \\ 0 & \mathbf{\Delta}_2\left(s\right) & \cdot & \cdot \\ 0 & \cdot & \mathbf{\Delta}_3\left(s\right) & \cdot \\ 0 & \cdot & \cdot & \mathbf{\Delta}_n\left(s\right) \end{bmatrix} \qquad (59)$$

where, n is the number of uncertainties and $\mathbf{\Delta}\left(s\right)$ represents the individual uncertainties applied to the plant. In the standard block diagram notation, the structured

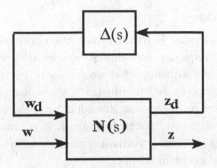

Fig. 14.   Unstructured uncertainty model for robustness

uncertainty, $\Delta(s)$ is represented in the same way as in Fig. 14. The uncertainty is scaled so that their infinity norms

$$\|\Delta_1\|_\infty \leq 1; \|\Delta_2\|_\infty \leq 1; \dots\dots; \|\Delta_n\|_\infty \leq 1 \Rightarrow \|\Delta\|_\infty \leq 1. \tag{60}$$

The general feedback system given in Fig. 14 is stable for all possible perturbations

$$\Delta(j\omega) \in \bar{\Delta}$$

and

$$\|\Delta(j\omega)\|_\infty \leq 1,$$

if and only if the nominal closed loop system is internally stable and

$$\sup_\omega \{\mu_{\bar{\Delta}} [N_{z_d w_d} (j\omega)]\} < 1 \tag{61}$$

where, $\mu_{\bar{\Delta}}$ is called the structured singular value and given by

$$\mu_{\bar{\Delta}} (N) = \frac{1}{\min_{\Delta \in \bar{\Delta}} [\bar{\sigma} (\Delta) | \quad \det (I + N\Delta) = 0]} \tag{62}$$

$$\mu_{\bar{\Delta}} (N) = 0 \quad \text{if} \quad \det(I + N\Delta) \neq 0 \quad \forall \quad \Delta \in \bar{\Delta}$$

The determination of robust stability is dependent on the computation of the structured singular value and can be impractical for a large number of cases. Hence, bounds on the structured singular values are generated and they provide good estimates of the structured singular value. In the case of performance robustness, the robust performance problem can be converted into an equivalent robust stability problem by appending an uncertainty block to the system in Fig. 14. The system meets the performance robustness objectives if and only if the new augmented system is robustly stable. Detailed description of robustness is beyond the scope of this chapter, and the readers are referred to books on robust optimal control[8,10,11] for a comprehensive discussion.

7 Concluding Remarks

The main idea of this chapter is to introduce the concepts of optimal structural control in the time and frequency domains. Augmentation techniques for structural control design where the frequency characteristics of excitations are introduced. Numerical examples are presented to illustrate the salient features of the control design. This chapter, by no means, is intended to provide an exhaustive review of the field of optimal structural control. Instead, it is aimed at providing a brief introduction to optimal structural control whose roots are strongly embedded in optimization and modern control theory.

8 Acknowledgments

The authors would like to acknowledge partial funding for this work by National Science Foundation, NSF CAREER GRANT-CMS 9996290.

References

1. B. F. Spencer and S. Nagarajaiah, "State of the art of structural control." *J. of Struc. Eng.*, ASCE, 129(7), 845-856 (2003).
2. F. Jabbari, W. E Schmitendorf and J. N. Yang, "H_∞ control for seismic excited buildings with acceleration feedback" *J. Eng. Mech.*, ASCE, 121(9), 994-1002 (1995).
3. S. Narasimhan, *Control of smart base isolated buildings with new semiactive devices and novel H_2/LQG, H_∞ and time-frequency controllers*. PhD thesis, Rice University (2004).
4. B. F. Spencer, J. Suhardjo and M. K. Sain, "Frequency domain optimal control strategies for aseismic protection". *ASCE Journal of Engineering Mechanics*, Vol. 120(1), 135-158 (1994).
5. T. T. Soong, *Active structural control: Theory and practice*. Longman Scientific and Technical, Essex, England(1990).
6. R. Stengel, *Optimal control and estimation*. Dover Publications, New York (1986).
7. J. S. Bendat and A. G. Piersol *Random Data: Analysis & Measurement Procedures*. Wiley-Interscience, 3rd edition (2000).
8. B. J. Burl, *Linear optimal control*. Addison Wesley Longman, Inc. (1999).
9. J.C. Doyle, K. Glover, P. P. Khargonekar and B. A. Francis, "State-space solutions to standard H_2 and H_∞ control problems" *IEEE Transactions on Automatic Control*, 34(8), 831-847 (1989).
10. B. A. Francis, *A course in H_∞ theory*. Springer-Verlag (1987).
11. M. Green and D. J. N Limebeer, *Linear robust control*. Prentice hall (1995).
12. M. G. Safanov and D. J. N Limebeer, "Simplifying the H_∞ theory via loop shifting". In *Proc., of the 27th IEEE Conference on Decision and Control*, Austin, Texas, 1399-1404 (1988).
13. MATLAB, The Math Works, Inc., Natick, Massachusetts (2000).

CHAPTER 9

OPTIMIZATION OF SYSTEMS FOR ACOUSTICS

Ashok D. Belegundu and Michael D. Grissom

The Pennsylvania State University, University Park, PA 16802
E-mail: abelegundu@psu.edu

An experimentally verified approach for the optimization of systems for acoustics with passive structural modifications is given. The method is general enough to handle a variety of structural modifications and structural impedances. Following some introductory acoustics and vibrations concepts, the optimization approach is formulated. Governing equations and solution methods are given, and finally several example applications are shown.

1. Introduction

This chapter discusses passive optimization techniques for minimizing or tuning acoustic response. The focus is on relating direct experience of our group, in the last ten years, on vibrating structures that are harmonically excited and which radiate sound into the open air. Work carried out here has been experimentally verified. A survey of all work done in this area is not attempted. The aim here is to share our important experiences in designing quiet structures. This chapter does not address noise in a cavity such as an automobile interior, nor flow induced noise such as jet noise, fan noise. Active noise cancellation techniques are also not addressed. Figure 1 shows some applications for noise reduction that involve vibrating panels. Other examples are radiated noise from engine valve covers, oil pans, timing chain cover plates, and cylindrical pressure vessels. Passive approaches involve attaching point masses, stiffeners, and vibration absorbers (point mass/stiffness/damper) to the structure (Fig. 2).

Recently, attaching thin acoustic cavities has shown potential (see Section 6). In high impedance structures made with thick metal plates for instance, we surround the noise source with a cover and attach absorbers to the cover (see gear box in Fig. 3). The cover is an air-tight enclosure made of thin sheet metal or

245

composite material. There are also passive devices to reduce noise in acoustic enclosures. Helmholtz resonators (Fig. 2) and wave guides are popular among these. Helmholtz resonators have been used in spacecraft fairings as also in motorcycle intake systems.

Computations are based on the use of finite element analysis for vibration analysis and a wave superposition-boundary element method for acoustic analysis. These codes are integrated with non-gradient optimizers (simulated annealing, differential evolution, and random search).   Adopted objective functions include kinetic energy, sound power, and a multiattribute value function.. This work is targeted up to medium frequency bands. At very high frequency bands with high modal density, it may be argued that techniques such as SEA (statistical energy analysis) are better suited than finite or boundary analysis.

Fig. 1. Examples of Noise Sources involving Vibrating Panels: (a) washing machine, (b) boat's motor housing or cowling, (c) Trim Panel in aircraft – current 120 dB interior noise levels must be reduced to about 70 dB with constraints on thickness and weight.

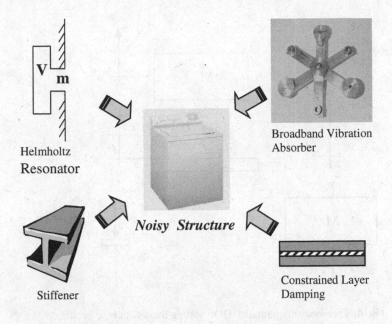

Helmholtz
Resonator

Broadband Vibration
Absorber

Noisy Structure

Stiffener

Constrained Layer
Damping

Fig. 2. Passive devices that may be attached to a vibrating structure.

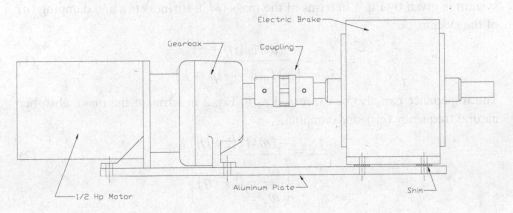

Fig. 3. Complex noise source application to be enclosed and then attaching absorbers to the cover.

2. Definitions and Introductory Concepts

The simplest vibrating structure is a single degree of freedom (1-DOF) sprung mass. A free-body diagram of a 1-DOF sprung mass with base excitation is shown in Fig. 4.

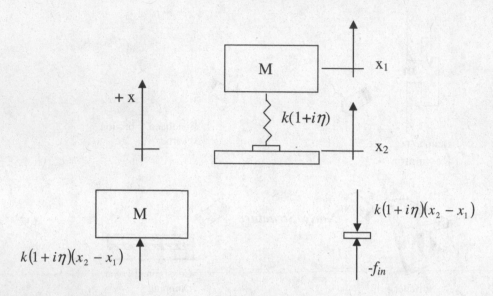

Fig. 4. Free-body diagram of 1-DOF sprung mass with base excitation.

The base impedance, transmitted force (f_{in}) divided by velocity at the base, of this system is given by Eq. 1 in terms of the mass (M_0), stiffness (k), and damping (η) of the system.

$$\frac{f_{in}}{\dot{x}_2} = \frac{i\omega M_0 (i\eta + k)}{\left(-M_0 \omega^2 + i\eta k + k\right)} \tag{1}$$

The impedance can also be expressed as in Eq. 2 in terms of the mass, absorber natural frequency (ω_n), and damping.

$$\frac{f_{in}}{\dot{x}_2} = \frac{-i\omega M_0 (1 + i\eta)}{\left(\dfrac{\omega^2}{\omega_n^2} - (1 + i\eta)\right)} \tag{2}$$

At $\omega \ll \omega_n$ the tuned absorber acts as a discrete mass on the base structure. At $\omega \gg \omega_n$ the tuned absorber acts as a discrete spring on the base structure. In the vicinity of ω_n the absorber has a spike in impedance that indicates that large forces can result even with a relatively low base velocity. At this resonance, the magnitude and width of the peak is determined by the mass and damping of the absorber. Greater mass increases the impedance linearly. Greater damping reduces the impedance but increases the bandwidth of the peak. It is these passive effects of mass, stiffness, and sprung masses that will be exploited to optimize systems for acoustics.

The simplest sound sources are the monopole and the dipole. While they are theoretical in nature, they form the basic patterns for more complex sound radiators. A monopole can be conceptualized as a small sphere, with its entire surface expanding and contracting in-phase. Acoustic waves radiate equally in all directions from a monopole. A dipole is two closely spaced monopoles, pulsating out of phase with one another. In contrast to the monopole, the sound radiated from a dipole is very directive and, at low frequencies, is not a very efficient acoustic radiator. The acoustic radiation of structures with complex geometries is evaluated by replacing the vibrating surfaces with monopoles and dipoles and solving their equivalent source strengths for the magnitude of the vibration. This concept is used to evaluate acoustic response which makes optimization of systems for acoustics possible for vibrating structures with complex geometries.

Acoustic optimization objectives vary widely from application to application. Practically, they are constrained more by the ability to measure them than the ability to calculate them. Several of the more common are mentioned here, but nearly all are based on sound pressure measurements. Sound pressure level (SPL) is measured at a point with a single microphone (p), and is generally reported in decibels relative to 2×10^{-5} Pascals in air (Eq. 3).

$$SPL\ (dB) = 20\log_{10}(\,p/2 \times 10^{-5})\qquad(3)$$

A single SPL measurement or calculation is usually not enough to characterize the acoustic effect of a sound producing system. The SPL measurement is also affected by surrounding structures, and it is often difficult to isolate a system from its surroundings to obtain a good measurement. Two objectives that are used to evaluate the overall effect due to acoustic energy flow are sound intensity ($\mathbf{I}$) and power (Π). Intensity is defined as the rate of acoustic energy flow through a unit area, the pressure p times the particle velocity ($\mathbf{v}$) (Eq. 4).

$$\mathbf{I} = p\,\mathbf{v}\qquad(4)$$

In practice, the real portion of the particle velocity is estimated by measuring the pressure at two closely spaced microphones, and the intensity is time averaged. The real part of the average intensity is reported in decibels relative to 1×10^{-12} W/m^2.

$$I_{avg}(dB) - 10\log_{10}(I_{ical}/1 \times 10^{-12})\qquad(5)$$

Sound power is defined as the rate of energy flow through a surface that completely surrounds the system of interest, or the real part of the intensity integrated over that surface (Eq. 6).

$$\Pi = \iint_S I_{real} \, dS \tag{6}$$

The sound power is reported in decibels relative to 1×10^{-12} W.

$$\Pi(dB) = 10 \log_{10}(\Pi/1 \times 10^{-12}) \tag{7}$$

There are many other sound metrics, but they are nearly all based on the previous three. In some situations, as in the aircraft trim panel in Fig. 1(c), the objective is to maximize transmission loss through the panel. This is defined as the ratio equal to the acoustic intensity incident on one side of the panel divided by the acoustic intensity transmitted on the other side.

Other metrics consist of weighting the sound power, pressure, or intensity in the frequency domain. The typical human ear responds more to some frequencies than others, and responds to some combinations of frequencies differently than others. The first model of a frequency dependence effect is to apply the A-weighting curve to the sound pressure measurements. Other attempts to measure sound quality include loudness, harshness, and annoyance metrics.

3. Optimization Problem Formulation

Optimization problems may be stated as minimizing an objective function $f(x)$ subject to constraints: $g(x) \leq 0$, $h(x) = 0$. As noted above, masses, vibration absorbers etc. can be attached to the structure (or a cover around the structure). Thus, design variables for optimization relate to mass (m), stiffness (k), or damping (c) of each attachment to the base structure. Often, variables are chosen which are related to these. Details will be given below in Section 6.

3.1. Objective Function Formulation

Regarding objective functions, kinetic energy (KE) is minimized first, to obtain a good starting point for sound power minimization. The KE is that of the original structure excluding the vibration absorbers. The physics of sound in this context is as follows. A structure can be vibrating significantly in a certain mode but still radiate very little sound power. This happens when the mode is a weak radiator where displaced volume velocity cancellation occurs as with the dipole source. In Fig. 5 below, mode (a) is a strong and (b) is a weak radiator, respectively. When the objective function is defined over a broad frequency band, the optimized designs resonate more as weak radiators.

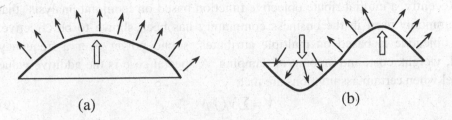

(a)   (b)

Fig. 5. Vibration Mode (a) is a strong and (b) is a weak -- radiator.

When KE is minimized by the optimizer, the energy is either transferred to the added absorbers, masses, and stiffeners or the structural impedance is increased at the forcing locations. Generally, total mass of attachments is constrained to be less than 10% of the original weight of structure.

Since all the metrics considered (KE, Π, ...) are frequency dependent (Fig. 6) each must be formed into a scalar objective function for optimization. Owing to light damping that is found in structures, we may add the response at resonant frequencies only.

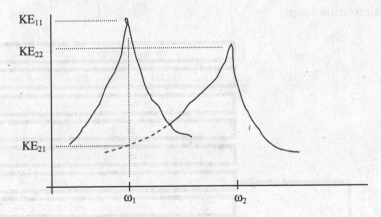

Fig. 6 Objective function for broadband frequency with resonant frequencies.

Thus, $KE_1 \equiv KE_{11} + KE_{21} + KE_{31} + ...$ will represent total energy at ω_1. Summing $KE_1 + KE_2 + ...$ over the band of interest will serve as a good measure for total, integrated, energy in an harmonically excited system. We may represent this as

$$W = \sum_{\omega} W(\omega_i) \tag{8}$$

From Parseval's theorem, this measure equals the average kinetic energy in a time period. Sound power calculations follow similar lines as KE as far as summing peaks.

Recently, a multiattribute objective function based on 'conjoint analysis' that is commonly used in the business community has been shown to be effective. The objective is based on multiple attributes: sound power over a frequency band, weight, cost, and amount of damping. A special case is the additive value model when certain assumptions are met:

$$V = \sum v_i(f_i) \tag{9}$$

where V is the objective, v_i is a consistently scaled function value associated with the level f_i of attribute i.

3.2. *Procedure for Optimal Design of Quiet Structures*

Most important in optimization is the need to perform reanalysis, i.e. analyze the structure with changing parameters in the optimization loop, <u>without</u> re-computing modes of the original structure (which is a large finite element model). Figure 7 below shows the design flow in computer-aided design of quiet structures including the multiple attribute objective function. Note the importance given to keeping a minimum amount of computations within the iterative optimization loop.

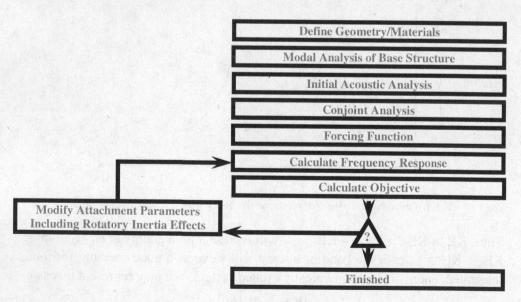

Fig. 7. Flowchart for computer-aided design of quiet structures.

Before getting into details of governing equations, measurements and examples, the general design approach, of which Fig. 7 above is only a part, is outlined below for a noise reduction problem. Note: "Structure" here is defined as the original structure or, in cases when the original structure has high impedance (e.g. made of very thick plates), of a cover structure around the original structure. Structure does not include the attachments (i.e. the absorbers).

Task 1. The power spectrum of the sound power radiating from the noise source is experimentally determined to identify the frequency band within which the sound power levels are high at the operating condition. Within this band, the kinetic energy KE and sound power Π may be discretized as $W = \sum W(\omega_i)$. Π is obtained by summing the power at each frequency over the frequency interval.
Task 2. Conduct modal analysis experiments of the structure.
Task 3. Develop a finite element model and validate the model by comparison with the modal analysis experiments.
Task 4. Develop a forcing function. White noise is a good choice when there is uncertainty. In this case, every node is excited normal to the surface with a load of c Newtons with random phase, where c may be determined to match sound power prediction with measured values. In the case when loading is due to acoustic excitation, nodal velocity measurements have to be taken and used to define an equivalent forcing function, through, say least-squares technique.
Task 5. To validate the acoustical model, compute the sound power radiated from the structure (or cover) based on results from the numerical model and the physical model to insure their agreement.
Task 6. Optimize the structure using tuned absorbers or other attachments.
Task 7. Experimentally validate the optimized design.

4. Governing Equations and Solution Methods

As discussed in the previous section, at the core of any method of optimization of systems for acoustics is the reanalysis method. In this case, the reanalysis method involves the recalculation of the acoustic radiation of a forced vibrating structure with modifications. The most general modification is a sprung mass (tuned absorber, Fig. 4) <u>as they add additional degrees of freedom</u>, so the reanalysis is presented in terms of adding tuned absorbers. This section begins with a description of the vibration analysis of a structure, then describes two possible vibration reanalysis methods, and finally gives the acoustic analysis method.

4.1. *Vibration Analysis of Base Structure (Without Modifications)*

The first step is to determine the eigenvalues and eigenvectors of the unmodified structure. We denote $\mathbf{\Phi}_0$ = matrix whose columns are eigenvectors and λ_o = diagonal matrix whose elements are eigenvalues. A natural frequency in rad/s is obtained from the eigenvalue as $\omega = \sqrt{\lambda}$. Dimension of the matrix $\mathbf{\Phi}_0$ is (number of degrees of freedom, number of modes in the basis). Modal information can be found from a finite element model or from experiment. Here we use finite elements to determine modal response. The basic equations for this are given below.

Equations of motion of the forced vibration for a finite element representation of a hysteritically damped base structure are

$$\mathbf{m}_0\ddot{\mathbf{x}} + \left[\mathbf{k}_0 + i\eta\mathbf{k}_0\right]\mathbf{x} = \mathbf{f}_0 \tag{10}$$

where $\mathbf{m}_0, \mathbf{k}_0, \eta, \mathbf{f}_0$, and $\mathbf{x}$ are the mass and stiffness matrices, the material loss factor, and the forcing and response vectors, respectively. Assuming harmonic excitation and response, we have

$$-\omega^2\mathbf{m}_0\mathbf{X} + \left[\mathbf{k}_0 + i\eta\mathbf{k}_0\right]\mathbf{X} = \mathbf{F}_0 \tag{11}$$

where $\mathbf{F}_0$ and $\mathbf{X}$ are the complex amplitudes of the force and response vectors. If the forcing vector and damping are set to zero, the normal modes can be found by solving the eigenvalue problem

$$\mathbf{k}_0\mathbf{\Phi}_0 = \lambda_0\mathbf{m}_0\mathbf{\Phi}_0 \tag{12}$$

The eigenvectors satisfy

$$\mathbf{\Phi}_0^T\mathbf{k}_0\mathbf{\Phi}_0 = \lambda_0 \qquad \mathbf{\Phi}_0^T\mathbf{m}_0\mathbf{\Phi}_0 = \mathbf{I} \tag{13}$$

Using mode superposition, the forced response of the structure can be given as

$$\mathbf{X} = \sum_{j=1}^{m} q_j\,\mathbf{\Phi}_o^j = \mathbf{\Phi}_0\mathbf{q} \tag{14}$$

where $\mathbf{q}$ is a vector of modal 'participation factors' or modal 'coordinates' given by

$$q_j = \frac{\mathbf{\Phi}_o^{j\,T}\mathbf{F}}{\left[-\omega^2 + (1+i\eta)\lambda_0^j\right]} \qquad j = 1,\ldots, m \tag{15}$$

At the *k*th resonance, whence $\omega = \omega_k$, and $\lambda^k = \omega_k^2$, we have $q_j = \dfrac{\Phi_o^{j^{\mathrm{T}}} \mathbf{F}}{i\eta\lambda_0^j}$:

Other quantities such as kinetic energy and radiated sound power can now be computed.

4.2. *Analysis of Modified Structure by the Impedance Method*

As discussed in the introduction, a few different methods exist for dynamic analysis of the structure with vibration absorbers attached to it. Of these, the impedance method and the reduced eigenvalue method are most attractive, since in each of these Eq. 12 is solved only once. We first discuss the impedance method. The reanalysis problem is formulated in terms of added impedances as

$$\mathbf{m}_0\ddot{\mathbf{x}} + \left[\mathbf{k}_0 + i\eta\mathbf{k}_0\right]\mathbf{x} = \mathbf{f}_0 = \mathbf{f}_{in} - \mathbf{z}\dot{\mathbf{x}} \tag{16}$$

where $\mathbf{f}_{in}$ is the forcing vector, and $\mathbf{z}$ is the impedance matrix of the modification. The impedance matrix is diagonal if each modification is independent and discrete as is the case with simple spring-mass absorbers. For example, impedance for a simple mass *m* takes the expression $z = i\omega m$, and for a spring-mass system with parameters *k*, *m* takes the form

$$z = \frac{i\omega m k}{k - m\omega^2}. \tag{17}$$

Replacing $\mathbf{f}_o$ by $\mathbf{f}_{in} - \mathbf{z}\dot{\mathbf{x}}$, we have

$$\mathbf{X} = \mathbf{\Phi}_0\left[-\omega^2 + (1 + i\eta)\lambda_0\right]^{-1}\mathbf{\Phi}_0^T\left(\mathbf{F}_{in} - i\omega\mathbf{z}\mathbf{X}\right) \tag{18}$$

Defining a diagonal matrix $\mathbf{A} = \left[-\omega^2 + \lambda_0(1 + i\eta)\right]^{-1}$, we can write the solution

$$\mathbf{X}_z = \left[\mathbf{I} + i\omega\mathbf{\Phi}_z\mathbf{A}\mathbf{\Phi}_z^T\mathbf{z}_z\right]^{-1}\mathbf{\Phi}_z\mathbf{A}\mathbf{\Phi}_0^T\mathbf{F}_{in} \tag{19}$$

where $\mathbf{z}_z$ is the matrix of impedances and $\mathbf{\Phi}_z$ is the matrix of eigenvectors corresponding only to (non-zero) impedance locations. Solution to Eq. 19 gives the response only at the impedance locations, $\mathbf{X}_z$. In Eq. 19, only a small *pxp* matrix, were *p* is the number of impedance (or absorber) locations, is inverted for each desired frequency. Response of the modified structure at a general degree of freedom (as opposed to where an absorber is attached) is obtained by

$$\mathbf{X} = \mathbf{\Phi}_0\mathbf{q}_z = \mathbf{\Phi}_0\mathbf{A}\mathbf{\Phi}_0^T\left(\mathbf{F}_0 - i\omega\mathbf{z}\mathbf{X}_m\right) \tag{20}$$

where $\mathbf{X}_m$ is the vector of $\mathbf{X}_z$ found through Eq. 18 augmented with zero values at the zero impedance locations, and $\mathbf{q}_z$ is the vector of modal coordinates of the modified structure.

The computational procedure may be summarized as follows. Given a set of absorbers with known locations and parameters, $\mathbf{z}$ is first defined. Then, for the specified frequency ω, Eqs. 18 and 19 are solved to obtain the displacement amplitude of the base structure $\mathbf{X}(\omega)$. Velocities are obtained from $\dot{\mathbf{X}} = i\omega\mathbf{X}$. Other quantities such as kinetic energy of the base structure are readily determined from the velocities.

4.3. *A Disadvantage with the Impedance Method For Estimating Broadband Response*

Two main difficulties exist with the impedance method. One is the derivation of expressions for impedance, $\mathbf{z}_z$, that incorporate rotatory inertia of the absorbers (as a result of base rotation). The other difficulty is as follows. The impedance method yields response at a specified frequency ω. Peak values of kinetic energy or other performance metric which occur at resonance frequencies not known *apriori* are not easily determined. The kinetic energy must be calculated at enough discrete frequencies that the peaks (or sum of peaks or an integral measure) over the broadband are accurately captured. The following figure illustrates the difficulty just mentioned. The kinetic energy of the base structure (Fig. 8) is computed and plotted at various frequencies for the given resolution. Only a single peak is included in the frequency band for illustration.

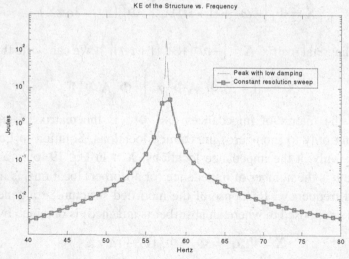

Fig. 8. A constant resolution sweep missing a peak in the kinetic energy.

Evaluation of kinetic energy at equal increments in frequency misses this peak value. For low structural damping, as is generally the case, extremely small steps must be used, and even this will not be accurate. Noting that small increments mean more computation, the problem of determining broadband response now becomes evident. This problem has not received much attention in the past as absorbers were used to target only a fixed frequency. Use of distributed tuned absorbers for broadband energy/sound reduction has exacerbated the problem of determining multiple response peaks.

4.4. *Reduced Eigenvalue Reanalysis Method*

As before, let $\mathbf{M}_0$ and $\mathbf{K}_0$ refer to the mass and stiffness matrices of the base structure without absorbers. An absorber is described by its own mass and stiffness matrices $\mathbf{M}_{abs}$ and $\mathbf{K}_{abs}$. Some degrees of freedom of these matrices coincide (shared) with those of base structure where they are attached, while other degrees of freedom are independent. Thus, the modification mass and stiffnesses of the absorbers may be partitioned as

$$\mathbf{M}_{abs} = \begin{bmatrix} \Delta\mathbf{m} & \mathbf{m}_a \\ \mathbf{m}_a^T & \mathbf{m}_z \end{bmatrix}, \quad \mathbf{K}_{abs} = \begin{bmatrix} \Delta\mathbf{k} & \mathbf{k}_a \\ \mathbf{k}_a^T & \mathbf{k}_z \end{bmatrix} \tag{21}$$

where $\Delta\mathbf{m}$ is the added mass matrix at the shared degrees of freedom, $\mathbf{m}_z$ is the added mass matrix at the new degrees of freedom, $\mathbf{m}_a$ is the coupling mass matrix, and similar descriptions for the stiffness submatrices. The modification element matrices are assembled into the base structure's mass and stiffness matrices as

$$\begin{bmatrix} \mathbf{m}_o + \Delta\mathbf{m} & \mathbf{m}_a \\ \mathbf{m}_a^T & \mathbf{m}_z \end{bmatrix}\begin{Bmatrix} \ddot{\mathbf{x}}_o \\ \ddot{\mathbf{x}}_z \end{Bmatrix} + \begin{bmatrix} (1+i\eta)\mathbf{k}_o + \Delta\mathbf{k} & \mathbf{k}_a \\ \mathbf{k}_a^T & \mathbf{k}_z \end{bmatrix}\begin{Bmatrix} \mathbf{x}_o \\ \mathbf{x}_z \end{Bmatrix} = \begin{Bmatrix} \mathbf{f}_o \\ 0 \end{Bmatrix} \tag{22}$$

Harmonic excitation, response, and modal superposition as defined for the unmodified structure is assumed. As stated earlier, attachment of small vibration absorbers allow us to assume that response of the base structure with additions can be represented in the original modes. Thus, modal superposition parallels Eq. 14 for the unmodified structure, but with added terms from the modifications:

$$\begin{Bmatrix} \mathbf{X} \\ \mathbf{X}_z \end{Bmatrix} = \begin{bmatrix} \Phi_0 & 0 \\ 0 & \mathbf{I} \end{bmatrix}\begin{Bmatrix} \mathbf{q} \\ \mathbf{X}_z \end{Bmatrix} \tag{23}$$

Equations 21 and 22 are combined to give Eq. 23

$$\left[-\omega^2\begin{bmatrix}\mathbf{m}_o+\Delta\mathbf{m} & \mathbf{m}_a \\ \mathbf{m}_a^T & \mathbf{m}_z\end{bmatrix}+\begin{bmatrix}(1+i\eta)\mathbf{k}_o+\Delta\mathbf{k} & \mathbf{k}_a \\ \mathbf{k}_a^T & \mathbf{k}_z\end{bmatrix}\begin{bmatrix}\boldsymbol{\Phi}_0 & 0 \\ 0 & \mathbf{I}\end{bmatrix}\right]\begin{Bmatrix}\mathbf{q} \\ \mathbf{X}_z\end{Bmatrix}=\begin{Bmatrix}\mathbf{F}_0 \\ 0\end{Bmatrix} \quad (24)$$

Both sides of Eq. 23 are pre-multiplied by the modal matrix in Eq. 22, and the result is simplified by taking advantage of the orthogonality conditions (Eq. 25):

$$\left[-\omega^2\begin{bmatrix}\mathbf{I}+\boldsymbol{\Phi}_0^T\Delta\mathbf{m}\boldsymbol{\Phi}_0 & \boldsymbol{\Phi}_0^T\mathbf{m}_a \\ \mathbf{m}_a^T\boldsymbol{\Phi}_0 & \mathbf{m}_z\end{bmatrix}+\begin{bmatrix}(1+i\eta)\lambda_o+\boldsymbol{\Phi}_0^T\Delta\mathbf{k}\boldsymbol{\Phi}_0 & \boldsymbol{\Phi}_0^T\mathbf{k}_a \\ \mathbf{k}_a^T\boldsymbol{\Phi}_0 & \mathbf{k}_z\end{bmatrix}\right]\begin{Bmatrix}\mathbf{q} \\ \mathbf{X}_z\end{Bmatrix}=\begin{Bmatrix}\boldsymbol{\Phi}_0^T\mathbf{F}_0 \\ 0\end{Bmatrix}$$

$$(25)$$

Equation 25 can be denoted as

$$\left[-\omega^2\hat{\mathbf{M}}+\hat{\mathbf{K}}\right]\hat{\mathbf{X}}=\hat{\mathbf{F}} \quad (26)$$

where

$$\hat{\mathbf{X}}=\begin{bmatrix}\mathbf{q} \\ \mathbf{X}_z\end{bmatrix} \quad (27)$$

Equation 25 involves inverting a smaller matrix. Dimension of $\hat{\mathbf{X}}$ equals m number of modes used in Eq. 14 plus the number of independent degrees of freedom associated with the absorbers. The solution of $\hat{\mathbf{X}}$ from the above equation can again be obtained using modal superposition. We set $\hat{\mathbf{F}}=0$ and solve for the modes from

$$\hat{\mathbf{K}}\,\hat{\boldsymbol{\varphi}}^j=\lambda_j\,\hat{\mathbf{M}}\,\hat{\boldsymbol{\varphi}}^j \quad j=1,...,\hat{m} \quad (28)$$

We then have

$$\hat{\mathbf{X}}=\sum_{j=1}^{\hat{m}}\psi_j\,\hat{\boldsymbol{\varphi}}^j \quad (29)$$

where

$$\hat{\boldsymbol{\varphi}}=\begin{bmatrix}\boldsymbol{\Phi}_q \\ \boldsymbol{\Phi}_z\end{bmatrix} \quad (30)$$

As in Section 2, we may use orthogonality properties to write the modal response as

$$\psi_j=\frac{\hat{\boldsymbol{\varphi}}^{j^T}\hat{\mathbf{F}}}{(-\omega^2+\lambda_m)} \quad (31)$$

where λ_m = (complex) eigenvalues of the modified structure with absorbers. From Eq. 22, we have $\mathbf{X}_m = \mathbf{\Phi}_0\,\mathbf{q}$, which together with Eqs. 27, 29-31 yields the response of the base structure degrees of freedom (i.e., excluding absorber degrees of freedom) as

$$\mathbf{X}_m = \sum_j \psi_j\,\mathbf{\Phi}_m^j \tag{32}$$

where the modes of the modified system are given by

$$\mathbf{\Phi}_m = \mathbf{\Phi}_0\,\mathbf{\Phi}_q \tag{33}$$

Eq. 22 can be written as

$$\mathbf{X}_m = \mathbf{\Phi}_m\left[-\omega^2 + \lambda_m\right]^{-1}\mathbf{\Phi}_m^{\mathrm{T}}\mathbf{F}_0 \tag{34}$$

which represents the forced response of the modified base structure.

While the impedance approach discussed earlier only provides $\mathbf{X}(\omega)$, and a search technique is needed to determine the peak responses, in the reduced eigenvalue approach each peak response is immediately obtained by setting the real part of $\lambda_m = \omega_m^2$ in Eq.34. Further, we have derived an efficient technique for generating absorber matrices, $\mathbf{M}_{\mathrm{abs}}$ and $\mathbf{K}_{\mathrm{abs}}$.

4.5. *Sound Power Calculations*

A full development of the boundary element / wave superposition method used here is given by Fahnline and Koopmann. The method replaces each of the (triangular) elements on a surface with point acoustic monopole and dipole sources. The strength of each of the sources is found through a volume velocity boundary condition:

$$\mathbf{u} = \mathbf{U}\mathbf{s} \tag{35}$$

where $\mathbf{s}$ is the vector of source strengths (one for each element on the structure), $\mathbf{U}$ is a matrix relating the source strengths to volume velocities and $\mathbf{u}$ is the vector of volume velocities. The volume velocity produced by a single element is defined as

$$u_n = \frac{1}{3}A_n\left(\dot{\mathbf{x}}_{1n} + \dot{\mathbf{x}}_{2n} + \dot{\mathbf{x}}_{3n}\right)\cdot\mathbf{n}_n \tag{36}$$

where A_n is the surface area of element n, $\mathbf{n}_n$ is the unit normal to the element, and $\dot{\mathbf{x}}_{in}$ ($i = 1, 2, 3$) are the nodal velocities at the corners of the triangular element. The nodal velocities are calculated as discussed in the previous section. The acoustic source strengths are found by inverting Eq. 35. We may write Eq. 36 in matrix form, solving for the entire volume velocity vector $\mathbf{u} = \mathbf{V}\dot{\mathbf{x}}$, where $\mathbf{V}$ is a matrix containing element surface areas, unit normals and connectivity information. Writing this in terms of modal participation factors we have $\mathbf{u} = i\omega\mathbf{V}\mathbf{\Phi}\mathbf{q}$ from which

$$\mathbf{s} = i\omega\mathbf{U}^{-1}\mathbf{V}\mathbf{\Phi}\mathbf{q} \tag{37}$$

The acoustic power is calculated by pre- and post-multiplying the source strength vector by a coupling matrix $\mathbf{S}$:

$$\Pi_{av} = \frac{1}{2}\operatorname{Re}\{\mathbf{s}^H\mathbf{S}\mathbf{s}\} \tag{38}$$

where the H superscript indicates complex conjugation and transposition. The matrix $\mathbf{S}$ is not given here for brevity.

Direct use of Eq. 38 above is not as efficient as the following procedure. Substituting Eq. 37 into Eq. 38 we have

$$\Pi_{av} = \frac{1}{2}\omega^2 \operatorname{Re}\{\mathbf{q}^H\mathbf{\Phi}^T\mathbf{V}^T\mathbf{U}^{-H}\mathbf{S}\mathbf{U}^{-1}\mathbf{V}\mathbf{\Phi}\mathbf{q}\} \tag{39}$$

or
$$\Pi_{av} = \frac{1}{2}\omega^2 \operatorname{Re}\{\mathbf{q}^H\mathbf{P}\mathbf{q}\} \tag{40}$$

where
$$\mathbf{P} = \mathbf{\Phi}^T\mathbf{V}^T\mathbf{U}^{-H}\mathbf{S}\mathbf{U}^{-1}\mathbf{V}\mathbf{\Phi} \tag{41}$$

Varying the parameters or locations of tuned absorbers on the structure will affect only the modal participation vector $\mathbf{q}$, while leaving the $\mathbf{P}$ matrix unchanged. *The $\mathbf{P}$ matrix may be calculated and stored before optimization begins*; calculation of sound power involves only the determination of the change in the modal participation vector through Eq. 20, and pre- and post-multiplication of this vector with the $\mathbf{P}$ matrix at each frequency. This new reanalysis procedure makes feasible the use of the sound power of a structure as an objective function in optimization with tuned absorbers.

4.6. *Determination of Initial Modal Participation Factors*

All of the analysis presented above are dependent on having knowledge of the modal participation factors (MPF's) of the unmodified structure, $\mathbf{q}_0$. Once the eigenvalues and eigenvectors of the structure are known, the modal participation factors for the structure can be had either experimentally or through a known forcing function. If the forcing function structure is known *a priori*, we may simply use $\mathbf{q}_0 = \mathbf{\Lambda}\mathbf{\Phi}^T\mathbf{F}$ to determine the MPF's. In many cases, however, the forcing function is unknown or is very complicated. For example, if the forcing function is an acoustic excitation induced by the noise source on the enclosure. In this situation, we may determine the MPF's experimentally as follows.

Examining $\dot{\mathbf{x}} = i\omega\mathbf{\Phi}\mathbf{\Lambda}\mathbf{\Phi}^{\mathrm{T}}\mathbf{f} = i\omega\mathbf{\Phi}\mathbf{q}_0$, we see that the MPF's are related to the nodal velocities through the matrix of eigenvectors. In theory, we could measure the nodal velocities at all points on the structure and invert the equation to solve for the MPF's. In practice, it is not necessary to measure all points to obtain good estimates of MPF's; only certain critical points (e.g. points located at antinodes of the mode shapes of the structure) need to be measured.

Let us denote the subset of measured velocities with an overbar, in which case we have

$$\bar{\dot{\mathbf{x}}} = i\omega\bar{\mathbf{\Phi}}\mathbf{y}_0 \tag{42}$$

where $\bar{\mathbf{\Phi}}$ is the submatrix of eigenvectors associated with the measured nodes. Note that this matrix is generally not square, but will have dimension $m \times r$ where m is the number of measured points and r is the number of known eigenvectors. Premultiplying Eq.42 by the transpose of the eigenvector matrix, we have

$$\bar{\mathbf{\Phi}}^{\mathrm{T}}\bar{\dot{\mathbf{x}}} = i\omega\bar{\mathbf{\Phi}}^{\mathrm{T}}\bar{\mathbf{\Phi}}\mathbf{q}_0 \tag{43}$$

The matrix preceding the $\mathbf{q}$ vector is now square, and we can invert this equation to solve for the modal participation factors without tuned absorbers.

$$\mathbf{q}_0 = -\frac{i}{\omega}\left(\bar{\mathbf{\Phi}}^{\mathrm{T}}\bar{\mathbf{\Phi}}\right)^{-1}\bar{\mathbf{\Phi}}^{\mathrm{T}}\bar{\dot{\mathbf{x}}} \tag{44}$$

For this method to be successful, the number of measured points must be equal to or greater than the number of modes used in the solution. Obviously, the method increases in accuracy as the number of measured points increases, and the measured points should include the locations of the largest displacements on the structure (modal antinodes).

5. Non-gradient Optimizers Used

Acoustic optimization does not usually lend itself to gradient optimizers. Objective functions generally involve many, closely spaced, sharp, resonance responses. Several non-gradient optimizers have been used in this effort. Simulated Annealing, random search, and Differential Evolution have been used.

6. Selected Example Problems

The preceding design method has been applied to a selection of three acoustic optimization applications. The three applications typify three design situations and the modifications that have proven to be effective for those applications. For similar applications to the ones given see the references. First, mass is added to a thin (low-impedance) structure to reduce acoustic radiation. Second, tuned absorbers are added to medium impedance structures. Third, a high impedance structure is surrounded by a close fitting cover.

6.1. *Optimization of a Half-Cylindrical Shell*

The optimization problem under consideration is an aluminum half-cylinder, rigidly mounted on a fixed, rigid plate. Figure 9 shows the finite element mesh of the half-cylinder. It is 12 inches (304.8 mm) long, with a 6.25 inch (158.75 mm) inside diameter. The cylinder was manufactured by reducing the thickness of a 6.25 inch ID aluminum pipe to 2 mm. The objective of the optimization problem is to minimize the sound power radiated by the half-cylindrical shell from its first five modes. To quiet the shell, two masses will be affixed to its surface. The locations of the masses will be determined by the optimization algorithm. Mathematically, the optimization problem can be written as:

$$\text{Minimize}\left\{ W_1 + W_2 + W_3 + W_4 + W_5 \right\}$$

where W_i is the sound power at the ith resonant frequency. The design variables are the locations of two masses, $\{x_1, x_2\}$. The two masses are small tungsten cylinders, each weighing 35.8 grams. The shell is driven at its top center point (41) with a harmonic shaker.

To model the half-cylindrical shell, a finite element mesh was created with 81 nodes and 128 elements. At the top center node a 19.0 g point mass was added, to simulate the inertia of the shaker assembly. The point masses were constrained to lie at nodal locations on the finite element mesh. A curvilinear coordinate system was mapped onto the finite element nodes so that each nodal location could be described by two coordinates: the distance along the long axis of the cylinder and circumferential location. Thus, each mass has two independent design variables, for a total of four ($n = 4$). It should be noted that it is possible to specify the location of each mass using only one coordinate: the node number. However, the use of this type of design variable leads to a highly discontinuous design space, which causes problems during optimization.

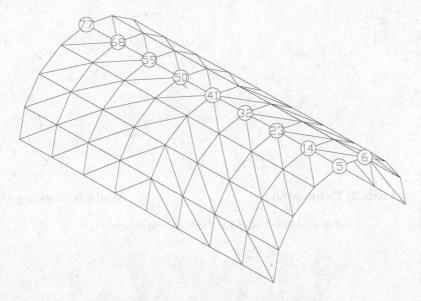

Fig. 9. Finite element mesh for half-cylinder.

It is noteworthy that the optimal mass placement (Fig. 10) is asymmetric with respect to the transverse axis of the cylinder. To understand the reason for the optimization results, some physical insight into the mode shapes of the shell is necessary. The mode shapes before optimization are shown in Fig. 11. The first mode shape is a 'swaying' mode and produces very little sound power, since the driving force is in the vertical direction. The second mode (a 'piston' type mode) produces the largest amount of sound power. The third mode (a 'rocking' or 'see-saw' type mode) creates only a small amount of noise, due to the effect of volume velocity cancellation. That is to say, one side of the shell moves up and compresses the air while the other side moves downward and decompresses the same amount of air like a dipole. The overall effect is that very little net volume velocity is created. The fourth and fifth modes (called 'rocking-2' and 'piston-2') produce little noise, again due to volume velocity cancellation.

Figure 12 shows the mode shapes of the half-cylinder with the optimal mass configuration. As can be seen, the piston and see-saw modes have both been converted into asymmetric 'rocking' type modes, both of which produce only a small amount of sound power. Thus, the shell has been converted into a weak radiator, through the addition of only 71.6 g of mass. The final result of this conversion is a 9.5 dB reduction in overall sound power at the first five modes.

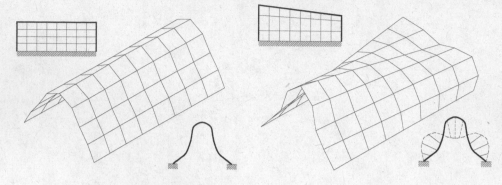

Mode 2: Piston Mode   **Mode 3: Rocking Mode**

Fig. 10. Mode Shapes for shell before optimization.

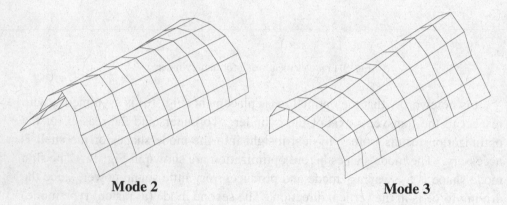

Mode 2   **Mode 3**

Fig. 11. Second and third shell mode shapes after optimization (Asymmetric rocking modes).

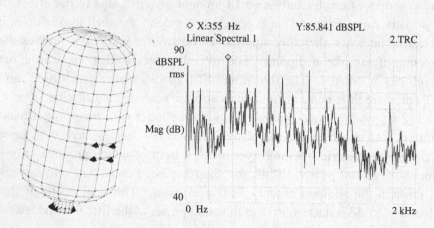

Fig. 12. Finite element depiction of curved pressure vessel and its sound pressure level in response to forcing.

6.2. *Multicriteria Optimization of a Pressure Vessel*

The concept of customer preference, or product value, prevalent in economics and management science, is just beginning to be used in engineering design. This concept and the associated measurement approaches offer us a theoretically appealing way to aggregating customer preferences for multiple product attributes into a single objective function, representing total product value, which may then be maximized. Among these methods, conjoint analysis has emerged as the most popular approach in marketing to estimate the value that customers attach to different features of a product that can be at different levels. We incorporate a designer's preferences for reducing noise in a curved pressure vessel excited with broadband noise (Fig. 12).

The shell is part of a large industrial machine. The 'product' here refers to a broadband vibration absorber(s) attached to the structure. Through direct interaction with the design engineer, we elicit his/her preferences for various alternative design configurations, and specify an aggregate value function. We then apply optimization techniques, interfaced to simulation codes, to maximize the value function. We show that this method provides more economical designs compared to certain conventional formulations.

The objective function represents total value of the product which is maximized using a non-gradient optimizer. Attributes considered are acoustic radiation (SPL), number of absorber beams (n_b), and damping material (η_i). The aggregate value of the attributes (v) to the designer is maximized, and the optimization problem is formulated in the following manner:

$$
\begin{aligned}
\text{Minimize:} \quad & -(v(\text{SPL}(m_i, L_i, \eta_i)) + v(n_b) + v(\eta_i)) \\
\text{Subject to:} \quad & 0 \text{ g} \leq m_i \leq 500 \text{ g} \\
& 1 \text{ mm} \leq L_i \leq 200 \text{ mm} \\
& 0.002 \leq \eta_i \leq 0.030 \\
& (i = 1, 2, \dots n_b)
\end{aligned}
$$

where m_i is the mass, L_i is the length, and η_i is the damping for the *ith* absorber. This formulation reduces the number of absorber beams by 40% with a negligible increase in the radiated sound over the commonly used single objective with constraint approach (Fig. 13).

Current trends in engineering design provide Pareto sets. While this is good as a tool for design space exploration, there remains the significant problem of choosing the best design. On the other hand, multiattribute design produces a single design, and modern multiattribute decision theory can provide the tradeoffs necessary to choose the optimal design. Here, lesser sound levels and easier to manufacture absorbers result when the tradeoffs are included in the optimization.

Attributes	Number of Beams	dB Reduction: Low Freq.	dB Reduction: Mid Freq.	dB Reduction: High Freq.
Kinetic Energy Objective	17	15	10	2
Sound Pressure Objective	10	14	6	7
Multiattribute Objective	6	13	6	7

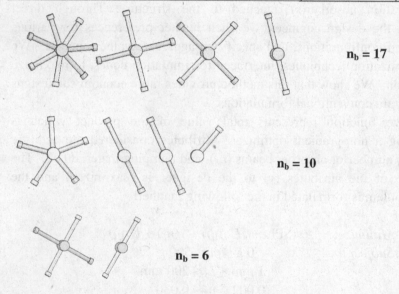

Fig. 13. Benefit of mulitattribute optimization compared to conventional formulation for acoustic pressure vessel.

Uncertainty in loading has been incorporated in arriving at the product design. However, the effect of variations in product manufacture (example, the beam lengths have some manufacturing tolerances) and certain other parameters have not been addressed. Details of this work will appear soon. There is need to generalize this work to design for robustness. It is assumed that the additive form of the value function is "reasonable". Many exciting developments are taking place in marketing science, in choice models and their applications, which are naturally relevant in engineering design.

6.3. *Thin Air Cavity Attached to a Structure*

A second plate attached with springs to an original vibrating plate requires modeling fluid-structure interaction for vibration analysis when the gap is small. This design concept has found to provide sound reduction even in the first, lowest vibration mode, where absorbers have had difficulty. An example problem demonstrates this. The plate is a 240 x 290 mm2, simply supported, 0.95 mm thick top cover plate attached with a 10 mm air gap (Fig. 14). The bottom plate is 8 mm thick which is a high impedance structure. When the spring sizes are optimized the cover plate reduces the sound power by 16 dB (Fig. 14).

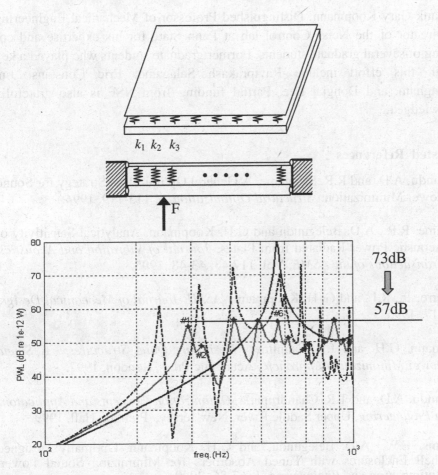

Fig. 14. Spring mounted cover plate sound power reduction plot with 25 springs. With optimized springs, 57 dB, no springs with only cavity, 73 dB, single plate (no air cavity), 80 dB.

7. Summary

The applications given in Section 6 show that passive structural modifications can be used to acoustically tailor a vibrating structure when applied with an effective optimization methodology. The applications cover a wide range of structural impedances, a variety of structural modifications, and multiple attributes. Finally, and most importantly, the optimized systems are significantly improved acoustically over the unmodified systems.

Acknowledgments

We thank Gary Koopmann, Distinguished Professor of Mechanical Engineering and Director of the Noise Control lab at Penn State for his expertise and co-advising of several graduate students. Former graduate students who played a key role in this effort include Raviprakash Salagame, Eric Constans, Jim Cunningham, and Dongjai Lee. Partial funding from NSF is also gratefully acknowledged.

Suggested References

Belegundu, A.D. and R.R. Salagame. A General Optimization Strategy for Sound Power Minimization. *Structural Optimization*, 8: 113-119, 1994.

Salagame, R.R., A.D. Belegundu and G.H. Koopmann. Analytical Sensitivity of Acoustic Power Radiated from Plates, *Journal of Vibration and Acoustics, Transactions of the ASME,* Vol. 117/43: 43-48, 1995.

St. Pierre, Jr., R.L. and G. H. Koopmann, *ASME Journal of Mechanical Design,* 117, 1995.

Koopmann, G.H. and J.B. Fahnline, *Designing Quiet Structures: A Sound Power Minimization Approach* , Academic Press, London, 1997.

Belegundu, A.D. and T.R. Chandrupatla, *Optimization Concepts and Applicatons in Engineering*, Upper Saddle River, New Jersey, Prentice-Hall, 1999.

Constans, E.W., A.D. Belegundu, and G.H. Koopmann. Optimally Designed Shell Enclosures with Tuned Absorbers for Minimizing Sound Power, *Optimization and Engineering*, 1, 67-86, 2000.

Cunningham, J.D. *Design and Optimization of Passive, Broadband Vibration Absorbers*, M.S. Thesis, The Pennsylvania State University, 2000.

Heinze, E.S., M.D. Grissom, and A.D. Belegundu, *Design and Optimization of Broadband Vibration Absorbers for Noise Control*, Proceedings of ASME IMECE 2001, New York, 2001.

Grissom, M.D. , A.D. Belegundu, and G.H. Koopmann, A Reduced Eigenvalue Method for Broadband Analysis of a Structure with Vibration Absorbers Possessing Rotatory Inertia, *Journal of Sound and Vibration*, 281, 869-886, 2005.

Lee, D., G.H. Koopmann, and A.D. Belegundu, Optimal design of acoustic enclosure for minimal broadband sound radiation, IMECE2005-82943, Proceedings 2005 ASME International Mechanical Engineering Congress and Exposition, Orlando, FL, Nov. 5-11, 2005.

Grissom, M.D., A.D. Belegundu, A. Rangaswamy, and G.H. Koopmann, Conjoint Analysis Based Multiattribute Optimization: Application in Acoustical Design, In Press, *Structural and Multidisciplinary Optimization*.

Christensen, S.T. and S.V. Sorokin On Analysis and Optimization in Structural Acoustics - Part I: Problem Formulation and Solution Techniques, Vol. 16, pp. 83-95, 1998. On Analysis and Optimization in Structural Acoustics - Part II: Exemplification for Axisymmetric Structures, *Structural Optimization*, Vol. 16, pp. 96-107, 1998.

Marburg S. and Hardtke, H. Investigation and Optimization of a Spare Wheel Well to Reduce Vehicle Interior Noise, *Journal of Computational Acoustics*, Vol. 11, No. 3, 425-449, 2003.

Dong, J., K. K. Choi, and N-H. Kim, N-H., Design Optimization for Structural-Acoustic Problems Using FEA-BEA with Adjoint Variable Method, *ASME Journal of Mechanical Design*, (126), 3, 2004.

CHAPTER 10

DESIGN OPTIMIZATION UNDER UNCERTAINTY

Sankaran Mahadevan

Department of Civil and Environmental Engineering, Vanderbilt University
VU Station B # 351831, 2301 Vanderbilt Place
Nashville, Tennessee, USA
E-mail: sankaran.mahadevan@vanderbilt.edu

Design optimization studies for mechanical systems have increasingly become concerned with mathematical treatment of uncertainties in system demands and capacity, boundary conditions, component interactions, and available resources. The problem of optimum design under uncertainty has been formulated as reliability-based design optimization (RBDO). Recent efforts in this context seek to integrate advances in two directions: computational reliability analysis methods and deterministic design optimization. Much current work is focused on developing computationally efficient strategies for such integration, using decoupled or single loop formulations instead of earlier nested formulations. The extension of reliability-based optimization to include robustness requirements leads to multi-objective optimization under uncertainty. Another important application concerns multidisciplinary problems, where the various reliability constraints are evaluated in different disciplinary analysis codes and there is feedback coupling between the codes. Applications of recently developed methods to automotive and aerospace design problems are discussed, and new directions for further study are outlined.

1. Introduction

The design of any engineering system requires the assurance of its reliability and quality. Uncertainties in the system characteristics and demand prevent such assurances from being given with absolute certainty. Traditional deterministic design methods have accounted for uncertainties through empirical safety factors. However such safety factors do not provide a quantitative measure of the safety margin in design, and are not quantitatively linked to the influence of different design variables and their uncertainties on the overall system performance. Therefore, in recent decades, a rational approach that quantifies the reliability of

performance or risk of failure in probabilistic terms, and includes these estimates directly in the design optimization, is gaining increased attention. For large systems in the aerospace and automotive
industries, reliability estimation based on expensive full-scale tests is not possible, and a model-based computational approach becomes valuable.

Deterministic optimization enhanced by reliability criteria and formulated within a probabilistic framework is referred to as reliability-based design optimization (RBDO). The aim of RBDO is to achieve adequate reliability with minimum cost. The reliability requirements for individual components as well as the entire system need to be included in the optimization formulation. In the optimization process, two types of variables are considered: deterministic and random variables. And the random variables may be further divided into random design variables and random system parameters. The design variables that appear in the objective function(s) of the RBDO problem may include the deterministic variables as well as the distribution parameters of the random design variables. For different RBDO formulations, the probability of failure or sometimes equivalently, the reliability index, may appear in either the objective function or the constraints or even both.

In RBDO, different objective functions have been used, such as minimization of weight, minimization of cost, and minimization of the probability of failure. The objective function can also be the life-cycle cost, in which the overall cost includes both the initial cost and the maintenance cost. Since any complex system has to satisfy many design criteria, resulting in multiple objectives and constraints, multi-objective optimization formulations have also been used[1,2].

Two types of reliability can be used in the optimization, namely, component reliability and system reliability[3,4]. The formulation of RBDO problems with both element-level and system-level reliability constraints may be expressed as:
Minimize $C(\mathbf{d}, \mathbf{X})$
Subject to:

$$P_{fi}(\mathbf{X}, \mathbf{p}) \leq P_{fia} \quad (i = 1, 2, ..., ncc)$$
$$P_{fsi}(\mathbf{X}, \mathbf{p}) \leq P_{fsia} \quad (i = 1, 2, ..., nsc)$$
$$d_{il} \leq d_i \leq d_{iu} \quad (i = 1, 2, ..., ndd) \tag{1}$$
$$X_{il} \leq X_i \leq X_{iu} \quad (i = 1, 2, ..., nrd)$$

where $C(\mathbf{d}, \mathbf{X})$ is the objective function; $\mathbf{d}$ and $\mathbf{X}$ are deterministic and random design vectors, respectively; $\mathbf{p}$ is the vector of random parameters; P_{fi} and P_{fsi} are the failure probability of the ith component and the system, respectively; P_{fia}

and P_{fsia} are the allowable failure probabilities for the ith component and the system, respectively; the deterministic and random design variables are subject to their lower and upper bounds, respectively. Other deterministic constraints may also be included in the above formulation.

This chapter examines available reliability analysis and design optimization methods, with particular focus on computational efficiency and applicability to practical problems. The issue of computational efficiency is addressed by techniques that decouple reliability analysis and optimization iterations. Practical applications pose challenges with respect to robust design, multiple objectives, multi-disciplinary coupled systems, model uncertainty etc. Recent developments in meeting these challenges are presented, and new directions for further research are outlined.

2. Reliability Analysis

For the reliability analysis of a single limit state, the first-order reliability method (FORM) is widely used due to its simplicity and speed. In FORM, the most probable point (MPP) is found by optimization methods such as Rackwitz and Fiessler's Newton-type method[5] sequential quadratic programming[6] (SQP), etc. The MPP is defined as the point on the limit state with the minimum distance to the origin in an uncorrelated reduced normal space. Well-established methods are available for transforming correlated, non-normal variables to an equivalent uncorrelated reduced normal space[7]. The probability of failure is then calculated using a first-order approximation to the limit state, as:

$$P_f = \Phi(-\beta) \tag{2}$$

Where the reliability index β is the distance from the origin to the MPP, and Φ is the cumulative standard normal distribution. Various second-order reliability methods[8,9] (SORM) have also been developed to improve the reliability estimation of FORM.

In the case of the system-level reliability constraint, the joint probability of multiple failure events often needs to be computed. For series systems, the system failure probability is computed through the union of the individual component failures. Analytical first-order and second-order bounds have been used to estimate the system probability of failure.

For problems where even second-order bounds are too wide, Monte Carlo simulation (MCS) could be used. MCS is simple to implement and can be applied to almost all problems at any desired accuracy; however, it is prohibitively expensive. The importance sampling method has been used to overcome this difficulty by concentrating most of the sampling in the failure region. Efficient adaptive importance sampling (AIS) techniques have been developed[10,11,12], which gradually refine the sampling density function to reflect the increasing knowledge of the failure domain. Multi-modal AIS[13,14] has been found particularly satisfactory for several applications[15,16].

In most RBDO studies, FORM has been commonly used for component-level reliability calculations in the objective and/or constraint functions[17,18]. For system-level reliability analysis in RBDO, various methods have been used, such as second-order bounds[19], multiple checking point method[20], PNET (probabilistic network evaluation technique)[21] and Monte Carlo simulation with response surface[22]. Formulations that include both component and system reliability requirements have been proposed by Mahadevan[23] and Pu et al.[24] The system reliability calculation has mostly been applied as a feasibility check, with the optimization iterations being based on component reliability, so that the computational effort is tremendously reduced.

One particular use of FORM is the computation of reliability sensitivity factors, given by the direction cosines of the MPP vector. In fact, the analyst may have more confidence in the sensitivity factors and finds them more useful in design decision-making than the failure probability estimates themselves, which are affected by numerous uncertainties. For reliability analysis using Monte Carlo simulation, approximate sensitivity factors have been derived in the case of Gaussian random variables[25].

3. Reliability-Based Optimization

The conventional RBDO approach is to employ nested optimization and reliability analysis loops, in which the reliability of the component(s) or the system is estimated inside the optimization loop, this makes the computational cost of RBDO prohibitive. Therefore, efforts to improve the efficiency of RBDO have been pursued in two directions: (1) improve the efficiency of the reliability analysis methods, and (2) develop equivalent formulations that use decoupling or single loop techniques.

Although RBDO is different from deterministic optimization, it is possible for the former to take advantage of the latter. It was found that it is more efficient to implement a two-level approach in which the RBDO process starts from a stationary point of the conventional deterministic optimization procedure, where the deterministic constraints are satisfied[26, 27].

Recent research has focused on developing decoupling techniques, such as by Royset et al. (2001)[28], in which the reliability terms in traditional RBDO are replaced by deterministic functions, and the SORA (sequential optimization and reliability analysis) approach of Du and Chen (2000)[29].

The reliability analysis used in SORA is based on the performance measure approach (PMA) suggested by Tu *et al.*[30] PMA is the inverse of the Reliability Index Approach (RIA), the traditional approach for implementing the First-order Reliability Method (FORM). SORA takes advantage of the PMA method in order to decouple the optimization and reliability analyses by separating each random design variable into a deterministic design component used for the optimization and a stochastic component used for the reliability analysis. The optimization is done independent of probabilistic analysis by controlling only the deterministic component of the random design variable (i.e. μ_x). During the optimization phase, the stochastic component η is kept constant, and during the reliability analysis phase, μ_x^k is kept constant and the value of η satisfying the reliability constraint is found. The algorithm terminates when successive cycles converge to consistent values of μ_x^k and η^k. The SORA concept for multiple constraints is outlined graphically in Fig. 1.

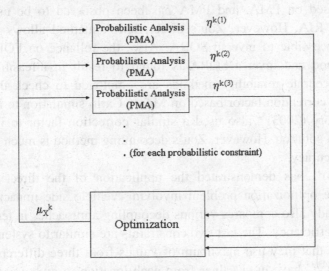

Fig. 1: SORA for multiple constraints

The SORA approach is based on inverse first-order reliability analysis (PMA). As a result, it may be inaccurate for nonlinear limit states. Due to the PMA strategy, there is no easy way to integrate more accurate higher order reliability approximations such as SORM, or more robust methods such as Monte Carlo simulation, within the SORA approach. SORA also presents a significant hurdle in including system reliability constraints in the optimization. A decoupling strategy that makes use of the direct reliability analysis formulation would be more useful in including different reliability analysis methods and system reliability requirements. Such a direct strategy has been developed by Zou (2004)[31], based on re-writing the reliability constraints using a first-order Taylor series approximation.

The reliability analysis provides the failure probability estimates and the derivatives for the approximated reliability constraints and makes them deterministic. Thus this is also a sequential approach, but it is based on direct reliability analysis as opposed to inverse reliability analysis. It is also possible to include a higher-order Taylor series approximation in this formulation. The important benefit is that this formulation works with direct reliability analysis and because of its modularity, it is able to include different reliability methods for different reliability requirements as dictated by the needs of accuracy, computational efficiency, and feasibility.

If FORM is used for reliability analysis, then Zou's direct SORA approach[31] may be referred to as RIA-based. When both the direct and inverse SORA methods use FORM, inverse SORA is likely to be computationally more efficient since it is based on PMA, and PMA has been observed to be usually more efficient than RIA. However, Zou's direct SORA method allows us to solve problems inaccessible to inverse SORA. Also, the reliance on FORM in both Royset's method and inverse SORA is found to result in infeasible solutions when more accurate reliability methods are employed to check the solution. Royset uses a correction factor based on Monte Carlo simulation to improve the solution; Sopory (2003)[32] also used a similar correction factor to improve the inverse SORA solution. However, Zou's decoupling method is much more direct in ensuring accuracy.

Zou (2004)[31] has demonstrated the application of the direct decoupling strategy to the optimization problem involving vehicle side impact, shown in Fig. 2. The study also compares various decoupling approaches in terms of both accuracy and efficiency. The last two constraints are similar to system reliability constraints, in that they use maximum of values from three different locations. Earlier studies[29,33] have used values from each location to give a constraint and avoided considering the maximum. However, guidelines from the Insurance Institute of Highway Safety require using the maximum, and the direct decoupling approach enables this.

Another recent development is a single loop strategy developed by Mourelatos (2003)[34], by exploiting the Kuhn-Tucker conditions at the optimum. This helps to adopt a well-known strategy for effectiveness in optimization, i.e., satisfying the constraints only at the optimum and allowing the solution to be infeasible before convergence. Initial results with several mathematical examples have shown this strategy to be computationally more efficient than the decoupling methods.

Minimize	Weight (W)
Subject to	$P(V_{\text{B-Pillar}} \leq 9.9 \ (\text{m/s})) \leq 0.1$
	$P(V_{\text{front door}} \leq 15.69 \ (\text{m/s})) \leq 0.1$
	$P(\text{Abdomen Load} \leq 1.0 \ (\text{KN})) \leq 0.1$
	$P(\text{Pubic Force} \leq 4.0 \ (\text{KN})) \leq 0.1$
	$P(\text{Max}(V*C) \leq 0.32 \ (\text{m/s})) \leq 0.1$
	$P(\text{Max}(D_{\text{rib}}) \leq 32 \ (\text{mm})) \leq 0.1$
where	V: Velocity, D: Deflection,
	V*C: Viscous Criterion

Fig. 2: RBDO with vehicle side impact constraints

4. Multi-objective Optimization

Two types of important problems lead to multi-objective optimization. The first type is straightforward, where multiple objectives need to be met; some of these objectives may be in conflict with each other. Another type is in reliability-based robust design (RBRD). Several alternative formulations are possible; one example is where we seek not only to minimize or maximize the mean of the objective but also to minimize the variance of the objective. Thus RBRD problems can also be solved through multi-objective optimization techniques.

The traditional approach in multi-objective optimization is to transform the problem into a single objective one by a generalized weighted sum as a scalar substitute. This approach includes subjective information and can be misleading concerning the nature of the optimum design[35]. Other approaches include ε-constraint method[36] which optimizes the most important objective subject to some pre-determined constraint values of other objectives, and goal programming[37] (GP) which tries to achieve the targets of all pre-determined objectives simultaneously. The Pareto frontier, an efficient frontier of solutions in the objective space, has the property that for any point in it, there is no other point in the set of possible outcomes with a better performance on all objectives simultaneously. Therefore, it can provide valuable information for decision-making. The Pareto frontier can be generated using the weighted-sum method[38], or other methods including genetic algorithms (GA)[39]. Tappeta and Renaud[40] used compromise programming followed by a local approximation to present the Pareto surface iteratively. Li and Fadel[41] proposed a hyper-ellipse method to approximate the bi-objective Pareto frontier.

When considering uncertainties in multi-objective optimization, the computational effort becomes even larger and as a result, very few studies have been completed. Frangopol and Fu[42] proposed a multi-objective optimization approach to deal with structural reliability-based design under multiple limit states. The weight of the structure and probabilities of failure with respect to collapse, first plastic yielding, and excessive elastic deformation, were selected as objectives. The probability of failure for the structural system was calculated by Ditlevsen's second-order upper bound[43].

For engineering problems with multiple objectives to be balanced under uncertainties, it is desirable that under different scenarios, different appropriate multi-objective optimization methods be used, and that for each sub-problem, a suitable reliability method be applied to ensure both accuracy and efficiency in RBDO. As seen earlier, the large computational effort of RBDO is a significant hurdle in applying multi-objective optimization under uncertainty to large engineering systems. In particular, MCS-based methods, although generally more accurate, are rarely used in RBDO for reliability estimation due to their inefficiency. These difficulties need to be overcome before a multi-objective RBDO problem can be solved successfully.

In the context of reliability-based robust design, Du and Chen (2003)[44] replace the bi-objective formulation, i.e., minimize/maximize the mean of the objective, and minimize its variance, with an equivalent single objective formulation, using the percentile approach. Minimizing the difference in objective function values corresponding to two percentiles (say 5% and 95%) also minimizes the variance. A more efficient procedure is to deal with only one of the percentile objectives. It minimizes the upper percentile objective if minimization of both mean and variance of the objective is required, and maximizes the lower percentile objective if maximization of the mean and minimization of variance of the objective is required. Sopory and Mahadevan (2003)[45] compared two single objective formulations -- weighted sum and percentile -- for several problems and found that the solutions and computational efforts of both formulations are similar. In the weighted sum formulation, one has to decide the weights for the two objectives w.r.t. mean and variance. However, the percentile formulation also indirectly assigns different weights to the two objectives by the choice of the percentile value.

Zou (2004)[31] illustrated the application of multi-objective RBDO to door quality optimization, considering closing effort and wind noise (Fig. 3).

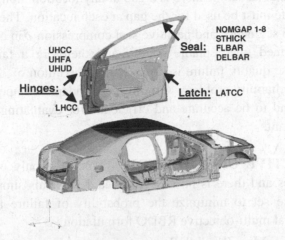

Fig. 3: Car body- door sub-system

DOOR CLOSING EFFORT – The design objective with respect to door closing effort is to keep the energy required to close the car door to be less than a predetermined amount. The energy required for the door to just reach the latch position is referred to as the door closing energy, and it is assumed to consist of three components. The first component, the energy loss to the air, is caused by the pressure rise in the vehicle when the door pushes the air ahead of itself. The second component, due to the seal compression, is the work that the door performs to compress the seal. The third component is due to door deformation, which takes into account the energy absorbed as the door deforms by the load and inertia from the hinge and the seal. The door closing effort is assumed to be unsatisfactory if total door closing effort is greater than 3.0 Joules. (Energy losses not included in this analysis are hinge and latch energy losses due to dynamic seal loads, and losses due to seals other than the primary door seal). The FORM method is found to be adequate for this single limit state problem.

WIND NOISE – The design objective with respect to wind noise is to have a positive seal compression $\delta = u_s - u_g$ at every location along the seal, where u_s represents its unreformed thickness, and u_g the operating gap between the body and door seal lines. If the seal compression δ, as defined by the above equation, is positive, there is a non-zero seal force F applied on both the body and the door. Otherwise, there is no seal force on the body and door since the seal is not compressed, thus leading to a wind noise problem.

Since wind noise occurs if there is a gap at any location along the seal line, a separate limit state must be used for the gap at each location. The continuous seal is divided into 16 segments, and negative seal compression δ in any one of the 16 segments is deemed unacceptable quality and therefore a failure. Thus, the overall wind noise quality failure is expressed by the union of 16 failure events, and is a system reliability problem. A multi-modal adaptive importance sampling method was found to be accurate and efficient[46] for evaluating the wind noise reliability constraint.

MULTI-OBJECTIVE OPTIMIZATION – Since the only concerns are the two quality issues and there is no cost associated with this simplified example, the objectives are set to minimize the probability of failure for both quality issues. The general multi-objective RBDO formulation is:

$$\text{Min } P_{f\_\text{CE}} \& P_{fs\_\text{WN}}$$
$$\text{s.t. } -1.7 \leq d_i \leq 1.9 \ (i = 1,..,4)$$

(3)

where P_{f_CE} is the probability of failure for the door closing effort problem calculated from FORM, P_{fs_WN} is the system-level probability of failure for wind noise problem calculated from multi-modal AIS, and $d_i's$ $(i = 1,..,4)$ represent the mean values of the design variables.

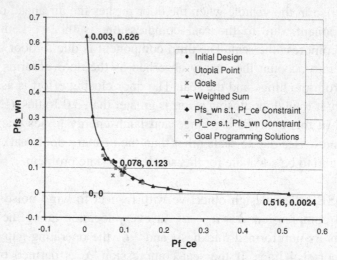

Fig. 4: Multi-objective optimization

Zou's study[31] used the traditional nested approach to solve this problem, since the inverse FORM-based decoupling methods then available could not be applied due to the use of Monte Carlo simulation to evaluate the wind noise probability. However, the recently developed direct decoupling method can overcome this hurdle. Since all objectives and/or constraints are in terms of failure probabilities, RBDO could not take advantage of the deterministic optimization either.

Three multi-objective optimization methods, the weighted sum method, the ε-constraint method and goal programming, were used to investigate the tradeoff between the two objectives, as shown in Fig. 4. An approximate Pareto frontier is constructed using a simple weighted sum approach, and then used for further optimization actions. Two other groups of candidate solutions are found by the ε-constraint method and goal programming. These optimal points are close to the approximate Pareto frontier, which validates their usage as alternative multi-objective RBDO methods.

Design of an aerospace vehicle is a complex process requiring analysis and optimization across multiple disciplines. In many cases, relatively mature (high and low fidelity) disciplinary analysis tools are available. These disciplinary analyses cannot be taken in isolation since they are coupled to one another through shared input and output. Furthermore, system design objectives and constraints may span several disciplines. Integrating disciplinary analyses into a multidisciplinary framework and finding practical ways to solve system optimization problems under uncertainty is a serious challenge.

RELIABILITY ANALYSIS – Multidisciplinary reliability analysis is particularly difficult when there is feedback coupling between the different disciplinary analyses. Consider the two-discipline system shown in Fig. 5. Variables $u_{1,2}$ and $u_{2,1}$ are the state variables (defined such that $u_{i,j}$ is an output of analysis i and an input to analysis j). It is seen that, regardless of which analysis is performed first; an unknown input state variable (either $u_{1,2}$ or $u_{2,1}$) is needed, indicating a feedback condition. Systems with feedback coupling are typically solved with fixed-point iteration. In other words, assumed values for the unknown state variables are initially used; then they are updated by performing the analyses from which they are derived; the analyses are performed again with the updated values; this process continues until convergence is reached.

Using fixed-point iteration within probabilistic analysis algorithms is not usually an ideal approach. For one, the number of probabilistic analysis loops multiplies the number of fixed-point iterations, which in turn multiplies the computational effort of a single set of the disciplinary analyses.

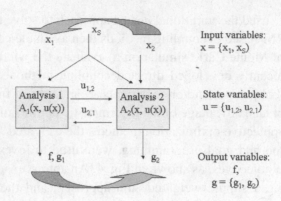

Fig. 5: Feedback coupling of a two-discipline system

Another difficulty is in obtaining gradient information, usually required for the more efficient analytical approximation algorithms. If a finite difference method were used, the fixed-point iteration process for convergence would need to be repeated for each variable. Furthermore, one might select a less stringent convergence criterion to reduce the number of fixed-point iterations, but this introduces 'noise' that can corrupt the gradient information and prevent the convergence of FORM-type reliability analysis algorithms. These problems become worse when probabilistic analysis and optimization are simultaneously attempted for multidisciplinary systems.

Smith (2002)[47] has developed two alternative strategies for multidisciplinary reliability analysis: (1) Markov Chain Monte Carlo simulation, and (2) performing first order second moment analysis first to obtain approximate statistics of the state variables, followed by either FORM or Monte Carlo simulation of individual disciplines.

MDO UNDER UNCERTAINTY – In the next step, optimization under uncertainty for a coupled multidisciplinary system can be achieved by combining the decoupled RBDO approach with deterministic multidisciplinary optimization (MDO) methods. Note that the decoupled RBDO methods (whether based on direct or inverse reliability analysis) replace the probabilistic constraints in the optimization problem with a deterministic approximate equivalent, and perform reliability analysis separately, not within the optimization loop. This makes it easy to apply existing deterministic MDO methods to the outer optimization. Chiralaksanakul et al. (2003)[48] have demonstrated this idea by combining the decoupled RBDO with three different deterministic MDO methods, namely, the multidisciplinary feasible (MDF), individual disciplinary feasible (IDF), and all-at-once (AAO) methods, thus successfully developing a probabilistic MDO methodology.

The probabilistic MDO methodology is taken another step forward by Smith and Mahadevan (2003)[49] for the design of an aerospace vehicle at two levels: a global geometry design and a component structural sizing application. The global geometry application is inter-disciplinary in that it considers a coupled analysis of geometry, weights, and aerodynamic disciplines. It is also a system-level design in terms of physical architecture, in that the geometry design variables define the global characteristics (length, radius, wing areas, etc.) of a vehicle comprised of many component parts (wings, fuel tanks, engines, etc.). The component sizing application involves a single disciplinary analysis, at the component level (in terms of physical architecture), analyzed in terms of multiple limit states. The two design processes are intrinsically linked, and an efficient iterative coupling process from the global to local design and vice versa given uncertainties in system parameters is necessary.

GLOBAL VEHICLE GEOMETRY DESIGN – The vehicle geometry, for illustration purposes, is shown in Fig. 6.

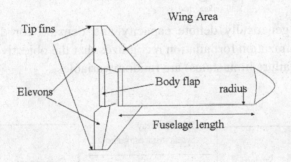

Fig. 6: Illustrative vehicle geometry concept

A vehicle geometry that minimizes mean dry weight is expected to minimize overall cost, so this is chosen as the objective function. For stability, the pitching moment (C_m) for the vehicle should be zero or extremely close to zero. In addition, C_m should decrease as the angle of attack increases. This is achieved by adjusting the control surfaces trim the vehicle as the angle of attack is increased. Thus the aerodynamic analysis for pitching moment constrains the vehicle geometry optimization.

The pitching moment constraint must hold during all flight conditions; nine flight scenarios (constructed with three velocity levels and three angles of attack) are used as a representative sample (Unal *et al.* (1998)[50]). The deterministic optimization problem is written as: *Minimize vehicle dry weight (W) such that the pitching moment coefficient (C_m) for each of 9 scenarios is within acceptable bounds [-0.01, +0.01]*. The problem is reformulated in probabilistic terms as: *Minimize <u>mean</u> weight such that the pitching moment coefficient for all 9 scenarios has a low <u>probability</u> (less than 0.1) of failing to be within the acceptable bounds [-0.01, +0.01]*. This is a multidisciplinary problem requiring the synthesis of information from three analysis codes: a geometry-scaling algorithm, a weights and sizing code and an aerodynamics code.

LOCAL TANK DESIGN – The design goal for the liquid hydrogen tank is to minimize the weight of the tank while meeting the requirements for fuel capacity and structural integrity. The fuel capacity requirement is maintained by choosing the appropriate tank geometry. With tank geometry dictated by the global design, an optimization problem may be formulated as:

$$\text{Min } \mu_{\text{tank weight}} = f\left(f_R(R)\right) \approx f(\mu_R)$$

$$\text{s.t. } P\left(\bigcup_{\text{all failure modes}} R < S\right) < P_{\text{required}} \tag{4}$$

where R and S generically denote capacity and demand for different failure modes. This optimization formulation recognizes that the objective (tank weight) and constraints (failure limit states) are random variables.

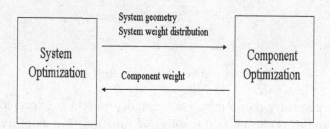

Fig.7: Iteration between system-level and component-level optimization

GLOBAL LOCAL INTEGRATION – An iterative process is needed to converge on optimal solutions for both the system and component designs.

Perhaps the most obvious iteration strategy is to use a brute force fixed-point iteration method; in other words to simply repeat the system–component–system design cycle and hope for ultimate convergence. This idea is depicted in Fig. 7.

This bi-level optimization is a common strategy for design; it does not require inter-level data flow during optimization and preserves a degree of autonomy for component-level designers. However, this strategy may not always work, and may not be able to find a converged solution to the bi-level system with a reasonable amount of computational effort. As more components are added, finding a feasible solution becomes more difficult.

An alternate approach is to integrate the two optimizations into a single problem, and solve the resulting probabilistic MDO problem by combining a decoupled RBDO approach with a deterministic MDO method. However, note that the local reliability constraint is expressed as a union of several failure modes. This presents a problem for inverse SORA, which is designed to handle only individual limit states; hence all three limit states were considered as three different constraints in the optimization[49]. On the other hand, Zou's new direct decoupling method[31] is easily able to handle system reliability constraints.

5. Concluding Remarks

The current state of the art in mechanical system design optimization under uncertainty is mainly focused on individual reliability requirements. System-level reliability requirements have not been widely considered, except with frame and truss structures. A newly developed direct decoupling strategy allows this consideration, and also facilitates a modular approach for including different methods for evaluating different reliability constraints. One particular issue is the availability of sensitivity information, particularly when Monte Carlo methods are used. The applications of various decoupling approaches to realistic problems and comparisons of their relative performances are yet to be done.

The extension of RBDO to robust design is quite recent, with several formulations being evaluated. The robustness requirement has been expressed either as single- or bi-objective optimization, and several approaches are being investigated. The inclusion of model uncertainty within RBDO is yet to be investigated.

The extension of RBDO methods to multi-disciplinary systems has recently been accomplished through the decoupling strategies. This has also been shown to be feasible for systems where the different disciplinary codes are connected through feedback coupling. The direct decoupling approach allows the use of different reliability analysis methods (such as extended FORM, Markov Chain Monte Carlo simulation etc.) for such coupled systems.

The application of RBDO to time-dependent problems has been indirect, with durability requirements expressed through time-independent measures[51]. Direct inclusion of time-dependent reliability in RBDO, with random process treatment of loads and system properties, is yet to be done, and could be computationally prohibitive. Time-dependent problems are particularly relevant to fluid dynamics in aerospace vehicle design. Previous studies with progressive failure were limited to trusses, frames, and composite laminates, where computational effort was not prohibitive. However, this is a challenge in the case of problems where functional evaluations are very time-consuming. A conservative first failure criterion has been used as a surrogate for ultimate system failure in the case of a composite aircraft wing (Liu and Mahadevan, 1998)[52], but requires further investigation.

An important issue in practical application of RBDO is model uncertainty, especially considering the frequent use of response surface models in many RBDO studies. Ongoing research at Vanderbilt University is developing Bayesian methodologies for model validation[53,54], model uncertainty quantification, and inclusion of model uncertainty in design optimization.

While RBDO methodology development marches ahead, with simple numerical examples as proofs of concept, practical implementation of RBDO still appears to be a distant goal, due to the lack of actual data on the random variables. Statistical data collection, which is expensive and time consuming, appears to be the single most difficult hurdle in the practical implementation of the RBDO methodology, and yet very important in developing confidence in the proposed methods.

Acknowledgments

The study on RBDO and robust design methods was funded by the General Motors Research and Development Center under project numbers PLMR-13106 (project manager: Dr. R. V. Lust) and ND0044200 (project manager: Dr. J. Tu). Professor Z. P. Mourelatos at Oakland University, formerly at General Motors, provided guidance with the door closing and wind noise analyses[46,55]. The study on multi-disciplinary systems was supported by funds from NASA Langley Research Center (LaRC), VA (Cooperative Agreements NCC 1-01031 and 1-03023) under the direction of the Multidisciplinary Optimization Branch (Monitors: Dr. Thomas A. Zang, Dr. Jean-Francois Barthelemy and Dr. Natalia Alexandrov). The support is gratefully acknowledged.

References

1. S. S. Rao, "Multi-objective Optimization in Structural Design with Uncertain Parameters and Stochastic Processes", *AIAA J.*, 22, 1670-1678, (1984).
2. D. M. Frangopol, "Multicriteria Reliability-Based Structural Optimization", *Struct. Safe.*, 2, 154-159, (1985).
3. Y. S. Feng and F. Moses, "Optimum Design, Redundancy and Reliability of Structural Systems", *Comput. Str.*, 24 (2), 239-251, (1986).
4. P. Thoft-Christensen and Y. Murotsu, *Application of Structural System Reliability Theory*, (Springer Verlag, Berlin, 1986).
5. R. Rackwitz and B. Fiessler, "Reliability under Combined Random Load Sequences", *Comput. Str.*, 9(5), 489-494, (1978).
6. C. T. Lawrence. and A. L. Tits, "Feasible Sequential Quadratic Programming for Finely Discretized Problems from SIP", *Semi-Infinite Programming, in the Series Nonconvex Optimization and Its Applications, (Eds. Reemtsen, R. and Ruckmann, J. –J.)*, Kluwer Academic Publishers, (1997).
7. A. Haldar and S. Mahadevan, *Probability Statistics and Reliability in Engineering Design*, John Wiley & Sons, (2000).
8. L. Tvedt, "Distribution of Quadratic Forms in Normal Space – Application to Structural Reliability", *J. Eng. Mech.* ASCE, 116(6), 1183-1197, (1990).
9. H. U. Köylüoğlu and S. R. K. Nielsen, "New Approximations for SORM Integrals", *Struct. Safe.*, 5, 119-126, (1988).
10. C. G. Bucher, "Adaptive Sampling – an Iterative Fast Mote Carlo Procedure", *Struct. Safe.*, 5, 119-126, (1988).
11. G. I. Schuëller, H. J. Pradlwarter and C. G. Bucher, "Efficient Computational Procedures for Reliability Estimate of MDOF-systems", *Int. J. Nonlinear Mech.*, 26(6), 961-974, (1991).
12. Y. T. Wu, *NESSUS/FPI User's and Theoretical Manuals – Version 2.4*, Southwest Research Institute, (San Antonio, TX, 1998).
13. R. E. Melchers, "Improved Importance Sampling Methods for Structural System Reliability Calculation", *P. Struct. Safe. Rel. (ICOSSAR)*, 1185-1192, (1989).
14. A. Karamchandani, P. Bjerager and C. A. Cornell, "Adaptive Importance Sampling", *5th Int. Conf. Struct. Safe. Rel. (Eds. Ang, A. H.-S., M. Shinozuka, and G. I. Schuëller), ASCE*, 855-862, (1989).
15. A. Dey and S. Mahadevan, "Ductile Structural System Reliability Analysis using Adaptive Importance Sampling", *Struct. Safe.*, 20, 137-154, (1998).
16. S. Mahadevan and P. Raghothamachar, "Adaptive Simulation for System Reliability Analysis of Large Structures", *Comput. Str.*, 77, 725-734, (2000).
17. S. Mahadevan and X. Liu, "Probabilistic Optimum Design of Composite Laminates", *J. Compos. Mater.*, 32(1), 68-82, (1998).
18. E. Aktas, F. Moses and M. Gohsn, "Cost and Safety Optimization of Structural Design Specifications", *Reliab. Eng. Syst. Safe.*, 73, 205-212, (2001).
19. A. S. Al-Harthy and D. M. Frangopol, "Integrating System Reliability and Optimization in Prestressed Concrete Design", *Comput. Str.*, 64(1-4), 729-35, (1997).
20. M. Miki, Y. Murotsu, T. Tanaka and S. Shao, "Reliability-Based Optimization of Fibrous Laminated Composites", *Reliab. Eng. Syst. Safe.*, 56, 285-290, (1997).
21. H. W. Leheta and A. E. Mansour, "Reliability-Based Method for Optimal Structural Design of Stiffened Panels", *Marine Str.*, 10, 323-352, (1997).
22. D. R. Oakley, R. H. Sues and G. S. Rhodes, "Performance Optimization of Multidisciplinary Mechanical Systems subject to Uncertainties", *Probabilist. Eng. Mech.*, 13(1), 15-26, (1998).
23. S. Mahadevan, "Probabilistic Optimum Design of Framed Structures", *Compu. Str.*, 42(3), 365-74, (1992).

24. Y. Pu, P.K. Das and D. Faulkner, "A Strategy for Reliability-Based Optimization", *Eng. Str.*, 19, 276-282, (1997).

25. Y. T. Wu, "Computational Methods for Efficient Structural Reliability and Reliability Sensitivity Analysis", *AIAA J.*, 32(8), 1717-23, (1994).

26. K. K. Choi, X. Yu and K. Chang, "A Mixed Design Approach for Probabilistic Structural Durability", *6th AIAA/USAF/NASA/ISSMO Symp. Multidiscip. Anal. Optim.*, 785-95, (Bellevue, WA, 1996).

27. Y. Tsompanakis and M. Papadrakakis, "Robust and Efficient Methods for Reliability-Based Structural Optimization", *Comput. Meth. Shell Spatial Str .IASS-IACM*, (Athens, Greece, 2000).

28. J. O. Royset, A. Der Kiureghian and E. Polak, "Reliability-based optimal structural design by the decoupling approach", *Reliab. Eng. Syst. Safe.*, 73, 213-221, (2001).

29. X. Du and W. Chen, "Sequential Optimization and Reliability Assessment Method for Efficient Probabilistic Design", *DETC2002/DAC-34127, Eng. Conf.*, (Montreal, Canada, Sep. 29-Oct. 2, 2002).

30. J. Tu, K. K. Choi and Y. H. Park, "A New Study on Reliability-Based Design Optimization", *J. Mech. Des.*, 121, 557-64, (1999).

31. T. Zou, "Efficient Methods for Reliability-Based Design Optimization", Ph.D. Dissertation, Vanderbilt University, (Nashville, Tennessee, 2004).

32. A. Sopory, J. Tu and A. Kloess, "Reliability-Based Design Optimization using a Decoupled Approach", GM Internship report, (2003).

33. R. J. Yang, L. Gu, C. H. Tho, K. K. Choi and B. D. Youn, "Reliability-Based Multidisciplinary Design Optimization of a Full Vehicle System", AIAA-2002-1758, *Proceedings, 43rd AIAA/ASME/ASCE/ AHS/ASC Str. Struct. Dyn. Mater. Conf. Exhibit*, (Atlanta, GA, 2002).

34. Z. Mourelatos, Personal Communication, (2003).

35. T. W. Athan and P. Y. Papalambros, "A Note on Weighted Criteria Methods for Compromise Solutions in Multi-Objective Optimization", *Eng. Optim.*, 27, 155-176, (1996).

36. R. E. Steuer, *Multiple Criteria Optimization: Theory, Computation, and Application*, John Wiley & Sons, Inc., (1986).

37. J. L. Cohon, *Multi-objective Programming and Planning*, Academic Press, Inc., (1978).

38. A. M. Geoffrion, "Proper Efficiency and the Theory of Vector Maximization", *J. of Math. Anal. Appl.*, 22, 618–630, (1968).

39. P. G. Busacca, M. Marseguerra and E. Zio, "Multi-objective Optimization by Genetic Algorithms: Application to Safety Systems", *Reliab. Eng. Syst. Safe.*, 72, 59-74, (2001).

40. R. Tappeta and J. Renaud, "Interactive Multi-objective Optimization Design Strategy for Decision Based Design", *ASME Des. Eng. Tech. Conf.*, DETC99/DAC-8581, (1999).

41. Y. Li and G. M. Fadel, "Approximating Pareto Curves using the Hyper-Ellipse", *AIAA/ISSMO Internet Conf. Multidiscip. Optim.*, (1998).

42. D. M. Frangopol and G. Fu, "Limit States Reliability Interaction in Optimum Design of Structural Systems", *5th Inter. Conf. Struct. Safe. Rel. (ICOSSAR '89)*, 1879-85, (1989).

43. O. Ditlevsen, "Narrow Reliability Bounds for Structural Systems", *J. Struct. Mech.* ASCE, 7 (4), 453-472, (1979).

44. X. Du, A. Sudjianto and W. Chen, "An Integrated Framework for Optimization using Inverse Reliability Strategy", DETC-DAC48706, *ASME Des. Auto. Conf.*, (Chicago, IL, September 2-6, 2003).

45. A. Sopory and S. Mahadevan, "Sequential Methods for Reliability-Based Optimization and Robust Design: Report to GM", Vanderbilt University, (Nashville, Tennessee, 2003).

46. T. Zou, Z. Mourelatos, S. Mahadevan and P. Meernik, "Reliability Analysis of Systems with Nonlinear Limit States; Application to Automotive Door Closing Effort", *SAE World Congr.*, 2003-01-0142, (Detroit, MI, March 3-6, 2003).

47. N. Smith, "Probabilistic Methods for Aerospace System Conceptual Design," M.S. thesis, Vanderbilt University, (Nashville, TN, 2002).

48. A. Chiralaksanakul, A. Copeland and S. Mahadevan, "Probabilistic Optimization of Multidisciplinary Systems: Report to NASA Langley Research Center", Vanderbilt University, (Nashville, Tennessee, 2003).

49. N. Smith and S. Mahadevan, "Probabilistic Coupling of System-Level and Component-Level Designs: Report to NASA Langley Research Center", Vanderbilt University, (Nashville, Tennessee, 2003).

50. R. Unal, R. A. Lepsch and M. L. McMillin, "Response Surface Model Building and Multidisciplinary Optimization Using D-Optimal Designs," *P. 7th AIAA/USAF/NASA/ISSMO Symp. Multidiscip. Anal. Optim.*, 405-411, AIAA 98-4759, (St. Louis, Sept. 1998).

51. K. K. Choi, B. D. Youn and J. Tang, "Structural Durability Design Optimization and Its Reliability Assessment", DETC-DAC48711, *ASME Des. Auto. Conf.*, Chicago, IL, September 2-6, 2003).

52. X. Liu and S. Mahadevan, "System Reliability-Based Optimization of Composite Structures," *P. 7th AIAA/USAF/NASA/ISSMO Symp. Multidiscip. Anal. Optim.*, St. Louis, Missouri, 856-860, Sept. (1998).

53. R. Zhang and S. Mahadevan, "Integration of Computation and Testing for Reliability Estimation," *Reliab. Eng. Syst. Safe.*, 74(1), 13-21, (2001).

54. R. Zhang and S. Mahadevan, "Bayesian Methodology for Reliability Model Acceptance", *Reliab. Eng. Syst. Safe.*, 80(1), 95-103, (2003).

55. T. Zou, Z. Mourelatos and S. Mahadevan, "Simulation-Based Reliability Analysis of Automotive Wind Noise Quality", *SAE World Congr.*, 2004-01-0238, (Detroit, MI, 2004).

CHAPTER 11

DESIGN OPTIMIZATION WITH UNCERTAINTY, LIFE-CYCLE PERFORMANCE AND COST CONSIDERATIONS

Dan M. Frangopol

Department of Civil, Environmental, and Architectural Engineering
University of Colorado, Boulder, CO 80309-0428, USA
E-mail: dan.frangopol@colorado.edu

Kurt Maute

Department of Aerospace Engineering Sciences
University of Colorado, Boulder, CO 80309-0429, USA
Email: maute@colorado.edu

Min Liu

Formerly, Department of Civil, Environmental, and Architectural Engineering,
University of Colorado, Boulder, CO 80309-0428, USA
Email: minliu@illinoisalumni.org

This chapter reviews design optimization approaches which account for uncertainty, life-cycle performance, and cost. State-of-the-art probabilistic methods for analyzing the stochastic response of components and structural systems are outlined and their integration into design optimization methods discussed. Formulations for including probabilistic design criteria into reliability-based optimization problems are presented. The importance of life-cycle optimization under uncertainty with multiple objectives is emphasized and optimization methods for such problems are presented. This chapter shows that accounting for uncertainty via probabilistic approaches is an important and powerful design tool in various fields of application.

1. Introduction

Our knowledge to design and optimize structural systems under uncertainty is continuously growing. The intended service lives of these systems are typically several decades and sometime centuries. During this time, structures are deteriorating and are usually exposed to abnormal loads of different types ranging from natural to man-made disasters. An integrated approach is necessary to optimize these systems taken into consideration uncertainty, life-cycle performance, and cost, among other factors. This chapter focuses on such an approach. It is assumed that all uncertainties are captured by probabilistic concepts and methods, that random processes are discretized by random variables, and that reliability analysis is the method to quantify the safety state of uncertain engineering systems under stochastic environments.

As indicated in Frangopol and Maute[1], the methodologies used in reliability-based structural optimization for taking into consideration uncertainties associated with engineering systems can be classified into two groups: robust design optimization (RDO) and reliability-based design optimization (RBDO). The purpose of the RDO approach is to simultaneously optimize the deterministic performance and to minimize the sensitivity of the performance with respect to random variations[2]. The RBDO approach allows the design for a specific target reliability level, accounting for the various sources of uncertainty in a quantitative manner. This approach operates on two sets of variables including both design and random variables.

Formally, RBDO and stochastic programming (SP) address design problems under uncertainty and the generic problem formulations are closely related. However, the research on SP has focused on decision-making processes in which the uncertainty can be gradually reduced as more information becomes available. The goal of SP is to find values of the initial decision variables as well as functions for updating these variables as additional information about the problem evolves. The solution of these kinds of problems requires the simulation and optimization of multi-stage processes in which decisions alternate with observations. The complexity of multi-stage problems has essentially limited the type of problems which can be solved today by SP methods. The main body of work on SP focuses on linear SP 2-stage problems. The reader is referred to Birge and Louveaux[3] for an introduction and overview over stochastic programming.

This chapter focuses on the RBDO approach. It presents a brief overview of formulation of reliability-based design optimization problems under uncertainty (Section 2), an in-depth description of reliability analysis (Section 3) and reliability-based design optimization (Section 4) methods, a discussion on life-cycle optimization under uncertainty with multiple objectives (Section 5), references to recent applications (Section 6), and conclusions (Section 7). Additional background, including the historical background of the RBDO approach can be found in Frangopol and Moses[4], Frangopol[5,6], and Frangopol and Maute[1,2].

2. Formulation of RBDO Problems

In the presence of uncertainties a design optimization problem typically includes deterministic and probabilistic design criteria. Probabilistic criteria are often embedded as constraints restricting the failure probability but can also be used to formulate objectives. Frangopol and Moses[4] identified several types of RBDO problems which are also summarized in Frangopol and Maute[1]. The objective function can be the total expected cost, the total expected utility, the system failure probability, and the probability of occurrence of a specified event, among others.

A generic reliability-based optimization problem can be formulated as follows with **s** being the set of N_S design variables and **y** the set of N_Y random variables:

$$\begin{aligned}
&\min_{\mathbf{s}} && P(c(\mathbf{s},\mathbf{y}) > \overline{c}) \\
&\text{subject to} && \overline{P}_i - P(g_i(\mathbf{s},\mathbf{y}) < 0) \geq 0 \quad i=1...N_P \\
&&& g_j^D(\mathbf{s}) \geq 0 \quad j=1...N_D \\
&&& \mathbf{s} \in \mathbf{S}
\end{aligned} \tag{1}$$

The objective is to minimize the probability of the cost function $c(\mathbf{s},\mathbf{y})$ being larger than a target value $\overline{c}$. One set of constraints limits the failure probabilities associated with the limit state functions $g_i(\mathbf{s},\mathbf{y})$, defining failure as $g_i(\mathbf{s},\mathbf{y}) < 0$. The maximum acceptable failure probability is denoted by $\overline{P}_i$ and the number of probabilistic constraints by N_P. A second set of inequalities, $g_j^D(\mathbf{s}) \geq 0$, contains all deterministic constraints. The number of deterministic constraints is denoted by N_D. The optimization problem (1) can be also formulated using different measures of probability, such as the reliability index. These formulations are discussed in Section 4 of this chapter. Recently, there has been considerable progress in formulating and solving RBDO with multiple objectives and in a time-dependent context. An overview of this progress is given in Frangopol and Liu[7], and a discussion is presented in Section 5.

The feasible set of design variables is denoted by **S**. A design variable can be either continuous or discrete. Typically the design variables and random variables define sizing variables (*e.g.*, cross-sectional dimensions), the geometry and the shape of the overall structure, the structural topology (*e.g.*, number of girders in a bridge, number of columns supporting a roof), the material distribution and material properties, the operating conditions (such as abnormal loadings and environments) and various inspection/maintenance parameters (such as the type and quality of inspection/maintenance methods). These structural properties, operating and inspection/maintenance conditions can be either deterministic or subject to stochastic variations. In the latter case, the design variables can be associated, for example, with mean, mode or median values.

The state equations governing the stochastic response of the system also depend on the design variables $\mathbf{s}$ and the random variables $\mathbf{y}$:

$$\mathbf{R}(\mathbf{s},\mathbf{y},\hat{\mathbf{v}}(\mathbf{s},\mathbf{y})) = \mathbf{0} \tag{2}$$

where $\mathbf{R}$ is the vector of residuals associated with a discretized system and $\hat{\mathbf{v}}$ is the vector of N_V state degrees of freedom. The state equations (2) can be either added to the set of constraints of the optimization problem (1) or explicitly satisfied when evaluating the objective and constraints. In the following presentation, the stochastic state equations (2) are treated separately, as their mathematical properties and meaning differ from other design constraints. The stochastic response and the reliability-based design criteria need to be evaluated at each iteration of the design optimization process. Although this approach allows calling the reliability analysis method in a black-box fashion, great care and insight is needed to select the appropriate reliability analysis method, depending on the formulation of the optimization problem and the type of probability criteria used.

3. Reliability Analysis Methods

The evaluation of the probabilistic design criteria is of pivotal importance for the overall design optimization procedure. These criteria characterize the stochastic response of the system being optimized. The choice of the stochastic analysis method depends on the expected failure probability, the nonlinearity of the failure criteria with respect to the random variables, and the number of random variables, among others. Reliability analysis methods are a subset of stochastic analyses approaches tailored to evaluate the probability of failure.

In the following, first the basic concepts of stochastic analysis are introduced and the reliability methods most often used within RBDO are presented. As stochastic analysis methods and RBDO have their roots in structural analysis and design, the presentation focuses on problems related to structural mechanics. A detailed presentation of structural reliability concepts can be found for example in the books by Ang and Tang[8], Madsen *et al.*[9], Thoft-Christensen and Murotsu[10], and Melchers[11], as well as in the review articles by Shinozuka[12], Rackwitz[13], and Schuëller[14,15]. The same concepts, however, can be applied to other disciplines and multi-disciplinary problems. An overview of applications of RBDO to multidisciplinary design problems is given, for example, in Agarwal and Renaud[16] and Allen and Maute[17].

3.1. *Probabilistic Measures*

The stochastic response of a system can be characterized by stochastic moments and probabilities of particular events occurring. Typically the first and second moments, that are the mean and the standard deviation, are used for formulating engineering design problems. Of particular interest in RBDO is the probability of failure P_f which is defined by the occurrence of the event $g(\mathbf{y}) < 0$, where $g(\mathbf{y})$ is the limit state function. The probability of failure P_f can be computed by evaluating the following N_y-dimensional convolution integral over the failure domain:

$$P_f = \int_{g<0} f(\mathbf{y}) \, d\mathbf{y} \qquad (3)$$

where $f(\mathbf{y})$ is the joint probability density function. Only in special cases this convolution integral has a closed-form solution, for example if the random variables $\mathbf{y}$ are normal (Gaussian) and the limit state function $g(\mathbf{y})$ is linear in the random variables $\mathbf{y}$. In this case the probability of failure P_f is:

$$P_f = \Phi(-\beta) \qquad (4)$$

where Φ is the standard normal distribution function and β the reliability index. The above concept can be generalized and the reliability level of a system can be characterized by the performance function P_p:

$$P_p = \int_{g<p_m} f(\mathbf{y}) \, d\mathbf{y} \qquad (5)$$

where p_m is the performance measure. If $p_m = 0$ the performance function P_p is identical with the failure probability P_f. Evaluating the performance function requires the solution of a N_y-dimensional convolution integral over the failure domain. The performance function is particularly useful in the context of design optimization and will be discussed in Section 4.

The fundamental reliability analysis problem introduced above involves a single failure mode of a single component. The reliability of a system, however, can only be correctly assessed by considering the full structural system as a single entity and can be accurately computed only if all its failure models are taken into account. Moses[18] and Frangopol and Moses[4] have shown for structural systems that the reliability of a system can be vastly different from the reliability of its components depending on the system topology and geometry, material behavior, statistical correlation, and variability in loads and strengths. The simplest models for system reliability are series and parallel failure modes. Ang and Tang[8] indicate practical expressions for system reliability based on lower and upper bounds for both series and parallel systems. Some of these bounds consider correlation between pairs of potential failure modes. More complex system models involving series systems of parallel systems have been used, for example, by Enevoldsen and Sørensen[19] and Estes and Frangopol[20].

In most reliability-based engineering studies the system, external loads and operating conditions are idealized as time invariant. However, loads and operating conditions often fluctuate in time and, under environmental stressors, system properties, such as the strength of materials, are time-variant. Therefore, reliability problems are typically time-variant. The study of time-variant structural system reliability is still under development. One option to characterize the stochastic nature of time-variant system is to evaluate the probability of failure at specific points in time, the so-called point-in-time failure probability. However more relevant for design problems but computationally more costly is to find the probability of system failure over a period of time t, that is the time-to-failure probability $P_f(t)$ given as:

$$P_f(t) = P\{ g(t) < 0 \ in \ (0,t] \} \tag{6}$$

where the limit state function $g(t)$ is a function of time t. This approach requires complex and costly integrations. For example, Mori and Ellingwood[21,22] Enright and Frangopol[23,24] and Kuschel and Rackwitz[25] applied this approach to both highly idealized and realistic systems.

3.2. Basic Analysis Methods

Determining the probability of occurrence of a single failure mode of a single component or the probability of a failure of an entire system, whether it is time-invariant or time-variant, requires the evaluation of the convolution integrals (1) or (3) at some level. As the failure domain is typically a nonlinear implicit function of the random variables, analytical solutions do not exist but numerical integration schemes are needed. Basic reliability analysis methods can be divided into sampling approaches, approximate reliability analysis methods, and stochastic perturbation and projection schemes for differential equations. These basic methods can be either directly applied to the model predicting the limit state $g(\mathbf{y})$ for a specific realization of the random variables $\mathbf{y}$ or to a meta-model approximating the limit state. Meta-models are frequently used in combination with sampling methods and large numerical simulation models for which direct sampling would lead to unacceptable large computations costs.

In general, a compromise between accuracy and numerical efficiency needs to be found. In addition, if the design optimization procedure is driven by a gradient-based optimization algorithm, reliability analysis methods need to be chosen which allow efficiently evaluating the design sensitivities of probabilistic design criteria. In the following, basic reliability analysis methods and meta-modeling approaches used within design optimization procedures will be presented.

3.2.1. *Sampling Methods*

The stochastic response of a system can be evaluated by sampling methods which allow evaluating not only the probability of failure but the entire stochastic response. By sampling repeatedly the joint probability density function $f(\mathbf{y})$ and evaluating the limit state function $g(\mathbf{y})$ for each sample, the probability of failure, the expected value and all statistical moments of $g(\mathbf{y})$ can be determined. The most well-known sampling method is Monte Carlo Simulation (MCS). While MCS is widely used due to its generality, simplicity, and accuracy on problems that are highly nonlinear with respect to the uncertainty parameters, it requires a large number of samples leading to often unacceptable numerical costs, in particular for reliability analysis of practical engineering problems with typically high levels of reliability. For example, if a constraint required a failure probability to be less than 10^{-4}, then to achieve an error of less than 20% based on a 95% confidence interval, MCS requires more than 10^{6} system evaluations. Therefore, Monte Carlo simulation is typically impractical for implicit systems solved, for example, by high-fidelity numerical simulation.

The efficiency of MCS can be improved by selective, constrained sampling. McKay *et al.*[26] proposed a Latin Hypercube sampling (LHS) approach which samples the entire space of the random variables but selects uniformly samples from intervals with equal probabilities. LHS requires significantly less samples than MCS but the sampling procedure is more involved. Importance sampling (IS) methods select samples from an area of interest, or important area, using an estimator of the following form:

$$P_f = \frac{1}{N_I} \sum_i^{N_I} I\left[g(\mathbf{y}^{(i)})\right] f(\mathbf{y}^{(i)}) / \tilde{f}(\mathbf{y}^{(i)}) \quad ; \quad I\left[g(\mathbf{y}^{(i)})\right] = \begin{cases} 0 \ for \ g(\mathbf{y}^{(i)}) > 0 \\ 1 \ for \ g(\mathbf{y}^{(i)}) \le 0 \end{cases} \quad (7)$$

where $\mathbf{y}^{(i)}$ are samples according the distribution $\tilde{f}(\mathbf{y}^{(i)})$ which is concentrated in the important area. The number of samples is denoted by N_I. For example, for determining the failure probability the area of interest is around the most probable point of failure along the limit state $g(\mathbf{y}) = 0$. IS methods were first introduced by Kahn and Marshall[27] and later adapted and refined for reliability analysis of engineering systems, for example, by Shinozuka[28], Schuëller and Stix[29], Wu et al.[30], Melchers[31], and Macke and Bucher[32]. Kim and Ra[33] report that the number of samples can be reduced by up to 20 times for typical engineering applications in comparison with Monte Carlo simulation. However, the authors note that the efficiency of IS techniques may strongly depend on the problem.

3.2.2. *Approximate Reliability Analysis Methods*

While sampling methods can characterize the overall stochastic response of a system, reliability analysis methods are tailored towards approximating the probability of failure. In order to roughly characterize the influence of random variables on the performance of the design, often first-order approaches such as First-order-second-moment (FOSM) methods are integrated into the design process. These methods may be sufficient to characterize the influence of small random perturbations about the mean design and are used within Robust Design Optimization (RDO) strategies. However, in order to consider the failure probability of engineering systems more sophisticated approximation methods need to be employed.

One of the most popular and well explored RBDO approaches is based on the concept of characterizing the probability of failure or survival by the reliability index and performing computations based on first-order reliability methods (FORM). The reliability index β is defined as the distance from the origin to the most probable point (MPP) of failure, also called design point, on the failure surface in the space of the standard normal variables $\mathbf{u}$:

$$\beta = \left|\mathbf{u}^*\right| \quad (8)$$

where $\mathbf{u}$ is the vector of N_U independent random variables and $\mathbf{u}^*$ denotes the MPP.

The limit state function $g(\mathbf{u})$ is assumed continuous and differentiable. Following Equation (3) the probability of failure $P(g < 0)$ is then found by integrating the first order approximation of the limit state function $g(\mathbf{u}) = 0$ at

Table 1: Transformations of standard normal variables into variables of common probability distribution functions.

Distribution Type	Transformation $y = T_u^{-1}(u)$
Normal (μ,σ)	$y = \mu + \sigma u$
Lognormal (μ,σ)	$y = e^{\mu+\sigma u}$
Uniform (a,b)	$y = a + \dfrac{(b-a)}{2}(1 + erf(u/\sqrt{2})$
Gamma (a,b)	$y = ab\left(\dfrac{u}{\sqrt{9a}} + 1 - \dfrac{1}{9a}\right)^3$

the MPP. The feasibility of the mean-value point needs to be checked, to identify on which side of the limit state surface the mean design point lies, that is whether $P(g < 0)$ is the probability of failure or safety.

To evaluate the reliability index by FORM, the random variables $\mathbf{y}$ are mapped into standard normal space $\mathbf{u}$:

$$\mathbf{u} = \mathbf{T}_u(\mathbf{y}) \tag{9}$$

where $\mathbf{T}_u : \mathbb{R}^{N_Y} \to \mathbb{R}^{N_U}$ is generally a non-linear mapping that depends on the type of random distribution of $\mathbf{y}$. For example, if the random variables $\mathbf{y}$ are correlated non-normal, the Rosenblatt transformation is typically used to map $\mathbf{y} \to \mathbf{u}$ [34,35]. While these transformations are in general not exact, the mapping errors are insignificant for the most common distributions. For uncorrelated variables $\mathbf{y}$ the transformation $\mathbf{T}_u^{-1}$ of the most common distributions are listed in Table 1. In the case of correlated variables $\mathbf{y}$, the random variables need to be transformed first into uncorrelated variables before the mappings listed in Table 1 can be applied.

In the standard normal space the Most Probable Point (MPP), or the closest point to the failure surface, is to be located by solving a nonlinear optimization problem subject to one equality constraint in the space of the standard normal variables $\mathbf{u} \in \mathbb{R}^{N_U}$:

$$\begin{aligned} &\min_{u} \quad |\mathbf{u}|^2 \\ &s.t. \quad g(\mathbf{u}) = 0 \end{aligned} \tag{10}$$

For convexification purposes, the above optimization problem can be augmented by a penalty term as follows:

$$\min_{u} \quad |\mathbf{u}|^2 + r_p |g(\mathbf{u})|^2$$

$$s.t. \quad g(\mathbf{u}) = 0 \tag{11}$$

where r_p is a penalty factor. Liu and Der Kiureghian[36] advocate formulation (11) if a dual solution method is employed requiring a convex optimization problem. In general, the optimization problems (10) and (11) can be either solved by nonlinear programming methods or by algorithms tailored to the specific structure of the problem, such as HL-RF iteration scheme by Hasofer and Lind[37] and Rackwitz and Fiessler[38]. Liu and Der Kiureghian[36] comparing generic optimization algorithms with the HL-RF scheme showed by numerical examples that both approaches are equally robust for FORM reliability analysis and recommend in particular a Sequential Quadratic Programming and the HL-RF methods due to their efficiency for problems with large nonlinear numerical simulation models. As the problems (10) and (11) are subject to only one constraint, the gradients of the limit state function with respect to the random variables $\mathbf{u}$ are preferably evaluated by the adjoint method, for efficiency purposes.

One of the shortcomings of FORM is the lack of an estimator of the approximation error. FORM has inherent errors in approximating the limit state function $g(\mathbf{u}) = 0$ as linear. Depending on the curvature of the limit state function, the probability will be underestimated or overestimated (see Figure 1). As long as the limit state surface $g(\mathbf{u}) = 0$ is approximately linear in the vicinity of the MPP, FORM leads to acceptable results. If the curvature is too large the accuracy of the probability can be improved by a second-order approximation which is discussed below. However, a FORM analysis does not provide any means of estimating the approximation error. In addition, FORM is prone to

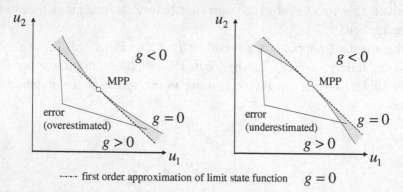

Fig. 1: Approximation error of FORM: probability of failure is overestimated (left), probability of failure is underestimated (right).

robustness issues. The FORM prediction strongly depends on how well the MPP is determined. In particular, for problems with a large number of random variables finding the MPP with high accuracy is costly and often not practical. In case of multiple local minima with comparable distances to the origin of the standard normal space, all local minima need to be determined. Der Kiureghian and Dakessian[39] propose the use of multiple linearization strategies by combining the individual reliability indices associated with each local minima. Systematic multiple starting point procedures, such as the hyperspace division strategy of Katsuki and Frangopol[40], can be used for finding the local minima.

In the context of design optimization, the basic FORM formulation may also suffer from the fact that for a particular design the problems (10) and (11) have no solution, that is there is no realization of the design for which the equality constraint is satisfied. This problem can be overcome by characterizing the stochastic response by the worst performance measure for a target reliability index $\overline{\beta}$. This approach leads to the following optimization problem:

$$\min_{u} \quad g(\mathbf{u})$$
$$s.t. \quad |\mathbf{u}|^2 - \overline{\beta}^2 = 0 \tag{12}$$

In contrast to the formulations (10) and (11) it is guaranteed that there exists a realization which satisfies the equality constraint. However, the meaning of the value of $g(\mathbf{u})$ at the solution of the formulation (12) provides only a qualitative rather than quantitative insight into the stochastic response. The usefulness of formulation (12) in the context of design optimization will be discussed in Section 4.

The accuracy of FORM can be improved by a second-order approximation of the failure surface, $g(\mathbf{u}) = 0$, at the MPP. In a transformed standard normal space $(\mathbf{u}', \mathbf{u}'')$ the approximation of the failure surface can written as follows[41]:

$$g(\mathbf{u}', \mathbf{u}'') = \beta - \mathbf{u}' + \frac{1}{2} \sum_{i=1}^{N_u - 1} \kappa_i \mathbf{u}''_i \tag{13}$$

where κ_i is the principal curvature and $\mathbf{u}''_i$ the principal directions of the paraboliod approximating the failure surface at the MPP. Tvedt[42] derived an exact solution of the second order approximation (13) for the special case where $\beta \kappa_i > -1$. For large β-values, Breitung[43] and Hohenbichler and Rackwitz[44] suggest the following asymptotic approximations:

$$P_f \approx \Phi(-\beta) \prod_{i=1}^{N_u - 1} \frac{1}{\sqrt{1 + \beta \kappa_i}} \approx \Phi(-\beta) \prod_{i=1}^{N_u - 1} \frac{1}{\sqrt{1 + (\varphi(\beta)/\Phi(\beta))\kappa_i}} \tag{14}$$

where φ is the standard normal density function.

The main disadvantage of the second-order reliability method (SORM) is the computational complexity. In addition to finding the MPP, SORM requires the evaluation of the second-order derivatives of the limit sate function:

$$H_{ij} = \frac{\partial^2 g}{\partial u_i \partial u_j}\bigg|_{MPP} \tag{15}$$

where H_{ij} is the Hessian of the limit state function. Often it is not practical to determine the second-order derivatives, either because analytical expressions are not available or because finite difference schemes lead to inaccurate results. Furthermore, the evaluation of H_{ij} is often prohibitively costly if the number of random variables is large. Alternatively, Der Kiureghian *et al.*[45] and Der Kiureghian and DeStefano[46] propose point-fitting and curve-fitting SORM algorithms, respectively, approximating the principal curvature without the need of computing second-order derivatives. However, the point-fitting approach requires finding $2(N_u - 1)$ points on the failure surface and the curve-fitting algorithm up to N_u MPP solutions.

While FORM may lead to inaccurate predictions of the probability of failure, numerous RBDO approaches are based on FORM. In addition to relative low computational costs in comparison to sampling methods, FORM allows the efficient evaluations of design sensitivities. In contrast, SORM is hardly ever used for RBDO problems due to the significantly increased computational costs over FORM.

3.2.3. *Stochastic Response Surface Methods*

For complex engineering systems with non-Gaussian random variables, sampling methods are often computationally too costly and approximate reliability methods lead to unacceptable errors. In these cases, Faravelli[47] and Rajashekhar and Ellingwood[48], among others, advocate stochastic analysis procedures based on response surfaces, also called surrogate models, providing a more efficient and accurate alternative. Stochastic response surface methods (SRSMs) approximate the limit state function $g(\mathbf{y})$ by typically simple explicit or implicit expressions, similar to response surface methods approximating the objective and constraints for design optimization purposes. Accurate sampling methods, for example, can be then applied to the explicitly approximated limit state function $\tilde{g}(\mathbf{y})$ at low computational costs. In the following a brief overview of SRSMs for reliability analysis is provided and their suitability for RBDO discussed. For details about stochastic SRSMs the reader is referred to Myers[49], Wang[50] and Bucher and Macke[51].

SRSMs for reliability analysis consist of the following three steps: (a) selection of the type of response surface, (b) fitting the response surface to exact limit state function, and (c) evaluating the probability of failure based on approximated limit state function. The type of response surface ranges from polynomials, exponential relationships, and radial basis functions up to Kriging models and neural networks. The response surface can be either formulated directly in the space of the random variables or in the standard normal space, in order to account for the stochastic distribution of random variables. For example, Isukapalli *et al.*[52] suggest a SRSM based on a polynomial chaos expansion (PCE), approximating the limit state function in the standard normal space by Hermite polynomials:

$$g(\mathbf{y}(\mathbf{u})) \approx a_0^p + \sum_{i=1}^{N_u} a_i^p \Gamma_1(u_i) + \sum_{i=1}^{N_u} \sum_{j=1}^{i} a_{ij}^p \Gamma_2(u_i, u_j) + \sum_{i=1}^{N_u} \sum_{j=1}^{i} \sum_{k=1}^{j} a_{ijk}^p \Gamma_3(u_i, u_j, u_k) + \dots \quad (16)$$

where $\mathbf{y}(\mathbf{u})$ is the inverse mapping of (9), p is the order of the polynomial, Γ_n are the multidimensional Hermite polynomials of degree $n = 1 \dots N_u$, and $\{a_i^p, a_{ij}^p, a_{ijk}^p, \dots\}$ are unknown coefficients to be determined. The accuracy of the approximation increases with the order of the polynomial and, for normal random variables $\mathbf{y}$, the approximation converges in a mean-square sense. However, as p increases the number of unknown coefficients N_{pc} and the computational costs for constructing the stochastic response surface increase exponentially. The number of coefficients of a p-th order approximation with N_u random variables can be computed as follows[53]:

$$N_{pc} = \sum_{i=0}^{p} \frac{(N_u + i - 1)!}{i!(N_u - 1)!} \quad (17)$$

The response surface is fitted to a particular limit state function by sampling the exact response at selected support points $\hat{\mathbf{y}}_i$ and performing a least-square fit. The selection scheme, also called designs-of-experiments, has a significant influence on the quality of the response surface and include saturated, full factorial, composite, Monte Carlo and Latin Hypercube sampling. In the context of reliability analysis, the samples are often constructed in a transformed space concentrated along the failure surface at the MPP. As the MPP is not known in advance, iterative adaptive schemes can be used to fine-tune the selection scheme. In order to reduce the numerical costs, random variables with low coefficient of variation and small influence on the failure surface can be eliminated, either a-priori or in an adaptive fashion[54]. As approximation error of the response surface may lead significant errors, the true response needs to be accurately represented in particular along the limit state surface $g(\mathbf{y}) = 0$ in the vicinity of the MPP. Therefore, the quality of the response surface needs to be checked, for example, by the analysis of variance (ANOVA) comparing the variation of the exact and approximated limit state functions.

SRSM can provide a powerful tool for characterizing the stochastic response of a system and for evaluating the probability of failure. The success of SRSMs strongly depends on the appropriate type of response surface and the design-of-experiment scheme. The main computational costs are associated with the evaluation of the exact limit state at the support points. In order to limit the number of support points, low order approximations are typically preferred.

The systematic integration of SRSMs into RBDO is still in its infancy. While SRMS are tailored for approximating the stochastic response with respect to the random variables, they need to be rebuilt for each design in the design optimization loop, leading to significant computational costs. Choi et al.[55] and Kim et al.[56] have integrated SRSMs based polynomial chaos expansions into design optimization problems with reliability constraints. In order to avoid rebuilding the response surface for approximating the reliability criteria, Eldred et al.[57] and Giunta et al.[58] have studied response surfaces approaches approximating either in a hierarchical or simultaneous fashion the response of the system in the space of random and design variables.

3.2.4. *Stochastic Projection Schemes*

In many engineering problems the system response is governed by partial differential equations (PDEs), such as in solid and fluid mechanics. Accounting for uncertainties in geometric and material parameters, as well as boundary and initial conditions, leads to stochastic PDEs. In order to determine the stochastic response of such problems, direct MCS is often too costly and even SRSMs can lead to unacceptable computational costs for problems with a large number of random variables. In cases where FORM and SORM are not sufficiently accurate or the entire stochastic response needs to be characterized, methods tailored to stochastic PDEs offer computationally efficient alternatives. Perturbation methods based on low-order approximations have been successfully applied to approximate the first two statistical moments of the response but become inaccurate for large variations of the input variables and are in general not appropriate for predicting the probability of failure[14]. In the past decade, computational methods for solving stochastic PDEs based on stochastic projection schemes have attracted substantial interest, in particular in the context of stochastic finite element methods. In the following a brief overview of the basic projection schemes is given and their potential for RBDO discussed. The reader is referred to Ghanem and Spanos[59], Schuëller[14], and Nair[53] for an introduction and overview of stochastic projection schemes.

The response of a static stochastic PDE $v(\mathbf{x}, \mathbf{y})$ depends on the spatial coordinate $\mathbf{x}$ and the random variables $\mathbf{y}$. First, the continuous spatial problem can be discretized by any standard technique, such as the finite element method, leading to the following semi-discrete form:

$$v(\mathbf{x}, \mathbf{y}) = \sum_{i=1}^{N_v} \phi_i(\mathbf{x}) \, \hat{v}_i(\mathbf{y}) = \boldsymbol{\varphi}(\mathbf{x}) \, \hat{\mathbf{v}}(\mathbf{y}) \tag{18}$$

where $\boldsymbol{\varphi}$ is a vector of local or global shape functions and $\hat{\mathbf{v}}$ is a vector of N_v discretized field variables which depends only on the random variables. For example, the discretization of a 1-D linear stochastic PDE governing the displacement of a linear elastic bar can be written as follows:

$$\mathbf{R}(\mathbf{y}) = \mathbf{K}(\mathbf{y}) \, \hat{\mathbf{v}}(\mathbf{y}) - \mathbf{P}(\mathbf{y}) = 0 \tag{19}$$

where $\mathbf{R}(\mathbf{y})$ is the vector of residuals. The stiffness matrix $\mathbf{K}(\mathbf{y})$ and the load vector $\mathbf{P}(\mathbf{y})$ are known functions of the random variables. Following a discretization scheme $\hat{\mathbf{v}}$ can be approximated in the space of the random variables as follows[59]:

$$\hat{\mathbf{v}}(\mathbf{y}) = \sum_{i=1}^{N_a} \boldsymbol{\psi}_i(\mathbf{y}) \, a_i = \boldsymbol{\Psi}(\mathbf{y}) \, \mathbf{a} \tag{20}$$

where $\boldsymbol{\Psi}$ is a $N_v \times N_a$ matrix of stochastic basis functions and $\mathbf{a}$ is a vector of deterministic coefficients. Introducing the approximation (20) in the semi-discrete formulation (19) the resulting error can be projected into the space of the stochastic basis functions. Depending on the definition of the inner product used to orthogonalize the residual, one obtains a deterministic or random algebraic equation. Ghanem and Spanos[59] introduced the following projection:

$$\left\langle \boldsymbol{\Psi}^T(\mathbf{y}) \, \mathbf{K}(\mathbf{y}) \, \boldsymbol{\Psi}(\mathbf{y}) \right\rangle \mathbf{a} - \left\langle \boldsymbol{\Psi}^T(\mathbf{y}) \, \mathbf{P}(\mathbf{y}) \right\rangle = 0 \tag{21}$$

where $\langle \cdot \rangle$ denotes the average operator which defines the inner product in the Hilbert space of the random variables. The reader may note that the unknown coefficients $\mathbf{a}$ are the solution of a deterministic system. Alternatively, the residual can be projected using an algebraic vector product leading to a system which needs to be solved for a particular realization of the random variables.

In the past decade, the above approach has gained substantial popularity applying a polynomial chaos expansion (PCE) approach to the solution vector $\hat{\mathbf{v}}$. Ghanem and Spanos[59] introduced a PCE scheme based on Hermite polynomials which form an orthogonal basis for normal random variables $\mathbf{u}$. The random variables $\mathbf{y}$ are transformed into the standard normal space, $\mathbf{y} \rightarrow \mathbf{u}$, for example by the same approaches as used for FORM. The coefficients of the solution vector $\hat{v}_i$ are approximated by a p-th order Hermite polynomial (16), leading to the number $N_a = N_v \times N_{pc}$ of unknown coefficients. Solving the resulting algebraic system for the coefficients a_i the solution of the stochastic PDE can be reconstructed with (18) and (20). The PCE approach features a stable and guaranteed convergence of mean response for an increasing order of the approximation. Xiu and Karniadakis[60] extended this approach to non-normal variables. The statistics of the stochastic field can be evaluated, either analytically or numerically. Sudret and Der Kiureghian[61] advocate evaluating the probability of failure by FORM applied to the PCE approximation. The PCE approach leads to very large problems requiring iterative solution techniques for realistic engineering problems and, therefore, has not been integrated into RBDO methods.

Instead of a polynomial chaos expansion of each coefficient of the solution vector v_i, Nair and Keane[62] introduced a reduced stochastic basis where $\boldsymbol{\Psi}(\mathbf{y})$ is formed by vectors spanning a Krylov subspace of the stochastic operator $\mathbf{K}(\mathbf{y})$.

$$\boldsymbol{\psi}_0(\mathbf{y}) = \mathbf{K}_0^{-1}\mathbf{P}(\mathbf{y}); \quad \boldsymbol{\psi}_i(\mathbf{y}) = \mathbf{K}_0^{-1}\mathbf{K}(\mathbf{y})\,\boldsymbol{\psi}_{i-1}(\mathbf{y}); \quad i = 1...N_a \tag{22}$$

where $\mathbf{K}_0$ is the mean operator. The essential difference to the PCE approach discussed previously is that the stochastic basis vectors in (22) expand the entire solution vector, not only a single component. Nair and Keane[62] showed that the statistics of linear elastic beam and bar structures can be approximated by only 3 reduced stochastic basis vectors, that is $N_a = 3$, reducing significantly the computational costs of PCE schemes.

The advantages of stochastic projection schemes based on reduced stochastic basis vectors have been recently exploited in RBDO of stochastic linear dynamic structures by Allen *et al.*[63] and extended to stochastic nonlinear dynamic problems by Maute *et al.*[64]. For illustration purposes, the following semi-discrete form of the linear equations of motion of a structural system is considered:

$$\mathbf{M}(\mathbf{y})\,\ddot{\hat{\mathbf{v}}}(\mathbf{y}) + \mathbf{C}(\mathbf{y})\,\dot{\hat{\mathbf{v}}}(\mathbf{y}) + \mathbf{K}(\mathbf{y})\,\hat{\mathbf{v}}(\mathbf{y}) = \mathbf{P}(t,\mathbf{y}) \tag{23}$$

where $\mathbf{M}, \mathbf{C}$, and $\mathbf{K}$ are the $N_v \times N_v$ mass, damping, and stiffness matrices and $\mathbf{P}(t,\mathbf{y})$ is the load vector. All matrices and the load vector depend on the random variables. Light damping is assumed. Following the ideas of Kirsch[65] developed for fast reanalysis techniques, the reduced stochastic basis vectors are formed by the eigenmodes and their derivatives with respect to the random variables.

$$\mathbf{\psi}_i(\mathbf{y}) = a_i^0(\mathbf{y})\, \mathbf{\theta}_i\big|_{\bar{y}} + \sum_{j=1}^{N_y} a_i^j(\mathbf{y})\, \frac{\partial \mathbf{\theta}_i}{\partial y_j}\bigg|_{\bar{y}} = \mathbf{\Theta}_i\, \mathbf{a}_i(\mathbf{y}); \quad i = 1\ldots N_m \tag{24}$$

where the matrix $\mathbf{\Theta}_i$ is composed of the i-th eigenvector $\mathbf{\theta}_i$ of the structure and the derivatives $\partial \mathbf{\theta}_i / \partial y_j$ with respect to the random variable at a reference configuration $\bar{\mathbf{y}}$. For structures undergoing forced harmonic vibrations, for example, only few eigenmodes $\mathbf{\psi}_i$ are needed to capture the system response. Therefore, the stochastic reduced basis is formed by the first N_m modes.

The coefficients $a_i^j(\mathbf{y})$, $j = 0 \ldots N_y$, forming the vector $\mathbf{a}_i(\mathbf{y})$ are computed by the combined approximation (CA) technique of Kirsch[66] as follows:

$$\left[\left(\mathbf{M}_i^C(\mathbf{y}) \right)^{-1} \mathbf{K}_i^C(\mathbf{y}) \right] \mathbf{a}_i = 0 \tag{25}$$

where the reduced matrices $\mathbf{K}_i^C(\mathbf{y})$ and $\mathbf{M}_i^C(\mathbf{y})$ are the stiffness and mass matrices projected onto the subspace $\mathbf{\Theta}_i$:

$$\mathbf{K}_i^C = \mathbf{\Theta}_i^T \mathbf{K}(\mathbf{y})\, \mathbf{\Theta}_i \quad ; \quad \mathbf{M}_i^C = \mathbf{\Theta}_i^T \mathbf{M}(\mathbf{y})\, \mathbf{\Theta}_i \tag{26}$$

The authors note that $\mathbf{K}(\mathbf{y})$ and $\mathbf{M}(\mathbf{y})$ are, respectively, the stiffness and mass matrices of the full-order finite element model evaluated at a particular realization of the random variables. Instead of the average operator in formulation (21), Allen *et al.*[63] follow a re-analysis concept leading to the following reduced system:

$$\mathbf{M}^R(\mathbf{y})\, \hat{\ddot{\mathbf{v}}}^R(\mathbf{y}) + \mathbf{C}^R(\mathbf{y})\, \hat{\dot{\mathbf{v}}}^R(\mathbf{y}) + \mathbf{K}^R(\mathbf{y})\, \hat{\mathbf{v}}^R(\mathbf{y}) = \mathbf{P}^R(t, \mathbf{y}) \tag{27}$$

where $\mathbf{M}^R, \mathbf{C}^R$ and $\mathbf{K}^R$ are $N_m \times N_m$ matrices and $\mathbf{P}^R$ is a vector of size N_m. The reduced stochastic response is denoted by $\hat{\mathbf{v}}^R$. The reduced terms in (27) are computed by projecting the full-order terms onto the subspace $\mathbf{\Psi}(\mathbf{y})$ as follows:

$$\begin{aligned} \mathbf{M}^R(\mathbf{y}) &= \mathbf{\Psi}^T(\mathbf{y})\, \mathbf{M}(\mathbf{y})\, \mathbf{\Psi}^T(\mathbf{y}) \\ \mathbf{C}^R(\mathbf{y}) &= \mathbf{\Psi}^T(\mathbf{y})\, \mathbf{C}(\mathbf{y})\, \mathbf{\Psi}^T(\mathbf{y}) \\ \mathbf{K}^R(\mathbf{y}) &= \mathbf{\Psi}^T(\mathbf{y})\, \mathbf{K}(\mathbf{y})\, \mathbf{\Psi}^T(\mathbf{y}) \\ \mathbf{P}^R(t, \mathbf{y}) &= \mathbf{\Psi}^T(\mathbf{y})\, \mathbf{P}(t, \mathbf{y}) \end{aligned} \tag{28}$$

As $N_m \ll N_v$ the reduced problem can be analyzed, either in the time or frequency domain, at significantly lower costs than the full-order system (22). The eigenmodes in (24) are computed once at the reference configuration $\bar{\mathbf{y}}$ by a modal analysis of the full-order system. The derivatives of the eigenmodes $\partial \mathbf{\theta}_i / \partial y_j$ can be efficiently evaluated by analytical sensitivity analysis techniques.

Haug *et al.*[67] and Kleiber *et al.*[68] provide a detailed introduction into parameter sensitivity analysis of mechanical systems; for coupled multi-physics problems the reader is referred to Maute *et al.*[69] and references therein. The $N_y + 1$ dimensional eigenvalue problem (25) needs to be solved for each realization of the random variables and each reduced stochastic basis. The reduction of the problem size significantly decreases the computational effort needed for computing the statistics of the transient structural response and the probability of failure of the system. Allen[70] showed that direct MCS, SRSM, and FORM can be applied to the reduced system approximating well the stochastic response.

The main differences between the scheme proposed by Nair and Keane[62] and the one of Allen *et al.*[63] are the definition of the inner product and the reduced basis vectors. While the averaging operator in (21) requires only the sampling of the projected operator and right hand side, the reduced response needs to be sampled following the approach of Allen *et al.*[63]. Assuming that the computational effort for solving the reduced order system is small in comparison to constructing the projection matrices, both schemes lead to similar computational costs. Constructing the reduced basis by preconditioned Krylov vectors features generality and provides mathematical insight gathered from research of iterative Krylov solvers for the reduced basis. For problems in linear structural dynamics an approximation scheme based on eigenmodes is a natural choice. Maute *et al.*[71] have extended this approach to nonlinear systems using empirical modes obtained, for example, by Proper Orthogonal Decomposition (POD) techniques.

In addition, the scheme of Allen *et al.*[63] offers an interesting extension regarding its integration into RBDO methods. As it is solely based on algebraic operations and not particularly tailored to stochastic problems, it is applicable to approximating the system response not only with respect to the random but also the design variables. Allen *et al.*[63] and Maute *et al.*[71] have successfully applied their approach to RBDO problems approximating simultaneously the system response with respect to both random variables and design variables. The derivatives of the reduced stochastic response $\hat{v}^R$ with respect to the design variables can be computed analytically without significant additional costs.

3.2.5. *Reliability Methods for RBDO*

The choice of reliability analysis methods for the successful integration into a computational RBDO framework depends on various factors: (a) the type of probabilistic design criteria of interest, that is for example the mean or the variation of the system response and the probability of failure, (b) the nonlinearity of the stochastic response with respect to the random variables, (c) the computational complexity of the system model, and (d) the characteristics of the design optimization problem and the appropriate optimization strategy. For sure, any of the above reliability methods can be integrated into a design optimization framework in a black-box fashion. Each function analysis within the iterative design optimization process simply calls for the evaluation of the stochastic response. For example, one can always solve the design optimization by a genetic algorithm (GA) and evaluate the probability of failure by direct MCS. While there are engineering design problems characterized by strong nonlinearities with respect to the random and design variables, for which such a brute-force strategy is the only choice, it leads in most cases to unacceptable computational costs.

Focusing on optimization problems with constraints on the probability of failure, the costs of the reliability analysis can often be reduced by improved sampling procedures and approximate reliability methods applied either directly to the system model or to an approximation of the stochastic response. However, for most realistic engineering problems, the computational costs of even advanced reliability analysis methods are too large when used within global design optimization search strategies, such as GAs and evolutionary strategies. Assuming that the objective and constraints are sufficiently smooth functions of the optimization variables, gradient-based optimization algorithm can significantly reduce the number of function evaluations but typically require the first-order gradients of the design criteria with respect to the optimization variables. For the sake of efficiency and robustness, the design sensitivities should be evaluated analytically rather than by finite difference schemes.

Only a small subset of reliability analysis methods is appropriate for efficiently evaluating low failure probabilities typically required for most engineering systems and allows for the efficient computation of the design sensitivities. The predominate representatives of this subset are reliability analysis methods based on approximate reliability analysis and a large body of work has been dedicated to the integration of FOSM and FORM into RBDO, which will be summarized in Section 4. Only few studies can be found in the literature reporting on the extension of stochastic response surface methods and stochastic projection schemes towards gradient-based optimization schemes.

4. RBDO Methods

The key issue that separates RBDO methods from optimization problems with only deterministic design criteria is the integration of the reliability analysis method. As discussed previously, design optimization algorithm and reliability analysis method need to be selected carefully in order to obtain an efficient and robust optimization strategy while accurately characterizing the stochastic response of the system. In general, design optimization algorithm and reliability analysis method share common features: both are iterative procedures calling for the repeated evaluation of the configuration of the system away from the initial (nominal) configuration and exploring the system response in a parameter space.

There are essentially three approaches for integrating reliability analysis into the design optimization process: nested-loop, sequential, and unilevel approaches. In a nested-loop approach the probabilistic design criteria are evaluated by an appropriate reliability analysis method whenever the optimization algorithm calls for the evaluation of the design criteria. The advantage of this concept is that optimization and reliability analysis methods can be treated in a black-box fashion simply linking the inputs and outputs of the algorithms. Gradient-based optimization algorithms can be only used if the reliability analysis method allows for the computation of the design sensitivities of the probabilistic design criteria, assuming they exist. This restriction is overcome by gradient-free optimization methods, such as evolutionary strategies or genetic algorithms (GAs), typically at cost of a larger number of iterations in the design optimization loop. In this section nested-loop approaches are discussed for combining gradient-based design optimization methods with approximate reliability methods based on FORM. The integration of MCS into a design optimization methods based on GAs is discussed in Section 5.

The fundamental disadvantage of the nested-loop approach is that the reliability analysis needs to be converged with high accuracy even far away from the optimal design point, leading to a large computational burden. This bottleneck can be overcome either by sequential or unilevel approaches. Sequential approaches construct a sequence of approximated design and/or reliability analysis problems which are solved in an alternate fashion[72,73]. While sequential approaches can significantly reduce computational costs, it can typically not be guaranteed that the sequence of subproblems eventually converges to the exact solution. Alternatively, the design optimization and reliability analysis problems can be merged and the design and random variables are simultaneously advanced. In the literature, methods following this concept are referred to as unilevel, hybrid, or one-shot methods. A short overview of unilevel methods is given at the end of this section.

4.1. *Reliability Index Approach (RIA)*

Approximate reliability methods and in particular FORM are an attractive tool for RBDO as they require only few evaluations of the deterministic system response and, in the case of FORM, the gradients of the failure probability with respect to the design variables can be conveniently determined. Using FORM, an RBDO problem formulation (see Section 2) can be rewritten in terms of reliability indices as follows:

$$\max_{s} \quad \beta_c\left(c(\mathbf{y}_c^*, \mathbf{s}) \geq \overline{c}\right)$$

$$s.t. \quad \beta_j\left(g_j(\mathbf{y}_j^*, \mathbf{s}) < 0\right) - \overline{\beta}_j \geq 0 \quad j = 1...N_P \quad (29)$$

$$s_i = \left\{s_i \in \mathfrak{R} \mid s_i^l \leq s_i \leq s_i^u\right\}$$

where β_c is the reliability index associated with the event that the costs c exceed the target cost $\overline{c}$. The constraints bounding the probability of failure $P(g_j < 0)$ can be expressed by imposing a lower bound $\overline{\beta}_j$ on the associated reliability index β_j. The solutions of the MPP searches for each reliability index are denoted by $\mathbf{y}_c^*$ and $\mathbf{y}_j^*$. In formulation (29) it is assumed that the optimization variables are continuous and the feasible set is constraint by lower and upper bounds s_i^l and s_i^u, respectively.

The above formulation is called Reliability Index Approach (RIA). Formulation (29) can be written in terms of the failure probability $P_f(g_j < 0)$, leading to the generic RBDO problem statement (1), by substituting the reliability index by the failure probability via the transformation (4). However, as the failure probability may vary over several orders of magnitude during the optimization process while the variations of the reliability index are usually small, from a numerical point of view, it is advantageous to use a formulation based on the reliability index.

Following a nested-loop approach, RIA calls for the evaluation of $(N_P + 1)$ reliability indices at each iteration of the outer design optimization loop. If the optimization process is driven by a gradient-based optimization algorithm, the derivatives of the reliability indices with respect to the design variables need to be provided.

One of the main advantages of FORM and RIA is the ease of analytically computing design sensitivities of β_c and β_j. In the following the basic steps for deriving and computing the derivative of the reliability index β with respect to a design variable s_i are summarized. The reader is referred to Hohenbichler and Rackwitz[74] and Karamchandani and Cornell[75] for detailed derivations.

Differentiating the definition of β (8) with respect to the design variable s_i yields the following expression:

$$\frac{d\beta}{ds_i} = \frac{1}{\beta} \left[\mathbf{u}^* \right]^T \frac{d\mathbf{u}^*}{ds_i}$$

(30)

Evaluating the Karush-Kuhn-Tucker (KKT) optimality conditions of the FORM minimization problem (10), the most probable point $\mathbf{u}^*$ can be written as follows:

$$\mathbf{u}^* = -\left[\frac{\beta}{|dg/d\mathbf{u}|} \frac{dg}{d\mathbf{u}} \right]_{\mathbf{u}^*}$$

(31)

Substituting this result into the total derivative (30) yields:

$$\frac{d\beta}{ds_i} = -\left[\frac{1}{|dg/d\mathbf{u}|} \frac{dg}{d\mathbf{u}} \right]_{\mathbf{u}^*}^T \frac{d\mathbf{u}^*}{ds_i}$$

(32)

Equation (32) can be simplified by exploiting the fact that the total variation of the equality constraint of the FORM minimization problem, $g(\mathbf{s}, \mathbf{u}(\mathbf{s})) = 0$, vanishes. In the case that the design variable s_i is associated with a deterministic quantity, one arrives at the following expression:

$$\frac{d\beta}{ds_i} = \frac{1}{|dg/d\mathbf{u}|_{\mathbf{u}^*}} \frac{dg}{ds_i} \bigg|_{\mathbf{u}^*}$$

(33)

where the second term on the right-hand side is the gradient of the limit state function with respect to the design variable s_i and can be computed by conventional sensitivity analysis.

A design variable can also be a probabilistic characteristic of a random variable. For example, the mean or the standard deviation of a random variable can be treated as a design variable. In this case, the gradient of the reliability index can be derived from (30) via the transformation (9), yielding the following expression:

$$\frac{d\beta}{ds_i} = \frac{1}{\beta} \left[\mathbf{u}^* \right]^T \frac{dT_u}{ds_i} \bigg|_{\mathbf{u}^*}$$

(34)

While the RIA approach allows for the efficient evaluation of the design criteria, FORM sometimes suffers from convergence issues and, as discussed previously, it can not be guaranteed that the RIA optimization problem has a solution for any given design during the design optimization process. If the mean design is far away from the failure surface, there may be no solution $\mathbf{u}^*$ such that the equality constraint $g = 0$ can be satisfied and the FORM MPP search is ill-posed. This situation may occur in the course of the design optimization process leading to serious robustness problems.

4.2. *Performance Measure Approach (PMA)*

The ill-posedness and ill-robustness issues of the FORM-based RIA approach can be overcome by reformulating the design optimization problem following the Performance Measure Approach (PMA) of Tu *et al.*[76], or also known as known as Fixed Norm Formulation introduced by Lee and Kwak[77], or Target Performance Approach discussed by Lee *et al.*[78]. These approaches characterize the safety of a system via the worst possible performance, that is the minimum value of the limit state function for a required reliability index $\bar{\beta}_j$, and call for the solution of the performance measure problem (12).

Since the reliability requirement is already included in the reliability analysis, the PMA based design optimization problem takes on the following form:

$$\max_s \quad g_c(\mathbf{y}_c^*, \mathbf{s})$$

$$s.t. \quad g_j(\mathbf{y}_j^*, \mathbf{s}) \geq 0 \qquad\qquad j = 1 \ldots N_{fc} \qquad (35)$$

$$s_i = \left\{ s_i \in \Re \mid s_i^l \leq s_i \leq s_i^u \right\}$$

where the performance g_c of the objective is measured with respect to a target reliability $\bar{\beta}_c$ and g_j are the performances associated with the probabilistic failure constraints. The random variables $\mathbf{y}_c^*$ and $\mathbf{y}_j^*$ are the solutions of each PMA problem (12).

Similar to the RIA approach, the design sensitivities of the performance measures g_c and g_j can be evaluated with minor additional effort. The total derivative of the performance measure g_j, for example, can be written as follows:

$$\frac{dg_j}{ds_i} = \frac{\partial g_j}{\partial s_i} + \underbrace{\frac{\partial g_j}{\partial \mathbf{u}}^T \frac{d\mathbf{u}}{ds_i}}_{=0} = \frac{\partial g_j}{\partial s_i} \qquad (36)$$

The partial derivative of g_j with respect to the design variables can be evaluated by differentiating either analytically or numerically the expressions in the analysis model. The second term in Equation (36) vanishes. The KKT conditions for the PMA optimization problem (12) are:

$$\frac{\partial g_j}{\partial \mathbf{u}} + \eta \ \frac{\partial \beta}{d\mathbf{u}} = 0 \tag{37}$$

where η is the Lagrange multiplier associated with the equality constraint $\beta_j - \overline{\beta}_j = 0$. Since the reliability index is constant and does not depend on the random variables $\mathbf{u}$, the second term in (37) and, therefore, the derivative of the performance measure with respect to $\mathbf{u}$ in (36) vanish.

4.3. *Univlevel RIA and PMA*

Essentially there are two approaches for merging the design optimization and the reliability analysis problems associated with the RIA and PMA formulations. Kuschel and Rackwitz[79] proposed to introduce the KKT conditions of the MPP search (10) into the RIA formulation (29) and to solve the augmented problem simultaneously for the design variables and standard normal variables. However, as the MPP problem may have no solution, this approach can lead to an ill-posed problem. Therefore, Agarwal *et al.*[80] applied the same idea to the PMA formulation and introduced the KKT conditions of problem (12) as additional equality constraints. While numerical studies for analytical and simple structural problems have shown that the above approaches can significantly reduce the total number of iterations, their applicability suffers from the need to compute second-order derivatives for solving the unilevel problem by gradient-based optimization methods.

Instead of introducing the optimality conditions of the reliability analysis problem as additional constraints, Kharmanda *et al.*[81] propose to reformulate the objective of the RIA problem and add zero-order constraints. While this approach does not require the evaluation of second-order derivatives, it can not be guaranteed that the solution of the augmented unilevel problem is also a solution of the original reliability problem.

5. Life-cycle Optimization with Multiple Objectives including Cost

Automation of civil engineering optimization problems, *e.g.* structural design and maintenance planning, rely heavily on numerical algorithms that lead the search process towards improved solutions in terms of better merit objective values. Many traditional optimization algorithms are problem-dependent and single-objective oriented. They usually make use of gradient information to guide the search process and continuous design variables are often assumed. In contrast, population-based natural evolution-inspired genetic algorithms (GAs) are general-purpose numerical tools, with which gradients are no longer needed and discrete-valued design variables can be handled without any difficulty. More importantly, multiple conflicting objective functions can be directly and simultaneously treated by GA through the concept of non-dominance. This section discusses features of different optimization algorithms with a particular focus on techniques of solving multiobjective optimization problems via GAs.

5.1. *Traditional Optimization Algorithms*

Traditional optimization techniques usually apply available mathematical programming algorithms to solve civil engineering optimization problems[82,83]. Optimality criteria are obtained through application of KKT conditions combined with Lagrange multipliers accounting for relevant constraints. With the linear programming approach, both the objective(s) and constraints must be linear functions of design variables. In contrast, nonlinear programming approaches deal with problems whose objective(s) and/or constraints are nonlinear differentiable functions of design variables. Note that traditional methods are usually problem-dependent in nature, that is, one method that is very efficient for a kind of problems may not be viable for another kind. It is therefore necessary for one to have knowledge of applicability of one particular optimization algorithm in order to use it properly.

Continuous design variables are often assumed in traditional optimization algorithms. To handle optimal design problems with discrete or mixed continuous-discrete design variables that are often encountered in real-world design practice, other numerical techniques are needed, including branch and bound method, integer programming, sequential linearization, rounding-off method, *etc.*[84]. With traditional methods, the optimization process usually starts with a single initial solution elected in the feasible solution space and the subsequent improved solutions evolve based on searching/updating rules that typically require sensitivity calculation. This point-by-point process may be trapped at local minima.

5.2. *Application to Multiobjective Optimization*

Most of traditional methods by themselves can only handle single-objective based optimization problems, for which a single final optimal solution will be obtained. In order to solve optimization problems with multiple (conflicting) objective functions, one has to convert, with nontrivial inconvenience, the original multiobjective problem into a series of equivalent single-objective optimization problems, which in total produce a set of solutions that exhibits optimal tradeoff among all competing objectives. The reader is referred to Stadler[85] and Eschenauer *et al.*[86] for an in-depth treatment of multicriteria design optimization in engineering and sciences. Pioneering research efforts in formulating multicriteria optimization in RBDO context were made by Fu and Frangopol[87,88].

5.3. *Genetic Algorithms*

A very brief introduction is provided in this section to the working mechanism of genetic algorithms (GAs). GA is global stochastic search and optimization heuristics that mimic natural evolution and are based on Darwin's survival-of-the-fittest principles[89]. GA belongs to the category of nature-inspired evolutionary algorithms that include as well ant-colony optimization, simulated evolution, DNA computing, and cultural algorithms[90]. Since their inception in the 1960's, GAs have been successfully used in a wide array of applications[91]. The growing popularity stems from GA's ease of implementation and robust performance for difficult engineering and science problems of vastly different natures.

GA usually operates on solutions that are encoded as genotypic representations (*e.g.*, binary strings), called chromosomes, from their original phenotypic representations (*i.e.*, actual data values), although there also exist real-coded GAs that work directly on original data with no coding operation[92].

Compared to most traditional optimization methods, GA has the following distinct advantages[91]: (a) GA works with a population of solutions instead of one solution at each iteration. By starting with a random set of solutions, the current population evolves to an offspring population at each iteration. Working with a number of solutions provides GAs with the ability of capturing multiple optimized tradeoff solutions in a single algorithm run. (b) Gradients are not required in guiding the genetic search, which makes GA insusceptible to pitfalls of traditional gradient-based hill-climbing searching procedures. This feature is especially useful for practical engineering problems where either objectives or constraints or both are non-differentiable with respect to discrete-valued design variables. (c) GA is a problem-independent universal search tool, which makes it more flexible as well as robust for application to problems of different natures; the requirement on users' capability is greatly reduced.

The most significant concern about GA is its extensive computational expenses. Usually a large population of solutions needs to be maintained in order to override the impact of errors due to stochastic transition operators. This entails a tremendous amount of computational time for evaluating objective functions as well as for performing genetic operations. Consequently, it is helpful to decide, before a GA based procedure is formally launched, if the benefits gained from GAs are worth the extra computational efforts spent. With the availability of high-speed computer facilities, however, computational demands might not be a very big obstacle for problems of small to moderate sizes.

5.4. *Multiobjective Optimization via Genetic Algorithms*

Multiobjective GAs have been comprehensively studied and fruitfully developed in the last decade[92]. Unlike problems with single objective functions for which one optimized solution is sought after, there is no unique optimized solution with respect to all conflicting objective functions for multiobjective optimization problems (unless all these objectives are strictly equivalent to one another). Instead, a set of optimized tradeoff solutions is expected among conflicting objectives. GA seeks for these multiple optimized tradeoff solutions based on the concept of dominance. A solution is said to dominate another solution, if two conditions are both met[90]: (a) is no worse than in all objectives; (b) is strictly better than in at least one objective. Among a set of solutions, the nondominated set contains those that are not dominated by any individual of the set. When the set is the entire search space, the resulting nondominated set is called the (global) Pareto optimal set (or front).

5.5. *Genetic Algorithm Based Multiobjective Structural Design*

GAs have been powerfully applied to a broad range of research areas in structural engineering due to their salient convenience over traditional optimization/search methods such as problem-independency that requires no gradient information to guide the search process, the global perspective, and easy treatment of discrete design variables. Goldberg and Samtani[93] first studied structural optimization with a ten-bar truss. Afterwards, researchers around the world applied GAs to many different structural systems. For example, Rajeev and Krishnamoorthy[94] investigated GA based optimal design of a 160-bar transmission tower system; Adeli and Cheng[95] used GAs to optimize large high-rise framing systems. GAs were also developed to optimize trusses by simultaneously considering size, shape, and topology[96,97]. In parallel to application, comparison studies were performed that recommended more efficient GA operators especially suited for structural optimization applications; for example, Leite and Topping[98] provided an extensive discussion on a wide range of GA applications as well as performance comparisons with other optimization methods. Ghaboussi[99] reviewed structural mechanics and engineering applications of biologically inspired soft computing methods (GAs and artificial neural network).

Applications of GA to steel moment frame designs have also received research attention. Member sizes of a practical steel frame can usually be selected from a catalog of commercially available sections only, which leads to optimization problems that are discrete and thus highly non-continuous, making gradient information very difficult to compute. GAs have no difficulty in solving these discrete optimization problems. Camp *et al.*[100] developed an automated algorithm that integrates GA, a finite element analysis program, and a complete database of AISC sections for the design of steel frames according to AISC-ASD provisions. Pezeshk *et al.*[101] incorporated into GAs a group selection scheme, an adaptive crossover scheme, and a finite element analysis to design frames according to AISC-LRFD as well as AISC-ASD. Foley and Schinler[102] used an advanced analysis and object-oriented evolutionary computation for optimizing both fully and partially restrained steel frame designs.

Applications of GA to structural design problems considering multiple merit objective functions have appeared only recently in literature. Cheng and Li[103] developed a GA based multiobjective procedure with an elitist strategy through niching and a "Pareto-set filter" technique that alleviate the effects of "genetic drift" or loss of diversity among final solutions; dominance characteristics and constraint violations were handled by a fuzzy-logic penalty function; three truss design examples were provided. Coello and Christiansen[104] proposed their GA-based multiobjective optimization technique and applied it to two truss design problems. Greiner *et al.*[105] used a nondominated sorting GA with an elitist strategy for a multiobjective problem that minimized both the number of different cross-section types (the largest number being 4) and the mass for design of frame structures. Cheng[106] summarized his previous research on multiobjective structural optimization emphasizing GAs as well as game theory; objective functions used in the numerical examples included structural weight, control performance index, seismic input energy, potential energy, and construction cost. Liu *et al.*[107,108] presented a GA-based seismic design optimization for steel moment frame structures with simultaneous consideration of multiple objectives including initial construction expenses and seismic structural performance.

5.6. *Life-cycle Cost-based Multiobjective Seismic Design*

Structural engineers tend to optimize the system they are designing based on different considerations such as cost-effectiveness, aesthetics, social/political issues, *etc*. In steel frame building designs, for example, widely used optimization objective functions include minimum structural weight, minimum compliance, a fully stressed state, desired component/system reliability levels, *etc*., many of which are related, either explicitly or implicitly, to present or future monetary expenses to some extent.

The consideration of both initial construction cost and lifetime cost (*e.g*., costs due to maintenance, operation, repair, damage, and/or failure consequences) leads to a so-called "life-cycle cost" analysis under which the optimal design is the one that balances these two general cost items appropriately according to pre-selected criteria. It is, however, very difficult at the early design stage to quantitatively express various sources of initial cost with comparable accuracies. For example, estimation of material usage related expenses may be relatively easier. In contrast, it is a demanding task to quantify the precise relationship between the complexity of a proposed design and its associated labor/erection cost. Furthermore, future costs due to maintenance, inspection, as well as direct/indirect economic losses due to environmental attacks (wind, earthquake, *etc*.) are uncertain in nature and can only be evaluated in a probabilistic sense. As previous research examples, Wen and Shinozuka[109] investigated cost-effective active structural control scenarios; Koh *et al*. (2000) evaluated cost-effectiveness for seismic-isolated bridges.

The future cost usually considered in earthquake-related studies comes from monetary-equivalent losses due to seismic events during a structure's lifetime; other future expenses (as well as possible benefits) are not taken into account due to their usually independency to seismic structural resistance. The primary criterion of traditional code provisions is to ensure life safety and prevent structural collapse. Recent earthquakes, however, revealed that economic losses induced by less drastic structural damages as well as functional disruptions can also be enormous comparable to the structure's initial cost. Therefore, the concept of damage control needs to be considered appropriately in the design stage in order to reduce future economic losses.

In the last decade, concepts of performance-based seismic design have been emerged and continuously developed as the new generation design methodologies. The most distinctive feature from conventional design practice is the explicit evaluation of actual structural performance under future seismic loading conditions expressed in probabilistic terms (*e.g*., ATC-40[111]; FEMA-273[112]; FEMA-350[113]). Permissible structural performances as well as damage states associated with each hazard level are both illustrated qualitatively based on previous earthquake-driven site inspections and are expressed quantitatively in terms of representative structural response indices (*e.g*., interstory drift ratio, axial column force). Structural designs conforming to these design guidelines with multiple limit states are expected not only to ensure life safety/collapse prevention under severe earthquakes but also to incur less damage-related direct/indirect consequences when subject to small to moderate seismic events.

By use of appropriate cost functions associated with varied damage states, designers have an opportunity to consider earthquake-related economic losses in a direct and explicit manner; seismic structural design based on life-cycle cost analysis then becomes a tractable alternative. In particular, the expected lifetime seismic damage cost could be derived by adding the product of each damage state cost and its associated expected failure probability during a lifetime. Minimization of the expected life-cycle cost, which is a direct sum of initial cost and the expected lifetime seismic damage cost, has primarily been the sole design criterion (i.e., objective function) that received fruitful research efforts. For example, Kang and Wen[114] developed a design methodology based on the minimum expected life-cycle cost and investigated its application in steel moment frame building design; using FEMA-released software HAZUS, Kohno and Collins[115] investigated minimum life-cycle cost design of reinforced concrete structures; Ang and Lee[116] analyzed reinforced concrete buildings built in Mexico based on cost functions in terms of the Park-Ang damage index. It should be pointed out that these pioneering research efforts were based on a series of conventional trial designs with varied base shear levels around the codified values; no formal optimization algorithms were actually involved. Their conclusions were that the codified base shear level should be increased appropriately in order to minimize the expected total life-cycle cost.

Liu *et al.*[117] presented a GA-based automated procedure for seismic design optimization of steel moment frame structures in accordance with 2000 NEHRP seismic provisions and AISC seismic steel design specifications. The life-cycle cost is considered through two separate objective functions: initial cost and lifetime seismic damage cost. The degree of design complexity is roughly accounted for by the number of different steel section types as the third objective function, which provides an additional dimension to the resulting tradeoff optimized design solutions. The damage cost is computed in this study with designer-specified confidence level related percentile limit state probabilities so that effects of randomness and uncertainty in seismic demand and capacity estimates as well as in seismic hazards are appropriately considered, following the SAC/FEMA guidelines[118].

5.7. *Life-cycle Bridge Maintenance Planning*

Satisfactory performance of civil infrastructure is of critical importance to sustained economic growth and social development of a modern society. In particular, the highway transportation system is considered to be one of the society's critical foundations. Among the many elements of this complex system, bridges are especially important. They are the most vulnerable elements in this system because of their distinct function of joining highways as the crucial nodes. The bridge infrastructure has been constantly exposed to aggressive environments and facing ever increasing traffic volumes and heavier truck-loads. These actions progressively degrade the long-term performance of highway bridges. Bridge managers urgently need to cost-effectively allocate limited budgets for maintaining and managing the large inventory of bridges with the goal of optimally balancing lifetime structure performance and whole-life maintenance cost.

Most existing methodologies for managing highway bridges are based on the least life-cycle cost criterion while enforcing relevant performance constraints[119,120]. There exist practical difficulties when applying such methodologies. For instance, if the available budgets are larger than the computed minimum life-cycle cost, the bridge performance can be maintained at a higher level than what is previously prescribed in deriving the minimum life-cycle cost solution. On the other hand, if the scarce financial resources are not enough to meet the computed minimum life-cycle cost, bridge managers have to look for another solution that can improve bridge performance to the highest possible level with the available budget. Indeed a satisfactory maintenance management solution should be determined by optimally balancing competing objectives of improving lifetime bridge performance and reducing the long-term costs of various origins.

In addition to the traditional life-cycle cost as a merit measure in making maintenance management decisions, bridge performance indicators include, for example, visual inspection-based condition and computation-based safety/reliability (*e.g.*, Frangopol and Das[121]). Optimizations of all these merit aspects are treated simultaneously in a multiobjective optimization based bridge management framework. This leads to a group of non-dominated solutions, each of which representing a unique optimized tradeoff between life-cycle cost reduction and bridge performance enhancement. These alternative solutions will significantly facilitate bridge managers' decision-making by actively selecting a solution that has the most desirable compromise between the conflicting life-cycle cost and lifetime bridge performance objectives.

Most existing research on bridge maintenance management can be categorized as project-level type because only individual bridges or a group of similar bridges are considered. In order to conduct realistic life-cycle bridge management, uncertainties associated with description of time-dependent bridge deterioration with and without maintenance interventions should be appropriately considered. In general there are two sources of uncertainty: the aleatory uncertainty due to inherent randomness and epistemic uncertainty stemming from lack of knowledge or imperfect modeling. The former can by no means be reduced while the latter can be lessened through, for example, Bayesian updating when additional information is gathered. The time-dependent performance of deteriorating bridges with and without maintenance interventions can be predicted by a continuous computational model. This model describes the performance profiles under no maintenance by a curve characterized by an initial performance level, time to damage initiation, and a deterioration curve governed by appropriate functions and in the simplest form, a linear function with a constant deterioration rate. Effects of a generic maintenance action, based on the computational model of Frangopol[122], discussed in Frangopol *et al.*[122] and van Noortwijk and Frangopol[123], include immediate performance enhancement, suppression of the deterioration process for a specified time interval, reduction in the deterioration rate, and another specified duration of maintenance effect beyond which the deterioration rate resumes to the original one. Uncertainties associated with the deterioration process are considered in terms of probabilistic distribution of the controlling parameters of this computational model. Monte Carlo simulation is used to account for these uncertainties by obtaining the statistical performance profiles of deteriorating structures.

Liu and Frangopol[124,125] applied multiobjective optimization techniques in life-cycle maintenance management of deteriorating bridge at project-level and network-level. Multiple and competing criteria in terms of condition, safety, and life-cycle cost are considered simultaneously. Uncertainties associated with the deterioration process with and without maintenance interventions are treated appropriately. GA is presented to solve the posed combinatorial optimization problems. As application examples, GA-based automated management procedures are applied to prioritize maintenance resources for deteriorating bridges over the specified time horizons. At project-level, the long-term performance of a large population of similar highway reinforced concrete crossheads is maintained. At network-level, a group of bridges that form the nodes of an existing highway network is considered.

5.7.1. *Project-level*

In Liu and Frangopol[124], maintenance needs are prioritized for deteriorating reinforced concrete highway crossheads through simultaneous optimization of both structure performance in terms of condition and safety states and life-cycle maintenance cost with a prescribed discount rate of money of 6% in 50 years. The deterioration processes with and without maintenance are simulated by the model as described above. Five maintenance strategies are considered: replacement of expansion joints, silane, cathodic protection, minor concrete repair, and rebuild. Three objective functions are considered: the condition index, the safety index, and the life-cycle maintenance cost. For reinforced concrete elements under corrosion attack, visual inspection-based condition is classified into four discrete levels, denoted as 0, 1, 2, and 3, that represent no chloride contamination, onset of corrosion, onset of cracking, and loose concrete/significant delamination, respectively. According to bridge specifications in the United Kingdom, the safety index is defined as the ratio of available to required live load capacity.

The GA-based maintenance prioritization procedure is carried out for deteriorating reinforced concrete crossheads. A total of 129 different optimized maintenance planning solutions are generated at the 30th generation. All these solutions are feasible in that the minimum performance levels are strictly satisfied, which are enforced as constraints. These solutions lead to, in a Pareto optimal sense, different levels of performance enhancement and maintenance needs. Based on this large set of solutions, civil infrastructure managers can actively select a solution that balances condition, safety, and life-cycle maintenance cost objectives in a preferred way.

5.7.2. *Network-level*

Maintenance management for deteriorating bridge infrastructure from a network perspective provides more rational solutions because the ultimate goal of bridge management is to improve the overall performance of a transportation network other than that of individual bridges in the network. In this section, the maintenance management of a group of bridges that form nodes of an existing highway network is discussed. Performance evaluation of deteriorating highway networks in the transportation engineering community has been focused on road performance in terms of travel time and capacity reliabilities[126]. Recently, Shinozuka *et al.*[127] evaluate the performance of highway bridge networks under earthquakes and demonstrate the importance of seismic retrofit in improving network performance in terms of drivers' travel delay reduction. Maintenance planning optimization for deteriorating bridge networks, which is important for daily operation other than under disastrous events such as earthquakes and floods, has not received adequate attention. Augusti *et al.*[128] investigate retrofitting efforts allocation for seismic protection of highway bridge networks. Adey *et al.*[129] develop a network bridge management system using a supply-and-demand system approach.

Liu and Frangopol[125] proposed network-level bridge maintenance management procedure. The overall performance of a bridge network was assessed in terms of the origin–destination connectivity reliability, which was evaluated by an event tree analysis, utilizing structural reliability index profiles of all deteriorating bridges in the network. Each maintenance solution was treated as a sequence of maintenance activities that are scheduled at discrete years and are applied to selected bridges. The conflicting objectives of maximizing the network connectivity reliability and minimizing the present value of total maintenance cost were considered in the posed combinatorial bi-objective optimization problem that was solved by GA. It is found that the proposed GA-based procedure is able to locate a set of different maintenance solutions that represent the optimized tradeoff between the two conflicting objectives. In particular, under given budgets, one can find the maintenance solution that best improves the network performance in terms of connectivity; alternatively, if a threshold is specified for the network performance, a maintenance solution that requires the least total maintenance cost can be found. In addition, the GA-based procedure can automatically prioritize maintenance needs among networked bridges and over the time horizon. It can effectively identify bridges that are more important to the functionality of the network and then prioritize limited resources to these bridges at selected years.

6. Applications

In recent publications, Frangopol and Maute[1,2] provided both the status of RBDO and reliability-based life-cycle cost optimization (RBLCCO) applications in the civil and aerospace engineering fields. For this reason, in this section only the most recent (i.e. 2004, 2005 and 2006) applications reported by the authors and their co-workers are mentioned. These applications, including also probability-based life-cycle design optimization (PBLCDO), are found in Kong and Frangopol[130,131], Stewart *et al.*[132], Estes and Frangopol[133], Liu *et al.*[107,108], Liu and Frangopol[124,125], *Yang et al.*[134,135], Allen *et al.*[63] and Maute *et al.*[64].

7. Conclusions

Concepts and methods of structural design optimization with uncertainty, life-cycle performance and cost considerations have greatly advanced over the past decade. Most of these advances were summarized in this chapter. However, there are still areas where improvement can be made. More widespread use of these concepts and methods in structural engineering design, maintenance and management practice will led to a better understanding of the limitations of existing RBDO approach.

The tools required performing RBDO including both aleatory and epistemic uncertainties[136], life-cycle performance and cost are rapidly becoming available. As indicated in Ellingwood[137], additional research is required to provide supporting models and data in several areas including deterioration of materials and structures, time-dependent reliability assessment, and infrastructure damage and loss estimation.

Recently developed techniques in the field of reliability-based multi-objective optimization of deteriorating structural systems and networks are useful for managing preferences in decision-making under uncertainty with conflicting criteria such as cost and probability of unsatisfactory performance.

Acknowledgments

The support by the National Science Foundation under grants CMS-9912525, CMS-0217290, CMS-0509772 and DMI-0300539 is gratefully acknowledged. Several postdoctoral researchers and a number of former and current graduate students at the University of Colorado at Boulder contributed to the results presented in this chapter. Their contributions and assistance are greatly appreciated. The opinions and conclusions presented in this chapter are those of the authors and do not necessarily reflect the views of the sponsoring organizations.

References

1. D.M. Frangopol and K. Maute, in *Engineering Design Reliability Handbook* (CRC Press, New York, 2004), p. 24.1.
2. D.M. Frangopol and K. Maute, *Computers & Structures*, 81 (2003).
3. J.R. Birge and F. Louveaux, *Introduction to stochastic programming* (Springer, New York, 1997).
4. D.M. Frangopol and F. Moses, in *Reliability-based structural optimization* (Chapman & Hall, London, 1994), p. 492.

5. D.M. Frangopol, in *Probabilistic Structural Mechanics Handbook* (Chapman & Hall, London, 1995), p. 352.
6. D.M. Frangopol, Progress in Structural Engineering and Materials, 1 (1998).
7. D.M. Frangopol and M. Liu, Structure and Infrastructure Engineering, in press.
8. A.H-S. Ang and W.H. Tang, *Probability Concepts in Engineering Planning and Design*, Vol. 2 (Wiley, New York, 1984).
9. H.O. Madsen, S. Krenk, and N.C. Lind, *Methods of Structural Safety* (Prentice-Hall, 1986).
10. P. Thoft-Christensen and Y. Murotsu, *Applications of Structural Systems Reliability Theory* (Springer, 1986).
11. R.E. Melchers, Structural Reliability Analysis and Prediction (Wiley, Chichester, 1999).
12. M. Shinozuka, *Structural Safety & Reliability*, Vol. 1 (ASCE, 1990).
13. R. Rackwitz, *Structural Safety*, 23 (2001).
14. G.I. Schuëller, Probabilistic Engineering Mechanics, 12 (1997).
15. G.I. Schuëller, *Computers & Structures*, 79 (2001).
16. H. Agarwal and J.H. Renaud, in Proceedings of the 43rd AIAA/ASME/ASCE/AHS/ASC Structures, Structural Dynamics, and Materials Conference (AIAA, 2002), AIAA-2002-1755.
17. M. Allen and K. Maute, Computer Methods in Applied Mechanics and Engineering, 194 (2005).
18. F. Moses, in *Optimum Structural Design* (Wiley, Chichester, 1973), p. 241.
19. I. Enevoldsen and J.D. Sørensen, *J. Structural Engineering*, ASCE, 119 (1993).
20. A.C. Estes and D.M. Frangopol, *J. Structural Engineering*, ASCE, 125 (1999).
21. Y. Mori and B. Ellingwood, *J. Structural Engineering*, ASCE, 119 (1993).
22. Y. Mori and B. Ellingwood, *J. Structural Engineering*, ASCE, 120 (1994).
23. M.P. Enright and D.M. Frangopol, *Journal of Structural Engineering*, ASCE, 124 (1998).
24. M.P. Enright and D.M. Frangopol, *Structural Safety*, 21 (1999).
25. N. Kuschel and R. Rackwitz, in *Proceedings of the 8th IFIPWG7.5 Working Conference*, (Sheridan Books, Chelsea, Michigan, 1998), p. 27.
26. M.D. McKay, W.J. Conover, and R.J. Beckman, *Technometrics*, 221 (1979).
27. H. Kahn and A.W. Marshall, Journal of the Operations Research Society, 1 (1953).
28. M. Shinozuka, *J. Structural Engineering*, 109 (1983).
29. G.I. Schuëller and R. Stix, *Structural Safety*, 4 (1987).
30. Y-T. Wu, H. Millwater, and T. Cruse, *AIAA Journal*, 12 (1990).
31. R.E. Melchers, *J. Engineering Mechanics*, 116 (1990).
32. M. Macke and C. Bucher, *Journal of Sound and Vibration*, 268 (2003).
33. S.-H. Kim and K.-W. Ra, in Proceedings of the 8th Joint Specialty Conference on Probabilistic Mechanics and Structural Reliability (ASCE, 2000).
34. M. Rosenblatt, M., *Annals Math. Stat.*, 23 (1952).
35. M. Hohenbichler and R. Rackwitz, *J. Engineering Mechanics*, ASCE 107 (1981).
36. P-L. Liu and A. Der Kiureghian, *Structural Safety*, 9 (1991).
37. A. Hasofer and N. Lind, *J. Engineering Mechanics*, 100 (1974).
38. R. Rackwitz and B. Fiessler, *Computers & Structures*, 9 (1978).
39. A. Der Kiureghian and T. Dakessian, *Structural Safety*, 20 (1998).
40. S. Katsuki and D.M. Frangopol, *J. Engineering Mechanics*, 120 (1994).
41. A. Der Kiureghian, in *Engineering Design Reliability Handbook* (CRC Press, New York 2004), p. 14.1.
42. L. Tvedt, J. Engineering Mechanics, 116 (1990).
43. K. Breitung, *J. Engineering Mechanics*, 110 (1984).

44. M. Hohenbichler and R. Rackwitz, *J. Engineering Mechanics*, 114 (1988).
45. A. Der Kiureghian, H.S. Lin, and S.-J. Hwang, *J. Engineering Mechanics*, 113 (1987).
46. A. Der Kiureghian and M. DeStefano, *J. Engineering Mechanics*, 117 (1991).
47. L. Faravelli, *J. Engineering Mechanics*, 115 (1989).
48. M. R. Rajashekhar and B. Ellingwood, *Struct. Safety,* 12 (1993).
49. R.H. Myers, *J. Quality Technology*, 31 (1999).
50. G.G. Wang, *J. Mechanical Design*, 125 (2003).
51. C.G. Bucher and M. Macke, in *Engineering Design Reliability Handbook*, (CRC Press, 2004) p. 19.1.
52. S.S. Isukapalli, A. Roy, and P.G. Georgopoulos, *Risk Analysis*, 20 (2000).
53. P.B. Nair, in *Engineering Design Reliability Handbook* (CRC Press, 2004), p. 21.1.
54. N.H. Kim, H. Wang, and N.V. Queipo, in Proceedings of 10th AIAA/ISSMO Multidisciplinary Analysis and Optimization Conference (AIAA, 2004). AIAA-2004-4514.
55. S.-K. Choi, R.V. Grandhi, and R.A. Canfield, in Proceedings of 10th AIAA/ISSMO Multidisciplinary Analysis and Optimization Conference (AIAA, 2004), AIAA-2004-4590.
56. N.H. Kim, H. Wang, and N.V. Queipo, in Proceedings of the 9th ASCE Joint Specialty Conference on Probabilistic Mechanics and Structural Reliability (2004).
57. M.S. Eldred, A.A. Giunta, S.F. Wojtkiewicz, Jr., and T.G. Trucano, in *Proceedings of the 9th AIAA/ISSMO Symposium on Multidisciplinary Analysis and Optimization* (AIAA, 2002), AIAA-2002-5585.
58. A.A. Giunta, M.S. Eldred, and J.P. Castro, in Proceedings of the 9th ASCE Joint Specialty Conference on Probabilistic Mechanics and Structural Reliability (2004).
59. R. Ghanem, and P.D. Spanos, *Stochastic finite element: a spectral approach* (Springer, New York, 1991).
60. D. Xiu and G.E. Karniadakis, *SIAM J. Sci. Comput.*, 24 (2002).
61. B. Sudret and A. Der Kiureghian, *Prob. Eng. Mech.*, 17 (2003).
62. P.B. Nair and A.J. Keane, *AIAA Journal*, 40 (2002).
63. M. Allen, G. Weickum, and K. Maute, in Proceedings of 10th AIAA/ISSMO Multidisciplinary Analysis and Optimization Conference (AIAA, 2004).
64. K. Maute, D.M. Frangopol, M. Allen, M. Liu, and G. Weickum, in Proceedings of the 2005 NSF Design, Service and Manufacturing Grantees and Research Conference (2005).
65. U. Kirsch, *AIAA Journal*. 39 (2001).
66. U. Kirsch, *Computers & Structures*, 78 (2000).
67. E.J. Haug, K.K. Choi, and V. Komkov, *Design sensitivity analysis of structural systems* (Academic Press, New York, 1986).
68. M. Kleiber, H. Antunez, H. Hien, and P. Kowalczyk, *Parameter Sensitivity in nonlinear mechanics* (Wiley, Chichester, 1997).
69. K. Maute, M. Nikbay, and C. Farhat, *Int. J. Num. Meth. Engrg.*, 56 (2003).
70. M. Allen, *Reliability-Based Design Optimization of Multiphysics, Aerospace Systems* (Ph.D. thesis, Department of Aerospace Engineering Sciences, University of Colorado, Boulder, 2004).
71. K. Maute, G. Weickum, and M. Allen, in Proceedings of the 9th International Conference on Structural Safety and Reliability ICOSSAR'05 (2005) , p. 2049.
72. W. Chen and X. Du, in Proceedings of the ASME Design Engineering Technical Conferences and Computers and Information in Engineering Conference (2002).

73. L. Wang and S. Kodiyalam, in Proceedings of the 43rd AIAA/ASME/ ASCE/AHS/ASC Structures, Structural Dynamics, and Materials Conference (AIAA, 2002), AIAA-2002-1754.
74. M. Hohenbichler and R. Rackwitz, *Civil Engineering Systems*, 3 (1986).
75. A. Karamchandani and C. Cornell, *Structural Safety*, 11 (1992).
76. J. Tu, K.K. Choi, and Y.H. Park, J. Mechanical Design, ASME, 121 (1999).
77. T. Lee and B. Kwak, Mechanics of Structures and Machines, 15 (1987).
78. J-O. Lee, Y-S. Yang, and W-S. Ruy, *Computers & Structures*, 80 (2002).
79. N. Kuschel and R. Rackwitz, Applications of Statistics and Probability, 2 (2000).
80. H. Agarwal, J.C. Lee, L.T. Watson, and J.E. Renaud, in Proceedings of the 45th AIAA/ASME/ASCE/AHS/ASC Structures, Structural Dynamics, and Materials Conference (AIAA, 2004), AIAA-2004-2029.
81. G. Kharmanda, A. Mohamed, and M. Lemaire, *Struct. Multidisc. Optim.*, 24 (2002).
82. R.H. Gallagher and O.C. Zienkiewicz, eds., *Optimal Structural Design: Theory and Application* (Wiley, Chichester, 1973).
83. J.S. Arora, *Introduction to Optimal Design.* (McGraw-Hill, New York, 1989).
84. J.S. Arora, in *Recent Advances in Optimal Structural Design*, ed. S.A. Burns (ASCE, 2002).
85. W. Stadler, Multicriteria Optimization in Engineering and in the Sciences (Plenum Press, London, 1988).
86. H.A. Eschenauer, J. Koski, and A. Osyczka, *Multicriteria Design Optimization* (Springer, Berlin, 1990).
87. G. Fu and D.M. Frangopol, *Structural Safety*, 7 (1990).
88. G. Fu and D.M. Frangopol, *Journal of Structural Engineering*, ASCE, 116 (1990).
89. J.H. Holland, *Adaptation in Natural and Artificial Systems* (University of Michigan Press, Ann Arbor, 1975).
90. K. Deb, Multi-objective Optimization Using Evolutionary Algorithms (Wiley, Chichester, 2001).
91. D.E. Goldberg, Genetic Algorithms in Search, Optimization and Machine Learning (Addison-Wesley, Reading, 1989).
92. C.A.C. Coello, Comp. Meth. Appl. Mech. Engrg., 191 (2002).
93. D.E. Goldberg and M.P. Samtani, in Proceedings of the 9th Conference on Electronic Computation (ASCE, 1986).
94. S. Rajeev and C.S. Krishnamoorthy, *J. Structural Engineering*, ASCE, 118 (1992).
95. H. Adeli and N.-T. Cheng, *J. Aerospace Engineering*, ASCE, 6 (1993).
96. S.D. Rajan, *J. Structural Engineering*, ASCE, 121 (1995).
97. S.M. Shrestha and J. Ghaboussi, *J. Structural Engineering*, ASCE, 124 (1998).
98. J.P.B. Leite and B.H.V. Topping, *Advances in Engineering Software*, 29 (1998).
99. J. Ghaboussi, Structural Engineering and Mechanics, 11 (2001).
100. C. Camp, S. Pezeshk, and G. Cao, *J. Structural Engineering*, ASCE, 124 (1998).
101. S. Pezeshk, C.V. Camp, and D. Chen, *J. Structural Engineering*, ASCE, 126 (2000).
102. C.M. Foley and D. Schinler, *J. Structural Engineering*, ASCE, 129 (2003).
103. F.Y. Cheng and D. Li, *J. Structural Engineering*, ASCE, 123 (1997).
104. C.A.C. Coello and A.D. Christiansen, *Computers & Structures*, 75 (2000).
105. D. Greiner, G. Winter, and J.M. Emperador, *Finite Elements in Analysis and Design*, 37 (2001).
106. F.Y. Cheng, in *Recent Advances in Optimal Structural Design*, ed., S.A. Burns (ASCE, 2002).

107. M. Liu, S.A. Burns, and Y.K. Wen, *Earthquake Engineering and Structural Dynamics*, 34 (2005).
108. M. Liu, S.A. Burns, and Y.K. Wen, *J. Structural Engineering*, ASCE, to appear.
109. Y.K. Wen and M. Shinozuka, *Engineering Structures*, 20 (1998).
110. H.M. Koh, J. Song, and D.-H. Ha, in Proceedings of the 12th World Conference on Earthquake Engineering (2000).
111. *ATC-40* (Applied Technology Council, Redwood City, CA, 1996).
112. *FEMA-273* (Federal Emergency Management Agency, Building Seismic Safety Council, Washington, D.C, 1997).
113. *FEMA-350* (Federal Emergency Management Agency, Building Seismic Safety Council, Washington, D.C, 2000).
114. Y.-J. Kang, and Y.K. Wen, *Structural Research Series,* No. 629 (Department of Civil and Environmental Engineering, University of Illinois at Urbana-Champaign, Urbana, 2000).
115. M. Kohno and K.R. Collins, *Research Report UMCEE 00/06* (Department of Civil and Environmental Engineering, University of Michigan, Ann Arbor, 2000).
116. A.H.-S. Ang and J.-C. Lee, Reliability Engineering and System Safety, 73 (2001).
117. M. Liu, Y.K. Wen, and S.A. Burns, *Engineering Structures*, 26 (2004).
118. C.A. Cornell, F. Jalayer, R.O. Hamburger, and D.A. Foutch, *J. Structural Engineering*, ASCE, 128 (2002).
119. H. Hawk and E.P. Small, Structural Engineering International, 8 (1998).
120. P.D. Thompson, E.P. Small, M. Johnson, and A.R. Marshall, *Structural Engineering International*, 8 (1998).
121. D.M. Frangopol and P.C. Das, in *Current and Future Trends in Bridge Design, Construction, and Maintenance*, eds., P.C. Das, D.M. Frangopol, and A.S. Nowak (Thomas Telford, London, 1999), p. 45.
122. D.M. Frangopol, J.S. Kong, and E.S. Gharaibeh, *J. Computing in Civil Engineering*, ASCE, 15 (2001).
123. J. van Noortwijk and D.M. Frangopol, *Probabilistic Engineering Mechanics*, 19 (2004).
124. M. Liu and D.M. Frangopol, Computer-aided Civil and Infrastructure Engineering, 20 (2005).
125. M. Liu and D.M. Frangopol, J. Bridge Engineering, ASCE, 10 (2005).
126. M.G.H. Bell and Y. Iida, *Transportation Network Analysis*, (Wiley, Chichester, 1997).
127. M. Shinozuka, Y. Murachi, X. Dong, Y. Zhou, and M.J. Orlikowski, in Proceedings of China-US Workshop on Protection of Urbana Infrastructure and Public Buildings against Earthquakes and Manmade Disasters (2003).
128. G. Augusti, M. Ciampoli, and D.M. Frangopol, *Engineering Structures*, 20 (1998).
129. B. Adey, R. Hajdin, and E. Bruhwiler, *Journal of Infrastructure Systems*, ASCE, 9 (2003).
130. J.S. Kong and D.M. Frangopol, *J. Structural Engineering*, ASCE, 130 (2004).
131. J.S. Kong and D.M. Frangopol, *J. Structural Engineering*, ASCE, 131 (2005).
132. M. Stewart, A.C. Estes, and D.M. Frangopol, J. Structural Engineering, ASCE, 130(2004).
133. A.C. Estes and D.M. Frangopol, in *Structural Engineering Handbook*, eds., W-F. Chen and E. M. Lui (CRC Press, 2005), p. 36-1.
134. S.-I. Yang, D.M. Frangopol, and L.C. Neves, *Engineering Structures*, 28 (2006).
135. S.-I. Yang, D.M. Frangopol, Y. Kawakami, and L.C. Neves, *Reliability Engineering and System Safety*, in press.
136. A.H-S. Ang and D. De Leon, Structure and Infrastructure Engineering, 1 (2005).
137. B.R. Ellingwood, Structure and Infrastructure Engineering, 1 (2005).

CHAPTER 12

OPTIMIZATION-BASED INVERSE KINEMATICS OF ARTICULATED LINKAGES

Karim Abdel-Malek and Jingzhou Yang

Center for Computer-Aided Design
The University of Iowa, Iowa City, IA 52242, USA
E-mail: jyang@engineering.uiowa.edu

Articulated linkages appear in many fields, among them, robotics, human modeling, and mechanism design. The inverse kinematics (IK) of articulated linkages forms the basic problem to solve in various scenarios. This chapter presents an optimization-based approach for IK of a specific articulated linkage, the human model. The problem is formulated as a single-objective optimization (SOO) problem with a single performance measure or as a multi-objective-optimization (MOO) problem with multiple combined performance measures. A human performance measure is a physics-based metric, such as delta potential energy, joint displacement, joint torque, or discomfort, and serves as an objective function (cost function) in an optimization formulation. The implementation of the presented approach is shown for three models: a 4-degree-of-freedom (DOF) finger model, 21-DOF torso-right hand model, and 31-DOF torso-both hands model. Preliminary validation using a motion capture system demonstrates the accuracy of the proposed method.

1. Introduction

This chapter presents an optimization-based method for the coordination of voluntary human motion (articulated human model), a special type of articulated linkages. Posture prediction means the estimation of joint variables that will allow the human body to assume a posture to achieve an objective. Note that the word prediction is often used instead of calculation since predicting realistic human motion is not an exact science, but rather a prediction of human behavior. For example, the prediction or calculation of upper extremity variables to achieve the grasping of an object is a posture prediction problem. Grasping, in this case, means the positioning of the hand that is also called the end-effector of a serial kinematic chain in the field of robotics. Similarly, predicting, i.e., calculating the joint variables of a lower extremity to allow the foot to be positioned on an object, a pedal, for example, is also a posture prediction problem. The challenge in predicting postures is evident in that there are a large (infinite) number of solutions.

In this chapter, the treatment of an open kinematic chain, as opposed to closed-loop systems, is addressed. For the human body, we consider a variety of articulated linkages, typically all beginning in the waist and extending to the hand, the foot, or the head. Indeed, a kinematic chain starts from the foot extending to the hand.

Many theories and methodologies have been reported for how the brain issues motor control commands to the central nervous system to achieve a task. This chapter will focus on the optimization-based method, which we believe is a natural process for completing a task.

One method for characterizing intelligent function in a digital human (avatar) is the ability to process a task using previous cost functions. The optimization methodology, introduced by the author and colleagues several years ago, first breaks a task into sub-tasks, then, breaks each sub-task into procedures. This process is called task planning and is well understood in the field of robotics. In order to accomplish a procedure, the procedure planner then selects one or more cost functions from a list of available human performance measures to be used in an optimization algorithm (one example is provided in Figure 1). Note that task planning and procedure planning are forms of intelligent engines. Selecting the appropriate type of cost functions and understanding their relative importance is itself a challenging problem that is not well understood. Nevertheless, optimizing for a combination of cost functions yields a behavior that is different than optimizing for a different set of cost functions. This aspect of selecting cost functions will likely continue to be a research area with great interest in that it produces behaviors that vary for the same procedure but different cost functions.

The main new idea here is the belief that humans execute procedures while minimizing or maximizing cost functions. This chapter presents an optimization-based inverse kinematics of the articulated linkages and is organized as follows: Section 2 introduces the Denavit-Hartenberg (DH) method;[1] Section 3 presents the formulation of inverse kinematics; Section 4 provides several examples; and the conclusions summarize the methodology and formulations.

2. Denavit-Hartenberg Method

The mathematical foundation for the human skeletal model used to control motion essentially relates a set of joint angles to the Cartesian coordinates of a point on the avatar. A human skeleton can be modeled as a kinematic system – a series of links with each pair of links connected by single degree-of-freedom (DOF) joints as shown in Figure 2. Therefore, a complete human body can be modeled as several open-loop chains, one of which is shown in Figure 2, where q_i is a generalized coordinate and represents the rotation of a single basic revolute joint.

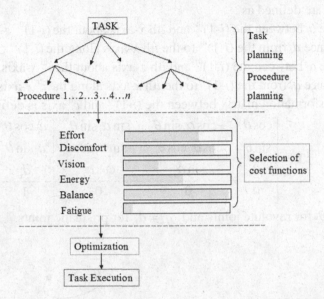

Figure 1. The overall planning and execution of a task using the theory of optimization

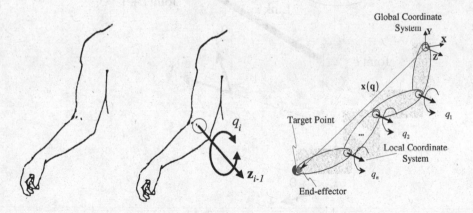

Figure 2. Definition of a kinematic pair

Each joint in the skeletal model is represented by one, two, or three revolute joints. $\mathbf{q} = [q_1, \ldots, q_n]^T \in R^n$ is an n-dimensional vector of generalized coordinates, where n is the number of DOF. The *end-effector* is a point on the virtual human, typically the tip of the index finger. $\mathbf{x(q)}$ represents the position of the end-effector in Cartesian space and is a function of the joint angles. Using the Denavit-Hartenberg (DH) method, the four parameters $(\theta_i, d_i, \alpha_i, a_i)$, shown in Figure 3, are defined as

(1) the angle θ_i between the $(i\text{-}1)^{\text{th}}$ and ith x-axis about the $(i\text{-}1)^{\text{th}}$ z-axis

(2) the distance d_i from the $(i\text{-}1)^{\text{th}}$ to the ith x-axis along the $(i\text{-}1)^{\text{th}}$ z-axis

(3) the angle α_i between the $(i\text{-}1)^{\text{th}}$ and ith z-axis about the ith x-axis

(4) the distance a_i from the $(i\text{-}1)^{\text{th}}$ to the ith x-axis along the ith x-axis

The transformation matrix between the $(i\text{-}1)^{\text{th}}$ and i^{th} axis is defined as

$$^{i-1}\mathbf{T}_i = \left[\begin{array}{ccc|c} \cos\theta_i & -\cos\alpha_i \sin\theta_i & \sin\alpha_i \sin\theta_i & a_i \cos\theta_i \\ \sin\theta_i & \cos\alpha_i \cos\theta_i & -\sin\alpha_i \cos\theta_i & a_i \sin\theta_i \\ 0 & \sin\alpha_i & \cos\alpha_i & d_i \\ \hline 0 & 0 & 0 & 1 \end{array}\right] \qquad (1)$$

where $q_i = \theta_i$ for revolute joints and $q_i = d_i$ for prismatic joints.

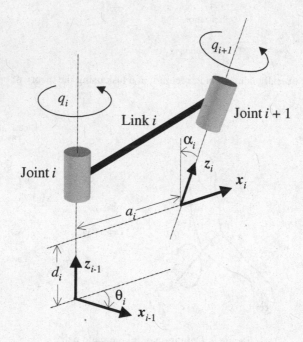

Figure 3. DH-Parameters

The global position-vector $\mathbf{x(q)}$ for the end-effector of the human model is given as:

$$\begin{bmatrix} \mathbf{x(q)} \\ 1 \end{bmatrix} = {}^0\mathbf{T}_n \begin{bmatrix} \mathbf{x}_n \\ 1 \end{bmatrix} \tag{2}$$

$$ {}^0\mathbf{T}_n = \left[\prod_{i=1}^{n} {}^{i-1}\mathbf{T}_i \right] = \begin{bmatrix} \mathbf{R} & \mathbf{P} \\ \mathbf{0} & 1 \end{bmatrix} \tag{3}$$

where $\mathbf{x}_n$ is the position of the end-effector with respect to the nth frame with $\mathbf{x}_n = \begin{bmatrix} x_n & y_n & z_n \end{bmatrix}^T$. Note that the 0th frame is the global reference frame (global coordinate system). $\mathbf{R}$ is the orientation matrix for the nth frame with respect to the 0th frame. $\mathbf{P}$ is the position vector of the nth frame.

3. Inverse Kinematics

Because of the apparent mathematical complexity of the problem of posture prediction and because the human body has many more than six DOF, methods for predicting postures are very involved. Posture prediction, at least on the surface, is equivalent to what is called Inverse Kinematics in the field of mechanisms and robotics (known as IK in the gaming and animation industries). This section addresses the benefits and shortcomings of IK:

Analytical and geometric IK methods. These are methods typically associated with arms of robotic manipulators that are no more than 6-DOF systems, where closed-form or numerical solutions are possible. Higher than 6-DOF systems become very redundant and require very complex numerical algorithms. Analytical or geometric IK methods for human posture prediction are almost impossible. One reason for this is the high number of DOF associated with the human model lead to severe difficulties in calculating solutions. Solutions are almost impossible for geometric methods and very difficult using numerical methods, as identifying and finding all solutions is an outstanding mathematical problem. A second reason is the need to calculate realistic postures. While some of the analytical methods may yield solutions, a choice of the "best" solution that looks most natural is difficult to achieve using analytic IK methods. The main benefit of analytic and geometric methods is that, if determined, they are determined very quickly. Typically, analytic methods, even though numerical in nature, are computationally efficient.

Empirically based posture prediction. This method is based on gathering much data of actual subjects while performing various postures. Anthropometry and posture data are captured and recorded. Statistical models, typically nonlinear regression, are then developed and used in posture prediction algorithms. A benefit of this method is its ability to predict postures that already have been recorded for the exact anthropometric model. Extrapolating postures and variations thereof are extremely difficult and become highly inaccurate, and this method completely fails to predict motion, the ultimate goal of posture prediction. If it is to be expanded to predict motion, the variability and many parameters associated with motion, including dynamic effects and inertia, are not only difficult to measure but impossible to correlate. The method requires an exhaustive and often very costly experimental setup involving thousands of subjects to generate a modest model for a small population.

This chapter introduces a framework with an associated algorithm for predicting postures that is based on an individual task. In order to better understand the motivation behind cost functions, consider first the case of a driver in a vehicle who is about to reach to a radio control button on the dashboard. It is believed that the driver will reach directly to the button while exerting minimum effort and perhaps expending minimum energy. Next, consider the same driver negotiating a curve. Here, he will have to place his/her hand on the steering wheel in such a way as to be able to exert the necessary force needed to turn the wheel. As a result, the driver will, involuntarily, select a posture that maximizes force at the hand, minimizes the torque at each joint, minimizes energy, and minimizes effort needed to accomplish this task (as illustrated in Figure 4).[2-9]

Therefore, our underlying plot is that each procedure is driven by the optimization of one or more cost functions, leading the person to position his/her body in the most natural manner. Note that simple logic has been implemented in the *Processor* module to correlate between the Task and the Cost Functions.

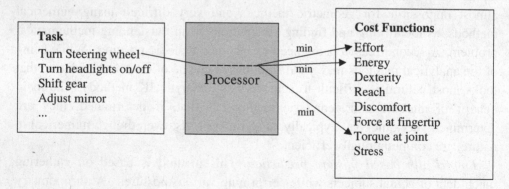

Figure 4. The task-based approach to selecting cost functions

The formulation contains three main components:

(1) *A cost function*: to be minimized or maximized. In our case, many cost functions thus form a multi-objective optimization problem.

(2) *Design variables*: the variables are individual degree-of-freedom joint displacements that will be calculated from the algorithm. In our case the joint variables that define the position and orientation of each segmental link are design variables.

(3) *Constraints*: are mathematical expressions that bound the problem. A constraint is imposed making sure the distance between the end-effector and the target point is within a specified tolerance. Furthermore, joint ranges of motion are necessary constraints but additional constraints are added as appropriate.

3.1. *Formulation of the Optimization Problem*

The optimum posture (set of q-values) is determined by solving the following optimization problem:

$$\text{Find: } \mathbf{q} \in R^{DOF} \tag{4}$$

$$\text{Minimize: } Discomfort, Effort, \text{etc.}$$

$$\text{Subject to: } \left[\mathbf{x}(\mathbf{q})^{end\text{-}effector} - \mathbf{x}^{target\ point} \right]^2 \leq \varepsilon$$

$$\text{and } q_i^L \leq q_i \leq q_i^U ; \ i = 1, 2, \ldots, DOF$$

The *feasible space* for a problem such as the one in (4) is defined as the set of all points $\mathbf{q}$ for which all of the constraints are satisfied.

3.3.1. *Design Variables*

As suggested earlier, the design variables are the generalized coordinates q_i, or in vector form as $\mathbf{q} = \begin{bmatrix} q_1 & \cdots & q_{DOF} \end{bmatrix}^T$. Since all joints are rotations, q_i have units of radians. Many optimization algorithms require an *initial guess*, which entails determining initial values for the design variables. The initial guess can be somewhat arbitrary, although it is helpful to use a feasible point. In this study, an initial guess is used that satisfies the joint limits constraint but does not necessarily satisfy the distance constraint, which is discussed below.

3.3.2. *Constraints*

The first constraint in (4) is the distance constraint and requires that the end-effector contact a predetermined target point (user specified) in Cartesian space, where ε is a small positive number that approximates zero. This constraint represents distance-squared. The DH-method is used to determine the position of the end-effector after a series of given displacements. In addition, each generalized coordinate is constrained to lie between lower and upper limits, represented by q_i^L and q_i^U, respectively. These limits ensure that the digital human does not assume an unrealistic position in an effort to contact the target point.

3.3.3. *Cost Function*

While many cost functions exist and others are being developed by many researchers and in several fields, we will attempt to demonstrate a number of such functions for posture prediction such as joint displacement, delta-potential energy, discomfort, effort, and joint torque. In this section we review all of the five cost functions and will demonstrate the use of some of these cost functions in the next section:

(1) *Joint Displacement*

Joint displacement is proportional to the deviation from the *neutral position*. The neutral potion is selected as a relatively comfortable posture, typically a standing position with arms at one's sides, where q_i^N is the neutral position of a joint, and $\mathbf{q}^N$ represents the overall posture. Technically, the displacement from the neutral position is given by $\left|q_i - q_i^N\right|$; however, for computational simplicity $\left(q_i - q_i^N\right)^2$ is used. Because motion in some joints contributes more significantly to the joint displacement, a weight w_i is introduced to stress the importance of a particular joint. Then, the cumulative joint displacement (of all joints) is characterized by the following objective function:

$$f_{Jo\text{int}-Displacement}\left(\mathbf{q}\right) = \sum_{i=1}^{n} w_i \left(q_i - q_i^N\right)^2 \tag{5}$$

(2) *Delta-Potential-Energy*

This section discusses a potential-energy function that is indirectly based on the weighted sum method for MOO. However, in this case, the weights are based on the mass of different segments of the body. With the previous (joint displacement) function, the weights are set based on intuition and experimentation, and although the postures obtained by minimizing joint displacement are realistic, there are other ways to assign relative importance to the components of the human performance measure. The idea of potential energy provides one such alternative.

We represent the primary segments of the upper body with nine lumped masses: one each for the lower, middle, and upper torso, respectively; one each for the right upper arm; right forearm; and right hand; and one each for the corresponding lumped masses of the left arm. We then determine the potential energy for each mass. The heights of these masses, rather than the joint displacements for the generalized coordinates, provide the components of the human performance measure. Mathematically, the weight (force of gravity) of a segment of the upper body provides a multiplier for movement of that segment in the vertical direction. The height of each segment is a function of generalized coordinates, so, in a sense, the weights of the lumped masses replace the scalar multipliers, w_i, which are used in the joint displacement function.

If the potential energy function were used directly, there would always be a tendency to bend over, thus reducing potential energy. Consequently, we actually minimize the *change* in potential energy. Each link in a segmented serial chain, as depicted in Figure 5 (e.g., the forearm), has a specified center of mass.

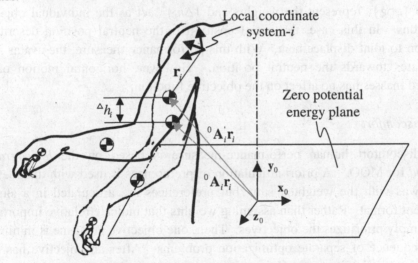

Figure 5. Illustration of the potential energy of the upper body

The vector from the origin of a link's local coordinate system to its center of mass is given by $^i\mathbf{r}_i$, where the subscript indicates the relevant local coordinate system. In order to determine the position and orientation of any part of the body, we use the transformation matrices $^{(i-1)}A_i$, which are 4×4 matrices that relate local coordinate system-i to local coordinate system-$i-1$. Consequently,

$\mathbf{r}_i$ is actually an augmented 4×1 vector with respect to local coordinate system i, rather than a 3×1 vector typically used with Cartesian space. $\mathbf{g} = \begin{bmatrix} 0 & 0 & -g \end{bmatrix}^T$ is the augmented gravity vector. When the human upper body moves from one configuration to another, there are two potential energies, P_i' which is associated with the initial configuration and P_i which is associated with the current configuration. Therefore, for the first body part in the chain (the lower torso), the potential energies are $P_1' = m_1 \mathbf{g}^T {}^0\mathbf{A}_1' \mathbf{r}_1$ and $P_1 = m_1 \mathbf{g}^T {}^0\mathbf{A}_1 \mathbf{r}_1$. The potential energies for the second body part are $P_2' = m_2 \mathbf{g}^T {}^0\mathbf{A}_1' {}^1\mathbf{A}_2' \mathbf{r}_2$ and $P_2 = m_2 \mathbf{g}^T {}^0\mathbf{A}_1 {}^1\mathbf{A}_2 \mathbf{r}_2$. The potential energies for the ith body part are $P_i' = m_i \mathbf{g}^T {}^0\mathbf{A}_1' \cdots {}^{i-1}\mathbf{A}_i' \mathbf{r}_i$ and $P_i = m_i \mathbf{g}^T {}^0\mathbf{A}_1 \cdots {}^{i-1}\mathbf{A}_i \mathbf{r}_i$. In Figure 5, Δh_i is the y-component of the vector ${}^0\mathbf{A}_1' \cdots {}^{i-1}\mathbf{A}_i' \mathbf{r}_i - {}^0\mathbf{A}_1 \cdots {}^{i-1}\mathbf{A}_i \mathbf{r}_i$. The final objective function, which is minimized, is defined as follows:

$$f_{Delta-Potential-Energy}(\mathbf{q}) = \sum_{i=1}^{9} \left(P_i - P_i' \right)^2 \tag{6}$$

Note that (6) can be written in the form of a weighted sum as follows:

$$f_{Delta-Potential-Energy}(\mathbf{q}) = \sum_{i=1}^{9} \left(m_i g \right)^2 \left(\Delta h_i \right)^2 \tag{7}$$

where $\left(m_i g \right)^2$ represent the weights and $\left(\Delta h_i \right)^2$ act as the individual objective functions. In this case, the initial position is the neutral position describe in relation to joint displacement. With this performance measure, the avatar again gravitates towards the neutral position. However, horizontal motion of the lumped masses has no affect on the objective function.

(3) Discomfort

The discomfort human performance-measure is based on the *lexicographic method* for MOO.[5] A priori articulation of preferences is used with this method, as it was with the weighted sum, but preferences are articulated in a slightly different format. Rather than assigning weights that indicate relative importance, one simply prioritizes the objectives. Then, one objective at a time is minimized in a sequence of separate optimization problems. After an objective has been minimized, it is incorporated as a constraint in the subsequent problems. The solution to this method is Pareto optimal, if it is unique.

The concept behind this new discomfort human performance measure is that groups of joints are utilized sequentially. That is, in an effort to reach a particular target point, one first uses one's arm. Then, if necessary, one bends the torso. Finally, if the target remains out of reach, one may extend the clavicle joint. The lexicographic method for MOO is designed to incorporate this type of preference structure. However, solving a sequence of optimization problems can be time consuming and impractical for real-time applications such as human simulation. The weighted sum method can be used to approximate results of the lexicographic method if the weights have infinitely different orders of magnitude.[10,11] This results in the weights shown in Table 1.

Although weights are used here, they do not need to be determined as indicators of the relative significance of their respective joints; they are simply fixed mathematical parameters. In addition, the exact values of the weights are irrelevant; they only need to have significantly different orders of magnitude. Some of the weights in Table 1 are discontinuous because movement in various directions can result in different degrees of acceptability. These discontinuities can lead to computational difficulties; however, with this discomfort objective, such discontinuities are avoided.

Table 1. Joint weights for discomfort

Joint Variable	Joint Weight	Comments
$q_1, \ldots, q_{12}$	1×10^4	Used with both positive and negative values of $q_i - q_i^N$
q_{13}, q_{14}	1×10^8	Used with both positive and negative values of $q_i - q_i^N$
$q_{15}, \ldots, q_{21}$	1	Used with both positive and negative values of $q_i - q_i^N$

The weights in Table 2 are used in a function that is similar to Eq. (5). However, prior to applying the weights, each term in Eq. (5) is normalized as follows:

$$\Delta q^{norm} = \frac{q_i - q_i^N}{q_i^U - q_i^L} \qquad (8)$$

With this normalization scheme, each term $\left(\Delta q_i^{norm}\right)^2$ acts as an individual objective function and has values between zero and one.

Although this approach generally works well it often results in postures with joints extended to their limits, and such postures can be uncomfortable. To rectify this problem, extra terms are added to the discomfort function such that the discomfort increases significantly as joint values approach their limits. The final discomfort function is given as follows:

$$f_{Discomfort}(q) = \frac{1}{G} \sum_{i=1}^{n} \gamma_i \left[\left(\Delta q_i^{norm} \right)^2 + G \times QU + G \times QL \right] \tag{9}$$

$$QU = \left(0.5 Sin \left(\frac{5.0 \left(q_i^U - q_i \right)}{q_i^U - q_i^L} + 1.571 \right) + 1 \right)^{100}$$

$$QL = \left(0.5 Sin \left(\frac{5.0 \left(q_i - q_i^L \right)}{q_i^U - q_i^L} + 1.571 \right) + 1 \right)^{100}$$

where $G \times QU$ is a penalty term associated with joint values that approach their upper limits, and $G \times QL$ is a penalty term associated with joint values that approach their lower limits. Each term varies between zero and G, as $\left(q_i^U - q_i \right) / \left(q_i^U - q_i^L \right)$ and $\left(q_i - q_i^L \right) / \left(q_i^U - q_i^L \right)$ vary between zero and one. In this case, $G = 1 \times 10^6$. Figure 6 illustrates the curve for the following function, which represents the basic structure of the penalty terms:

$$Q = \left(0.5 Sin \left(5.0r + 1.571 \right) + 1 \right)^{100} \tag{10}$$

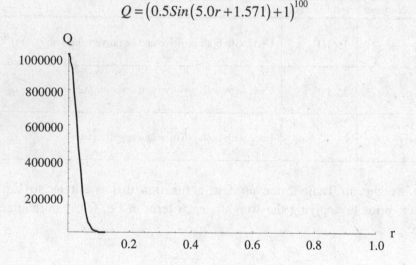

Figure 6. Graph of discomfort joint-limit penalty term

r represents either $\left(q_i^U - q_i \right) / \left(q_i^U - q_i^L \right)$ or $\left(q_i - q_i^L \right) / \left(q_i^U - q_i^L \right)$. Thus, as Figure 6 illustrates, the penalty term has a value of zero until the joint value reaches the upper or lower 10% of its range, where either $\left(q_i^U - q_i \right) / \left(q_i^U - q_i^L \right) \leq 0.1$ or $\left(q_i - q_i^L \right) / \left(q_i^U - q_i^L \right) \leq 0.1$. The curve for the penalty term is differentiable, and reaches its maximum of $G = 1 \times 10^6$ when $r = 0$. The final function in (9) is multiplied by $1/G$ to avoid extremely high function-values.

(4) *Effort*

Effort is measured as the displacement of a joint from its original position. Effort will greatly depend on the initial configuration of the limb, i.e., prior to the joint moving to another position. For an initial set of joint variables $q_i^{initial}$ and for a final set of joint variables q_i, a simple measure of the effort is expressed by

$$f_{effort}(\mathbf{q}) = \sum_{i=1}^{n} \alpha_i \left(q_i - q_i^{initial} \right)^2 \tag{11}$$

where α_i is a weight function assigned to each joint. Note that f_{effort} depends on the initial configuration of each joint.

(5) *Torque*

Stress induced at a joint is a function of torque imposed at that joint due to the biomechanical interaction. A person will generate the torque at a given joint to overcome a load by exerting muscle forces, but torque is also a function of the position and orientation of the joint during loading. In order to account for all of the elements that enter into calculating the torque at a given joint, we must employ a systematic formulation. To develop a mathematical expression for the torque, we first introduce a few preliminary concepts.

The velocity of a point on the hand is obtained by differentiating the position vector as

$$\dot{\mathbf{x}} = \mathbf{J}_x \dot{\mathbf{q}} \tag{12}$$

where the position Jacobian $\mathbf{J}_x(\mathbf{q}) = [\partial \mathbf{x}/\partial \mathbf{q}]$ is a $(3 \times n)$ matrix and $\dot{\mathbf{q}}$ is the vector of joint velocities. Note that the reach envelope can be determined from analytically stratifying the Jacobian.[2] Similarly, the angular velocity can be obtained as

$$\boldsymbol{\omega} = \mathbf{J}_w \dot{\mathbf{q}} \tag{13}$$

where the orientation Jacobian $\mathbf{J}_\omega$ is a $(3 \times n)$ matrix. Combining Eqs. (12) and (13) into one vector yields

$$\mathbf{v} = \begin{bmatrix} \dot{\mathbf{x}} \\ \boldsymbol{\omega} \end{bmatrix} = \mathbf{J}(\mathbf{q})\dot{\mathbf{q}} \tag{14}$$

where $\mathbf{J}(\mathbf{q})$ is the Jacobian of the limb or kinematic structure defined by

$$\mathbf{J}(\mathbf{q}) = \begin{bmatrix} \mathbf{J}_x \\ \mathbf{J}_\omega \end{bmatrix} \tag{15}$$

The goal is to determine the relationship between the generalized forces applied to the end-effortor (hand), for example, carrying a load, and generalized forces applied to the joints. Let $\boldsymbol{\tau}$ denote the $(n \times 1)$ vector of joint torques and

$\mathbf{F}$ the $(m \times 1)$ vector of hand forces applied at $\mathbf{p}$, where m is the dimension of the operational space of interest (typically six).

Using the principle of virtual work, we can determine a relationship of joint torques and forces at the end-effector. Since the upper and lower body form a kinematic system with time-invariant, holonomic constraints, its configuration only depends on the joint variables $\mathbf{q}$ (not explicitly on time). Consider the virtual work performed by the two force systems. As for the joint torques, the associated virtual work is

$$dW_\tau = \tau^T d\mathbf{q} \tag{16}$$

For the hand forces $\mathbf{F} = \begin{bmatrix} \mathbf{f}^T & \mathbf{m}^T \end{bmatrix}^T$, comprised of a force vector $\mathbf{f}$ and moment vector $\mathbf{m}$, the virtual work performed is

$$dW_F = \mathbf{f}^T d\mathbf{x} + \mathbf{m}^T \omega dt \tag{17}$$

where $d\mathbf{x}$ is the linear displacement and ωdt is the angular displacement.

Substituting Eqs. (14) and (15) into Eq. (17) yields

$$dW_F = \mathbf{f}^T \mathbf{J}_x d\mathbf{q} + \mathbf{m}^T \mathbf{J}_\omega d\mathbf{q}$$
$$dW_F = \mathbf{F}^T \mathbf{J} d\mathbf{q} \tag{18}$$

Since virtual and elementary displacements coincide, virtual works associated with the two systems are

$$\delta W_\tau = \tau^T \delta\mathbf{q} \tag{19}$$
$$\delta W_F = \mathbf{F}^T \mathbf{J} \delta\mathbf{q} \tag{20}$$

where δ denotes a virtual quantity. The system is under static equilibrium if and only if

$$dW_F = dW_\tau \qquad \forall \delta\mathbf{q} \tag{21}$$

which means that the difference between the virtual work of the joint torques and the virtual work of the hand forces shall be null for all joint displacements. Substituting Eqs. (16) and (18) into Eq. (21) yields

$$\tau^T \delta\mathbf{q} = \mathbf{F}^T \mathbf{J}(\mathbf{q}) \delta\mathbf{q} \quad \forall \mathbf{q} \tag{22}$$

Therefore, the relationship between the joint torques and forces on the hand is given by

$$\tau = \mathbf{J}^T \mathbf{F} \tag{23}$$

where the torque vector is $\tau = \begin{bmatrix} \tau_1, \tau_2, ..., \tau_n \end{bmatrix}^T$. The torque cost function is comprised of the weighted summation of all joint torques

$$f_{torque} = \sum_{i=1}^{n} \lambda_i |\tau_i| \tag{24}$$

where λ_i is a weight function used to distribute the importance of the cost function among all joints.

4. Examples

This section provides several examples to illustrate the optimization-based method for inverse kinematics of accumulated linkages. The first example is a simple 4-DOF index finger model, where, given the finger-tip position, we use the optimization-based method to find the joint angles. The second example demonstrates a 21-DOF upper body model, and orientation is also considered for inverse kinematics. The third example is a 30-DOF model for upper body with dual arms.

4.1. *A 4-DOF Finger Example*

An index finger may be modeled as a 4 DOF system shown in Figure 7.[12] The joint ranges of motion are $13^o \le q_1 \le 42^o$, $0^o \le q_2 \le 80^o$, $0^o \le q_3 \le 100^o$, $-10^o \le q_4 \le 90^o$, and the neural configuration is $q_1 = 0^o$, $q_2 = 30^o$, $q_3 = 30^o$, $q_4 = 10^o$. In Eq. (5) the weights are $w_1 = 100$, $w_2 = 10$, $w_3 = 5$, $w_4 = 5$. The inverse kinematic problem is defined as: given finger-tip position $\mathbf{x} = \begin{bmatrix} x & y & z \end{bmatrix}^T$ to find the joint angles.

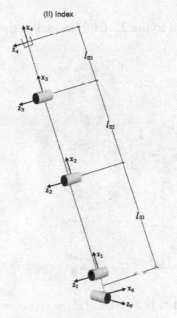

Figure 7. Index finger DH model

The optimization problem will be defined as follows:

$$\text{Find: } \mathbf{q} = \begin{bmatrix} q_1 & q_2 & q_3 & q_4 \end{bmatrix}^T$$

$$\text{to minimize: } f_{Joint-Displacement}(\mathbf{q}) = \sum_{i=1}^{4} w_i \left(q_i - q_i^N \right)^2$$

$$\text{subject to: } \left\| \mathbf{x}^{end-effector} - \mathbf{x}^p \right\| \leq \varepsilon$$

When the target point is at $\begin{bmatrix} -1.0 & 6.68 & -6.95 \end{bmatrix}^T$ joint angles are $\mathbf{q} = \begin{bmatrix} 8.4292 & 30.5679 & 28.511 & 9.0965 \end{bmatrix}^T$ in degrees, and joint displacement is 0.0233, where $\varepsilon = 0.0001$. If the target point is $\begin{bmatrix} -1.0 & 5.25 & -4.45 \end{bmatrix}^T$, then joint angles are $\mathbf{q} = \begin{bmatrix} 10.6792 & 0.7002 & 69.2643 & 48.8281 \end{bmatrix}^T$, and joint displacement is 0.0659.

4.2. A 21-DOF Model

A 21-DOF model is developed to represent realistically the movement of a *skinned* 3-dimensional model of a human, where the term skinned implies that a visually realistic outer surface is illustrated and attached to the kinemtic model (skeleton). Skinning is a term typically used by animators to describe the operation needed to lay the skin over the skeletal structure of a digital character.

4.2.1. DH Model

The DH method is applied to the 21-DOF model of the upper torso and right arm as illustrated in Figure 8.

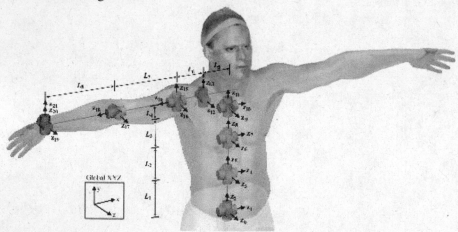

Figure 8. Detailed 21-DOF model (21st frame coincides with 20th frame)

Although the neutral posture remains a standing position with arms at the sides, the values are such that the position appears more relaxed. This posture is chosen based on observation of the skinned model rather than the skeletal model. The resulting vector $\mathbf{q}^N$ is defined as

$$q_i^N \quad 0; \ i \quad 1,...,12,19,20,$$

$$q_{13}^N \quad 15.0, \ q_{14}^N \quad 20.0, \ q_{15}^N \quad 100.0, \ q_{16}^N \quad 10.0, \ q_{17}^N \quad 80.0, \ q_{18}^N \quad 35.0, \ q_{21}^N \quad 15.0$$

Notice that the global coordinate system, selected to match this model's *zero*[th] coordinate frame, has also changed. The end-effector for this model is given by the position $(d, 0, 0)$, where d is a distance along the x-axis of the last coordinate frame.

4.2.2. Results

Figure 9 shows the results with the 21-DOF model and the target point $(-57.0, 31.5, 41.2)$. Discomfort = 1.6317, and the distance from the end-effector to the target point is 0.003 cm. The resulting q-vector is

$$\mathbf{q} = [-2.27, -.35, .39, -2.33, -.37, .34, -2.39, -.39, .30, -2.47, -.41,$$

$$.24, -15.82, 11.16, 36.24, 1.39, -79.38, -45.86, 1.07, -4.11, -.95]^T$$

In this case, all of the generalized coordinates have units of degrees.

Target Point

Figure 9. 21-DOF posture prediction for a target point of (−57.0, 31.5, 41.2)

The results with a target point of (-5.0, -20.0, 55.0) are shown in Figure 10. Discomfort = 1.7751. The distance from the end-effector to the target point is 0.003 cm. The resulting q-vector is

$$\mathbf{q} = [1.41, 5.37, 2.54, 1.51, 4.30, 2.53, 1.54, 3.20, 2.53, 1.50, 1.94,$$

$$2.53, -15.0, 25.0, 68.23, 36.05, -79.81, -24.13, -.57, 2.17, 2.97]^T$$

Figure 10. 21-DOF posture prediction for a target point of (-5.0, -20.0, 55.0)

As shown in Figure 10, there is a slight extension in the shoulder that was not possible with the previous model. Although the figure does not indicate it clearly, there is also a slight natural bend in the torso.

4.2.3. *Specifying the Orientation*

The rotation matrix **R** controls the orientation of frame-*i*. Therefore, by constraining portions of **R**, we can constrain the orientation of frame-*i*. Specifically, we can dictate in what direction (in terms of the global coordinate system) each axis of a local coordinate system points. This is done by specifying values for each column of **R**. The first column represents the direction in which the x-axis points, the second column represents the y-axis, and the third column represents the z-axis. Considering that the axes are orthogonal, only two axes can be constrained at a time.

For the 21-DOF model, there are 21 independent transformation matrices $^{i-1}\mathbf{T}_i$, and each one includes an independent rotation matrix $^{i-1}\mathbf{R}_i$. The cumulative transformation matrix $^0\mathbf{T}_n$ determines the position of the end-effector in terms of the global coordinate system. $^0\mathbf{R}_n$ is the rotation matrix incorporated in this cumulative matrix, and it determines the orientation of the nth reference frame with respect to the global reference frame. The nth reference frame is attached to the hand. Thus, to constrain the orientation of the hand, we constrain the components of $^0\mathbf{R}_n$.

To write an independent constraint for each component of $^0\mathbf{R}_n$ can be cumbersome when it comes to developing a general user-interface and can unnecessarily complicate the optimization problem. Instead, the components are combined into a single inequality constraint as follows:

$$\left\{\left[^0R_n(1,1)-l_{11}\right]^2+\left[^0R_n(2,1)-l_{21}\right]^2+\left[^0R_n(3,1)-l_{31}\right]^2\right\}+$$
$$\left\{\left[^0R_n(1,2)-l_{12}\right]^2+\left[^0R_n(2,2)-l_{22}\right]^2+\left[^0R_n(3,2)-l_{32}\right]^2\right\}\leq\gamma \tag{25}$$

where γ is a small positive number approximating zero. For this study $\gamma=1\times10^{-8}$. This constraint is incorporated in the formulation in (4). l_{ij} indicates the constrained value for $^0R_n(i,j)$. For example, with the following values for l_{ij}, the local x-axis for the hand points in the global **Z**-direction, and the local y-axis for the hand points in the global negative **Y**-direction:

$$l_{11}=0, l_{21}=0, l_{31}=1$$
$$l_{12}=0, l_{22}=-1, l_{32}=0 \tag{26}$$

The first three terms in (25) relate to the x-axis, while the second three terms relate to the y-axis. The z-axis is orthogonal to the x-axis and the y-axis. It is possible to simplify (25) and only constrain the orientation of one axis. In addition, the local axes do not necessarily have to be constrained such that they are parallel with a global axis; they may be given any direction.

4.2.4. *Example of an Orientation Constraint*

Figure 11 shows a standard posture with a target point of $(-57, 31.5, 41.2)$ and no orientation constraint.

Figure 11. Standard posture without orientation constraint

Figure 12. Orientation constraint on one axis

Figure 12 illustrates a posture with the same target point as shown in Figure 11 but with the local x-axis of the hand constrained to be aligned with the global $\mathbf{Z}$-axis. Such a posture may be required, for instance, when one reaches to push a button.

In this case, $l_{11} = 0$, $l_{21} = 0$, and $l_{31} = 1$. $^{0}R_{n}(1,1) = -9.7 \times 10^{-5}$, $^{0}R_{n}(2,1) = -2.5 \times 10^{-5}$, and $^{0}R_{n}(3,1) = 1.0$. The results are acceptable numerically and visually.

In Figure 13, the local x-axis of the hand is again aligned with the global $\mathbf{Z}$-axis, and the local y-axis of the hand is constrained so that it is aligned with the global $\mathbf{Y}$-axis but in the negative direction. Technically, the local y-axis is perpendicular to the hand, so in this case, the palm of the hand is parallel with the ground.

Figure 13. Orientation constraint on two axes

With this second example, $l_{11} = 0$, $l_{21} = 0$, $l_{31} = 1$, $l_{11} = 0$, $l_{21} = 0$, and $l_{31} = 1$. $^{0}R_{n}(1,1) = -3.7 \times 10^{-4}$, $^{0}R_{n}(2,1) = 5.4 \times 10^{-6}$, $^{0}R_{n}(3,1) = 1.0$, $^{0}R_{n}(1,2) = 1.0 \times 10^{-4}$, $^{0}R_{n}(2,2) = -1.0$, and $^{0}R_{n}(3,2) = 5.4 \times 10^{-6}$. There is a slight decrease in accuracy with some of the components of the rotation matrix along with a slight increase in the accuracy of others. Nonetheless, the results are reasonable. With both examples, the final distance from the target point is 0.003 cm.

4.3. A 30-DOF Model

The 21-DOF model proved to be a realistic representation of human movement; however, it is limited to single-arm tasks. This model has been reflected to the left arm for an additional 9 DOF. The result is the 30-DOF model shown in Figure 14.

4.3.1. DH Model

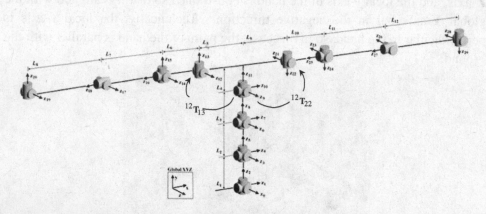

Figure 14. Detailed 30-DOF model

Although the development of the left arm is conceptually the same as that of the right, it is important to note that this addition leads to a *double dependence* on the spine. This means that there must be two transformation matrices from the DOF corresponding to z_{11}: one which represents the transformation from the *12*th DOF to the *13*th, called $^{12}\mathbf{T}_{13}$, and one that represents the transformation from the *12*th DOF to the *22*nd, called $^{12}\mathbf{T}_{22}$. The 21-DOF model also requires two end-effectors, one on each hand. The positions of these end-effectors, $\mathbf{x}_R$ and $\mathbf{x}_L$, are given by the following equations:

$$\mathbf{x}_R = \left(\mathbf{T}_{body}\right)\left(^{12}\mathbf{T}_{13}\right)\left(\prod_{i=13}^{21} {}^{i-1}\mathbf{T}_i\right)\mathbf{x}_{R_n} \tag{27}$$

$$\mathbf{x}_L = \left(\mathbf{T}_{body}\right)\left(^{12}\mathbf{T}_{22}\right)\left(\prod_{i=23}^{31} {}^{i-1}\mathbf{T}_i\right)\mathbf{x}_{L_n} \tag{28}$$

$$\mathbf{T}_{body} = \prod_{i=1}^{12} {}^{i-1}\mathbf{T}_i \tag{29}$$

$\mathbf{x}_{Rn}$ and $\mathbf{x}_{Ln}$ are the position vectors of the end-effectors with respect to their respective local coordinate frames. Because there are two transformations from the *12*th DOF, there are 31 transformation matrices, which correspond to 30 DOF. The transformation matrices are given in (1), where $i = 1, \ldots, 21, 23, \ldots, 31$ and

$$
{}^{12}\mathbf{T}_{22} = \begin{pmatrix} \cos\theta & -\cos\alpha\sin\theta & \sin\alpha\sin\theta & a\cos\theta \\ \sin\theta & \cos\alpha\cos\theta & -\sin\alpha\cos\theta & a\sin\theta \\ 0 & \sin\alpha & \cos\alpha & d \\ 0 & 0 & 0 & 1 \end{pmatrix} \tag{30}
$$

where the parameters θ, d, α, and a are measured from *12*th DOF to the *22*nd.

The optimum posture is determined using a formulation similar to (4). However, with two arms, there are two distance-constraints given as follows:

$$
\left[\mathbf{x}_R(\mathbf{q})^{\textit{right end-effector}} - \mathbf{x}_R^{\textit{right target point}} \right]^2 \leq \varepsilon \tag{31}
$$

$$
\left[\mathbf{x}_L(\mathbf{q})^{\textit{left end-effector}} - \mathbf{x}_L^{\textit{left target point}} \right]^2 \leq \varepsilon \tag{32}
$$

SNOPT[13] is used to solve this problem directly but without analytical gradients. The limits for the additional degrees of freedom are chosen so that the left joints are constrained by the same limits as the right joints. Furthermore, weights in the joint displacement function are added to reflect the new degrees of freedom. The updated weights are shown in Table 2.

Table 2. Joint weights used for 30 DOF

Joint Variable	Joint Weight	Comments
$q_1, \ldots, q_{12}$	100	Used with both positive and negative values of $q_i - q_i^N$
q_2	100 1000	When $q_i - q_i^N > 0$ When $q_i - q_i^N < 0$
q_{13}, q_{22}	75	Used with both positive and negative values of $q_i - q_i^N$
q_{17}, q_{26}	50	When $q_i - q_i^N > 0$

4.3.2. *Results*

The repercussions of incorporating multiple limbs that share common DOF can be demonstrated by comparing dual-arm results with single-arm results from posture prediction. Figure 15 (a) shows a single-arm posture, whereas Figure 15

(b) shows a dual-arm posture using the same target point for the right end-effector. Note how the shared DOF in the spine shift to facilitate reaching both targets.

(a) 21-DOF, single-arm model   (b) 30-DOF, dual-arm model
Figure 15. Posture prediction results for the same right-arm target point

This dual-arm coordination makes it possible to analyze workspace design and prototype evaluation over a wider variety of human postures. For example, the 30-DOF model can be used to consider the workspace of a vehicle for dual-arm reaches (Figure 16).

One benefit of the optimization-based approach to posture prediction is computational efficiency. Posture prediction feedback can thus be obtained in real-time or near real-time speeds. This might be especially useful to quickly evaluate workspace or compare a variety of postures over different anthropometries. In fact, the new approach, incorporating multiple limbs and shared DOF, maintains computational speed. The dual-arm posture prediction on the 30-DOF model took only approximately 0.15 sec for feasible targets on a 2.6GHz Pentium4 CPU with 512MB RAM. Single-arm posture prediction on the 21-DOF model takes approximately 0.10 sec on the same machine.

Figure 16. Evaluation of a vehicle workspace for dual-arm reaches

4.3.3. *Validation*

In order to validate the optimization-based approach for inverse kinematcis of articulated linkages, the Virtual Solder Research (VSR) lab at The University of Iowa provided use of their motion capture system, which consists of eight infrared miniature cameras from Vicon Motion Systems. For comparable interpretation of the captured motion, a skeletal model corresponding to the 30-DOF model was created within the motion capture software (Figure 17). Subjects wore several markers and were then tracked by the motion capture system over a number of reaching tasks (Figure 18). The motion capture system mapped the motions of the subjects to the skeletal model and recorded the results to a file. Finally, the results were subject to post-processing, which involves characterizing parameters such as subject anthropometry and local joint rotations with respect to the skeletal model.

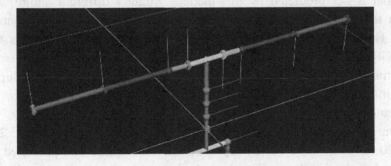

Figure 17. Skeletal model developed for motion capture corresponds to 30-DOF model

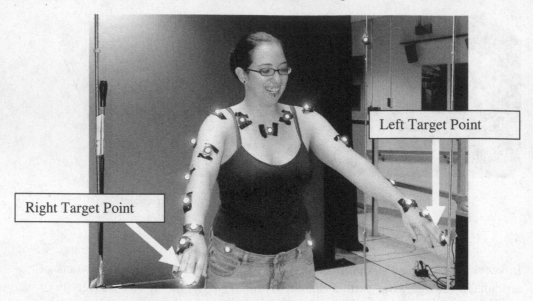

Figure 18. Subject prepared for motion capture using several markers

The predicted postures for three sets of target points are compared to motion capture results from a male subject. Target positions are given with respect to a global coordinate frame located in the torso, coincident with the zero[th] frame in the 30-DOF model, and are measured in centimeters. For target set #1, the right end-effector target, $\mathbf{x}_R^{target}$, is (-41.7, -4.3, 38.7); the left end-effector target, $\mathbf{x}_L^{target}$, is (39.1, -4.4, 40.1). The motion capture result for this target set is shown in Figure 19 (a), while the predicted posture is visualized in Figure 19 (b). Both postures are similar; however, the motion capture shows a slight bending at the elbow that is not predicted by this model. Slightly different anthropometries between the model and the motion capture subject are a possible contributing factor. However, minimizing joint displacement conceptually means that the model will tend toward the neutral posture. Since the neutral posture is defined with a straight arm, the result of the optimization will tend toward a straight arm. Hence, more realistic results should be possible with a more inclusive human performance measure(s).

For target set #2, $\mathbf{x}_R^{target}$ is (-65.3, 44.7, -41.0) and $\mathbf{x}_L^{target}$ is (39.4, -5.2, 40.6). Figure 20 (a) and Figure 20 (b) depict the motion capture and predicted results, respectively. Again, the predicted result shows less bending in the elbow, and also less twisting in the arm. For target set #3, $\mathbf{x}_R^{target}$ is (-41.3, 44.5, 60.9) and $\mathbf{x}_L^{target}$ is (-36.4, 44.4, 63.8). The motion capture result is shown in Figure 21 (a), and the predicted result is shown on the model in Figure 21 (b). The predicted result more closely resembles the motion capture result in this case.

Although these results provide an indication that the postures are accurate, there exists a need for additional results. Most notably, the subject in this study and the model had different anthropometries, and this necessarily affects the resulting posture. The validation study described below uses a 30-DOF model whose anthropometry is varied to match that of the subject.

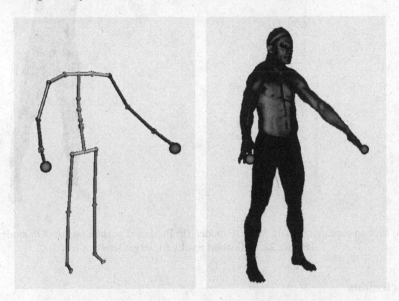

(a) Motion capture result on 30-DOF model (b) Predicted posture on 30-DOF model
Figure 19. Validation results for target set #1

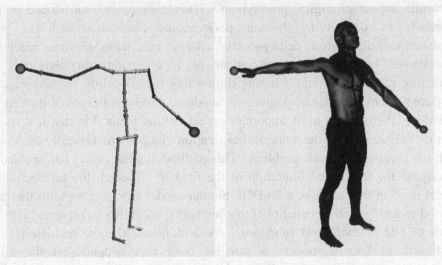

(a) Motion capture result on 30-DOF model (b) Predicted posture on 30-DOF model
Figure 20. Validation results for target set #2

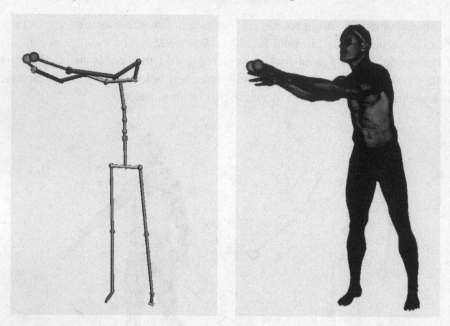

(a) Motion capture result on 30-DOF model (b) Predicted posture on 30-DOF model
Figure 21. Validation results for target set #3

5. Conclusions

A general mathematical formulation for inverse kinematics of articulated linkages has been proposed and demonstrated. Predicting human motion is an important aspect of digital prototyping. This chapter has introduced a new approach to quantifying human performance measures such as joint displacement, discomfort, delta-potential energy, etc., using rigorous analytical expressions, towards the development of a computational algorithm for predicting realistic postures. It was shown that the modeling method was not restricted to any number of degrees of freedom, and joint ranges of motion are included. Perhaps the most important aspect of this method is that it does not employ the Jacobian in the numerical evaluation of a posture, typically associated with the inverse kinematic problem. This methodological aspect has enabled us to surpass the traditional limitation of the 6 DOF. Indeed, the biomechanical model used in this work is a 30 DOF human model from the torso to the hand (seated reach). Another benefits of this method is its ability to represent tasks in terms of one or more cost functions. As demonstrated, more realistic posture prediction of human models is possible, one that depends on the initial configuration, the range of motion, and the exact dimensions. Validation of the method with motion capture results was presented. It is evident that the proposed method yields postures that minimize the specified cost function and render a realistic posture. Our method can predict realistic postures and can be applied to general. However, many more cost functions are needed, and more elaborate mathematical descriptions of human performance measures are required for various tasks. Nevertheless, this method provides a robust approach to realistic posture prediction that can handle a biomechanically accurate model.

Acknowledgments

This work was supported by partly funded by the US Army TACOM project: Digital Humans and Virtual Reality for Future Combat Systems (FCS) (Contract No.: DAAE07-03-D-L003) and by Caterpillar Inc. project: Digital Human Modeling and Simulation for Safety and Serviceability. The authors would like to thank all other Virtual Soldier Research (VSR) research members for their contributions.

References

1. Denavit, J., Hartenberg, R.S. A kinematic notation for lower-pair mechanisms based on matrices, *Journal of Applied Mechanics,* 1955, 77: 215-221.
2. Abdel-Malek, K., Yu, W., and Jaber, M., 2001, "Realistic posture prediction," *2001 SAE Digital Human Modeling and Simulation.*
3. Yang, J., Marler, R.T., Kim, H.J., Arora, J.S., and Abdel-Malek, K., 2004, "Multi-objective optimization for upper body posture prediciton", *10th AIAA/ISSMO Multidisciplinary Analysis and Optimization Conference* (August 30-September 1, 2004), Albany, NY.
4. Mi, Z., 2004, *Task-Based Prediction of Upper Body Motion,* Ph.D. Thesis, University of Iowa, Iowa City, IA.
5. Marler, R.T., 2005, *A Study of Multi-Objective Optimization Methods for Engineering Applications*, Ph.D. Thesis, University of Iowa, Iowa City, IA.
6. Marler, T., Yang, J., Arora, J., and Abdel-Malek, K., 2005a, "A Study of A Bi-Criterion Posture Prediction Problem Using Pareto-Optimal Sets," *The Fifth IASTED International Conference on Modelling, Simulation, and Optimization*, August 29-31, 2005, Oranjestad, Aruba.
7. Marler, R.T., Rahmatalla, S., Shanahan, M., and Abdel-Malek, K., 2005b, "A new discomfort function for optimization-based posture prediction", *SAE Digital Human Modeling for Design and Engineering* (June 14-16, 2005), Iowa City, IA.
8. Farrell, K., 2005, *Kinematic Human Modeling and Simulation using Optimization-based Posture Prediction*, M.S. Thesis, University of Iowa, Iowa City, IA.
9. Yang, J., Abdel-Malek, K, Marler, R.T., and Kim, J., "Real-Time Optimal Reach Posture Prediction in a New Interactive Virtual Environment," *Journal of Computer Science and Technology*, Vol. 21, No. 2, 2006, pp. 189-198.
10. Miettinen, K., 1999, *Nonlinear Multiobjective Optimization*, Kluwer Academic Publishers, Boston.

11. Romero, C., 2000, "Bi-criteria Utility Functions: Analytical Considerations and Implications in The Short-run Labor Market", *European Journal of Operations Research*, Vol. 122, No. 1, pp. 91-100.
12. Yang, J., Pena Pitarch, E., Kim, J., and Abdel-Malek, K., "Posture Prediction and Force/Torque Analysis for Human Hands," *Proceedings of SAE Digital Human Modeling for Design and Engineering*, July 4-6, 2006, 2005, Lyon, France.
13. Gill, P., Murray, W., and Saunders, A, 2002, "SNOPT: An SQP Algorithm for Large-Scale Constrained Optimization", *SIAM Journal of Optimization*, Vol. 12, No. 4, pp. 979-1006.

CHAPTER 13

MULTIDISCIPLINARY DESIGN OPTIMIZATION

Gyung-Jin Park

Department of Mechanical Engineering, Hanyang University
1271 Sa Dong, Ansan City, Gyeonggi Do, Korea 426-791
E-mail: gjpark@hanyang.ac.kr

The concept of multidisciplinary design optimization (MDO) has been addressed to solve optimization problems with multiple disciplines. Conventional optimization generally solves the problems with a single discipline. Disciplines are coupled in MDO problems. Many MDO methods have been proposed. The methods are classified and some representative methods are introduced. The advantages and drawbacks of each method are described and discussed.

1. Introduction

Engineering systems are fairly large and complicated these days. Design requirements are rigorous and stringent for such systems. Accordingly, design engineers are seeking new methods and one of them is multidisciplinary design optimization (MDO). MDO is a design optimization method. Generally, optimization has been applied by considering only one discipline. However, it is quite difficult to solve modern engineering problems with a single discipline. Therefore, we need design methodologies which can cover multiple disciplines. Diverse MDO methods have been proposed to overcome the difficulties.[1]

MDO usually has many design variables, objective functions and constraints. Analysis can be independently performed for a discipline. And the disciplines can be linked through analyses. That is, output from a discipline can be input to other disciplines. This phenomenon is called the coupled process. In MDO, it is important to efficiently manage the coupled process and to optimize large scale problems.

The need for MDO has emerged in the design of airplane wings. Pressures are imposed on the airplane wings by the flow of the surrounding air. The pressures act on the wings as the lift force and the drag force. The pressures are external forces on the wing structure and the shape of the wing is changed due to the external forces. Then the air flow is changed and the pressures are changed accordingly. Equilibrium is obtained when the coupled process converges. This phenomenon is investigated in aeroelasticity and computer simulation is adopted to solve this problem. The pressure distribution is solved by computational fluid dynamics and the deformation of the wing is analyzed by the finite element method. The analysis method considering the coupled process is called multidisciplinary analysis (MDA).

When we design airplane wings, the above two disciplines should be simultaneously considered. A design problem is separately defined for each discipline. For example, the design to minimize the drag force is performed by the output from computational fluid dynamics and the weight of the wing structure is minimized by using the output of the finite element method. When the optimization technology is employed in this case, it is called MDO. The two disciplines have common design variables, objective functions and constraints. They can also be independently defined in each discipline.

Coupling is predominant in the analysis and design of airplane wings. The problem is not yet thoroughly solved from an MDO viewpoint so designers examine various methods. Sometimes, a discipline of the control theory is involved in the design of airplane wings and there are many references for the design of wings. Therefore, the aspects for the airplane wings are not explained in this chapter. Instead, general methods for MDO are discussed and the application of them is investigated.[1]

Overall, there are seven methods of MDO. The formulation of the Multidisciplinary Feasible (MDF) method is basic. It is easy to use even though complicated system analysis should be carried out for each step. The coupled relationship is solved in the system analysis. The formulation of the Individual Feasible (IDF) method has been developed to eliminate the system analysis, and each discipline is independently solved. In the All-at-Once (AAO) method, even the analysis process is eliminated in the disciplines. Each discipline does not determine the design in the above methods. Analysis is only carried out in the disciplines.[2]

Various methods have been developed to allocate the design process to the disciplines. They are Concurrent Subspace Optimization (CSSO)[3], Collaborative Optimization (CO)[4], Bi-Level Integrated System Synthesis (BLISS)[5] and Multidisciplinary Optimization Based on Independent Subspaces (MDOIS).[6]

The characteristics of MDO are investigated and the above seven methods are compared. The methods are constantly evolving and being improved upon. In this chapter, they are explained based on the initial proposition of the methods.

They can still solve small scale problems or problems with weak couplings. Approximation methods are frequently utilized for large scale problems or problems with strong couplings.[7-8] Currently, it seems that no method can solve all the MDO problems. For large scale problems, an improved solution is sought with specific approximation methods. Therefore, none of the introduced methods is said to be the ultimate one.

2. Multidisciplinary Design Optimization

Design is the process to make the responses of a target object as a designer wants while analysis is the process to evaluate the responses. Currently, the analysis process is mostly carried out by computers. In an MDO process, analysis of a discipline is the base of a design. Various analyses are required in MDO. The analyses of the disciplines are related and this phenomenon is called coupling.

Coupling is the relationship between disciplines. It can be a direct one or an indirect one. A change in a discipline may return to the discipline again or it may not. Between independent disciplines, a change in a discipline does not affect the other one. More attention is given to the solving process of the coupling in analysis while coupling in the design process is also considered.

The couplings in analysis or in design can be explained as follows:

Coupling in analysis: When results (responses) of an analysis are input for the analysis of other disciplines, coupling in analysis exists. Generally, coupling in MDO means the coupling in analysis.

Coupling in design: When design variables are shared by multiple disciplines, a coupled relationship exists.

Relationship between analysis and design: The process which uses the analysis results in design is called Nested Analysis and Design (NAND). Meanwhile, analysis may not be performed and an evaluator is used instead. This process is called Simultaneous Analysis and Design (SAND).[9]

In this chapter, methods are explained based on the assumption that there are two disciplines. When there are more than two disciplines, the derivation is valid as well.

2.1. *Coupling in Analysis*

In MDO, there are multiple disciplines which can have independent analysis and design procedures. The coupling for analysis can be expressed as[10]

$$\mathbf{z}_1 = \mathbf{h}_1(\mathbf{x}_1, \mathbf{x}_c, \mathbf{z}_2^c) \tag{1}$$

$$\mathbf{z}_2 = \mathbf{h}_2(\mathbf{x}_2, \mathbf{x}_c, \mathbf{z}_1^c) \tag{2}$$

where $\mathbf{h}_1$ and $\mathbf{h}_2$ are the analyzers for disciplines 1 and 2, respectively. Each analyzer can have its own design variables and coupling variables. $\mathbf{x}_1 \in R^{nd1}$ and $\mathbf{x}_2 \in R^{nd2}$ are local design variable vectors for disciplines 1 and 2, respectively. $\mathbf{x}_c \in R^{ndc}$ is the common design variable vector shared by the two disciplines. The state variable vectors $\mathbf{z}_1 \in R^{ns1}$ and $\mathbf{z}_2 \in R^{ns2}$ are the results (responses) from the two disciplines, respectively.

When we have couplings between disciplines, some or all of the state variables of a discipline are used as input to the other discipline. Such state variables are called coupling variable vectors $\mathbf{z}_1^c \in R^{nc1}$ and $\mathbf{z}_2^c \in R^{nc2}$. The state variables which are not coupling variables are non-coupling variables $\mathbf{z}_1^{nc} \in R^{nnc1}$ and $\mathbf{z}_2^{nc} \in R^{nnc2}$. The state variable vectors are expressed by $\mathbf{z}_1 = [\mathbf{z}_1^{nc^T} \quad \mathbf{z}_1^{c^T}]^T$ and $\mathbf{z}_2 = [\mathbf{z}_2^{nc^T} \quad \mathbf{z}_2^{c^T}]^T$.

Eqs. (1)-(2) can be expressed as

$$\mathbf{z}_1^{nc} = \mathbf{h}_{11}(\mathbf{x}_1, \mathbf{x}_c, \mathbf{z}_2^c) \tag{3}$$

$$\mathbf{z}_1^c = \mathbf{h}_{12}(\mathbf{x}_1, \mathbf{x}_c, \mathbf{z}_2^c) \tag{4}$$

$$\mathbf{z}_2^{nc} = \mathbf{h}_{21}(\mathbf{x}_2, \mathbf{x}_c, \mathbf{z}_1^c) \tag{5}$$

$$\mathbf{z}_2^c = \mathbf{h}_{22}(\mathbf{x}_2, \mathbf{x}_c, \mathbf{z}_1^c) \tag{6}$$

If we have couplings in analysis, the coupling variables should be evaluated. A representative method is the fixed point iteration method. It is difficult to obtain coupling variables in practice. Approximation methods are frequently adopted for calculation of the coupling variables. The state where all the coupled relationships are satisfied is called Multidisciplinary Feasibility (MF). That is, the equilibrium between disciplines is satisfied. The state where the equilibrium of a discipline is satisfied is called Individual Feasibility (IF). In IF, the coupled relationship may not be satisfied. Achieving MF or IF is a fairly important problem in MDO.

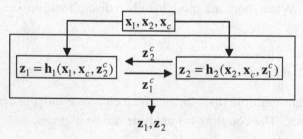

Fig. 1. Coupled relationship in analysis

The coupled relationship of analysis involved is presented in Fig. 1. In Fig. 1, the design variable vectors $\mathbf{x}_1, \mathbf{x}_2$ and $\mathbf{x}_c$ are fixed and the state variable vectors $\mathbf{z}_1$ and $\mathbf{z}_2$ are obtained from $\mathbf{h}_1$ and $\mathbf{h}_2$. Output of a discipline is used as input of the other discipline. The calculation process of Fig. 1 is called the system analysis. The system analysis process terminates when $\mathbf{z}_1^c$ and $\mathbf{z}_2^c$ are not changed. MF is the state when the system analysis terminates.

2.2. *Formulation of MDO*

MDO formulation is defined based on the coupled aspects of the design variables, the objective function and constraints. An MDO problem is formulated as follows:

$$\text{Find} \quad \mathbf{x}_1, \mathbf{x}_2, \mathbf{x}_c \tag{7}$$

$$\text{to minimize} \quad \mathbf{f}(\mathbf{x}_1, \mathbf{x}_2, \mathbf{x}_c, \mathbf{z}_1, \mathbf{z}_2) \tag{8}$$

$$\text{subject to} \quad \mathbf{g}(\mathbf{x}_1, \mathbf{x}_2, \mathbf{x}_c, \mathbf{z}_1, \mathbf{z}_2) \leq 0 \tag{9}$$

$$\mathbf{z}_1 = \mathbf{h}_1(\mathbf{x}_1, \mathbf{x}_c, \mathbf{z}_2^c) \tag{10}$$

$$\mathbf{z}_2 = \mathbf{h}_2(\mathbf{x}_2, \mathbf{x}_c, \mathbf{z}_1^c) \tag{11}$$

where $\mathbf{x}_1$, $\mathbf{x}_2$ and $\mathbf{x}_c$ are local design variable vectors of disciplines 1 and 2, and the common design variable vector shared by the two disciplines, respectively. $\mathbf{f} = [f_1 \, f_2 \, f_c]$ is the local objective functions for disciplines 1 and 2, and the common objective function. $\mathbf{g} \in R^m$ is the constraint vector.

2.3. *Classification of the MDO Methods*

The relationships between disciplines are represented in Fig. 2(a). MDO methods solve the problem which has the structure of Fig. 2(a).[1] The methods are classified into single level methods and multi-level methods. The single level methods generally have a single optimizer. On the other hand, multi-level methods modify the structure in Fig. 2(a) into a hierarchical structure as illustrated in Fig. 2(b). Each level has an optimizer. Generally the multi-level has two levels.

The disciplines may or may not be decomposed in the single level method. When they are decomposed, each discipline is handled separately. They are always decomposed in the multi-level methods.

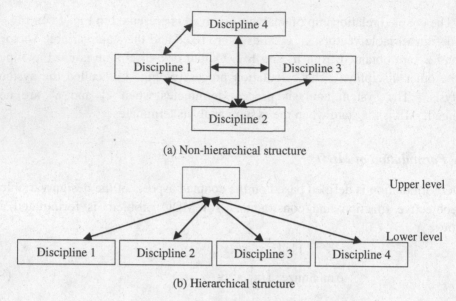

(a) Non-hierarchical structure

(b) Hierarchical structure

Fig. 2. Structures of the MDO processes

3. Linear Decomposition and Global Sensitivity Equation

The optimization problem in Eqs. (7)-(11) may be difficult to solve for large scale problems. A large scale problem can be decomposed with respect to each discipline.[11] The decomposition is usually carried out based on the type of analysis. Using linear addition for objective functions in Eq. (8) ($f = f_1 + f_2 + f_c$), Eqs. (7)-(9) can be modified as

$$\text{Find} \quad \mathbf{x}_1, \mathbf{x}_2, \mathbf{x}_c \tag{12}$$

$$\text{to minimize} \quad f(\mathbf{x}_1, \mathbf{x}_2, \mathbf{x}_c, \mathbf{z}_1, \mathbf{z}_2) \tag{13}$$

$$\text{subject to} \quad \mathbf{g}(\mathbf{x}_1, \mathbf{x}_2, \mathbf{x}_c, \mathbf{z}_1, \mathbf{z}_2) \leq \mathbf{0} \tag{14}$$

In Eqs. (12)-(14), the objective function and constraints are coupled with respect to the common design variables. The objective function and constraints can be linearly expanded as follows:

$$f \cong f_0 + \frac{df}{d\mathbf{x}_1} \cdot \Delta\mathbf{x}_1 + \frac{df}{d\mathbf{x}_2} \cdot \Delta\mathbf{x}_2 + \frac{df}{d\mathbf{x}_c} \cdot \Delta\mathbf{x}_c \tag{15}$$

$$g_i \cong g_{i0} + \frac{dg_i}{d\mathbf{x}_1} \cdot \Delta\mathbf{x}_1 + \frac{dg_i}{d\mathbf{x}_2} \cdot \Delta\mathbf{x}_2 + \frac{dg_i}{d\mathbf{x}_c} \cdot \Delta\mathbf{x}_c \quad (i = 1, 2, ..., m) \tag{16}$$

where $\dfrac{d}{d\mathbf{x}_l} = \left[\dfrac{d}{dx_{l1}} \quad \dfrac{d}{dx_{l2}} \quad \cdots \quad \dfrac{d}{dx_{lr}} \right]^T$, $\Delta\mathbf{x}_l = [\Delta x_{l1} \quad \Delta x_{l2} \quad \cdots \quad \Delta x_{lr}]^T$ $l \in \{1, 2, c\}$

and $r \in \{nd1, nd2, ndc\}$. Subscript 0 means the current design and Δ is perturbation from the current design.

Using Eqs. (15)-(16), an optimization problem of each discipline can be formulated by its own design variables. Suppose $\mathbf{x}_1$ and $\mathbf{x}_c$ are allocated to discipline 1 and $\mathbf{x}_2$ is allocated to discipline 2. Then the optimization problem for each discipline is defined as

(1) Discipline 1

$$\text{Find} \quad \mathbf{x}_1, \mathbf{x}_c \tag{17}$$

$$\text{to minimize} \quad f_0 + \frac{df}{d\mathbf{x}_1} \cdot \Delta\mathbf{x}_1 + \frac{df}{d\mathbf{x}_c} \cdot \Delta\mathbf{x}_c \tag{18}$$

$$\text{subject to} \quad g_{i0} + \frac{dg_i}{d\mathbf{x}_1} \cdot \Delta\mathbf{x}_1 + \frac{dg_i}{d\mathbf{x}_c} \cdot \Delta\mathbf{x}_c \leq 0 \quad (i = 1, 2, ..., m) \tag{19}$$

(2) Discipline 2

$$\text{Find} \quad \mathbf{x}_2 \tag{20}$$

$$\text{to minimize} \quad f_0 + \frac{df}{d\mathbf{x}_2} \cdot \Delta\mathbf{x}_2 \tag{21}$$

$$\text{subject to} \quad g_{i0} + \frac{dg_i}{d\mathbf{x}_2} \cdot \Delta\mathbf{x}_2 \leq 0 \quad (i = 1, 2, ..., m) \tag{22}$$

In Eqs. (17)-(22), the optimization problem in Eqs. (7)-(11) is linearized and decomposed. This process is called linear decomposition. Derivatives (sensitivity information) of the objective function and constraints are required for the decomposition.

The derivative terms in Eqs. (15)-(16) are derived by the chain rule and implicit differentiation.[11] The results are

$$\begin{bmatrix} \mathbf{I} & 0 & 0 & -\dfrac{\partial\mathbf{h}_{11}}{\partial\mathbf{z}_2^c} \\ 0 & \mathbf{I} & 0 & -\dfrac{\partial\mathbf{h}_{12}}{\partial\mathbf{z}_2^c} \\ 0 & -\dfrac{\partial\mathbf{h}_{21}}{\partial\mathbf{z}_1^c} & \mathbf{I} & 0 \\ 0 & -\dfrac{\partial\mathbf{h}_{22}}{\partial\mathbf{z}_1^c} & 0 & \mathbf{I} \end{bmatrix} \begin{bmatrix} \dfrac{d\mathbf{z}_1^{nc}}{d\mathbf{x}_1} \\ \dfrac{d\mathbf{z}_1^c}{d\mathbf{x}_1} \\ \dfrac{d\mathbf{z}_2^{nc}}{d\mathbf{x}_1} \\ \dfrac{d\mathbf{z}_2^c}{d\mathbf{x}_1} \end{bmatrix} = \begin{bmatrix} \dfrac{\partial\mathbf{h}_{11}}{\partial\mathbf{x}_1} \\ \dfrac{\partial\mathbf{h}_{12}}{\partial\mathbf{x}_1} \\ 0 \\ 0 \end{bmatrix} \tag{23}$$

The terms related to the coupling variables in Eq. (23) are expressed as

$$
\begin{bmatrix} \mathbf{I} & -\dfrac{\partial \mathbf{h}_{12}}{\partial \mathbf{z}_2^c} \\[2mm] -\dfrac{\partial \mathbf{h}_{22}}{\partial \mathbf{z}_1^c} & \mathbf{I} \end{bmatrix} \begin{bmatrix} \dfrac{d\mathbf{z}_1^c}{d\mathbf{x}_1} \\[2mm] \dfrac{d\mathbf{z}_2^c}{d\mathbf{x}_1} \end{bmatrix} = \begin{bmatrix} \dfrac{\partial \mathbf{h}_{12}}{\partial \mathbf{x}_1} \\[2mm] \mathbf{0} \end{bmatrix}
\tag{24}
$$

Eqs. (23)-(24) are called global sensitivity equations (GSE).[12-14] The solution of them is the derivative of coupling variables with respect to design variables. It is used in various MDO methods.

4. Multidisciplinary Design Optimization Methods

The aforementioned seven MDO methods are explained.[2]

4.1. *Multidisciplinary Feasible (MDF)*

The MDF method is a single level method which has the system analysis. It is also called the All-in-One (AiO) method. In the MDF method, the system analysis is treated as the analyzer of the single optimization.

Fig. 3 presents the general structure of the MDF method. The lower part of the figure is the system analysis. The system analysis usually uses the fixed point iteration method. Meanwhile, sensitivity information is needed in the optimization process. The sensitivity information is obtained from the aforementioned global sensitivity equations or the finite difference method on the system analysis. When the process converges, the MDF method can find a mathematical optimum. Therefore, the solution from the MDF method is considered as the standard solution when MDO methods are compared.

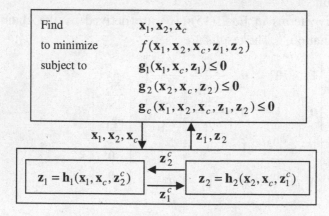

Fig. 3. Structure of the MDF method

The MDF method directly solves the problem in Eqs. (7)-(11). Multiple objective functions are modified into a single objective function. Eqs. (10)-(11) are handled in the system analysis. The MDF method does not decompose the disciplines. The number of optimizers is one and the number of design variables is the sum of the number of local variables ($nd1+nd2$) and the number of common variables (ndc).

4.2. *Individual Discipline Feasible (IDF)*

The IDF method is a single level method which decomposes the disciplines. It does not have the system analysis. Complimentary vectors (s_1^c and s_2^c) are adopted for the coupling variable vectors (z_1^c and z_2^c). Instead of MF, compatibility conditions are utilized and they are included as equality constraints in the optimization process. The formulation of the IDF method is

$$\text{Find} \quad \mathbf{x}_1, \mathbf{x}_2, \mathbf{x}_c, \mathbf{s}_1^c, \mathbf{s}_2^c \tag{25}$$

$$\text{to minimize} \quad f(\mathbf{x}_1, \mathbf{x}_2, \mathbf{x}_c, \mathbf{z}_1, \mathbf{z}_2) \tag{26}$$

$$\text{subject to} \quad \mathbf{g}_1(\mathbf{x}_1, \mathbf{x}_c, \mathbf{z}_1) \leq 0 \tag{27}$$

$$\mathbf{g}_2(\mathbf{x}_2, \mathbf{x}_c, \mathbf{z}_2) \leq 0 \tag{28}$$

$$\mathbf{g}_c(\mathbf{x}_1, \mathbf{x}_2, \mathbf{x}_c, \mathbf{z}_1, \mathbf{z}_2) \leq 0 \tag{29}$$

$$\mathbf{c}_1 = \mathbf{s}_1^c - \mathbf{z}_1^c = 0 \tag{30}$$

$$\mathbf{c}_2 = \mathbf{s}_2^c - \mathbf{z}_2^c = 0 \tag{31}$$

$$\mathbf{z}_1 = \mathbf{h}_1(\mathbf{x}_1, \mathbf{x}_c, \mathbf{s}_2^c) \tag{32}$$

$$\mathbf{z}_2 = \mathbf{h}_2(\mathbf{x}_2, \mathbf{x}_c, \mathbf{s}_1^c) \tag{33}$$

An analyzer of a discipline does not wait for the output from the other analyzers. Instead, it performs the analysis (IF) by using the complimentary vectors (s_1^c and s_2^c). Eventually, the coupling variables (z_1^c and z_2^c) should be the same (MF) as the complimentary vectors (s_1^c and s_2^c). Thus, compatibility conditions (c_1 and c_2) are defined as Eqs. (30)-(31). The overall flow of the IDF method is presented in Fig. 4.

When the IDF method is used, the number of design variables is increased by the number of complimentary variables. The number of design variables is $nd1+nd2+ndc$ plus $N(s_1^c)+N(s_2^c)$. $N(s_1^c)$ means the number of the elements in vector s_1^c. The number of equality constraints is the number of the coupling variables. IF is always satisfied during the optimization process and MF is satisfied when the optimum design is obtained.

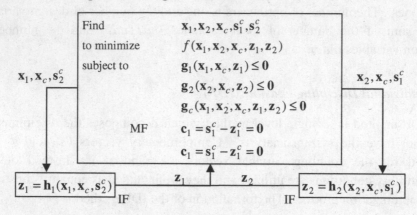

$$x_1, x_c, s_2^c$$

Find	$x_1, x_2, x_c, s_1^c, s_2^c$
to minimize	$f(x_1, x_2, x_c, z_1, z_2)$
subject to	$g_1(x_1, x_c, z_1) \le 0$
	$g_2(x_2, x_c, z_2) \le 0$
	$g_c(x_1, x_2, x_c, z_1, z_2) \le 0$
MF	$c_1 = s_1^c - z_1^c = 0$
	$c_1 = s_1^c - z_1^c = 0$

$$x_2, x_c, s_1^c$$

$z_1 = h_1(x_1, x_c, s_2^c)$ IF z_1 z_2 IF $z_2 = h_2(x_2, x_c, s_1^c)$

Fig. 4. Structure of the IDF method

4.3. *All-at-Once (AAO)*

The AAO method is a single level method that performs neither the system analysis nor individual analysis for each discipline. Analysis is generally carried out many times for a design and is quite expensive. The analysis results during the design process are meaningless when the final solution is obtained. The AAO method has been developed to abolish the expensive analysis process. Evaluators are utilized instead of analyzers in the AAO method.

State variables are calculated with respect to design variables in analysis. The analysis process can be viewed as illustrated in Fig. 5.

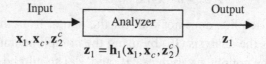

Input → Analyzer → Output

x_1, x_c, z_2^c

$z_1 = h_1(x_1, x_c, z_2^c)$ z_1

Fig. 5. Analysis process

Figure 5 represents the analysis process of discipline 1. State variable vector z_1 is obtained from design variable vectors x_1 and x_c and coupling variable vector z_2^c.

The analysis process can also be viewed from a different perspective. Fig. 6 presents an evaluator. The design and state variable vectors are inputs. When the results of the evaluator are the zero vector, it is regarded that the analysis is finished.

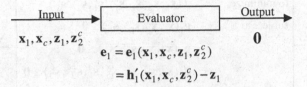

$$e_1 = e_1(\mathbf{x}_1, \mathbf{x}_c, \mathbf{z}_1, \mathbf{z}_2^c)$$

$$= \mathbf{h}_1'(\mathbf{x}_1, \mathbf{x}_c, \mathbf{z}_2^c) - \mathbf{z}_1$$

Fig. 6. The role of the evaluator

The concept in Fig. 6 is expanded to MDO by using the SAND concept. Complimentary variable vectors ($\mathbf{s}_1^c$ and $\mathbf{s}_2^c$) are adopted for state variable vectors ($\mathbf{z}_1^c$ and $\mathbf{z}_2^c$). The process of the evaluator is considered as equality constraints. It is similar to the IDF method. In the IDF method, the complimentary variables for coupling variables are used as design variables in the optimization process. On the other hand, the complimentary variables for coupling and non-coupling variables are included in the design variable set. That is, all the state variables are regarded as design variables.

The optimization formulation of the AAO method is

$$\text{Find} \quad \mathbf{x}_1, \mathbf{x}_2, \mathbf{x}_c, \mathbf{s}_1^{nc}, \mathbf{s}_2^{nc}, \mathbf{s}_1^c, \mathbf{s}_2^c \tag{34}$$

$$\text{to minimize} \quad f(\mathbf{x}_1, \mathbf{x}_2, \mathbf{x}_c, \mathbf{s}_1^{nc}, \mathbf{s}_1^c, \mathbf{s}_2^{nc}, \mathbf{s}_2^c) \tag{35}$$

$$\text{subject to} \quad \mathbf{g}_1(\mathbf{x}_1, \mathbf{x}_c, \mathbf{s}_1^{nc}, \mathbf{s}_1^c) \leq 0 \tag{36}$$

$$\mathbf{g}_2(\mathbf{x}_2, \mathbf{x}_c, \mathbf{s}_2^{nc}, \mathbf{s}_2^c) \leq 0 \tag{37}$$

$$\mathbf{g}_c(\mathbf{x}_1, \mathbf{x}_2, \mathbf{x}_c, \mathbf{s}_1^{nc}, \mathbf{s}_1^c, \mathbf{s}_2^{nc}, \mathbf{s}_2^c) \leq 0 \tag{38}$$

$$\mathbf{e}_1 = 0 \tag{39}$$

$$\mathbf{e}_2 = 0 \tag{40}$$

$$\mathbf{e}_1 = \mathbf{h}_1'(\mathbf{x}_1, \mathbf{x}_c, \mathbf{s}_2^c) - \mathbf{s}_1 \tag{41}$$

$$\mathbf{e}_2 = \mathbf{h}_2'(\mathbf{x}_2, \mathbf{x}_c, \mathbf{s}_1^c) - \mathbf{s}_2 \tag{42}$$

Eqs. (41)-(42) represent the evaluator.

The overall flow of the AAO method is illustrated in Fig. 7. Each discipline evaluates the validity of the given input. When the optimization process ends, IF and MF are simultaneously satisfied. The number of equality constraints is the same as the number of state variables. The number of design variables is $nd1+nd2+ndc$ plus $N(\mathbf{s}_1^{nc}) + N(\mathbf{s}_2^{nc}) + N(\mathbf{s}_1^c) + N(\mathbf{s}_2^c)$.

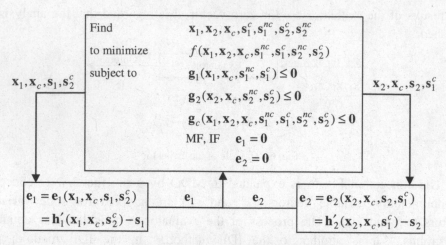

$$\mathbf{x}_1, \mathbf{x}_c, \mathbf{s}_1, \mathbf{s}_2^c$$

$$\mathbf{x}_2, \mathbf{x}_c, \mathbf{s}_2, \mathbf{s}_1^c$$

Find $\quad \mathbf{x}_1, \mathbf{x}_2, \mathbf{x}_c, \mathbf{s}_1^c, \mathbf{s}_1^{nc}, \mathbf{s}_2^c, \mathbf{s}_2^{nc}$

to minimize $\quad f(\mathbf{x}_1, \mathbf{x}_2, \mathbf{x}_c, \mathbf{s}_1^{nc}, \mathbf{s}_1^c, \mathbf{s}_2^{nc}, \mathbf{s}_2^c)$

subject to $\quad \mathbf{g}_1(\mathbf{x}_1, \mathbf{x}_c, \mathbf{s}_1^{nc}, \mathbf{s}_1^c) \le \mathbf{0}$

$$\mathbf{g}_2(\mathbf{x}_2, \mathbf{x}_c, \mathbf{s}_2^{nc}, \mathbf{s}_2^c) \le \mathbf{0}$$

$$\mathbf{g}_c(\mathbf{x}_1, \mathbf{x}_2, \mathbf{x}_c, \mathbf{s}_1^{nc}, \mathbf{s}_1^c, \mathbf{s}_2^{nc}, \mathbf{s}_2^c) \le \mathbf{0}$$

MF, IF $\quad \mathbf{e}_1 = \mathbf{0}$

$$\mathbf{e}_2 = \mathbf{0}$$

$$\mathbf{e}_1 = \mathbf{e}_1(\mathbf{x}_1, \mathbf{x}_c, \mathbf{s}_1, \mathbf{s}_2^c)$$
$$= \mathbf{h}_1'(\mathbf{x}_1, \mathbf{x}_c, \mathbf{s}_2^c) - \mathbf{s}_1$$

$$\mathbf{e}_1 \qquad \mathbf{e}_2$$

$$\mathbf{e}_2 = \mathbf{e}_2(\mathbf{x}_2, \mathbf{x}_c, \mathbf{s}_2, \mathbf{s}_1^c)$$
$$= \mathbf{h}_2'(\mathbf{x}_2, \mathbf{x}_c, \mathbf{s}_1^c) - \mathbf{s}_2$$

Fig. 7. Structure of the AAO method

The above three methods have the following characteristics:

1. They are single level methods.
2. An analyzer or an evaluator is used in each discipline.
3. The formulations are simple and easy to understand.
4. Common objective function and constraints are handled in the same manner as the independent ones.

Some MDO methods give the design process to each discipline. They are introduced in the subsequent sections.

4.4. *Concurrent Subspace Optimization (CSSO)*

CSSO is a multi-level method, which has the system analysis and decomposes disciplines. The design process is given to the disciplines. CSSO divides an MDO problem into upper and lower levels. In the upper level, some parameters are determined by an optimization process for the design process of the lower level. The parameters enable a discipline to consider the objective function and constraints of other disciplines. The optimization problem of the upper level is called the Coordinate Optimization Problem (COP). In the lower level (it is also called a subspace) for each discipline, optimization is carried out with the parameters from the upper level. The functions in a discipline are defined by using the linear decomposition. The lower level provides the optimization results and the optimum sensitivity. Optimum sensitivity is the sensitivity of the objective function with respect to a parameter which is considered as a constant in the optimization process.[15-16] The optimization results are used in the system analysis and the optimum sensitivity is used in the optimization of the upper level. The process proceeds in an iterative fashion.

CSSO first performs the system analysis in Eqs. (10)-(11). A method such as the fixed point iteration method is employed for the system analysis until the coupling variable vectors z_1^c and z_2^c are not changed. Details are in Eqs. (3)-(6).

Sensitivity information is needed to decompose the entire problem into optimization problems for the disciplines. Global sensitivity information in Eqs. (23)-(24) is needed for the coupling variables and used for linear approximation of the coupling variables. When the number of constraints in a discipline is large, all the constraints can be transformed to a constraint which is a cumulative constraint using the Kreisselmeier-Steinhouser (KS) function. It is

$$C = \frac{1}{\rho}\ln\left[\sum_{i=1}^{m}\exp(\rho g_i)\right] \tag{43}$$

where ρ is the adjustment coefficient, m is the number of constraints and g_i is the ith constraint. If any one of the constraints is violated, the value of Eq. (43) is positive. When all the constraints are satisfied, it is zero or negative. Thus, if the value of (43) is zero or negative, all the constraints are satisfied.

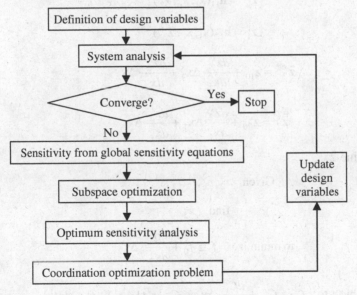

Fig. 8. Flow of CSSO

The flow of CSSO is illustrated in Fig. 8. At the first cycle, the system analysis is performed and the global sensitivity are calculated with respect to x_1, x_2 and x_c. With the results of the upper level, optimization is carried out for each discipline in the lower level and optimum sensitivity is evaluated. In the upper level, the COP problem is defined by the optimum sensitivity information. Then the coefficients for the next cycle are defined for the lower level. The convergence criteria are checked from the second cycle.

The problem in the lower level is investigated. In CSSO, the common variable $\mathbf{x}_c$ should be considered in one discipline. The optimization problems of the two disciplines are defined as follows:

(1) Discipline 1

$$\text{Given} \quad r_1^1, r_1^2, t_1^1, t_1^2, s^1, s^2 \tag{44}$$

$$\text{find} \quad \mathbf{x}_1, \mathbf{x}_c \tag{45}$$

$$\text{to minimize} \quad f \cong f_0 + \frac{df}{d\mathbf{x}_1} \cdot \Delta\mathbf{x}_1 + \frac{df}{d\mathbf{x}_c} \cdot \Delta\mathbf{x}_c \tag{46}$$

$$\text{subject to} \quad C_1 \le C_{10}[s^1(1-r_1^1)+(1-s^1)t_1^1] \tag{47}$$

$$\tilde{C}_2 \cong C_{20} + \frac{dC_2}{d\mathbf{x}_1} \cdot \Delta\mathbf{x}_1 + \frac{dC_2}{d\mathbf{x}_c} \cdot \Delta\mathbf{x}_c \le C_{20}[s^2(1-r_1^2)+(1-s^2)t_1^2] \tag{48}$$

$$\mathbf{z}_1^{nc} = \mathbf{h}_{11}(\mathbf{x}_1, \mathbf{x}_c, \tilde{\mathbf{z}}_2^c) \tag{49}$$

$$\mathbf{z}_1^c = \mathbf{h}_{12}(\mathbf{x}_1, \mathbf{x}_c, \tilde{\mathbf{z}}_2^c) \tag{50}$$

$$\tilde{\mathbf{z}}_2^{nc} = \mathbf{z}_{10}^{nc} + \frac{d\mathbf{z}_2^{nc}}{d\mathbf{x}_1} \Delta\mathbf{x}_1 + \frac{d\mathbf{z}_2^{nc}}{d\mathbf{x}_c} \Delta\mathbf{x}_c \tag{51}$$

$$\tilde{\mathbf{z}}_2^c = \mathbf{z}_{20}^c + \frac{d\mathbf{z}_2^c}{d\mathbf{x}_1} \Delta\mathbf{x}_1 + \frac{d\mathbf{z}_2^c}{d\mathbf{x}_c} \Delta\mathbf{x}_c \tag{52}$$

(2) Discipline 2

$$\text{Given} \quad r_2^1, r_2^2, t_2^1, t_2^2, s^1, s^2 \tag{53}$$

$$\text{find} \quad \mathbf{x}_2 \tag{54}$$

$$\text{to minimize} \quad f \cong f_0 + \frac{df}{d\mathbf{x}_2} \cdot \Delta\mathbf{x}_2 \tag{55}$$

$$\text{subject to} \quad \tilde{C}_1 \cong C_{10} + \frac{dC_1}{d\mathbf{x}_2} \cdot \Delta\mathbf{x}_2 \le C_{10}[s^1(1-r_2^1)+(1-s^1)t_2^1] \tag{56}$$

$$C_2 \le C_{20}[s^2(1-r_2^2)+(1-s^2)t_2^2] \tag{57}$$

$$\tilde{\mathbf{z}}_1^{nc} = \mathbf{z}_{10}^{nc} + \frac{d\mathbf{z}_1^{nc}}{d\mathbf{x}_2} \Delta\mathbf{x}_2 \tag{58}$$

$$\tilde{\mathbf{z}}_1^c = \mathbf{z}_{10}^c + \frac{d\mathbf{z}_1^c}{d\mathbf{x}_2}\Delta\mathbf{x}_2 \tag{59}$$

$$\mathbf{z}_2^{nc} = \mathbf{h}_{21}(\mathbf{x}_2, \mathbf{x}_c, \tilde{\mathbf{z}}_1^c) \tag{60}$$

$$\mathbf{z}_2^c = \mathbf{h}_{22}(\mathbf{x}_2, \mathbf{x}_c, \tilde{\mathbf{z}}_1^c) \tag{61}$$

s^p, r_k^p and t_k^p are constants in the lower level and are determined in the COP of the upper level. k and p represent the discipline number, and s^p, r_k^p and t_k^p .are given by the user at the first cycle. When the objective function of a discipline is a function of the local variables of the corresponding discipline, the above approximation is not needed. Otherwise, the objective function is linearly approximated. A cumulative constraint is employed to include all the constraints. The cumulative constraint of a discipline is considered through approximation by another discipline. This process enables to consider the constraints of the other discipline because a change of design in a discipline may affect the other discipline through coupling. It is shown in Eqs. (48) and (56). $dC/d\mathbf{x}_i$ ($i=1, 2, c$) is the information obtained from the global sensitivity equation. The constraints of the other discipline are considered by s^p, r_k^p and t_k^p.

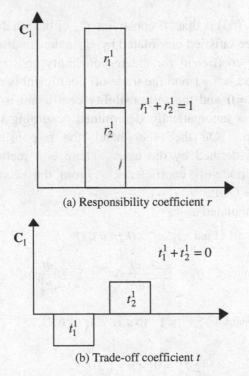

(a) Responsibility coefficient r

(b) Trade-off coefficient t

Fig. 9. Responsibility and trade-off coefficients

The coefficients are illustrated in Fig. 9. Fig. 9(a) explains r_k^p. Suppose that the constraint of discipline 1, C_1 is violated. It is corrected by r_1^1 in discipline 1 and by r_2^1 in discipline 2. Thus, r_2^1 is the amount of responsibility by discipline 2 for the constraint violation in discipline 1. It is called the responsibility coefficient and has the following relationship:

$$\sum_{k=1}^{NSS} r_k^p = 1 \qquad (p = 1, 2, ..., NSS) \tag{62}$$

where NSS is the number of disciplines. The meaning of Eq. (62) is that "if the constraint violation of the pth discipline, C_p is 1, r_k^p is the amount of responsibility for the kth discipline."

Trade-off coefficient t is schematically drawn in Fig. 9. Suppose that constraint C_1 in discipline 1 is satisfied. t_1^1 is negative, thus, discipline 1 tries to satisfy C_1 in surplus by t_1^1. t_2^1 is positive and discipline 2 tries to further reduce the objective function although $\tilde{C}_1$ is violated by t_2^1. t_k^p is the trade-off coefficient of constraint C_p in the kth discipline. The trade-off coefficients have the following relationship:

$$\sum_{k=1}^{NSS} t_k^p = 0 \qquad (p = 1, 2, ..., NSS) \tag{63}$$

The meaning of Eq. (63) is that "if constraint C_p of the pth discipline is satisfied, the constraint is over-satisfied or violated by t_k^p in the kth discipline."

s^p is a switch coefficient for the responsibility and trade-off coefficients. When C_p is violated, $s^p = 1$ and the trade-off coefficient is not activated. When C_p is satisfied, $s^p = 0$ and the responsibility coefficient is not activated. The switch coefficient is automatically determined according to the status of the constraint violation. On the other hand, the responsibility and trade-off coefficients can be defined by the user. There is a method to initialize the responsibility and trade-off coefficients.[3] From the second cycle, they are determined from the COP.

The COP is formulated as

$$\text{Find} \quad r_k^p, t_k^p \quad (k, p = 1, 2) \tag{64}$$

$$\text{to minimize} \quad f_0 + \sum_{p=1}^{2} \sum_{k=1}^{2} \frac{df}{dr_k^p} \Delta r_k^p + \sum_{p=1}^{2} \sum_{k=1}^{2} \frac{df}{dt_k^p} \Delta t_k^p \tag{65}$$

$$\text{subject to} \quad \sum_{k=1}^{2} r_k^p = 1 \quad (0 \le r_k^p \le 1) \quad (p = 1, 2) \tag{66}$$

$$\sum_{k=1}^{2} t_k^p = 0 \quad (p = 1, 2) \tag{67}$$

Eq. (65) needs the optimum sensitivity of the objective function with respect to the responsibility and trade-off coefficients. Since the coefficients are treated as constants in each discipline, the optimum sensitivity is obtained from each discipline.

CSSO is a two-level optimization method. Each discipline of the lower level (subspace) needs an optimizer and the upper level also needs an optimizer. IF and MF are satisfied by the system analysis. Constraints are indirectly shared by all the disciplines and common design variables are specifically handled by one discipline.

4.5. *Bi-Level Integrated System Synthesis (BLISS)*

BLISS is a multi-level method, which has the system analysis and decomposes disciplines.[5] It divides an MDO problem into an upper level and a lower level. Each discipline in the lower level has local design variables while the common variables are considered as constants. On the other hand, the upper level has common variables as design variables while the local variables of the disciplines are regarded as constants.

BLISS can handle an MDO problem without common constraints. Using the linear decomposition, the problem is formulated as follows:

(1) Discipline 1

$$\text{Given} \quad \mathbf{x}_1, \mathbf{x}_c, \mathbf{z}_2^c \tag{68}$$

$$\text{find} \quad \Delta \mathbf{x}_1 \tag{69}$$

$$\text{to minimize} \quad \frac{df}{d\mathbf{x}_1} \cdot \Delta \mathbf{x}_1 \tag{70}$$

$$\text{subject to} \quad \mathbf{g}_1(\mathbf{x}_1, \mathbf{x}_c, \mathbf{z}_1) \leq \mathbf{0} \tag{71}$$

$$\mathbf{z}_1^{nc} = \mathbf{h}_{11}(\mathbf{x}_1, \mathbf{x}_c, \mathbf{z}_2^c) \tag{72}$$

$$\mathbf{z}_1^c = \mathbf{h}_{12}(\mathbf{x}_1, \mathbf{x}_c, \mathbf{z}_2^c) \tag{73}$$

(2) Discipline 2

$$\text{Given} \quad \mathbf{x}_2, \mathbf{x}_c, \mathbf{z}_1^c \tag{74}$$

$$\text{find} \quad \Delta \mathbf{x}_2 \tag{75}$$

$$\text{to minimize} \quad \frac{df}{d\mathbf{x}_2} \cdot \Delta \mathbf{x}_2 \tag{76}$$

$$\mathbf{g}_2(\mathbf{x}_2, \mathbf{x}_c, \mathbf{z}_2) \leq \mathbf{0} \tag{77}$$

$$\mathbf{z}_2^{nc} = \mathbf{h}_{21}(\mathbf{x}_2, \mathbf{x}_c, \mathbf{z}_1^c) \tag{78}$$

$$\mathbf{z}_2^c = \mathbf{h}_{22}(\mathbf{x}_2, \mathbf{x}_c, \mathbf{z}_1^c) \tag{79}$$

(3) Upper level

$$\text{Given} \quad \mathbf{x}_c \tag{80}$$

$$\text{find} \quad \Delta \mathbf{x}_c \tag{81}$$

$$\text{to minimize} \quad \frac{df}{d\mathbf{x}_c} \cdot \Delta \mathbf{x}_c \tag{82}$$

The objective functions in the lower level are linearized with respect to the local variables as shown in Eqs. (70) and (76). The global sensitivity is utilized for the linearization.[12-14] For the upper level, the optimum sensitivity of the objective function with respect to common variables is required as shown in Eq. (82). The optimum sensitivity is obtained from the lower level. Since there are two disciplines in the lower level, different optimum sensitivities can be obtained from different disciplines. Therefore, it is calculated based on satisfaction of the system analysis. BLISS uses GSE/OS (GSE/Optimized Subsystems) for this. GSE/OS is a union of the global sensitivity equation and the optimum sensitivity equations in Eqs. (68)-(79).[5] It is as follows:

$$
\begin{bmatrix}
\mathbf{I} & 0 & 0 & -\dfrac{\partial \mathbf{h}_{11}}{\partial \mathbf{z}_2^c} & -\dfrac{\partial \mathbf{h}_{11}}{\partial \mathbf{x}_1} & 0 \\
0 & \mathbf{I} & 0 & -\dfrac{\partial \mathbf{h}_{12}}{\partial \mathbf{z}_2^c} & -\dfrac{\partial \mathbf{h}_{12}}{\partial \mathbf{x}_1} & 0 \\
0 & -\dfrac{\partial \mathbf{h}_{21}}{\partial \mathbf{z}_1^c} & \mathbf{I} & 0 & 0 & -\dfrac{\partial \mathbf{h}_{21}}{\partial \mathbf{x}_2} \\
0 & -\dfrac{\partial \mathbf{h}_{22}}{\partial \mathbf{z}_1^c} & 0 & \mathbf{I} & 0 & -\dfrac{\partial \mathbf{h}_{22}}{\partial \mathbf{x}_2} \\
0 & 0 & 0 & -\dfrac{\partial \mathbf{x}_1}{\partial \mathbf{z}_2^c} & \mathbf{I} & 0 \\
0 & -\dfrac{\partial \mathbf{x}_2}{\partial \mathbf{z}_1^c} & 0 & 0 & 0 & \mathbf{I}
\end{bmatrix}
\begin{bmatrix}
\dfrac{d\mathbf{z}_1^{nc}}{d\mathbf{x}_c} \\
\dfrac{d\mathbf{z}_1^c}{d\mathbf{x}_c} \\
\dfrac{d\mathbf{z}_2^{nc}}{d\mathbf{x}_c} \\
\dfrac{d\mathbf{z}_2^c}{d\mathbf{x}_c} \\
\dfrac{d\mathbf{x}_1}{d\mathbf{x}_c} \\
\dfrac{d\mathbf{x}_2}{d\mathbf{x}_c}
\end{bmatrix}
=
\begin{bmatrix}
\dfrac{\partial \mathbf{h}_{11}}{\partial \mathbf{x}_c} \\
\dfrac{\partial \mathbf{h}_{12}}{\partial \mathbf{x}_c} \\
\dfrac{\partial \mathbf{h}_{21}}{\partial \mathbf{x}_c} \\
\dfrac{\partial \mathbf{h}_{22}}{\partial \mathbf{x}_c} \\
\dfrac{\partial \mathbf{x}_1}{\partial \mathbf{x}_c} \\
\dfrac{\partial \mathbf{x}_2}{\partial \mathbf{x}_c}
\end{bmatrix}
\tag{83}
$$

where $\partial \mathbf{x}_1 / \partial \mathbf{z}_2^c$ and $\partial \mathbf{x}_1 / \partial \mathbf{x}_c$ are the optimum sensitivities of discipline 1, and $\partial \mathbf{x}_2 / \partial \mathbf{z}_1^c$ and $\partial \mathbf{x}_2 / \partial \mathbf{x}_c$ are those of discipline 2. The solution of Eq. (83) is used for the linearization in Eq. (82).

Steps of BLISS are as follows:

Step 1. Design variable vectors $\mathbf{x}_1, \mathbf{x}_2$ and $\mathbf{x}_c$ are initialized.

Step 2. The system analysis is performed to calculate z_1 and z_2.

Step 3. Convergence is checked from the second cycle.

Step 4. Global sensitivities are evaluated.

Step 5. Problems in Eqs. (68)-(79) are solved in the lower level.

Step 6. Eq. (83) for GSE/OS is established and solved.

Step 7. In the upper level, the problem in Eqs. (80)-(82) is solved.

Step 8. Update the design variables and go to Step 2.

In BLISS, IF and MF are satisfied during optimization because the system analysis is performed first.

4.6. *Collaborative Optimization (CO)*

Collaborative Optimization (CO) is a multi-level method, which does not perform the system analysis and decomposes disciplines.[4, 17] It has two levels such as the upper level which manages the overall process and the lower level which manipulates the design for each discipline. System analysis is replaced by adopting complimentary variables and compatibility conditions. The upper level minimizes the objective function with the results from the lower level and determines target values for disciplines in the lower level. The lower level tries to find a design to satisfy the constraints and the target values from the upper level.

The first step of CO is defining the complimentary variables. It can be different according to the user. Here, they are defined for local variables (x_1 and x_2), common design variables (x_c) and coupling variables as follows:

$$x_1 = s_1, \quad x_2 = s_2, \quad x_c = s_c, \quad z_1^c = s_1^c, \quad z_2^c = s_2^c \tag{84}$$

The left terms of Eq. (84) are design variable vectors of the lower level or the coupling variable vectors. The right terms are vectors determined in the upper level and used as target values in the lower level.

The problem in Eqs. (7)-(11) is modified for each discipline as follows:

(1) Discipline 1

$$\text{Given} \quad s_1, s_c, s_1^c, s_2^c \tag{85}$$

$$\text{find} \quad x_1, x_c, z_2^c \tag{86}$$

$$\text{to minimize} \quad c_1 = \| x_1 - s_1 \|^2 + \| x_c - s_c \|^2 + \| z_1^c - s_1^c \|^2 + \| z_2^c - s_2^c \|^2 \tag{87}$$

$$\text{subject to} \quad g_1(x_1, x_c, z_1) \leq 0 \tag{88}$$

$$z_1^{nc} = h_{11}(x_1, x_c, z_2^c) \tag{89}$$

$$z_1^c = \mathbf{h}_{12}(\mathbf{x}_1, \mathbf{x}_c, \mathbf{z}_2^c) \tag{90}$$

(2) Discipline 2

$$\text{Given} \quad s_2, s_c, s_1^c, s_2^c \tag{91}$$

$$\text{find} \quad \mathbf{x}_2, \mathbf{x}_c, \mathbf{z}_1^c \tag{92}$$

$$\text{to minimize} \quad c_2 = \| \mathbf{x}_2 - \mathbf{s}_2 \|^2 + \| \mathbf{x}_c - \mathbf{s}_c \|^2 + \| \mathbf{z}_1^c - \mathbf{s}_1^c \|^2 + \| \mathbf{z}_2^c - \mathbf{s}_2^c \|^2 \tag{93}$$

$$\text{subject to} \quad \mathbf{g}_2(\mathbf{x}_2, \mathbf{x}_c, \mathbf{z}_2) \leq \mathbf{0} \tag{94}$$

$$\mathbf{z}_2^{nc} = \mathbf{h}_{21}(\mathbf{x}_2, \mathbf{x}_c, \mathbf{z}_1^c) \tag{95}$$

$$\mathbf{z}_2^c = \mathbf{h}_{22}(\mathbf{x}_2, \mathbf{x}_c, \mathbf{z}_1^c) \tag{96}$$

where $\| \cdot \|$ means the L_2 norm which generally represents the magnitude of a vector.

Each discipline receives target values from the upper level. At the first cycle, the targets are arbitrarily defined and the optimization for each discipline of the lower level is initiated. The objective function of a discipline is the difference between the target values and the local design variables (state variables and design variables). CO can only consider local constraints. It is noted that the common and coupling variables are considered as design variables in all the disciplines.

The following formulation is utilized in the upper level:

$$\text{Find} \quad s_1, s_2, s_c, s_1^c, s_2^c \tag{97}$$

$$\text{to minimize} \quad f(s_1, s_2, s_c, \mathbf{z}_1^{nc}, \mathbf{z}_1^c, \mathbf{z}_2^{nc}, \mathbf{z}_2^c) \tag{98}$$

$$\text{subject to} \quad c_1 = c_{10} + \frac{dc_1}{d\mathbf{s}} \cdot \Delta \mathbf{s} = 0 \tag{99}$$

$$c_2 = c_{20} + \frac{dc_2}{d\mathbf{s}} \cdot \Delta \mathbf{s} = 0 \tag{100}$$

$$\mathbf{z}_1^{nc} = \mathbf{h}_{11}(s_1, s_c, s_2^c) \tag{101}$$

$$\mathbf{z}_1^c = \mathbf{h}_{12}(s_1, s_c, s_2^c) \tag{102}$$

$$\mathbf{z}_2^{nc} = \mathbf{h}_{21}(s_2, s_c, s_1^c) \tag{103}$$

$$\mathbf{z}_2^c = \mathbf{h}_{22}(s_2, s_c, s_1^c) \tag{104}$$

The sensitivity of the objective function in the upper level can be obtained analytically or by the finite difference method as following:

$$\frac{df}{ds} \cong \frac{\Delta f}{\Delta s} = \frac{f(s+\Delta s) - f(s-\Delta s)}{2\Delta s} \quad s \in \{s_1, s_2, s_c, s_1^c, s_2^c\} \tag{105}$$

The equality constraints in Eqs. (99)-(100) are objective functions of the disciplines. The derivative in Eqs. (99)-(100) is the optimum sensitivity. The optimum sensitivity is[4]

$$\frac{dc_j}{ds} = \frac{\partial c_j}{\partial s} + \lambda_j^T \frac{\partial g_j}{\partial s} \quad (j = 1, 2 \quad s \in \{s_1, s_2, s_c, s_1^c, s_2^c\}) \tag{106}$$

Eq. (106) is investigated. The local constraints are implicit functions with respect to the target values, therefore, the second term of the right hand side is zero. Since the first term is a partial derivative with respect to the target values, Eq. (106) yields

$$\frac{dc_1}{ds} = -2(d_1 - s) \quad (d_1 \in \{x_1, x_c, z_1^c, z_2^c\}, \quad s \in \{s_1, s_c, s_1^c, s_2^c\}) \tag{107}$$

$$\frac{dc_2}{ds} = -2(d_2 - s) \quad (d_2 \in \{x_2, x_c, z_1^c, z_2^c\}, \quad s \in \{s_2, s_c, s_1^c, s_2^c\}) \tag{108}$$

As shown in Eqs. (107)-(108), the optimum sensitivity is easily obtained.

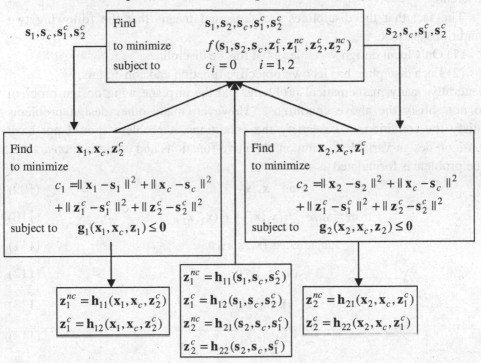

Fig. 10. Flow of CO

The overall flow of CO is illustrated in Fig. 10.

CO is basically an expansion of the IDF method and is a two-level optimization method. It does not have the system analysis. The objective functions of the disciplines are closely related to the constraints of the upper level in order to satisfy MF. Common constraints cannot be handled. It can be easily applied to problems where disciplines are obviously distinguishable. Analysis and design are independently performed in a discipline.

4.7. *Multidisciplinary Design Optimization Based on Independent Subspaces (MDOIS)*

MDOIS is a single level method, which has the system analysis and decomposes disciplines. In many MDO problems, the disciplines are physically separated and a separate analyzer is used for a discipline. Disciplines are coupled only in the analysis process. An independent design problem is defined for each discipline (subspace) of such problems. They are relatively simpler problems than a general MDO problem. MDOIS has been developed to solve such problems efficiently.

The fact that the disciplines are separated means that the following two conditions are met:

(1) Only local design variables exist for the disciplines.

(2) Each discipline has its own objective function and constraints.

Generally, many mathematical problems and the airplane wing design problem do not satisfy the above conditions. However, many other design problems satisfy the conditions. To satisfy the conditions, a problem should not have common design variables, common objective functions and common constraints. The problem is formulated as

$$\text{Find} \quad \mathbf{x}_1, \mathbf{x}_2 \tag{109}$$

$$\text{to minimize} \quad f_1(\mathbf{x}_1, \mathbf{z}_1) + f_2(\mathbf{x}_2, \mathbf{z}_2) \tag{110}$$

$$\text{subjec to} \quad \mathbf{g}_1(\mathbf{x}_1, \mathbf{z}_1) \leq \mathbf{0} \tag{111}$$

$$\mathbf{g}_2(\mathbf{x}_2, \mathbf{z}_2) \leq \mathbf{0} \tag{112}$$

$$\mathbf{z}_1 = \mathbf{h}_1(\mathbf{x}_1, \mathbf{z}_2^c) \tag{113}$$

$$\mathbf{z}_2 = \mathbf{h}_2(\mathbf{x}_2, \mathbf{z}_1^c) \tag{114}$$

The system analysis should be performed to solve Eqs. (113)-(114). The global sensitivity information may or may not be needed. The design problem for each discipline is formulated as follows:

(1) Discipline 1

$$\text{Given} \quad \mathbf{z}_2^c \tag{115}$$

$$\text{find} \quad \mathbf{x}_1 \tag{116}$$

$$\text{to minimize} \quad f_1(\mathbf{x}_1, \mathbf{z}_1) \tag{117}$$

$$\text{subject to} \quad \mathbf{g}_1(\mathbf{x}_1, \mathbf{z}_1) \leq 0 \tag{118}$$

$$\mathbf{z}_1 = \mathbf{h}_1(\mathbf{x}_1, \mathbf{z}_2^c) \tag{119}$$

(2) Discipline 2

$$\text{Given} \quad \mathbf{z}_1^c \tag{120}$$

$$\text{find} \quad \mathbf{x}_2 \tag{121}$$

$$\text{to minimize} \quad f_2(\mathbf{x}_2, \mathbf{z}_2) \tag{122}$$

$$\text{subject to} \quad \mathbf{g}_2(\mathbf{x}_2, \mathbf{z}_2) \leq 0 \tag{123}$$

$$\mathbf{z}_2 = \mathbf{h}_2(\mathbf{x}_2, \mathbf{z}_1^c) \tag{124}$$

Eqs. (115) and (120) represent the coupling variables obtained from the system analysis. If the global sensitivity is available, the coupling variables in Eqs. (119) and (124) can be replaced to consider the changes of the coupling variables with respect to the changes of the design variables by the following equations:

$$\mathbf{z}_1^c = \mathbf{z}_{10}^c + \frac{d\mathbf{z}_1^c}{d\mathbf{x}_2} \Delta \mathbf{x}_2 \tag{125}$$

$$\mathbf{z}_2^c = \mathbf{z}_{20}^c + \frac{d\mathbf{z}_2^c}{d\mathbf{x}_1} \Delta \mathbf{x}_1 \tag{126}$$

where $\mathbf{z}_{10}^c$ and $\mathbf{z}_{20}^c$ are the outputs of the system analysis for $\mathbf{z}_1^c$ and $\mathbf{z}_2^c$. Eqs. (125)-(126) may not be used. That is, the coupling variables can be considered as constants in the optimization process of a discipline.

The flow of MDOIS is illustrated in Fig. 11. Fig. 11 represents the case where the global sensitivities are utilized. MDOIS is a practical method in design practice. The design proceeds when IF and MF are satisfied. Design is carried out for each discipline.

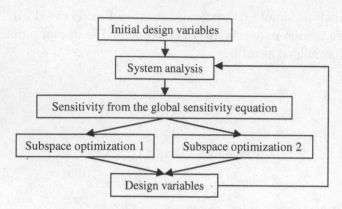

Fig. 11. Flow of MDOIS

4.8. *Comparison of MDO Methods*

In the previous sections, various MDO methods are introduced. MDO methods basically solve a non-hierarchical system in Fig. 2(a) directly or by transforming it to a hierarchical system in Fig. 2(b). Problem formulation and the solving process can be newly defined according to the characteristics of the problem. Therefore, the introduced methods are several selected methods out of many methods. There are many other MDO methods.

MDO methods are classified into single level methods and multi-level methods. MDF, IDF, AAO and MDOIS belong to the single level methods while CSSO, BLISS and CO belong to the multi-level methods. The single level methods are simple. However, the characteristics of multi-disciplines are not well exploited in the single level methods such as the MDF, IDF and AAO methods. The multi-level methods are complicated to use.

MDO methods are also classified based on the existence of the system analysis. The system analysis finds the coupling variables between analyses. MDF, CSSO, BLISS and MDOIS have the system analysis. They have an advantage in that MF (multidisciplinary feasibility) is always satisfied during the design process. However, many system analyses are required in MDF. In some multi-level methods, the results of the system analysis and the global sensitivity in the upper level are transmitted to disciplines in the lower level. The global sensitivity is generally fairly expensive to calculate.

IDF and CO have an advantage in that they do not need the system analysis. IF (individual feasibility) is satisfied, but MF is not satisfied during the design process. AAO uses evaluators instead of analyzers. It is the simplest one to use.

In the application of AAO, if we have only analyzers, we have the disadvantage in that an analyzer must be transformed to the evaluator. Neither MF nor IF is satisfied during the design process.

MDO methods handle the common constraints and the common design variables in different ways. Single level methods such as MDF, IDF and AAO can handle them in unified ways. MDOIS cannot have common variables or functions. CSSO cannot have common constraints and common design variables. Constraints of other disciplines are considered by the global sensitivity. BLISS can have common design variables, but cannot handle the common constraints. Common design variables are considered in the upper level. The effects of the common design variables in the lower level are transmitted to the upper level via optimum sensitivity. Calculation of the optimum sensitivity is quite expensive except for that in CO. The characteristics of the methods are compared in Table 1. O means 'yes' and × means 'no.' In this chapter, all the methods are explained to obtain an exact solution. In design practice, it may not be easy to obtain an exact optimum solution. In many cases, approximation is appropriately adopted in the processes of the MDO methods.[7-8]

5. Discussion

Multidisciplinary design optimization (MDO) has been created to manage design problems which are defined throughout multiple disciplines. The formulation of MDO can be relatively easily defined. However, it is quite difficult to find a universal method to solve MDO problems. Efforts are still needed to solve such problems.

The first difficulty in MDO is the coupling aspect between analyses. For example, fluid mechanics and solid mechanics are coupled in aeroelasticity analysis for airplane wings. It may not be impossible, but difficult. Moreover, it is extremely difficult to include it in optimization since it requires a large number of analyses. Some MDO methods need a complicated mathematical process such as global sensitivity or optimum sensitivity. Therefore, an MDO method is rarely applied to aeroelasticity problems.

There are engineering problems which have weak couplings between disciplines. It is relatively easy to apply an MDO method to such problems. Some problems need large scale analysis. Even single disciplinary optimization is not easy to use in such problems. As the analysis tools are developed significantly, many problems are enormous in size. Therefore, the exact process of the introduced MDO methods may not be easy to use in practical design.

Table 1 Comparison of MDO methods

Method	Level	Common constraints	Common design variables	System analysis	Global sensitivity	Optimum sensitivity	Satisfaction of IF	Satisfaction of MF
MDF	single	O	O	O	X	X	always	always
IDF	single	O	O	X	X	X	always	At optimum
AAO	single	O	O	X	X	X	At optimum	At optimum
CSSO	multi	X	X	O	O	O	always	always
BLISS	multi	X	O	O	O	O	always	always
CO	multi	X	O	X	X	O	always	At optimum
MDOIS	single	X	X	O	OX	X	always	always

Although there are no coupled aspects between analyses, multiple analyzers can be utilized in some problems. Recently, such problems are considered as MDO problems. Also a problem which has common design variables or common constraints can be regarded as an MDO problem.

Due to the above difficulties, a different MDO method can be utilized for a different application. Sometimes, a combination of the methods is used or a new method is defined using the ideas of the existing methods. Approximation is frequently adopted to solve complicated problems. Especially, when analysis is complicated the response surface method is frequently employed.

MDO is used to solve large scale design problems. There is still not a universal method. Therefore, more researches are anticipated in this area.

References

1. Sobieszczanski-Sobieski J, Haftka RT (1996) Multidisciplinary Aerospace Design Optimization:Survey of Recent Developments. AIAA Paper 96-0711
2. Cramer EJ, Dennis JE Jr., Frank PD, Lewis RM, Shubin GR (1993) Problem Formulation for Multidisciplinary Optimization. Center for Research on Parallel Computation Rice Univ. Rept. CRPC-TR93334
3. Sobieszczanski-Sobieski J (1988) Optimization by Decomposition: A Step from Hierarchic to Non-Hierarchic Systems. Recent Advances in Multidisciplinary Analysis and Optimization. NASA TM 101494
4. Braun RD (1996) Collaborative Optimization: An Architecture for Large-Scale Distributed Design. Ph.D. Dissertation, Stanford University
5. Sobieszczanski-Sobieski J, Agte J, Sandusky R Jr. (1998) Bi-Level Integrated System Synthesis. Proceedings of AIAA/USAF/NASA/ISSMO Symposium on Multidisciplinary Analysis and Optimization, AIAA-98-4916
6. Shin MK, Park GJ (2005) Multidisciplinary Design Optimization Based on Independent Subspaces. International Journal for Numerical Methods in Engineering 64:599-617
7. Renaud JE, Gabriele GA (1994) Approximation in Nonhierarchic System Optimization. AIAA Journal 32(1):198-205
8. Sellar RS, Renaud JE, Batille SM (1994) Optimization of Mixed Discrete/Continuous Design Variables System Using Neural Networks. 5th AIAA/NASA/USAF/ISSMO Symposium on Multidisplinary Analysis and Optimization, Panama City, Florida, USA
9. Balling RJ, Sobieszczanski-Sobieski J (1996) Optimization of Coupled Systems: A Critical Overview of Approach. AIAA Journal 34(1):6-17
10. Park GJ (2006) Analytic Methods for Design Practice. Springer Verlag, Germany, to be published
11. Sobieszczanski-Sobieski J (1982) A Linear Decomposition Method for Large Optimization Problems-Blueprint for Development. NASA TM 83248
12. Sobieszczanski-Sobieski J (1990) On the Sensitivity of Complex, Internally Coupled Systems. AIAA Journal 28(1):153-160
13. Sobieszczanski-Sobieski J (1990) Sensitivity Analysis and Multidisciplinary Design Optimization for Aircraft Design: Recent Advances and Results. Journal of Aircraft 27(12):993-1001

G. J. Park

14. Hajela P, Bloebaum CL, Sobieszczanski-Sobieski J (1990) Application of Global Sensitivity Equations in Multidisciplinary Aircraft Synthesis. Journal of Aircraft 27(12):1002-1010
15. Sobieszczanski-Sobieski J, Barthelemy JF, Riley KM (1982) Sensitivity of Optimum Solutions to Problem Parameters. AIAA Journal 20(9):1291-1299
16. Vanderplaats GN (1984) Numerical Optimization Techniques for Engineering Design. McGraw-Hill, New York
17. Sobieski IP, Kroo I (1996) Collaborative Optimization Applied to an Aircraft Design Problem. Proceedings of AIAA Aerospace Sciences Meeting AIAA-96-0715

CHAPTER 14

MESHFREE METHOD AND APPLICATION
TO SHAPE OPTIMIZATION

J. S. Chen

Civil & Environmental Engineering Department
University of California, Los Angeles
Los Angeles, CA 90095-1593
E-mail: jschen@seas.ucla.edu

Nam Ho Kim

Mechanical & Aerospace Engineering Department
University of Florida
Gainesville, Florida 32611-6250
E-mail: nkim@ufl.edu

Recent developments in meshfree method and its application to shape optimization are presented. The approximation theory of the Reproducing Kernel Particle Method is first introduced. The computational issues in domain integration and imposition of boundary conditions are discussed. A stabilization of nodal integration in meshfree discretization of boundary value problems is presented. Shape optimization based on meshfree method is presented, and the treatment of essential boundary conditions as well as the dependence of the shape function on the design variation is discussed. The proposed meshfree based shape design optimization yields a significantly reduced number of design iterations due to the meshfree approximation of sensitivity information without the need of remeshing. It is shown through numerical examples that the mesh distortion difficulty exists in the finite element–based design approach for design problems with large shape changes is effectively resolved.

1. Introduction

Meshfree methods developed in recent years introduced new approximation methods that are less restrictive in meeting the regularity requirement in the approximation and discretization of partial differential equations.[1-10] These methods are more flexible in embedding special enrichment functions in the approximation for solving problems with known characteristics, such as fracture problems,[11] more straightforward in constructing h– or p– adaptive refinement,[12-14] and less sensitive to large geometry changes such as those in large deformation problems[5,15] and shape optimization problems.[16,17]

Primary computational challenges involved in shape optimization using finite element methods (FEM) arise from the excessive mesh distortion that occurs during large shape changes and mesh–dependent solution accuracy. Numerous difficulties were encountered in finite element analysis, such as those involving mesh distortion, mesh adaptation, and those with the need for a large number of re-meshing during shape optimization.[18,19] Meshfree method is ideal for shape optimization because it allows field variables to be interpolated at the global level, therefore avoiding the use of a mesh. The main purpose of this chapter is to introduce special features of the meshfree method from a design sensitivity analysis (DSA) and optimization viewpoint, as well as the associated numerical aspects. Mesh distortion and re-meshing problems encountered in FEM-based shape optimization can be avoided and the design costs can be significantly reduced as a result of the accurate and efficient computation of design sensitivity. An important aspect to be considered in the shape optimization using meshfree method is the design derivation of meshfree shape functions. Unlike the finite element shape functions which are independent to the design variation due to the local construction of shape functions using natural coordinates, the meshfree shape functions depend on a global coordinate of material points that are related to the design parameters in shape DSA. Thus, the design derivative of the meshfree shape functions needs to be considered in DSA.

This Chapter is organized as follows. In Section 2, the reproducing kernel approximation for solving boundary value problems under the framework of reproducing kernel particle method (RKPM) is first introduced. Methods to impose boundary conditions and issues associated with the domain integration of Galerkin approximation are discussed, and an example demonstrating the accuracy and convergence property of RKPM is presented. In Section 3, shape design parameterization and design velocity are first defined. The shape sensitivity derivation is then introduced, and design derivation of meshfree shape functions and RKPM discretization of sensitivity equation are discussed. Two shape design optimization problems solved using the proposed methods are presented in Section 4. Concluding remarks are given in Section 5.

2. Reproducing Kernel Particle Method

In meshfree methods, the approximation of unknowns in the partial differential equations are constructed entirely based on a set of discrete points without using a structured mesh topology. Approximation methods such as moving least-squares,[20] reproducing kernel approximation,[4] partition of unity,[7] radial basis functions,[21] among others, have been introduced in formulating meshfree discrete equations. For demonstration purposes, the reproducing kernel approximation is presented herein. Other methods can also be employed under this framework.

2.1. *Reproducing Kernel Approximation*

The reproducing kernel approximation of a function $u(\mathbf{x})$, denoted by $u^h(\mathbf{x})$, is expressed as

$$u^h(\mathbf{x}) = \sum_{I=1}^{NP} \Psi_I(\mathbf{x})d_I , \tag{1}$$

where NP is the number of points used in the discretization of the problem domain Ω, d_I is the coefficient of the approximation at point I, and $\Psi_I(\mathbf{x})$ is called the reproducing kernel shape function. The reproducing kernel shape function is formed by a multiplication of two functions

$$\Psi_I(\mathbf{x}) = C(\mathbf{x};\mathbf{x} - \mathbf{x}_I)\Phi_a(\mathbf{x} - \mathbf{x}_I), \tag{2}$$

where $\Phi_a(\mathbf{x} - \mathbf{x}_I)$ is a kernel function that defines the continuity (smoothness) and the locality of the approximation with compact support (cover) Ω_I measured by the parameter a. The order of continuity in this approximation can be introduced without complexity. For example, the box function gives C^{-1} continuity, the hat function leads to C^0 continuity, the quadratic spline function results in C^1 continuity, and the cubic B-spline function yields C^2 continuity, etc. A commonly used kernel function is the cubic B-spline function given as

$$\Phi_a(x - x_I) = \begin{cases} \frac{2}{3} - 4z^2 + 4z^3 & for \ z \leq \frac{1}{2} \\ \frac{4}{3} - 4z + 4z^2 - \frac{4}{3}z^3 & for \ \frac{1}{2} < z \leq 1, \\ 0 & for \ z > 1 \end{cases} \tag{3}$$

where $z = |x - x_I|/a$. In multi-dimension, the kernel function can be constructed by using the distance function to yield an isotropic kernel function,

$$\Phi_a(\mathbf{x} - \mathbf{x}_I) = \Phi_a(z), \quad z = \|\mathbf{x} - \mathbf{x}_I\|/a, \tag{4}$$

or by the tensor product of the one-dimensional kernel functions to yield a anisotropic kernel

$$\Phi_a(\mathbf{x} - \mathbf{x}_I) = \Phi_{a_1}(x_1 - x_{1I})\Phi_{a_2}(x_2 - x_{2I}). \tag{5}$$

The union of all the kernel supports (covers) should cover the entire problem domain, i.e., $\bigcup_I \Omega_I \supset \Omega$ as shown in Figure 1. The term $C(\mathbf{x}, \mathbf{x} - \mathbf{x}_I)$ in Eq. (2) is the correction function or enrichment function. This function determines the completeness of the approximation and the order of consistency in solving PDE's.

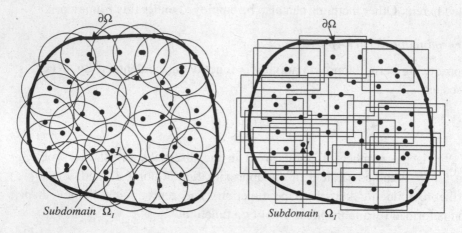

(a) Isotropic kernel supports (covers)   (b) Anisotropic kernel supports (covers)
Figure 1. Domain discretization and kernel supports

In general, $C(\mathbf{x}, \mathbf{x} - \mathbf{x}_I)$ is constructed by the monomial bases:

$$C(\mathbf{x}; \mathbf{x} - \mathbf{x}_I) = \sum_{i+j=0}^{n} (x_1 - x_{1I})^i (x_2 - x_{2I})^j b_{ij}(\mathbf{x})$$

$$= \mathbf{H}^T(\mathbf{x} - \mathbf{x}_I)\mathbf{b}(\mathbf{x}), \qquad i, j \geq 0 \tag{6}$$

$$\mathbf{H}^T(\mathbf{x} - \mathbf{x}_I) = \begin{bmatrix} 1 & x_1 - x_{1I} & x_2 - x_{2I} & \cdots & (x_2 - x_{2I})^n \end{bmatrix} \tag{7}$$

$$\mathbf{b}(\mathbf{x}) = \begin{bmatrix} b_{00} & b_{10} & b_{01} & \cdots & b_{0n} \end{bmatrix}, \tag{8}$$

where n is the order of the complete monomial bases, and this number defines the completeness of the approximation. The unknown coefficients $\mathbf{b}(\mathbf{x})$ can be determined by imposing the n-th order reproducing conditions

$$\sum_{I=1}^{NP} \Psi_I(\mathbf{x}) x_{1I}^i x_{2I}^j = x_1^i x_2^j \quad i + j = 0, \cdots, n, \tag{9}$$

or equivalently in the following form:

$$\sum_{I=1}^{NP} \Psi_I(\mathbf{x})(x_1 - x_{1I})^i (x_2 - x_{2I})^j = \delta_{i0}\delta_{j0} \quad i + j = 0, \cdots, n. \tag{10}$$

Substituting Eqs. (6) and (2) into Eq. (10) results in

$$\mathbf{M}(\mathbf{x})\mathbf{b}(\mathbf{x}) = \mathbf{H}(0), \tag{11}$$

where $\mathbf{M}(\mathbf{x})$ is the moment matrix of the kernel function $\Phi_a(\mathbf{x} - \mathbf{x}_I)$

$$\mathbf{M}(\mathbf{x}) = \sum_{I=1}^{NP} \mathbf{H}(\mathbf{x} - \mathbf{x}_I)\mathbf{H}^T(\mathbf{x} - \mathbf{x}_I)\Phi_a(\mathbf{x} - \mathbf{x}_I). \tag{12}$$

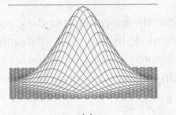 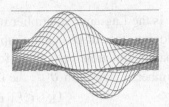

(a)   (b)

Figure 2. (a) 2-dimensional meshfree shape function Ψ_I (b) Shape function derivative $\Psi_{I,x}$

Therefore, the coefficient function $\mathbf{b}(\mathbf{x})$ is solved by

$$\mathbf{b}(\mathbf{x}) = \mathbf{M}^{-1}(\mathbf{x})\mathbf{H}(\mathbf{0}). \tag{13}$$

Notice that for the moment matrix $\mathbf{M}(\mathbf{x})$ to be non-singular, any spatial position $\mathbf{x}$ should be covered at least by n linearly independent kernel functions such that the n reproducing equations are solvable. Subsequently, the correction function and the discrete reproducing kernel shape function can be obtained as

$$C(\mathbf{x}; \mathbf{x} - \mathbf{x}_I) = \mathbf{H}^T(\mathbf{0})\mathbf{M}^{-1}(\mathbf{x})\mathbf{H}(\mathbf{x} - \mathbf{x}_I) \tag{14}$$

$$\Psi_I(\mathbf{x}) = \mathbf{H}^T(\mathbf{0})\mathbf{M}^{-1}(\mathbf{x})\mathbf{H}(\mathbf{x} - \mathbf{x}_I)\Phi_a(\mathbf{x} - \mathbf{x}_I). \tag{15}$$

The plots of the shape function and its derivatives are given in Figure 2, where linear basis and cubic B-spline kernel functions are employed. The meshfree shape function $\Psi_I(\mathbf{x})$ does not possess the Kronecker delta property; therefore, additional treatments are required to enforce the essential boundary conditions.

2.2. *Galerkin Approximation and Discretization*

For demonstration purposes, consider the following elastostatic problem:

$$(\sigma_{ij})_{,j} + b_i = 0 \quad in \quad \Omega \tag{16}$$

$$u_i = g_i \quad on \quad \partial\Omega^g \tag{17}$$

$$\sigma_{ij}n_j = h_i \quad on \quad \partial\Omega^h \tag{18}$$

where $\sigma_{ij} = C_{ijkl}\epsilon_{kl}$, $\epsilon_{ij} = \frac{1}{2}(u_{i,j} + u_{j,i}) = (\nabla^S u)_{ij}$ in Ω, Ω is the problem domain with essential boundary $\partial\Omega^g$ and natural boundary $\partial\Omega^h$, u_i is displacement, σ_{ij}

is stress, ε_{ij} is strain, C_{ijkl} is the elasticity tensor, h_i is the surface traction, and b_i is the body force. The weak form of the above problem is

$$\int_{\Omega} \delta(\nabla^S u)_{ij} C_{ijkl} (\nabla^S u)_{kl} d\Omega - \int_{\Omega} \delta u_i b_i d\Omega - \int_{\partial\Omega^h} \delta u_i h_i d\Gamma - \delta \int_{\partial\Omega^g} \lambda_i (u_i - g_i) d\Gamma = 0 \quad (19)$$

where λ_i is the Lagrange multiplier to impose the essential boundary conditions. By employing the displacement approximation in Eq. (1), and introducing approximation function $\{\phi_I(\mathbf{x})\}_{I=1}^{Ng}$ for the approximation of λ_i on $\partial\Omega^g$, where Ng is number of points on $\partial\Omega^g$, the following discrete equation is obtained

$$\begin{bmatrix} \mathbf{K} & \mathbf{G}^T \\ \mathbf{G} & \mathbf{0} \end{bmatrix} \begin{bmatrix} \mathbf{d} \\ \boldsymbol{\lambda} \end{bmatrix} = \begin{bmatrix} \mathbf{f} \\ \mathbf{q} \end{bmatrix} \quad (20)$$

$$\mathbf{K}_{IJ} = \int_{\Omega} \mathbf{B}_I^T \mathbf{C} \mathbf{B}_J d\Omega, \ \mathbf{G}_{IJ} = \mathbf{I} \int_{\partial\Omega^g} \phi_I \Psi_J d\Gamma, \ \mathbf{f}_I = \int_{\Omega} \Psi_I \mathbf{b} d\Omega + \int_{\partial\Omega^h} \Psi_I \mathbf{h} d\Gamma,$$

$$\mathbf{q}_I = \int_{\partial\Omega^\gamma} \phi_I \mathbf{g} d\Gamma, \ \mathbf{B}_I = \begin{bmatrix} \Psi_{I,1} & 0 \\ 0 & \Psi_{I,2} \\ \Psi_{I,2} & \Psi_{I,1} \end{bmatrix} \quad (21)$$

Compared with the standard finite element Galerkin approximation, two major disadvantages in computational efficiency are observed. First, additional degrees of freedom for Lagrange multiplier for imposition of essential boundary conditions are needed. Second, domain integration requires a background grid if Gauss integration is to be employed. It has been observed by Belytschko *et al.*[22] that very high order quadrature rule and very fine integration grids are required to reduce the integration error, and thus yields poor efficiency.

2.3. *Two Alternative Methods for Imposition of Essential Boundary Conditions*

2.3.1. *Transformation Method*

Recall reproducing kernel approximation of displacement

$$u_i^h(\mathbf{x}) = \sum_{I=1}^{NP} \Psi_I(\mathbf{x}) d_{iI}. \quad (22)$$

The nodal value of u_i^h at node J, $\hat{d}_{iJ}$, is obtained by

$$\hat{d}_{iJ} = u_i^h(\mathbf{x}_J) = \sum_{I=1}^{NP} \Psi_I(\mathbf{x}_J) d_{iI}, \ \text{or } \hat{\mathbf{d}} = \mathbf{\Lambda} \mathbf{d}, \quad (23)$$

where

$$
\hat{\mathbf{d}}_I = \begin{bmatrix} \hat{d}_{1I} \\ \hat{d}_{2I} \end{bmatrix}, \quad \mathbf{d}_I = \begin{bmatrix} d_{1I} \\ d_{2I} \end{bmatrix}, \quad \Lambda_{IJ} = \begin{bmatrix} \Psi_J(\mathbf{x}_I) & 0 \\ 0 & \Psi_J(\mathbf{x}_I) \end{bmatrix} = \Psi_J(\mathbf{x}_I)\mathbf{I} \tag{24}
$$

and Λ is the transformation matrix between generalized displacement vector $\mathbf{d}$ and nodal displacement vector $\hat{\mathbf{d}}$.

For computational efficiency, only the degrees of freedom associated with the essential boundaries are expressed in the nodal coordinate. Following Chen and Wang[23], the discrete points are first partitioned into two groups: a boundary group G^B containing all points on $\partial\Omega^g$, and an interior group G^I containing the rest of the points. Further partitioning the displacement vectors into boundary and interior components, $\mathbf{d}^T = \begin{bmatrix} \mathbf{d}^{B^T} & \mathbf{d}^{I^T} \end{bmatrix}$, $\hat{\mathbf{d}}^T = \begin{bmatrix} \hat{\mathbf{d}}^{B^T} & \hat{\mathbf{d}}^{I^T} \end{bmatrix}$, we rewrite Eq. (23) as:

$$
\hat{\mathbf{d}} = \begin{bmatrix} \hat{\mathbf{d}}^B \\ \hat{\mathbf{d}}^I \end{bmatrix} = \begin{bmatrix} \Lambda^{BB} & \Lambda^{BI} \\ \Lambda^{IB} & \Lambda^{II} \end{bmatrix} \begin{bmatrix} \mathbf{d}^B \\ \mathbf{d}^I \end{bmatrix} \equiv \hat{\Lambda}\mathbf{d}. \tag{25}
$$

Next, define a mixed displacement vector $\boldsymbol{d}^*$

$$
\mathbf{d}^* = \begin{bmatrix} \hat{\mathbf{d}}^B \\ \mathbf{d}^I \end{bmatrix} = \begin{bmatrix} \Lambda^{BB} & \Lambda^{BI} \\ \mathbf{0} & \mathbf{I} \end{bmatrix} \begin{bmatrix} \mathbf{d}^B \\ \mathbf{d}^I \end{bmatrix} = \Lambda^*\mathbf{d} \tag{26}
$$

or

$$
\mathbf{d} = \Lambda^{*-1}\mathbf{d}^*, \quad \Lambda^{*-1} = \begin{bmatrix} \Lambda^{BB-1} & -\Lambda^{BB-1}\Lambda^{BI} \\ \mathbf{0} & \mathbf{I} \end{bmatrix}. \tag{27}
$$

The displacement approximation can now be approximated as

$$
\mathbf{u}^h(\mathbf{x}) = \Psi(\mathbf{x})\mathbf{d} = \Psi(\mathbf{x})\Lambda^{*-1}\mathbf{d}^*, \tag{28}
$$

where

$$
\Psi(\mathbf{x}) = \begin{bmatrix} \Psi_I(\mathbf{x}) & \Psi_2(\mathbf{x}) & \cdots & \Psi_{NP}(\mathbf{x}) \end{bmatrix}, \quad \Psi_I(\mathbf{x}) = \begin{bmatrix} \Psi_I(\mathbf{x}) & 0 \\ 0 & \Psi_I(\mathbf{x}) \end{bmatrix}. \tag{29}
$$

With Eq. (28), kinematically admissible approximation of $u_i^h \in H_g^1$ and $\delta u_i^h \in H_0^1$ can be constructed, and the Galerkin approximation of weak form can be stated as: Find $u_i^h \in H_g^1$, $\forall \delta u_i^h \in H_0^1$, such that the following equation is satisfied:

$$
\int_\Omega \delta(\nabla^S u^h)_{ij} C_{ijkl} (\nabla^S u^h)_{kl}\, d\Omega - \int_\Omega \delta u_i^h b_i\, d\Omega - \int_{\partial\Omega^h} \delta u_i^h h_i\, d\Gamma = 0. \tag{30}
$$

Consider taking $\delta\mathbf{u}^h(\mathbf{x}) = \Psi(\mathbf{x})\Lambda^{*-1}\delta\mathbf{d}^*$ and $\mathbf{u}^h(\mathbf{x}) = \Psi(\mathbf{x})\mathbf{d}$ in Eq. (30) to yield

$$
\delta\mathbf{d}^{*T}\Lambda^{*-T}\mathbf{K}\mathbf{d} = \delta\mathbf{d}^{*T}\Lambda^{*-T}\mathbf{f} \quad \text{or} \quad \delta\mathbf{d}^{*T}(\mathbf{K}^*\mathbf{d} - \mathbf{f}^*) = 0, \tag{31}
$$

where

$$\mathbf{K}^* = \mathbf{\Lambda}^{*-T} \mathbf{K} = \begin{bmatrix} \mathbf{K}^{*BB} & \mathbf{K}^{*BI} \\ \mathbf{K}^{*IB} & \mathbf{K}^{*II} \end{bmatrix}, \quad \mathbf{f}^* = \mathbf{\Lambda}^{*-T} \mathbf{f} = \begin{bmatrix} \mathbf{f}^{*B} \\ \mathbf{f}^{*I} \end{bmatrix}. \tag{32}$$

Let Nb be the total number of degrees of freedom associated with essential boundary conditions on $\partial\Omega^g$. By considering the essential boundary conditions, we have $\hat{\mathbf{d}}^B = \mathbf{g}$, $\delta\hat{\mathbf{d}}^B = \mathbf{0}$, and $\delta\mathbf{d}^{*T} = \begin{bmatrix} \mathbf{0}^T & \mathbf{d}^{I^T} \end{bmatrix}$. By examining the discrete weak form, $\delta\mathbf{d}^{*T}\left(\mathbf{K}^*\mathbf{d} - \mathbf{f}^*\right) = 0$, it is apparent that the first Nb equations of $\mathbf{K}^*\mathbf{d} - \mathbf{f}^* = \mathbf{0}$ become redundant and can be replaced by the Nb equations of essential boundary conditions $\begin{bmatrix} \mathbf{\Lambda}^{BB} & \mathbf{\Lambda}^{BI} \end{bmatrix}\mathbf{d} = \mathbf{g}$ (Eq. (26)) to yield

$$\begin{bmatrix} \mathbf{\Lambda}^{BB} & \mathbf{\Lambda}^{BI} \\ \mathbf{K}^{*IB} & \mathbf{K}^{*II} \end{bmatrix}\begin{bmatrix} \mathbf{d}^B \\ \mathbf{d}^I \end{bmatrix} = \begin{bmatrix} \mathbf{g} \\ \mathbf{f}^{*I} \end{bmatrix}. \tag{33}$$

2.3.2. Modified Reproducing Kernel with Nodal Interpolation Properties

In this section we introduce an alternative method to construct a reproducing kernel approximation with Kronecker delta properties so that the essential boundary conditions can be imposed directly. Chen et al.[24] proposed a general formulation for developing reproducing kernel approximation with nodal interpolation properties.

Consider a modified reproducing kernel approximation of $u(x)$ as follows:

$$u_i^h(\mathbf{x}) = \sum_I \Psi_I(\mathbf{x})d_{iI} = \sum_I \left(\hat{\Psi}_I(\mathbf{x}) + \bar{\Psi}_I(\mathbf{x})\right)d_{iI}. \tag{34}$$

In Eq. (34), $\hat{\Psi}_I(\mathbf{x})$ is a primitive function used to introduce discrete Kronecker delta properties, and $\bar{\Psi}_I(\mathbf{x})$ is an enrichment function for imposing n-th order reproducing conditions. Consider the following construction of $\hat{\Psi}_I(\mathbf{x})$:

$$\hat{\Psi}_I(\mathbf{x}) = \frac{\hat{\Phi}_{\hat{a}_I}(\mathbf{x} - \mathbf{x}_I)}{\hat{\Phi}_{\hat{a}_I}(\mathbf{0})}, \quad \hat{a}_I < \min\{\|\mathbf{x}_I - \mathbf{x}_J\|, \forall J \neq I\}. \tag{35}$$

The support size $\hat{a}_I$ of $\hat{\Phi}_{\hat{a}_I}(\mathbf{x} - \mathbf{x}_I)$ is so chosen that it does not cover any neighbor points, and thus Kronecker delta conditions are satisfied in $\hat{\Psi}_I(\mathbf{x})$. The enrichment function is taken as the standard reproducing kernel form as

$$\bar{\Psi}_I(\mathbf{x}) = \mathbf{H}^T(\mathbf{x} - \mathbf{x}_I)\mathbf{a}(\mathbf{x})\bar{\Phi}_{\bar{a}_I}(\mathbf{x} - \mathbf{x}_I). \tag{36}$$

The coefficients $\mathbf{a}(\mathbf{x})$ in $\bar{\Psi}_I(\mathbf{x})$ are obtained by the following reproducing conditions:

$$\sum_I [\frac{\hat{\Phi}_{\hat{a}_I}(\mathbf{x} - \mathbf{x}_I)}{\hat{\Phi}_{\hat{a}_I}(\mathbf{0})} + \mathbf{H}^T(\mathbf{x} - \mathbf{x}_I)\mathbf{a}(\mathbf{x})\bar{\Phi}_{\bar{a}_I}(\mathbf{x} - \mathbf{x}_I)]x_{1I}^i x_{2I}^j = x_1^i x_2^j, \quad 0 \leq i + j \leq n. \tag{37}$$

Equation (37) can be rewritten as:

$$\sum_I [\frac{\hat{\Phi}_{\hat{a}_I}(\mathbf{x} - \mathbf{x}_I)}{\hat{\Phi}_{\hat{a}_I}(0)} + \mathbf{H}^T(\mathbf{x} - \mathbf{x}_I)\mathbf{a}(\mathbf{x})\bar{\Phi}_{\bar{a}_I}(\mathbf{x} - \mathbf{x}_I)(x_1 - x_{1I})^i(x_2 - x_{2I})^j] = \delta_{0i}\delta_{0j}, \tag{38}$$

$$0 \leq i + j \leq n$$

or

$$\sum_I [\frac{\hat{\Phi}_{\hat{a}_I}(\mathbf{x} - \mathbf{x}_I)}{\hat{\Phi}_{\hat{a}_I}(0)} + \mathbf{H}^T(\mathbf{x} - \mathbf{x}_I)\mathbf{a}(\mathbf{x})\bar{\Phi}_{\bar{a}_I}(\mathbf{x} - \mathbf{x}_I)\mathbf{H}(\mathbf{x} - \mathbf{x}_I)] = \mathbf{H}(0). \tag{39}$$

The coefficient vector $\mathbf{a}(\mathbf{x})$ is obtained from Eq. (39) by

$$\mathbf{a}(\mathbf{x}) = \mathbf{Q}^{-1}(\mathbf{x})[\mathbf{H}(0) - \hat{\mathbf{F}}(\mathbf{x})] \tag{40}$$

$$\hat{\mathbf{F}}(\mathbf{x}) = \sum_I \mathbf{H}(\mathbf{x} - \mathbf{x}_I)\frac{\hat{\Phi}_{\hat{a}_I}(\mathbf{x} - \mathbf{x}_I)}{\hat{\Phi}_{\hat{a}_I}(0)} \tag{41}$$

Finally, the reproducing kernel interpolation function is obtained:

$$\Psi_I(\mathbf{x}) = \frac{\hat{\Phi}_{\hat{a}_I}(\mathbf{x} - \mathbf{x}_I)}{\hat{\Phi}_{\hat{a}_I}(0)} + \mathbf{H}^T(\mathbf{x} - \mathbf{x}_I)\mathbf{Q}^{-1}(\mathbf{x})[\mathbf{H}(0) - \hat{\mathbf{F}}(\mathbf{x})]\bar{\Phi}_{\bar{a}_I}(\mathbf{x} - \mathbf{x}_I) \tag{42}$$

The Kronecker delta properties can be easily shown:

$$\Psi_I(\mathbf{x}_J) = \frac{\hat{\Phi}_{\hat{a}_I}(\mathbf{x}_J - \mathbf{x}_I)}{\hat{\Phi}_{\hat{a}_I}(0)} + \mathbf{H}^T(\mathbf{x}_J - \mathbf{x}_I)\mathbf{Q}^{-1}(\mathbf{x}_J)[\mathbf{H}(0) - \hat{\mathbf{F}}(\mathbf{x}_J)]\bar{\Phi}_{\bar{a}_I}(\mathbf{x}_J - \mathbf{x}_I) \tag{43}$$

$$= \delta_{IJ} + \mathbf{H}^T(\mathbf{x}_J - \mathbf{x}_I)\mathbf{Q}^{-1}(\mathbf{x}_J)[\mathbf{H}(0) - \hat{\mathbf{F}}(\mathbf{x}_J)]\bar{\Phi}_{\bar{a}_I}(\mathbf{x}_J - \mathbf{x}_I) = \delta_{IJ}$$

Note that in Eq. (43) the property $\hat{a}_I < \min\{\|\mathbf{x}_I - \mathbf{x}_J\|, \forall J \neq I\}$ has been used. The RK interpolation function in Eq. (42) bares the following properties:

(i) $Supp(\Psi_I(\mathbf{x})) = \max(\bar{a}_I, \hat{a}_I)$

(ii) If $\bar{\Phi}_{\bar{a}_I} \in C^{\bar{m}}$, $\hat{\Phi}_{\hat{a}_I} \in C^{\hat{m}}$, then $\Psi_I \in C^k$, $k = \min(\bar{m}, \hat{m})$

(iii) The singularity of $\mathbf{Q}(\mathbf{x})$ is only dependent on $\bar{a}_I$ and the order of basis function in $\mathbf{G}(\mathbf{x} - \mathbf{x}_I)$, and is independent to $\hat{a}_I$.

(iv) For better accuracy in solving PDE's, the primitive functions are included only in the shape functions associated with the nodes on the essential boundary. Following Chen *et al.*[24], it can be shown that the coefficients of shape functions with primitive functions included are nodal values and essential boundary conditions can be imposed directly.

2.4. *Stabilized Conformation Nodal Integration (SCNI)*

2.4.1. *Integration Constraints*

The traditional approach to perform domain integration is Gauss integration. However, if Gauss quadrature is employed for integrating the weak form, an additional background grid is required, and higher order quadrature rule is required to reduce the integration error. Another drawback of Gauss integration for the meshfree weak form is that it does not satisfy the integration constraints,[25] therefore the first order accuracy is not guaranteed even if the approximation of test and trial functions is linearly complete. Integration constraints are necessary conditions for linear exactness in the Galerkin approximation as identified by Chen *et al.*[25] There are two requirements for linear exactness in the 2nd order differential equations. The first condition, related to the approximation, requires the shape function to satisfy the linear consistency conditions given by

$$\begin{cases} \displaystyle\sum_{I=1}^{NP} \Psi_I(\mathbf{x}) = 1 \\ \displaystyle\sum_{I=1}^{NP} \Psi_I(\mathbf{x})\mathbf{x}_I = \mathbf{x} \end{cases} \tag{44}$$

Note that $\mathbf{x}$ and $\mathbf{x}_I$ are vectors. These conditions are automatically satisfied in the reproducing kernel shape functions if complete linear basis functions are used. The second condition requires the integration of the gradient of the shape function to vanish if the shape function does not intersect with the boundary,[25] i.e.,

$$\sum_{L=1}^{NIT} \nabla \Psi_I(\mathbf{x}_L) w_L = 0 \quad \text{if} \quad \text{supp}(\Psi_I) \cap \partial\Omega = \varnothing, \tag{45}$$

where NIT is the number of integration points. If Gauss integration is employed, $\mathbf{x}_L$ are the spatial coordinates of the Gauss points and w_L are the weights of integration. If nodal integration is applied, $\mathbf{x}_L$ are the coordinates of the discrete integration points and w_L are the associated weights at the discrete points.

For shape function that intersects with the natural boundary, the integration of the gradient of the shape function should satisfy the divergence equation

$$\sum_{L=1}^{NIT} \nabla \Psi_I(\mathbf{x}_L) w_L = \sum_{K=1}^{NITB} \mathbf{n} \Psi_I(\mathbf{x}_K) s_K \quad \text{if} \quad \text{supp}(\Psi_I) \cap \Omega^h \neq \varnothing, \tag{46}$$

where $NITB$ is the number of integration points on the natural boundary that are covered by the support of node I, $\mathbf{n}$ is the outward normal of the natural boundary, and s_K are the weights of boundary integration.

2.4.2. *Strain Smoothing*

To satisfy the integration constraints as stated in Eqs. (45) and (46), a strain smoothing[25] is introduced

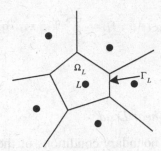

Figure 3. Nodal representative domain for SCNI

$$\bar{\varepsilon}_{ij}^h(\mathbf{x}_L) = \frac{1}{2A_L}\int_{\Omega_L}\left(u_{i,j}^h + u_{j,i}^h\right)d\Omega = \frac{1}{2A_L}\int_{\Gamma_L}\left(u_i^h n_j + u_j^h n_i\right)d\Gamma$$

$$= \frac{1}{2}[\bar{b}_{iI}(\mathbf{x}_L)d_{jI} + \bar{b}_{jI}(\mathbf{x}_L)d_{iI}]$$

(47)

$$\bar{b}_{iI}(\mathbf{x}_L) = \frac{1}{A_L}\int_{\Gamma_L}\Psi_I(\mathbf{x})n_i d\Gamma.$$

(48)

Here Ω_L is the nodal representative domain for node L as shown in Fig. 3, Γ_L is the boundary of the representative domain, and A_L is the volume (for 3D) or area (for 2D) of the representative domain. A Voronoi diagram at particle L as shown in Fig. 3 can be employed to generate the nodal representative domain.

The smoothed strain approximation can be expressed as

$$\bar{\varepsilon}^h(\mathbf{x}_L) = \sum_{I=1}^{NP}\bar{\mathbf{B}}_I(\mathbf{x}_L)\mathbf{d}_I$$

(49)

$$\bar{\mathbf{B}}_I(\mathbf{x}_L) = \begin{bmatrix} \bar{b}_{I1}(\mathbf{x}_L) & 0 \\ 0 & \bar{b}_{I2}(\mathbf{x}_L) \\ \bar{b}_{I2}(\mathbf{x}_L) & \bar{b}_{I1}(\mathbf{x}_L) \end{bmatrix},$$

(50)

where NP is the number of nodes whose support covers node $\mathbf{x}_L$. It has been shown that the smoothed gradient matrix $\bar{\mathbf{B}}$ satisfies the integration constraints.[25]

To introduce the smoothed strain into strain approximation, consider the following assumed strain variational equation:

$$\int_{\Omega_x}\delta\bar{\varepsilon}_{ij}C_{ijkl}\bar{\varepsilon}_{kl}d\Omega - \int_{\Omega}\delta u_i b_i d\Omega - \int_{\partial\Omega^h}\delta u_i h_i d\Gamma = 0.$$

(51)

By employing the smoothed strain approximation in Eq. (49) and the displacement approximation in Eq. (1), we have the discrete equation:

$$\mathbf{Kd} = \mathbf{f} \tag{52}$$

$$\mathbf{K}_{IJ} = \sum_{M=1}^{NP} \bar{\mathbf{B}}_I^T (\mathbf{x}_M) \mathbf{C} \bar{\mathbf{B}}_J (\mathbf{x}_M) A_M \,, \ \mathbf{f}_I = \sum_{M=1}^{NP} \Psi_I (\mathbf{x}_M) \mathbf{b} A_M + \sum_{L=1}^{Nh} \Psi_I (\mathbf{x}_L) \mathbf{h} s_L \tag{53}$$

2.5. Numerical Examples

2.5.1. Beam Subjected to a Shear Load

The problem statement and boundary conditions of the beam problem are given in Fig. 4(a). The numerical solution obtained from SCNI is compared with the solutions obtained by Gauss integration with 5x5 quadrature rule and the direct

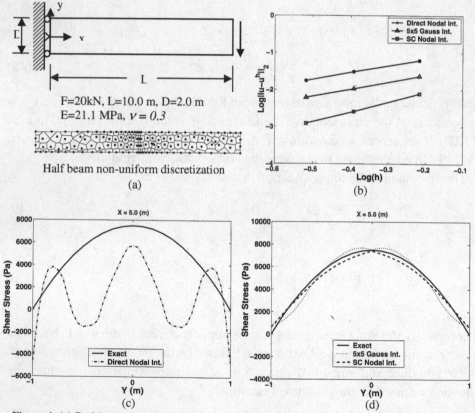

Figure 4. (a) Problem statement and discretization, (5) displacement L2 error norm, (c) shear stress distribution along x=0.5 L obtained by a direct nodal integration, (d) shear stress distribution along x=0.5 L obtained by the SCNI and 5x5 Gauss integration

Table 1 Tip displacement accuracy (%) using highly irregular discretization

Discrete Model	5x5 Gauss Int.	Direct Nodal Int.	SC Nodal Int.
124 nodes	94.99	192.82	99.25

nodal integration. Linear basis functions and a normalized support size of 2.01 are used in all three uniform discretizations. The comparison of displacement L_2 error norm is shown in Fig. 4(b). The solution of the direct nodal integration presents lower accuracy than that obtained from Gauss integration. SCNI not only enhances accuracy of the direct nodal integration, the method performs better than the Gauss integration method. Shear stress distributions along regular nodes at $x = 0.5L$ in Figs. 4(c) and 4(d) clearly demonstrate the superior performance of SCNI. A very non uniform 124-node model as shown in Fig. 4(a) is created to test the performance of three methods under highly non uniform discretization. The tip displacement solution of the direct nodal integration method deteriorates significantly in this case as shown in Table 1. On the contrary, SCNI still maintains a 99.25 % accuracy in the tip displacement; much better than direct nodal and Gauss integration methods.

3. Structural Shape Optimization

Structural design problems can be categorized based on the type of design variables. While the sizing design is related to parameters of the structure, the shape design is related to the structure's geometry. In the shape design problem, the structural domain or its boundary is defined as design variables. Since the domain itself is part of a design, the structural geometry appears implicitly as the design parameter. This fact makes the shape design problem more difficult than the conventional sizing design problem.

3.1. *Shape Design Parameterization and Design Velocity*

Shape design parameterization, which describes the boundary shape of a structure as a function of the design variables, is an essential step in the shape design process. Inappropriate parameterization can lead to unacceptable shapes. To parameterize the structural boundaries and to achieve optimum shape design, boundary shape can be described in three ways: (1) by using boundary nodal coordinates, (2) by using polynomials,[26-28] and (3) by using spline blending functions.[19,29-32] All these methods describe how the design variable changes the shape of the boundary.

In the meshfree method, the structural domain is discretized by a set of particles. When the boundary of the structure is changed according to the shape

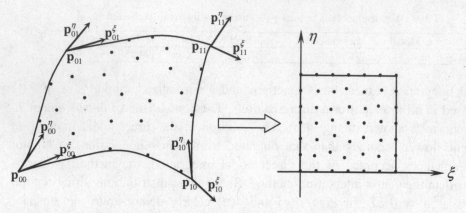

Figure 5. Parametric representation of a surface geometry. Corner points and their tangent vectors can be served as shape design variables. The parametric coordinates remain constant

design variable, the location of meshfree particles is changed accordingly. The direction that each particle moves with respect to the shape design variable is called the design velocity. Let $\mathbf{x}$ be the location of a particle in the domain and the location at the perturbed design be given as

$$\mathbf{x}_\tau = \mathbf{x} + \tau \mathbf{V}(\mathbf{x}), \tag{54}$$

where $\mathbf{V}(\mathbf{x})$ is the design velocity and the parameter τ controls the magnitude of design change. The process is similar to the dynamic process by considering τ as time. Because of this analogy, the direction $\mathbf{V}(\mathbf{x})$ is called the design velocity.

In order to illustrate the shape design change, we consider a simple geometric representation as an example. In many geometric modelers, the location of particles is often represented using a parametric technique. For example, the location $\mathbf{x}$ in two–dimensional space can be represented using two parameters as

$$\mathbf{x}(\xi,\eta) = \mathbf{U}(\xi)^T \mathbf{M} \mathbf{G} \mathbf{M}^T \mathbf{W}(\eta), \tag{55}$$

where $\mathbf{U}(\xi) = [\xi^3 \quad \xi^2 \quad \xi \quad 1]^T$ and $\mathbf{W}(\eta) = [\eta^3 \quad \eta^2 \cdot \quad \eta \quad 1]^T$ are vectors in the parametric coordinates, and $\mathbf{M}$ is a constant matrix defined as

$$\mathbf{M} = \begin{bmatrix} 2 & -2 & 1 & 1 \\ -3 & 3 & -2 & -1 \\ 0 & 0 & 1 & 0 \\ 1 & 0 & 0 & 0 \end{bmatrix} \tag{56}$$

and $\mathbf{G}$ is the geometric matrix defined as

$$G = \begin{bmatrix} \mathbf{p}_{00} & \mathbf{p}_{01} & \mathbf{p}_{00}^{\eta} & \mathbf{p}_{01}^{\eta} \\ \mathbf{p}_{10} & \mathbf{p}_{11} & \mathbf{p}_{10}^{\eta} & \mathbf{p}_{11}^{\eta} \\ \mathbf{p}_{00}^{\xi} & \mathbf{p}_{01}^{\xi} & \mathbf{p}_{00}^{\xi\eta} & \mathbf{p}_{01}^{\xi\eta} \\ \mathbf{p}_{10}^{\xi} & \mathbf{p}_{11}^{\xi} & \mathbf{p}_{10}^{\xi\eta} & \mathbf{p}_{11}^{\xi\eta} \end{bmatrix}_{4 \times 4 \times 3} \tag{57}$$

where $\mathbf{p}_{ij}$ are coordinates of the corner points on the surface, $\mathbf{p}_{ij}^{\xi}$ and $\mathbf{p}_{ij}^{\eta}$ are the tangent vectors in ξ and η directions, respectively, and $\mathbf{p}_{ij}^{\xi\eta}$ is the twist vectors. All components or combination of them can be served as shape design variables. Figure 5 shows the geometry and its transformation into the parametric coordinate.

The computation of $\mathbf{V}(\mathbf{x}) = \mathbf{V}(\xi, \eta)$ is directly related to the parametric representation of the neutral surface, as given in Eq. (55). For the purpose of explanation, let us consider one design variable b. Equation (55) is rewritten with design dependence as

$$\mathbf{x}(b; \xi, \eta) = \mathbf{U}(\xi)^T \mathbf{M} \mathbf{G}(b) \mathbf{M}^T \mathbf{W}(\eta). \tag{58}$$

Since geometric matrix $\mathbf{G}(b)$ is a function of the design, the design velocity $\mathbf{V}(\xi, \eta)$ can be obtained by perturbing b to $b + \tau \delta b$, and then differentiating with respect to τ as

$$\mathbf{V}(\xi, \eta) = \left. \frac{d\mathbf{x}(b + \tau \delta b; \xi, \eta)}{d\tau} \right|_{\tau=0} = \mathbf{U}(\xi)^T \mathbf{M} \left(\frac{\partial \mathbf{G}}{\partial b} \delta b \right) \mathbf{M}^T \mathbf{W}(\eta). \tag{59}$$

For example, when the x–component of $\mathbf{p}_{00}$ is chosen as the design variable, matrix $\partial \mathbf{G} / \partial b$ has all zero components except for the component at $(1,1)$ that has a value of $[1, 0, 0]$. The design velocity field must be obtained per each shape design variable.

An advantage of the design velocity computation in Eq. (59) is that it is unnecessary to store design velocity for all particles. It is sufficient to simply store matrix $\partial \mathbf{G} / \partial b$ for each design variable. Note that $\mathbf{V}(\xi, \eta)$ remains constant during the optimization process.

3.2. *Shape Sensitivity Analysis*

Design sensitivity analysis computes the rate of performance measure change with respect to design variable changes.[33] With the structural analysis, the design sensitivity analysis generates a critical information, gradient, for optimization. In this text, performance measures are presumed to be differentiable with respect to design, at least in the neighborhood of the current design point. For complex engineering applications, however, it is not simple to prove the differentiability.[34]

In general, a structural performance measure depends on the design parameters. For example, a change in the cross-sectional area of a beam would affect the structural weight. This type of dependence is simple if the expression of weight in terms of the design variables is known. This type of function is *explicitly dependent* on the design. Consequently, only algebraic calculation is involved to obtain the design sensitivity of an explicitly dependent performance measure.

However, in many cases, performance measures implicitly depend on the design. For example, there is no explicit way to express the stress of a structure explicitly in terms of the design variable b. Consider the general performance measure $\psi = \psi(\mathbf{u}(b), b)$ that depends on the design explicitly and implicitly. The sensitivity of ψ can be expressed as

$$\frac{d\psi(\mathbf{u}(b), b)}{db} = \frac{\partial \psi}{\partial b}\bigg|_{\mathbf{u}=\text{const}} + \frac{\partial \psi}{\partial \mathbf{u}}\bigg|_{b=\text{const}}^{T} \frac{d\mathbf{u}}{db}. \tag{60}$$

From the expression of $\psi(\mathbf{u}(b), b)$, the explicitly dependent term, $d\psi/db$, and the derivative, $d\psi/d\mathbf{u}$, can easily be obtained. The only unknown term in Eq. (60) is $d\mathbf{u}/db$, which is the sensitivity of the state variable with respect to the design variable. The key procedure of design sensitivity analysis is to calculate the sensitivity of the state variable by differentiating the structural equation. For a given shape design velocity field $\mathbf{V}(\mathbf{x})$, the shape sensitivity formulation expresses the sensitivity of state variable in terms of the design velocity. In this text, only linear problem is considered. The nonlinear sensitivity analysis is presented in Chapter 8.

Four approaches are used to obtain design sensitivity: the finite difference, discrete, continuum, and computational derivatives. In the finite difference approach, design sensitivity is obtained by either the *forward* or *central finite difference method*. In the discrete method, design sensitivity is obtained by taking design derivatives of the discrete governing equation. In the continuum approach, the design derivative of the variational equation is taken before discretization. Finally, computational or automatic differentiation refers to a differentiation of the computer code itself. The continuum approach is employed in this text because this formulation is independent of discretization methods. The particular discretization using the meshfree method will be discussed in the next section.

First, the sensitivity of the state variable is defined using the material derivative concept in continuum mechanics, as

$$\dot{\mathbf{u}}(\mathbf{x}) = \lim_{\tau \to 0} \left[\frac{\mathbf{u}_\tau(\mathbf{x} + \tau\mathbf{V}(\mathbf{x})) - \mathbf{u}(\mathbf{x})}{\tau} \right]$$

$$= \lim_{\tau \to 0} \left[\frac{\mathbf{u}_\tau(\mathbf{x}) - \mathbf{u}(\mathbf{x})}{\tau} \right] + \lim_{\tau \to 0} \left[\frac{\mathbf{u}_\tau(\mathbf{x} + \tau\mathbf{V}(\mathbf{x})) - \mathbf{u}_\tau(\mathbf{x})}{\tau} \right], \qquad (61)$$

$$= \mathbf{u}'(\mathbf{x}) + \nabla\mathbf{u} \cdot \mathbf{V}(\mathbf{x}),$$

where $\mathbf{u}'$ is the partial derivative. The above material derivative can be applied to general functions. The order of differentiation can be changed between the partial derivative and the spatial derivative, such that $(\nabla\mathbf{u})' = \nabla(\mathbf{u}')$. However, it is not true for the material derivative in Eq. (61). In such a case,

$$(\nabla\mathbf{u})^\cdot = \nabla\dot{\mathbf{u}} - \nabla\mathbf{u} \cdot \nabla\mathbf{V}. \qquad (62)$$

Since the structural equation is expressed in terms of functionals, the following two formulas are useful for deriving the sensitivity equation:

$$\psi_1' = \left[\int_\Omega f(\mathbf{x}) d\Omega \right]' = \int_\Omega [\dot{f}(\mathbf{x}) + f(\mathbf{x}) div\mathbf{V}] d\Omega \qquad (63)$$

$$\psi_2' = \left[\int_{\partial\Omega} g(\mathbf{x}) d\Gamma \right]' = \int_{\partial\Omega} [\dot{g}(\mathbf{x}) + \kappa g(\mathbf{x})V_n] d\Gamma. \qquad (64)$$

In Eq. (64), κ is the curvature of the boundary and V_n is the normal component of the design velocity on the boundary.

The variational equation in Eq. (30) is used for deriving the sensitivity equation. For the illustration, Eq. (30) is rewritten in the following form:

$$a_\Omega(\mathbf{u}^h, \delta\mathbf{u}^h) := \int_\Omega \delta(\nabla^S u^h)_{ij} C_{ijkl} (\nabla^S u^h)_{kl} d\Omega$$

$$= \int_\Omega \delta u_i^h b_i d\Omega + \int_{\partial\Omega^h} \delta u_i^h h_i d\Gamma := \ell_\Omega(\delta\mathbf{u}^h) \qquad (65)$$

The notation is selected such that $a_\Omega(\mathbf{u}, \delta\mathbf{u})$ is bilinear with respect to its two arguments, while $\ell_\Omega(\delta\mathbf{u})$ is linear. The above variational equation must satisfy for all kinematically admissible fields $u_i^h \in H_g^1$ and $\delta u_i^h \in H_0^1$.

Using the formulas in Eqs. (63) and (64), the above variational equation is differentiated to obtain the sensitivity equation:

$$a_\Omega(\dot{\mathbf{u}}^h, \delta\mathbf{u}^h) = \ell_\mathbf{V}'(\delta\mathbf{u}^h) - a_\mathbf{V}'(\mathbf{u}^h, \delta\mathbf{u}^h), \qquad (66)$$

for all $\delta u_i^h \in H_0^1$. In Eq. (66), the left-hand side is identical with that of Eq. (65) if $\mathbf{u}^h$ is substituted with its sensitivity $\dot{\mathbf{u}}^h$, and two terms on the right-hand side are defined as

$$\ell'_{\mathbf{V}}(\delta \mathbf{u}^h) = \int_{\Omega} [\delta u_i^h (\nabla b)_{ij} V_j + \delta u_i^h b_i \, div \mathbf{V}] d\Omega$$

$$+ \int_{\partial \Omega^h} [\delta u_i^h (\nabla h)_{ij} V_j + \kappa \delta u_i^h h_i V_n] d\Gamma \qquad (67)$$

and

$$a'_{\mathbf{V}}(\mathbf{u}^h, \delta \mathbf{u}^h) = \int_{\Omega} [\varepsilon_{ij}^{\mathbf{V}}(\delta \mathbf{u}^h) C_{ijkl} (\nabla^S u^h)_{kl} + \delta(\nabla^S u^h)_{ij} C_{ijkl} \varepsilon_{kl}^{\mathbf{V}}(\mathbf{u}^h)$$

$$+ \delta(\nabla^S u^h)_{ij} C_{ijkl} (\nabla^S u^h)_{kl} \, div \mathbf{V}] d\Omega \qquad , \qquad (68)$$

where $div \mathbf{V} = \partial V_i / \partial x_i$, and

$$\varepsilon_{ij}^{\mathbf{V}}(\mathbf{u}^h) = -\frac{1}{2} \left(\frac{\partial u_i^h}{\partial x_k} \frac{\partial V_k}{\partial x_j} + \frac{\partial u_j^h}{\partial x_k} \frac{\partial V_k}{\partial x_i} \right). \qquad (69)$$

The detailed derivations can be found in Choi and Seong.[35]

It is well known that the adjoint variable method is more efficient than solving the design sensitivity equation (66) when the number of design variables is greater than the number of performance functions. However, since this paper aims to address the dependence between shape design variables and the meshfree approximation function, the discussion will be limited to the direct differentiation method as in Eq. (66).

The shape sensitivity equation (66) is independent of discretization method. Either finite element[36] or meshfree method[37] can be used for numerically calculating the sensitivity of the state variable. In the following section, the implementation using the meshfree method is discussed.

3.3. Meshfree Discretization of Sensitivity Equation

3.3.1. Material Derivative of Meshfree Shape Function

Since the main unknown variable of the meshfree method is generalized displacement d_{iI}, the design sensitivity equation (66) in the continuum form, which is written in terms of $\dot{u}_i^h$, has to be discretized using $\dot{d}_{iI}$. Since u_i^h is approximated using the meshfree shape function in Eq. (22), $\dot{u}_i^h$ can be approximated by differentiating Eq. (22), as

$$\dot{u}_i^h(\mathbf{x}) = \sum_{I=1}^{NP} (\Psi_I(\mathbf{x}) \dot{d}_{iI} + \dot{\Psi}_I(\mathbf{x}) d_{iI}). \qquad (70)$$

This decomposition is quite different from the finite element method in which the shape function is independent of the design. The first term constitutes the main unknown $\dot{d}_{iI}$ of the sensitivity equation, while the second represents the dependence of the shape function on design, which is explicit in $\mathbf{V}(\mathbf{x})$.

From the observation that u_i^h and $\dot{u}_i^h$ belong to the same space,[†] $\dot{u}_i^h$ can be approximated directly using the meshfree shape function[38] as

$$\dot{u}_i^h(\mathbf{x}) = \sum_{I=1}^{NP} \Psi_I(\mathbf{x})\tilde{d}_{iI} . \tag{71}$$

By comparing Eq. (70) with Eq. (71), the latter seems to provide simpler approximation than the former. However, the former will yield numerical results that are more consistent than the latter. In addition, since $\tilde{d}_{iI}$ is not the material derivative of d_{iI}, the penalty function must be used for imposing the essential boundary conditions. In this text, the approximation in Eq. (70) will be used.

A numerical method to compute $\dot{\Psi}_I(\mathbf{x})$ will now be introduced. From the relation $\mathbf{x}_\tau = \mathbf{x} + \tau \mathbf{V}(\mathbf{x})$, the derivative of the material point $\mathbf{x}$ is nothing but the design velocity $\mathbf{V}(\mathbf{x})$. Consider the material derivative of the kernel function in Eq. (3) for a one–dimensional problem,

$$\dot{\Phi}_a(x - x_I) = \frac{4(V_I - V)}{a} \begin{cases} 2z - 3z^2, & z \leq \frac{1}{2} \\ (1 - z)^2, & \frac{1}{2} < z \leq 1 \\ 0, & \text{otherwise} \end{cases} \tag{72}$$

where V_I is the design velocity at x_I, and V is the design velocity at x. For a multi–dimensional problem, the product rule in Eq. (5) can be used.

To compute $\dot{\Psi}_I(\mathbf{x})$, the material derivative of the reproducing condition in Eq. (11) has to be taken, to obtain

$$\dot{\mathbf{b}}(\mathbf{x}) = -\mathbf{M}(\mathbf{x})^{-1}\dot{\mathbf{M}}(\mathbf{x})\mathbf{b}(\mathbf{x}) , \tag{73}$$

where,

$$\dot{\mathbf{M}}(\mathbf{x}) = \sum_{I=1}^{NP} [\dot{\mathbf{H}}\mathbf{H}^T\Phi_a + \mathbf{H}\dot{\mathbf{H}}^T\Phi_a + \mathbf{H}\mathbf{H}^T\dot{\Phi}_a] \tag{74}$$

$$\dot{\mathbf{H}}(\mathbf{x} - \mathbf{x}_I) = \begin{bmatrix} 0 & V_1 - V_{1I} & V_2 - V_{2I} & \cdots & n(x_2 - x_{2I})^{n-1}(V_2 - V_{2I}) \end{bmatrix}^T . \tag{75}$$

Thus, from the definition of the meshfree shape function in Eq. (15), we have

$$\dot{\Psi}_I(\mathbf{x}) = \dot{\mathbf{b}}^T\mathbf{H}\Phi_a + \mathbf{b}^T\dot{\mathbf{H}}\Phi_a + \mathbf{b}^T\mathbf{H}\dot{\Phi}_a . \tag{76}$$

For given design velocity $\mathbf{V}(\mathbf{x})$, Eq. (76) can be explicitly calculated even before any sensitivity analysis. The material derivative of $d\Psi_I / dx$ can also be calculated using a similar procedure.

[†] This can be observed by comparing Eq. (65) with Eq. (66).

3.3.2. *Discrete Form of Sensitivity Equation*

In developing sensitivity formulation, it is necessary to take the material derivative of strain or, equivalently, the gradient of displacement, $u_{i,j}^h = \partial u_i^h / \partial x_j$. Choi and Kim[33] uses the concept of a partial derivative that is commutable to a spatial gradient. By using Eqs. (62) and (70), a meshfree approximation of the material derivative of $u_{i,j}^h$ can be expressed as

$$\left(u_{i,j}^h\right)^{\cdot} = \sum_{I=1}^{NP} \left(\Psi_{I,j}\dot{d}_{iI} + \dot{\Psi}_{I,j}d_{iI} - \Psi_{I,k}d_{iI}V_{k,j}\right). \tag{77}$$

Note that the last two terms are explicitly dependent on the design velocity. The only unknown term is $\dot{d}_{iI}$ which will be computed from the sensitivity equation. To simplify the approximation of Eq. (77), the following relation can be used:

$$\left(\Psi_{I,j}\right)^{\cdot} = \dot{\Psi}_{I,j} - \Psi_{I,k}V_{k,j}. \tag{78}$$

Thus, the last two terms of Eq. (77) can be combined to represent an explicitly dependent term on $\mathbf{V}(\mathbf{x})$ through $(\Psi_{I,i})^{\cdot}$. Using Eq. (78), Eq. (77) is simplified to

$$\left(u_{i,j}^h\right)^{\cdot} = \sum_{I=1}^{NP} \Psi_{I,j}\dot{d}_{iI} + \sum_{I=1}^{NP} \left(\Psi_{I,j}\right)^{\cdot} d_{iI}. \tag{79}$$

Note that the two summations of Eq. (79) have a similar format. The first term on the right–hand side has to be solved using a design sensitivity equation, and the second term can be computed explicitly using the design velocity.

In sensitivity analysis, it is often assumed that the space H_0^1 of kinematically admissible displacements is independent of shape design, i.e., $\delta u_i^h = 0$. Even if the assumption of $\delta u_i^h = 0$ is not used, since $\delta \dot{u}_i^h \in H_0^1$, the following relation is satisfied

$$a_\Omega(\mathbf{u}^h, \delta \dot{\mathbf{u}}^h) = \ell_\Omega(\delta \dot{\mathbf{u}}^h). \tag{80}$$

Because of Eq. (80), the contribution of $\delta \dot{u}_i^h$ will be ignored in the derivation of sensitivity equation. In addition, from the relation in Eq. (77),

$$\left(\delta u_{i,j}^h\right)^{\cdot} = -\delta u_{i,k}^h V_{k,j} = -\sum_{I=1}^{NP} \Psi_{I,k}\delta d_{iI}V_{k,j}. \tag{81}$$

The approximation of the sensitivity equation (66) follows the same method as meshfree analysis. For a given meshfree shape function, using its material derivatives from Eq. (76), as well as using the relation in Eq. (79), the following approximation can be obtained:

$$\epsilon^{\mathbf{V}}(\mathbf{u}^h) = \sum_{I=1}^{NP} \dot{\mathbf{B}}_I \mathbf{d}_I, \tag{82}$$

where $\dot{\mathbf{B}}_I$ is the material derivative of $\mathbf{B}_I$, defined by

$$\dot{\mathbf{B}}_I = \begin{bmatrix} (\Psi_{I,1})^{\bullet} & (\Psi_{I,2})^{\bullet} & 0 \\ 0 & (\Psi_{I,1})^{\bullet} & (\Psi_{I,2})^{\bullet} \end{bmatrix}^T \tag{83}$$

In contrast, the approximation of $\varepsilon^V(\delta\mathbf{u}^h)$ has a different expression because of Eq. (81),

$$\varepsilon^V(\delta\mathbf{u}^h) = \sum_{I=1}^{NP} \mathbf{B}_I^V \delta\mathbf{d}_I \tag{84}$$

$$\mathbf{B}_I^V = -\begin{bmatrix} \Psi_{I,k}V_{k,1} & \Psi_{I,k}V_{k,2} & 0 \\ 0 & \Psi_{I,k}V_{k,1} & \Psi_{I,k}V_{k,2} \end{bmatrix}^T. \tag{85}$$

Now, the right-hand sides of sensitivity equation, Eqs. (67) and (68), can be approximated by

$$\ell'_V(\delta\mathbf{u}^h) \approx \int_\Omega \sum_{I=1}^{NP} \delta d_{iI} \Psi_I [(\nabla b)_{ij} V_j + b_i div\mathbf{V})] d\Omega$$

$$+ \int_{\partial\Omega^h} \sum_{I=1}^{NP} \delta d_{iI} \Psi_I [(\nabla h)_{ij} V_j + \kappa h_i V_n)] d\Omega \equiv \delta\mathbf{d}^T \mathbf{F}^\ell \tag{86}$$

$$a'_V(\mathbf{u}^h, \delta\mathbf{u}^h) \approx \int_\Omega \sum_{I=1}^{NP} \delta\mathbf{d}_I^T [\mathbf{B}_I^{V^T} \sigma + \mathbf{B}_I^T \mathbf{C} \varepsilon^V(\mathbf{u}^h) + \mathbf{B}_I^T \sigma div\mathbf{V}] d\Omega \equiv \delta\mathbf{d}^T \mathbf{F}^a. \tag{87}$$

Thus, the discrete form of the sensitivity equation becomes

$$\delta\mathbf{d}^T \mathbf{K} \dot{\mathbf{d}} = \delta\mathbf{d}^T (\mathbf{F}^\ell - \mathbf{F}^a), \tag{88}$$

for all $\delta\mathbf{d}$ whose counterparts $\delta\mathbf{u}$ belong to the space H_0^1 of kinematically admissible displacements.

3.3.3 Imposing Essential Boundary Conditions

The discrete sensitivity equation (88) cannot be solved directly because it is not trivial to construct the kinematically admissible $\delta\mathbf{d}$ from Eq. (88). As discussed in Section 2, the Lagrange multiplier method in Eq. (20) can be used for the purpose of sensitivity analysis. In such a case, the sensitivity of the Lagrange multiplier also needs to be calculated. In addition, the coefficient matrix becomes positive semi-definite, which requires a special treatment in solving the matrix equation. When the modified reproducing kernel approximation is used, Eq. (88) can directly be used because the modified meshfree shape functions for the boundary particles satisfy the interpolation property. Thus, the transformation method will be discussed in the following.

By following the same response analysis procedure to construct kinematically admissible displacements, the following linear matrix equation is solved:

$$\mathbf{K}^*\dot{\mathbf{d}} = \mathbf{\Lambda}^{*-T}(\mathbf{F}^\ell - \mathbf{F}^a), \tag{89}$$

where $\mathbf{K}^*$ represents the same stiffness matrix with meshfree analysis as in Eq. (32), which is already factorized. Thus, it is very efficient to solve (89) with different right-hand sides.

Consideration of the essential boundary conditions is somewhat different from that of the analysis undertaken in Eq. (33), since the transformation matrix $\mathbf{\Lambda}^*$, which is composed of a meshfree shape function, depends on the shape design. Let the prescribed displacement $\mathbf{g}$ at $\mathbf{x} \in \partial\Omega^g$ be independent of design, which is true in most cases. Then, from $\hat{\mathbf{d}}^B = \mathbf{g}$ and Eq. (26), we have

$$\mathbf{\Lambda}^B\dot{\mathbf{d}} = -\dot{\mathbf{\Lambda}}^B\mathbf{d}, \tag{90}$$

where $\mathbf{\Lambda}^B = [\mathbf{\Lambda}^{BB} \quad \mathbf{\Lambda}^{BI}]$ and $\dot{\mathbf{\Lambda}}^B_{IJ} = \dot{\Psi}_I(\mathbf{x}_J)$ is obtained from Eq. (70) with $\mathbf{x}_J \in \partial\Omega^g$. Equation (90) is substituted into Eq. (89) for those rows that correspond to the particles on the essential boundary, and by the use of Eq. (33), we have

$$\begin{bmatrix} \mathbf{\Lambda}^{BB} & \mathbf{\Lambda}^{BI} \\ \mathbf{K}^{*IB} & \mathbf{K}^{*II} \end{bmatrix} \begin{bmatrix} \dot{\mathbf{d}}^B \\ \dot{\mathbf{d}}^I \end{bmatrix} = \begin{bmatrix} -\dot{\mathbf{\Lambda}}^B\mathbf{d} \\ \left(\mathbf{\Lambda}^{*-T}(\mathbf{F}^\ell - \mathbf{F}^a)\right)^I \end{bmatrix}. \tag{91}$$

Equation (91) is solved for each design parameter with the same coefficient matrix with the meshfree analysis. After solving $\dot{\mathbf{d}}$, the material derivative of physical displacement can be calculated from the relation in Eq. (70).

4. Numerical Examples

4.1. *Torque-Arm Model*

The shape of the torque-arm model in Fig. 6 is optimized according to the eight shape design variables that control the boundary curves. In each design variable, the design velocity is calculated using Eq. (59). The torque-arm is modeled using 239 meshfree particles. Figure 6(a) shows meshfree particles and analysis results. The sensitivity computation requires only 10% of meshfree analysis computation per design variable due to the use of the same tangent operator as shown in Eqs. (33) and (91).

The design optimization problem is formulated to minimize the structural mass, with the effective stress constraint, as

$$\text{minimize} \quad \text{mass}$$
$$\text{subject to} \quad \sigma_{MAX} \leq 800 \quad \text{MPa} \tag{92}$$

The sequential quadratic programming method is used in a commercially available optimization program.[39] Figure 6(b) shows the meshfree analysis results at optimum design where the stress constraints along the upper side of torque arm is active. No re-modeling is used during the design optimization procedure. Through optimization, the structural mass is reduced by 48%. A total of 41 meshfree analyses and 20 sensitivity analyses are carried out during 20 optimization iterations. When finite element analysis is used with a re-meshing process,[40] the optimization process converged at 45 iterations with eight re-meshing processes. Thus, this approach reduces the cost of design optimization more than 50%, leaves along the cost related to the re-meshing process.

Since the initial particle distribution is used throughout the optimization process, a very non uniform particle distribution is resulted in the optimum design. The analysis result from evenly distributed particles at the optimum design confirms that the solution accuracy is insensitive to the particle distribution.

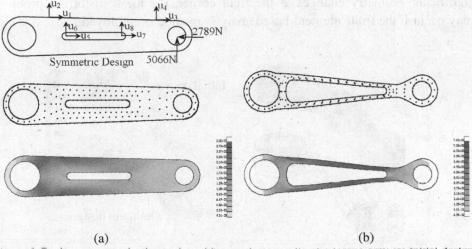

(a)   (b)

Figure 6. Design parameterization and meshfree analysis results of a torque-arm: (a) Initial design and (b) optimum design

4.2. *Road-Arm Model*

The advantage of meshfree analysis for structural optimization is more significant in the case of three-dimensional problem. Figure 7 shows a road arm model that is discretized with 1,455 meshfree particles. Eight shape design variables are defined to optimize the shape of the boundary. Since the geometries in the corner are so complicated, it is challenging to construct a regular-shaped finite element mesh. In addition to the complicated initial geometry, the structural shape further changes during design optimization process, which will cause a mesh distortion problem if a finite element method is used.

As is illustrated in Figure 7, the stress concentration appears in the left corner of the road arm. If the highest stress level in the left corner is considered as a reference value, then the dimension of the right corner cross-section can be reduced, because this region has a large amount of safety margin.

The design optimization is carried out to minimize the structural weight of the road arm, while maintaining the maximum stress level. Design optimization problem converges after eight iterations. Figure 7 also compares the meshfree analysis result at the initial and optimum designs. The structural weight at the optimum design is reduced by 23% compared to the initial weight. Since the stress concentration appears at the left corner in the initial design, the optimization algorithm intends to reduce the cross-sectional area of the right corner so that both parts may have the same level of stress values. Because of the significant geometry changes at the right corner, the mesh distortion problem may occur if the finite element-based analysis method is employed.

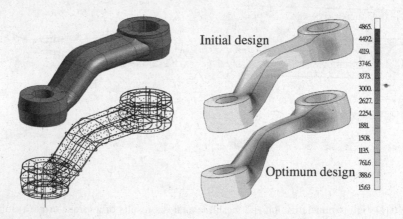

Figure 7. Meshfree discretization and analysis results of a road arm

5. Summary and Conclusions

Design sensitivity analysis (DSA) and optimization based on meshfree method have been proposed. Unlike finite element and boundary element methods, in meshfree approach the shape function of the meshfree approximation depends on shape design parameterization, and this effect has been discussed in detail. DSA based on stabilized conforming nodal integration completely removes background mesh, and the integration of the shape DSA and optimization capability has been effectively carried out. It has also been shown that shape design optimization of structures undergoing large shape changes can be effectively carried out using meshfree methods without re-meshing. Fast convergence of the design optimization algorithm has been accomplished using the accurate sensitivity information.

References

1. R. A. Gingold and J. J. Monaghan, *Monthly Notices Royal Astro. Soc.* **181,** 375-389 (1977).
2. B. Nayroles, G. Touzot, and P. Villon, *Comp. Mech.* **10** (1992).
3. T. Belytschko, Y. Y. Lu, and L. Gu, *Int. J. Numer. Methods Eng.* **37,** 229-256 (1994).
4. W. K. Liu, S. Jun, and Y. F. Zhang, *Int. J. Numer. Methods Fluids* **20,** 1081-1106 (1995).
5. J. S. Chen, C. Pan, C. T. Wu, and W. K. Liu, *Comp. Methods Appl. Mech. Eng.* **139,** 195-227 (1996).
6. C. A. M. Duarte and J. T. Oden, *Comp. Methods Appl. Mech. Eng.* **139,** 237-262 (1996).
7. J. M. Melenk and I. Babuska, *Comp. Methods Appl. Mech. Eng.* **139,** 289-314 (1996).
8. N. Sukumar, B. Moran, and T. Belytschko, *Int. J. Numer. Methods Eng.* **43,** 839-887 (1998).
9. S. N. Atluri and T. Zhu, *Comp. Mech.* **25,** 169-179 (2000).
10. S. De and K. J. Bathe, *Comp. Mech.* **25,** 329-345 (2000).
11. T. Belytschko, Y. Y. Lu, and L. Gu, *Eng. Fracture Mech.* **51,** 295-315 (1995).
12. C. A. M. Duarte and J. T. Oden, *Numer. Methods Partial Diff. Equa.* **12,** 673-705 (1996).
13. I. Babuska and J. M. Melenk, *Int. J. Numer. Methods Eng.* **40,** 727-758 (1997).
14. Y. You, J. S. Chen, and H. Lu, *Comp. Mech.* **31,** 316-326 (2003).
15. J. S. Chen, C. Pan, C. M. O. Roque, and H. P. Wang, *Comp. Mech.* **22,** 289-307 (1998).
16. I. Grindeanu, K. K. Choi, and J. S. Chen, *AIAA J.* **37,** 990-1016 (1999).
17. N. H. Kim, K. K. Choi, and J. S. Chen, *AIAA J.* **38,** 1742-1753 (2000).
18. M. E. Botkin, *AIAA J.* **20,** 268-273 (1982).
19. T. M. Yao and K. K. Choi, *ASME J. Struct. Eng.* **115,** 2401-2405 (1989).
20. P. Lancaster and K. Salkauskas, *Math. Comp.* **37,** 141-158 (1981).
21. R. L. Hardy, *J. Geophysics Res.* **176,** 1905-1915 (1971).
22. J. Dolbow and T. Belytschko, *Comp. Mech.* **23,** 219-230 (1999).
23. J. S. Chen and H. P. Wang, *Comp. Methods Appl. Mech. Eng.* **187,** 441-468 (2000).
24. J. S. Chen, W. Han, Y. You, and X. Meng, *Int. J. Numer. Methods Eng.* **56,** 935-960 (2003).
25. J. S. Chen, C. T. Wu, S. Yoon, and Y. You, *Int. J. Numer. Methods Eng.* **50,** 435-466 (2001).
26. S. S. Bhavikatti and C. V. Ramakrishnan, *Comp. Struct.* **11,** 397-401 (1980).

27. E. S. Kristensen and N. F. Madsen, *Int. J. Numer. Methods Eng.* **10,** 1007-1019 (1976).
28. P. Pedersen and C. L. Laursen, *J. Struct. Mech.* **10,** 375-391 (1982-83).
29. R. J. Yang and K. K. Choi, *J. Struct. Mech.* **13,** 223-239 (1985).
30. M. L. Luchi, A. Poggialini, and F. Persiani, *Comp. Struct.* **11,** 629-637 (1980).
31. M. Weck and P. Steinke, *J. Struct. Mech.* **11,** 433-449 (1983-4).
32. T. M. Yao and K. K. Choi, *Int. J. Numer. Methods Eng.* **28,** 369-384 (1989).
33. K. K. Choi and N. H. Kim, *Structural Sensitivity Analysis and Optimization 1: Linear Systems* (Springer, New York, 2004).
34. E. J. Haug, K. K. Choi, and V. Komkov, *Design Sensitivity Analysis of Structural Systems* (Academic Press, London, 1986).
35. K. K. Choi and H. G. Seong, *Comp. Methods Appl. Mech. Eng.* **57,** 1-15 (1986).
36. K. H. Chang, K. K. Choi, C. S. Tsai, B. S. Choi, and X. M. Yu, *Comp. Sys. Eng.* **6,** 151-175 (1995).
37. N. H. Kim, K. K. Choi, and M. Botkin, *Struct. Multidiscipl. Optim.* **24,** 418-429 (2003).
38. F. Bobaru and S. Mukherjee, *Comp. Methods Appl. Mech. Eng.* **190,** 4319-4337 (2001).
39. G. N. Vanderplaats, *Numerical Optimization Techniques for Engineering Design with Applications* (Vanderplaats Research & Development Inc., Colorado Springs, CO, 1999).
40. J. A. Bennett and M. E. Botkin, *AIAA J.* **23,** 458-464 (1985).

CHAPTER 15

SENSITIVITY-FREE FORMULATIONS FOR STRUCTURAL AND MECHANICAL SYSTEM OPTIMIZATION

Jasbir S. Arora and Qian Wang

Department of Civil and Environmental Engineering
Department of Mechanical and Industrial Engineering
Center for Computer Aided Design
The University of Iowa
Iowa City, Iowa, U.S.A.
E-mail: Jasbir-Arora@uiowa.edu

Sensitivity-free formulations do not require design sensitivity analysis of problem functions during optimization iterations. These formulations include some of the state variables of the problem as optimization variables in addition to the real design variables. This gives explicit dependence of the problem functions on the optimization variables. Therefore gradients of the functions with respect to the optimization variables can be calculated easily. Sensitivity-free formulations include the simultaneous analysis and design (SAND) approaches, mathematical programs with equilibrium constraints (MPEC), and partial differential equations (PDE)-constrained optimization problems. In addition to the sensitivity-free formulations, the conventional formulations for optimization of structural and mechanical systems are described. Advantages and disadvantages of the formulations are discussed. Some recent evaluations of the formulations are also described.

1. Introduction

Formulations for optimization of structural and mechanical systems can be divided into three broad categories. The first one, called the *conventional formulation* or the *nested analysis and design* (NAND) approach, treats only the real design variables as the optimization variables. The problem functions in these formulations are implicit functions of the design variables. Gradient evaluation of the functions can be tedious and difficult to implement. The second category is known as the *simultaneous analysis and design* (SAND) formulations. In these formulations, some of the state variables, such as the nodal

415

displacements, are also treated as optimization variables in addition to the real design variables. With these additional variables, the problem functions become explicit in terms of the optimization variables, making it easier to evaluate and implement the gradients of the functions. We shall refer to these as the *sensitivity-free* formulations. The third category of formulations is known as the displacement-based two-phase approach where the displacements are treated as optimization variables in the outer loop and the real design variables as the optimization variables in the inner loop. This chapter describes the SAND-type formulations along with their advantages and disadvantages. Some recent evaluations of these formulations are also discussed. The material for the chapter is taken from several recent papers by the authors.[1-5]

1.1. *Overview of Alternative Formulations*

Several alternative formulations for the structural optimization problem have been discussed based on different independent variables, analysis methods and forms of the resulting constraints.[6] These include design variable space (conventional), SAND, optimality criteria (OC), and some simplified SAND formulations. A more recent review by Arora and Wang[1] covers various SAND formulations for sizing, shape and topology optimization as well as displacement-based formulations, mathematical programming with equilibrium constraints, partial-differential-equations-constrained (PDE-constrained) optimization problems, optimal control, and multidisciplinary design optimization.

The SAND formulation is a major class of alternative formulations that has been discussed in the structural optimization literature. Some of the earliest attempts to include state variables in the structural optimization problem were by Schmit and Fox,[7,8] and Fuchs.[9,10] A SAND formulation based on an element-by-element preconditioned conjugate gradient technique was developed by Haftka,[11] and Haftka and Kamat.[12] Shin *et al.*[13] considered the SAND approach to solve the problem with eigenvalue constraints. Ringertz[14-16] presented SAND methods for the optimal design of nonlinear structures. The SAND formulation usually has large numbers of optimization variables and constraints. However, the matrix sparsity in the constraint Jacobian can be exploited for numerical efficiency.[4,16,17] Orozco and Ghattas[18] developed a reduced SQP method using SAND ideas to optimize geometrically nonlinear truss structures.

The SAND formulations have also been successfully applied to the configuration and topology design of structures.[19] It is well-known that a crucial step for success of the SAND formulations is the solution of very large scale optimization problems. Therefore considerable focus has been put on the development of new algorithms to solve large-scale optimization problems.[16,18-24] The SAND-type optimization formulations have also been discussed in other fields. A class of formulations known as *mathematical programs with equilibrium constraints* (MPEC), has been developed and studied. The word "equilibrium" in MPEC refers to the variational equalities or inequalities that model the equilibrium phenomenon in engineering and other applications. Another class of formulations that has been presented and analyzed recently is known as the *partial differential equations* (PDE)-constrained optimization. In these formulations, the equilibrium equations are expressed in a continuum form, the PDEs. SAND-type approaches have also been used to solve dynamic response optimization and optimal control problems.

A *displacement-based two-phase approach* was introduced by McKeown for optimal design of multilaminar, fiber-reinforced continua.[25] The procedure divided the problem into two subproblems, the inner and outer subproblems. In the inner subproblem, the cost was minimized subject to satisfaction of the equilibrium equations. The displacement field was specified and the design variables were the optimization variables. In the outer subproblem, the displacements were treated as optimization variables to minimize the cost function subject to the stress and displacement constraints. The two-phase algorithm was analyzed and applied to optimize trusses.[26] The outer subproblem was solved using either sequential linear programming (SLP) or another nonlinear programming (NLP) algorithm. McKeown expanded the two-phase optimization procedure to geometry and layout design of trusses.[27] Instead of using a complex ground structure, the author considered growing least-volume trusses, starting from the simplest possible layout. Wang *et al.* also presented a two-stage linear programming (LP) procedure for the minimum weight design of trusses.[28] An optimality criterion method (the maximum total strain energy criterion, as an LP) was used in the first stage to determine the joint displacements. Striz and Sobieszczanski-Sobieski proposed a displacement-based multilevel approach for structural optimization.[29] Both the subsystems level optimizations and the system level FE analysis and optimization could be performed in parallel.[30,31] In a paper by Missoum and Gürdal, the two-phase optimization procedure was applied to optimum design of static and dynamic trusses.[32] LP and SLP algorithms were used to solve the inner and outer subproblems, respectively. Since the weight was an implicit function of the displacements, a procedure was presented to calculate derivatives of the weight function with respect to the displacements. The approach was extended to the design of trusses with nonlinear material behavior,[33] and both geometric and material nonlinearities.[34] Path-independent material models were used. Slack variables were added to the equilibrium equation to define a relaxed problem that guarantees feasible solutions for the inner problem. It was shown that the displacement-based approach was quite efficient compared to the conventional NAND approach. Missoum *et al.* also extended their work to the optimization of geometrically nonlinear frames.[35] However, the displacement-based approach encountered some convergence difficulties for that problem. This approach is not discussed further in this chapter; for more details, papers by Arora and Wang[1] and others can be consulted.

1.2. *Overview of the Chapter*

The main purpose of this chapter is to describe various sensitivity-free formulations for optimization of structural and mechanical systems. The conventional NAND formulation is also described in order to compare and contrast it with the sensitivity-free formulations. Section 2 describes the conventional NAND formulation and Section 3 covers SAND formulations. Literature on linear and nonlinear problems is covered, and optimization algorithms that have been used for SAND formulations are discussed. Section 4 presents a comparative evaluation of the conventional and SAND formulations. Section 5 covers the literature on configuration and topology optimization of structures. Section 6 describes the PDE-constrained optimization formulation where the equilibrium equations are kept in the continuum form. Section 7 covers the formulation of MPEC. Section 8 covers the literature on optimal control problems, and Section 9 covers the literature on multidisciplinary design optimization. Section 10 presents some recent evaluations of the alternative formulations for truss and transient dynamic response optimization problems. Finally, some concluding remarks are given in Section 11.

2. Conventional Formulation

We start with a brief description of the conventional formulation for optimization of structural and mechanical systems. This is the most commonly used approach that has been developed for various types of problems, such as linear, nonlinear, transient dynamics, optimal control, etc. In the formulations, only the real design variables for the problem are treated as optimization variables. All other response quantities, such as displacements, stresses, strains and internal forces are treated as implicit functions of the design variables. Since most of the constraint functions depend implicitly on the design variables, their gradient evaluation requires the so-called design sensitivity analysis. This is described later in this section. Therefore these are not classified as sensitivity-free formulations. To keep presentation of the ideas clear and straightforward, only linear analysis problem (small displacements and linearly elastic material model) are considered in the discretized form. The approach can also be described for nonlinear analysis, using a continuum form of the analysis equations that is more general because it is not tied to any particular discretization for numerical calculations.[36,37]

2.1. *Formulation*

To describe the conventional formulation, let us define the following notation:

$\quad$ **x** = a k-dimensional vector of design variables that describes design of the system.

$\quad$ **z** = an n-dimensional vector of generalized displacements for a discretized model of the system.

For linear small displacement analysis, the governing equilibrium equation for the system is discretized using the finite element procedure as follows:

$$\mathbf{K}(\mathbf{x})\mathbf{z} = \mathbf{F}(\mathbf{x}) \tag{1}$$

where

$\quad$ $\mathbf{K}(\mathbf{x})$ = is a nonsingular $n \times n$ stiffness matrix that depends on the design of the system

$\quad$ $\mathbf{F}(\mathbf{x})$ = an n dimensional vector of equivalent external loads applied at the nodes of the discretized model for the system.

For a given design **x** and boundary conditions, Eq. (1) is assembled using contributions from each finite element, and solved for the state variable vector **z**. Using the vector **z**, strains and stresses at all points of the structure can be evaluated. Equation (1) has been implemented into many computer programs to analyze various structural systems. These programs are now widely used in practice. It is important to note that when the system is nonlinear (large displacements, elastoplastic material), Eq. (1) becomes nonlinear because $\mathbf{K}(\mathbf{x})$ and $\mathbf{F}(\mathbf{x})$ depend on the state variables **z** for the system. This complicates the solution process for Eq. (1) because it requires incremental and iterative procedures, such as the Newton-Raphson approach.

The optimal design problem is defined as follows:

Find the design variable vector **x** to minimize a cost function,

$$f = f(\mathbf{x}, \mathbf{z}) \tag{2}$$

subject to the inequality constraints

$$g(x, z) \leq 0 \tag{3}$$

Equality constraints, if present in the formulation, can be treated quite routinely. Note that in the above formulation, variables x and z are not independent, they are related by the equilibrium condition in Eq. (1). Only the design variables x are treated as the independent optimization variables, while the state variables z as the dependent variables. This is the central idea of the conventional formulation; i.e., to treat x as the only optimization variable and treat z as an implicit function of x, $z = z(x)$. Therefore, it is natural to set up a nested analysis and optimization process where (i) given x, z is calculated using an analysis code, and then (ii) given z, x is updated using an optimization algorithm. In the MPEC literature, this procedure is called the *implicit programming* approach. Since the displacement based finite element analysis programs are readily available, the conventional formulation has been the usual approach to solve optimization problems. Another reason for the popularity of the conventional formulation is that the widely-used approximate resizing rules can be obtained based on the optimality criteria.[11] Such rules can be quite efficient in terms of calculations as well as easy to implement though they are usually not robust. Other analysis methods are also available for the conventional formulation, such as the force method,[38] the boundary element method or the meshfree method.[39] Equation (1) therefore needs to be consistent with the corresponding analysis method. Structural analysis techniques based on conjugate gradient minimization of the energy functional have also been used for design optimization.[40]

2.2. *Gradient Evaluation*

Numerical values for z can be obtained from the state equation (1) once x is specified. However, an explicit functional form for z in terms of x cannot be obtained. In other words, z cannot be eliminated from the optimization problem by substitution. In the gradient-based optimization process, derivatives of the cost function $f(x, z)$ and the constraint functions $g(x, z)$ with respect to x are needed. The explicit expressions for these derivatives in terms of x cannot be obtained, since z is an implicit function of x. Therefore, usually the finite difference methods have been used to calculate the gradients since they are easy to implement and explicit expressions for the cost and constraint functions are not needed. However, the finite difference methods have accuracy problems.[41] Another drawback is that they are slow because many solutions of the state equation (1) are required.

To derive analytical expressions for gradients of the functions, implicit differentiation procedures need to be used, which is called *design sensitivity analysis*. The formulations that do not require this sensitivity analysis are called *sensitivity-free formulations*. To explain the design sensitivity analysis, calculation of the derivatives of one of the functions, say $f(\mathbf{x}, \mathbf{z})$, is briefly described. Other functions can be treated similarly. Taking total derivative of $f(\mathbf{x}, \mathbf{z})$ with respect to $\mathbf{x}$, we get

$$\left.\frac{df(\mathbf{x}, \mathbf{z}(\mathbf{x}))}{d\mathbf{x}}\right|_{k \times 1} = \left.\frac{\partial f(\mathbf{x}, \mathbf{z})}{\partial \mathbf{x}}\right|_{k \times 1} + \left.\frac{d\mathbf{z}(\mathbf{x})}{d\mathbf{x}}\right|_{k \times n} \left.\frac{\partial f(\mathbf{x}, \mathbf{z})}{\partial \mathbf{z}}\right|_{n \times 1} \tag{4}$$

Calculation of the partial derivatives of $f(\mathbf{x}, \mathbf{z})$ with respect to $\mathbf{x}$ and $\mathbf{z}$ presents no particular difficulty because explicit dependence of the function on $\mathbf{x}$ and $\mathbf{z}$ is known. However, calculation of $d\mathbf{z}/d\mathbf{x}$ in Eq. (4) needs further analysis and explanation. To calculate this $k \times n$ matrix, we take a total derivative of the state equation (1) with respect to the design variables $\mathbf{x}$ and rearrange the resulting equation to obtain:

$$\mathbf{KZ} = \mathbf{S} \tag{5}$$

where

$$\left.\mathbf{Z}\right|_{n \times k} = \frac{d\mathbf{z}(\mathbf{x})^T}{d\mathbf{x}} \; ; \quad \left.\mathbf{S}\right|_{n \times k} = \frac{\partial}{\partial \mathbf{x}}\left(\mathbf{F}(\mathbf{x}) - \mathbf{K}(\mathbf{x})\mathbf{z}\right)^T \tag{6}$$

Equation (5) looks deceptively simple and similar to the state equation (1). However, its solution variable $\mathbf{Z}$ is not a vector but a matrix of dimension $n \times k$. The right side $\mathbf{S}$ is also a matrix of the same dimension. Once the right side has been calculated, Eq. (5) can be solved using the same process that was used for solving Eq. (1). The decomposed matrix $\mathbf{K}$ needs to be saved for re-use with Eq. (5), requiring certain amount of data manipulation and storage. If iterative methods are used to solve state equation (1), then the decomposed $\mathbf{K}$ is not available to solve Eq. (5). Therefore the sensitivity equation (5) must also be solved using the iterative procedure which is more time consuming than the foregoing procedure where the decomposed $\mathbf{K}$ is available.

Calculation of the matrix $\mathbf{S}$ in Eq. (6) requires partial differentiation of the equilibrium equation for each finite element with respect to the design variables $\mathbf{x}$ and then the assembly process to form matrix $\mathbf{S}$. This process requires additional programming to extend an existing analysis code to implement the design sensitivity analysis capability. In addition, if new finite elements are added or the current ones are updated, the code for the design sensitivity analysis needs to be updated accordingly. Further, implementation of design sensitivity analysis for nonlinear and multi-physics problems becomes more complex and computationally more expensive because $\mathbf{K}$ and $\mathbf{F}$ depend on the state of the system as well. This is one of the stumbling blocks for engineering applications of optimization.

The above procedure for design sensitivity analysis is called the *direct differentiation method*. There is an alternate approach of design sensitivity analysis called the *adjoint variable method*. To derive that method, Eq. (5) is substituted into Eq. (4) as $\mathbf{Z} = \mathbf{K}^{-1}\mathbf{S}$, and an adjoint problem is defined with adjoint load as $\partial f / \partial \mathbf{z}$. The adjoint displacement vector is substituted into Eq. (4) to obtain an expression for the design gradient. Under certain circumstances, this method is more efficient than the direct differentiation method. However, the method is even more difficult to implement into existing analysis codes, especially for nonlinear and transient dynamic problems. Substantial literature is available that describes theoretical as well as implementation aspects of the design sensitivity analysis approaches.[36,37]

Conventional optimization formulations and solution methods for structural and mechanical systems can be difficult to use for design of practical structural and mechanical systems due to the following two main reasons: (i) Many practical applications are complex requiring interaction between several disciplines; i.e., require the use of different analysis software that are discipline-specific. Since they are independent programs, it is difficult to integrate them into the conventional design optimization formulations and algorithms. (ii) The conventional formulation requires design sensitivity analysis which is difficult to implement and maintain with existing analysis software.

To alleviate some of the difficulties noted above, several different research avenues have been explored in the literature. First, efficient structural reanalysis methods for analyzing a modified structure have been developed.[42-45] These methods can be useful for efficient analysis of updated designs and for calculation of the design derivatives during the optimization process. Second, various methods to develop approximate models, the so-called *meta-models*, such as the response surface approximations, have been proposed and evaluated for optimization of complex structural and mechanical systems.[46,47] Third, some alternative formulations have also been proposed and evaluated for optimization of structural and mechanical systems since 1960s. These formulations are discussed next.

3. Sensitivity-Free (SF) Formulations

The sensitivity-free (SF) formulations do not require design sensitivity analysis procedures described in the previous section to evaluate gradients of the problem functions with respect to the optimization variables. In other words, such formulations do not have any functions that depend implicitly on the optimization variables. There are optimization methods that do not use gradients of functions to solve the optimization problem. These are also sensitivity-free methods that are discussed elsewhere in this book. Here, we discuss sensitivity-free formulations that are suitable for gradient-based optimization methods. In the literature, the sensitivity-free formulations have been called *simultaneous analysis and design* approaches, of in short SAND approaches. Such approaches basically formulate the optimization problem in a space of mixed design and state variables. The structural analysis equations get embedded as equality constraints in one single optimization problem. Therefore explicit structural analysis and design sensitivity analysis are not needed. Note that there are many interesting formulations that can be derived using the SAND concept. Many of these formulations are described later in this chapter. In this section, the most common approach is described to discuss the basic concepts of sensitivity-free formulations.

3.1. *Formulation*

In this approach, the formulation of the problem is modified by treating the state and design variables z and x as independent optimization variables. To describe the approach, let us define a composite vector of optimization variables as

$$\mathbf{X} = \begin{bmatrix} \mathbf{x} \\ \mathbf{z} \end{bmatrix} \tag{7}$$

Note that if the structure is subjected to multiple loading conditions, the vector $\mathbf{X}$ will include multiple z vectors, one for each loading condition. In terms of the vector $\mathbf{X}$, the optimization problem is now defined as follows:

Find $\mathbf{X}$ to minimize a cost function

$$f = f(\mathbf{X}) \tag{8}$$

subject to the constraints

$$\mathbf{h}(\mathbf{X}) = \mathbf{K}(\mathbf{b})\mathbf{z} - \mathbf{F}(\mathbf{b}) = 0 \tag{9}$$

$$\mathbf{g}(\mathbf{X}) \leq 0 \tag{10}$$

Equation (9) is the equilibrium condition for the structure that is treated as an equality constraint.

3.2. *Gradient Evaluation*

For the alternate formulation in Eqs. (8) to (10) all the problem functions are explicit in terms of the variable $\mathbf{X}$. In the optimization process, partial derivatives of the functions with respect to $\mathbf{X}$ are needed; i.e., with respect to $\mathbf{x}$ and $\mathbf{z}$. Partial derivatives of f and $\mathbf{g}$ with respect to $\mathbf{x}$ and $\mathbf{z}$ can be easily calculated as noted before. Partial derivative of $\mathbf{h}$ with respect to $\mathbf{z}$ gives the stiffness matrix $\mathbf{K}$ and the partial derivative of $\mathbf{h}$ with respect to $\mathbf{x}$ gives the matrix $\mathbf{S}$ defined in Eq. (6). However, $d\mathbf{z}/d\mathbf{x}$ is not needed and no system of equations needs to be solved in the numerical solution process.

It is interesting to note that the sensitivity-free formulation does not require $\mathbf{h}(\mathbf{X})=\mathbf{0}$ be satisfied exactly at each iteration of the optimization process, i.e., the equilibrium equation need not be satisfied at every iteration, which can be advantageous for nonlinear problems. It needs to be satisfied only at the final solution point. This actually implies that the equation $\mathbf{h}(\mathbf{X})=\mathbf{0}$ never needs to be solved for $\mathbf{z}$ because $\mathbf{z}$ is treated as an independent optimization variable. The element level equilibrium equations can be used in the solution process. Thus the SF formulation is ideally suited for implementation on a parallel computer where each finite element can be assigned to one processor. All processors can be used to generate the element level quantities and thus speed-up the optimization process considerably.[11,12]

It is noted that the equilibrium equation (9) may not be the displacement based FEM equation, even though it is the most commonly used approach. Most work in the literature has used displacements as optimization variables. However, the force method or the mixed method of analysis can also be combined with the SF formulations.[6,48,49] The basic idea is to treat the redundant forces as optimization variables in addition to the design variables, and the compatibility conditions are treated as equality constraints. The SF formulation can also be combined with the optimality criteria methods.[6] A more recent analysis model - cellular automata (CA) has also been imbedded into SF formulations as equality constraints.[50] Besides displacements, other state quantities, such as forces and stresses can also be used as optimization variables.[2,6,9,51-60]

3.3. *Literature for Linear Problems*

Saka[61,62] presented SF formulations for the optimum shape design of trusses, and minimum weight design of rigid frames. Fuchs[9,10] presented an explicit optimum design method for linear elastic trusses. Three techniques were presented, according to the three classical analysis methods – force, displacement and hybrid (or mixed) methods. Haftka[11] presented an element-by-element formulation and showed that substantial computational savings of the SF formulations compared to the conventional nested approach. Ringertz[63] presented a branch and bound algorithm for topology design of truss structures, based on a ground structure approach. Ben-Tal and Bendsøe[64] proposed two alternative approaches for topology design of trusses for maximum stiffness with a prescribed volume. The optimization problem could be solved by an SF approach. Alternatively, this large, nonconvex formulation was transformed to an equivalent, unconstrained and convex problem in terms of nodal displacements only. This new formulation was mathematically equivalent to the original problem, and solved by a nonsmooth, steepest descent algorithm. Topology optimization of trusses for direct minimum weight design using the SF formulation was presented by Sankaranarayanan *et al.*[65] The SF formulation was compared with the minimum compliance formulation. It was also concluded that the minimum compliance method might not get the true optimal design.

3.4. *Literature for Nonlinear Problems*

In the conventional formulation, nonlinear analysis equations must be solved for any design update. However, the SF approach does not require repeated solution of the nonlinear analysis equations, since they need to be satisfied only at the optimal solution. The equilibrium equations in Eq. (9) become nonlinear that are treated as equality constraints. This is an additional advantage of the SF formulation for nonlinear problems. For such problems, Eq. (9) is written as

$$\mathbf{h}(\mathbf{X}) = \mathbf{P}(\mathbf{x}, \mathbf{z}) - \mathbf{F}(\mathbf{x}) = \mathbf{0} \tag{11}$$

where $\mathbf{P}(\mathbf{x},\mathbf{z})$ is the internal force vector. The evaluation of $\mathbf{h}$ in Eq. (11) is quite straightforward, as no matrix decomposition is needed. Assuming $\mathbf{F}$ is not a function of $\mathbf{z}$, the derivatives of Eq. (11) with respect to $\mathbf{x}$ and $\mathbf{z}$ are given as

$$\frac{\partial \mathbf{h}(\mathbf{X})}{\partial \mathbf{x}} = \frac{\partial \mathbf{P}(\mathbf{x}, \mathbf{z})}{\partial \mathbf{x}} - \frac{\partial \mathbf{F}(\mathbf{x})}{\partial \mathbf{x}} \tag{12}$$

$$\frac{\partial \mathbf{h}(\mathbf{X})}{\partial \mathbf{z}} = \frac{\partial \mathbf{P}(\mathbf{x}, \mathbf{z})}{\partial \mathbf{z}} = \mathbf{K}_T(\mathbf{x}, \mathbf{z}) \tag{13}$$

where $\mathbf{K}_T(\mathbf{x}, \mathbf{z})$ is the tangent stiffness matrix, and $\partial \mathbf{P}(\mathbf{x}, \mathbf{z})/\partial \mathbf{x}$ in Eq. (12) can be calculated in an element-by-element manner. If the equilibrium equation (11) is derived from the minimum potential energy, the tangent stiffness matrix in Eq. (13) can also be obtained as Hessian of the strain energy U as $\nabla^T \nabla U(\mathbf{X})$.[15,16]

Schmit and Fox[7,8] included state variables in the sizing and shape optimization problem of materially nonlinear trusses that was called "an integrated approach to structural synthesis". Smaoui and Schmit[66] presented an integrated approach to the minimum weight design of geometrically nonlinear static truss structures with geometric imperfections. The SF and nested approaches were compared for three geometrically nonlinear truss optimization problems by Haftka and Kamat.[12] It was concluded that the SF approach was competitive with the conventional nested approach, and that it was more efficient for large-scale problems. Ringertz[14] formulated the minimum weight design of truss structures with geometrically nonlinear behavior. In later papers, Ringertz[15,16] presented optimal design of geometrically nonlinear shell structures. The SF formulation was large but sparse. To make it tractable, sparse matrix approaches must be used.[16,17] A reduced SQP method was presented to solve geometrically nonlinear trusses by Orozco and Ghattas.[18] It was desired to utilize the structure of the problem functions so that the existing finite element analysis programs might be utilized. It was concluded that the reduced SF formulation required fewer structural analyses but the same amount of storage as NAND. Tin-Loi[58] discussed optimum shakedown design of discretized elastoplastic structures subjected to variable repeated loads and residual displacement constraints, which was formulated according to the classical lower bound theorem of shakedown.

3.5. *Optimization Techniques for SF Formulations*

The SF formulations have been solved successfully by various methods in the literature. New solution techniques have been developed in recent years. SUMT based on the penalty function techniques were used by Schmit and Fox,[7] Fuchs,[10] Haftka,[11] Haftka and Kamat,[12] and Ringertz.[15] Augmented Lagrangian methods were used by Sankaranarayanan et al.,[65] and Larsson and Rönnqvist.[67] Saka[61,62] and Achtziger[54,55] used the sequential linear programming (SLP) approach. A generalized reduced gradient (GRG) algorithm was used to solve the integrated problem by Smaoui and Schmit,[66] and Tin-Loi.[57,58,68] Haftka and Kamat,[12] and Orozco and Ghattas[17] used the projected Lagrangian algorithm. Various SQP methods were used by Ringertz,[14,16,63] Orozco and Ghattas,[18] Schulz and Bock,[69] Dreyer et al.,[70] Stolpe and Svanberg,[59] Schulz,[71] and Wang and Arora.[2] Ben-Tal and Roth,[19] Jarre et al.,[21] Maar and Schulz,[22] Herskovits et al.,[23] Hoppe et al.,[24] Herskovits,[72] and Hoppe and Petrova[73] used newly-developed interior point (IP) algorithms to solve SF formulations. Multigrid methods combined with SQP or interior point method have been successfully applied to SF formulations.[22,70] An

SF approach based on cellular automata (CA) and genetic algorithm (GA) for sizing and shape design of discrete structural systems was presented by Canyurt and Hajela.[50] Parallelization potential for the GA based SF approach was noted as a major advantage of the approach.[50]

4. Comparison of Conventional and SF Formulations

Table 1 lists the sizes of all the conventional and SF formulations. The following symbols are used: k = dimension of design variables vector $\mathbf{x}$; n = dimension of state variables (e.g., displacement) vector $\mathbf{z}$; m = number of inequality constraints in $\mathbf{g} \leq \mathbf{0}$ (e.g., stress constraints, excluding bounds on variables). Assume that there are $2k$ bound constraints on the design variable vector $\mathbf{x}$, and $2n$ bound constraints on the state variable vector $\mathbf{z}$. Table 2 lists comparison of the two formulations - the conventional and SF. Advantages and disadvantages of the formulation are discussed.

Table 1. Number of variables and constraints for the formulations

	Conventional Formulation	SF Formulation
No. of Variables	k	$k+n$
No. of Equality Constraints	0	n
No. of Inequality Constraints	$m+2n$	m
No. of Simple Bounds	$2k$	$2k+2n$

In SF formulations, displacements are chosen as optimization variables, and the analysis equations are treated as equality constraints. The inclusion of displacements as variables simplifies the constraint expressions and computer implementations. The reason is that they lead to a simpler form for the constraints which the optimization algorithm can treat more efficiently. Also the SF formulations avoid repeated analysis of the structure; therefore, they can be more efficient. This will also be the case for nonlinear structures where the conventional formulations need to solve the equilibrium equations at each iteration, which is expensive.

In the SF formulations, the optimization problem is very large because there are more variables in a single optimization process. It can easily exceed the capacity of current optimization codes and computers. In addition, variable and constraint scaling is needed in the SF formulations to reduce numerical ill-conditioning, since they include variables and constraints of different orders of magnitude. However, SF formulations simplify the forms of constraints and their

Jacobians, which are advantageous for numerical algorithms and implementations.

Table 2. Advantages and disadvantages of the two formulations

Formulation	Advantages	Disadvantages
Conventional	1. Least number of optimization variables. 2. Equilibrium equation is satisfied at each iteration. 3. Intermediate solutions may be usable.	1. Equilibrium equation must be solved at each iteration, which can be expensive. 2. Constraints are implicit functions of the variables; their evaluation requires analysis. 3. Design sensitivity analysis must be performed. 4. Implementation is tedious. 5. Dense Jacobian and Hessian matrices; difficult to treat large-scale problems
SF	1. Formulations are explicit in terms of variables. 2. Equilibrium equation is not solved at each iteration. 3. Many constraints become linear in variables. 4. Jacobians and Hessian are sparse. 5. Design sensitivity analysis is not needed. 6. Implementation is relatively straightforward. 7. Multi-physics problems are easier to optimize. 8. Lagrange multipliers for more constraints become available which may give further insights for practical applications.	1. Numbers of variables and constraints are large. 2. Intermediate solutions may not be usable. 3. Optimization algorithms for large-scale problems must be used, utilizing sparsity of the Jacobians and Hessians. 4. Optimization variables need to be normalized.

5. Configuration and Topology Design

Many formulations for configuration and topology design optimization have been presented in the literature. These include the ground structure approach for discrete element structures and a more general continuum topology optimization formulation. The literature on the subject of topology optimization is vast and many good references are available that describe various formulations and solution algorithms.[74-78] Here we focus on describing only some recent work related to the SF formulations and the corresponding computational algorithms for topology optimization. It turns out that the SF formulation is an important foundation for these design problems. A typical approach for topology optimization is to minimize the external work (compliance), where design variables together with the nodal displacements are the optimization variables.[79]

These problems are not convex when the equilibrium equations are included in the formulation. However, they may be reformulated as convex problems in different ways. Bendsøe *et al.*[80] and Bendsøe and Sigmund[78] reviewed different formulations for minimizing the compliance for the truss geometry and topology design. They noted that the compliance could be expressed in a number of equivalent potential or complementary energy formulations using the member forces, displacements and bar areas. Using the duality principles and nonsmooth analysis it was shown how displacement-only and stress-only formulations could be obtained. The equilibrium equations were part of these formulations even though they might be in the dual problem or other simplified forms. Based on duality principles, Muralidhar and Rao[52] also presented several new equivalent formulations for optimal truss topology design for limit states based on a unified elastic/plastic analysis. The strictly plastic and elastic limit design models were reduced to LP problems, and were shown to be equivalent to the widely studied model for minimum compliance topology design of elastic trusses. Topology optimization has also been re-formulated into some alternative formulations, such as semidefinate programming (SDP)[81-83] (also refer to Chapter 20) and linear programming (LP) problems.[52,84] Some detailed SF formulations and references to the convex re-formulations mentioned above can be found in the literature.[74,78]

The most common way to formulate a structural topology optimization is the *minimization of compliance*, defined as:

$$\tfrac{1}{2}\mathbf{F}^T\mathbf{z} \tag{14}$$

subjected to the state equations (1), and the constraints on the total volume and each element volume:

$$\sum \mathbf{v} = V_{total} \tag{15}$$

$$\mathbf{v}^L \le \mathbf{v} \le \mathbf{v}^U \tag{16}$$

where $\mathbf{F}$ and $\mathbf{z}$ are the same as defined in Eq. (1). $\mathbf{v}$ is the vector of element volumes and $\mathbf{v}^L$ and $\mathbf{v}^U$ are the corresponding lower and upper bounds. V_{total} is the total given volume of the structure. Although the problem of minimization of compliance can be solved by the SF approach,[64] direct minimization of the weight for truss topology design is also possible.[53-55,59,60,63,65] Oberndorfer *et al.*[85] discussed the advantages and disadvantages of these two formulations. They showed that for the condition where the allowable stresses for tension and compression members of trusses were identical, the two formulations became equivalent.

The beauty of the SF formulations for topology design is that both cross-sectional areas and displacements are treated as independent variables; therefore it is possible for member cross-sectional areas to reach zero value without causing singularity or non-differentiability. If the ground structure method is used in topology design, there are a very large number of cross-sectional areas and a relatively small number of displacement variables; therefore, the SF formulation has an advantage, since the size of the problem is not increased substantially. The use of various SF formulations for configuration and topology design can be found in References 19-22, 24, 51, 52, 54-56, 59-61, 63-65, 74, 78, 80, 82-84, and 86-88. Displacement-based two-phase approaches for configuration and topology design have also been used by McKeown,[26,27] Wang et al.,[28] Missoum and Gürdal,[32] and Gu et al.[33]

In recent years, there have been extensive developments of efficient numerical algorithms to solve large-scale structural or PDE-based topology optimization problems. These works include various interior point (IP) methods developed by Ben-Tal and Roth,[19] Jarre et al.,[21] Kočvara et al.,[82] Ben-Tal et al.,[83] Maar and Schulz,[22] Hoppe et al.,[24] and Hoppe and Petrova.[73] Dreyer et al.[70] and Schulz[71] studied efficient simultaneous solution based on SQP and reduced SQP. Multigrid solution techniques for topology optimization combined with IP and SQP were studied by Maar and Schulz,[22] and Dreyer et al.,[70] respectively. Some PDE-based models for topology optimization were studied.[24,71] The SF formulation is also a key component when formulating structural design problems with integer design variables. Some related works can be found in the literature.[60,86,89]

6. PDE-Constrained Optimization

Recently, a general class of formulations known as PDE-constrained optimization has been presented and discussed. In this formulation, the equilibrium equations are kept in the continuum form instead of the discretized form given in Eq. (1). Use of the continuum form offers flexibility in terms of the range of applications of optimization to many different fields including multidisciplinary applications. Also, many PDE solution algorithms and solvers, including the finite element method, can be used to perform optimization of complex systems.[24,71] The design or the control variables may also be described in the distributed parameter form and discretized for numerical calculations. In the PDE-constrained optimization literature, the term "decision variables" is used to represent design or control variables, or both of them. Problems of optimal design, optimal control, and parameter estimation of systems from many diverse application areas can be formulated in this way.

It is clear that the PDE-constrained optimization formulation is a generalization of the discretized optimization formulations discussed in the previous sections. Therefore it is important to note that the conventional NAND and the SF approaches discussed previously are applicable directly to the PDE-constrained optimization formulations. Thus all the advantages and disadvantages discussed previously for NAND and SF approaches apply to this formulation as well.

The *First Sandia Workshop on Large-Scale PDE-Constrained Optimization* was held in 2003 to focus on the issues related to this topic. The basic idea was to bring researchers in the fields of PDE and optimization together to foster greater synergy and collaboration between these communities. The proceeding of this workshop is an excellent source of references that describe the state-of-the-art on this subject.[90] The major topics discussed at the workshop included: large-scale computational fluid dynamics (CFD) applications, multifidelity models and inexactness of simulations, sensitivities for PDE-based optimization, NLP algorithms and inequality constraints, time-dependent problems, and software frameworks for PDE-constrained optimization. Several papers on these topics are included in the proceedings. Seven challenging issues needing further research and collaboration between the PDE and optimization communities were identified:

- Problem size in PDE-constrained optimization
- Integration of NLP and PDE-solvers
- Physics-based globalizations and inexact solution
- Approximate Jacobians
- Implicitly-defined and nonsmooth PDE residuals
- Treatment of inequalities
- Time-dependent problems

More details of these aspects can be found in Biegler *et al.*,[90] and Arora and Wang.[1]

7. Mathematical Programs with Equilibrium Constraints (MPEC)

Mathematical programming with equilibrium constraints (MPEC) is a general class of optimization problems in which some of the constraints are defined by a parametric variational inequality or the so-called complementarity system.[91] The variational equality or inequality constraints model the equilibrium requirements. The MPEC formulation is an extension of the so-called *bilevel programs*, also known as the *mathematical programs with optimization constraints*. The complementarity system of equations mentioned above is a result of the optimality conditions for the optimization constraints. It turns out that the SF formulations discussed earlier can be viewed as a special case of the MPEC.

The general MPEC is a nonconvex and nondifferentiable optimization problem which is computationally difficult to solve.[91] Various formulations of MPEC have been studied by Lou et al.,[91] Outrata et al.,[92] and others. Existence of optimal solutions has been discussed. Exact penalty functions for the complementarity system have been employed to obtain the first order optimality conditions for the MPEC. Examples of MPEC problems discussed in the literature can be found in Refs. 1, 91, and 93.

In engineering applications, the MPEC problems can be formulated in a continuum form where a variational principle governs the equilibrium state of the system, such as the principle of minimum potential energy or Hamilton's principle. An advantage of the continuum formulation is that the solution procedure is not tied to any particular numerical discretization approach. Thus it offers more flexibility for numerical solution of the problem. However, to keep the presentation of the basic ideas clearer, we stay with the discretized models of the system. To present and discuss an MPEC problem, consider an elastic body that comes into contact with a rigid smooth object. The problem is to design the body such that an objective function is minimized subject to equilibrium and other requirements, such as non-penetration of bodies, stress and displacement constraints. Using the notations defined earlier, the problem is defined as follows:

$$\underset{\mathbf{x}}{\text{minimize}} \; f(\mathbf{x}, \mathbf{z}) \tag{17}$$

subject to

$$\underset{\mathbf{z}}{\text{minimize}} \; V(\mathbf{x}, \mathbf{z}) \tag{18}$$

$$\mathbf{g}(\mathbf{x}, \mathbf{z}) \leq \mathbf{0} \tag{19}$$

$$\mathbf{c}(\mathbf{x}) \leq \mathbf{0} \tag{20}$$

In the outer problem, $f(\mathbf{x}, \mathbf{z})$ is the overall objective function to be minimized over the design variables $\mathbf{x}$. In the inner subproblem, the total potential energy function $V(\mathbf{x}, \mathbf{z})$ is minimized over the state variable $\mathbf{z}$. Some of the constraints in Eq. (19) may be imposed in the inner optimization subproblem while others may be imposed in the outer problem. For example, contact and non-penetration constraints may be imposed while solving the inner subproblem while the stress and displacement constraints may be imposed in the outer problem. Also some of these constraints may be equalities. The constraints in Eq. (20) that depend only on the design variables are imposed in the outer problem. The foregoing formulation is an instance of the bilevel optimization problems.

With the assumption of linearly elastic behavior under small displacements, the total potential energy function is given as

$$V(\mathbf{x}, \mathbf{z}) = \tfrac{1}{2} \mathbf{z}^T \mathbf{K}(\mathbf{x}) \mathbf{z} - \mathbf{z}^T \mathbf{F}(\mathbf{x}) \tag{21}$$

where $\mathbf{K}(\mathbf{x})$ is the structural stiffness matrix, and $\mathbf{F}(\mathbf{x})$ is the equivalent external force vector. Many times the constraints in Eq. (19) can be written as linear function of $\mathbf{z}$ as

$$\mathbf{g}(\mathbf{x}, \mathbf{z}) = \mathbf{A}(\mathbf{x}) \mathbf{z} - \mathbf{z}_0 \leq \mathbf{0} \tag{22}$$

where $\mathbf{A}(\mathbf{x})$ is a matrix of appropriate dimension and $\mathbf{z}_0$ is a specified vector. Now, writing the KKT optimality conditions for the inner subproblem, Eq. (18) and Eq. (19) in the formulation can be replaced by the following conditions:

$$\mathbf{K}(\mathbf{x}) \mathbf{z} - \mathbf{F}(\mathbf{x}) + \mathbf{A}(\mathbf{x})^T \mathbf{p} = \mathbf{0} \tag{23}$$

$$\mathbf{p} \geq \mathbf{0}; \quad \mathbf{g}(\mathbf{x}, \mathbf{z}) = \mathbf{A}(\mathbf{x}) \mathbf{z} - \mathbf{z}_0 \leq \mathbf{0}; \quad \mathbf{p}^T \mathbf{g}(\mathbf{x}, \mathbf{z}) = 0 \tag{24}$$

where $\mathbf{p}$ is the Lagrange multiplier vector for the constraints in Eq. (22). $\mathbf{p}$ is interpreted as the forces required to impose the constraints; e.g., if the constraints in Eq. (22) represent the non-penetration contact conditions then $\mathbf{p}$ represents the vector of contact forces. For the frictionless contact case, it represents the normal contact forces between the deformable body and the rigid object. The conditions in Eq. (24) represent the complementarity problem.

Hilding et al.[87] presented a detailed review of optimization of structures in unilateral mechanical contact. Emphasis was put on linear elastic structures in frictionless contact. They explained that, in general, structural optimization problems involving contact could not be treated by classical smooth optimization theory; instead, modern fields such as nonsmooth optimization and MPECs needed to be used. Various formulations and algorithms for contact analysis problems have also been studied by Mijar and Arora.[94,95] Variational equality and inequality formulations were studied for frictionless and frictional contact problems. Although the contact analysis problems can be formulated and solved using standard optimization algorithms, they can also be formulated as MPECs. Such formulations are nondifferentiable and generalized Newton method must be used to solve them.[95] Recently, Mijar and Arora[96,97] have also presented an augmented Lagrangian algorithm for frictional contact problems where the solution does not depend on the user-specified penalty parameter or the load step size.

MPEC formulations are suitable for inverse problems, such as system identifications. Ferris and Tin-Loi[98] and Pang and Tin-Loi[99] studied identification of the yield limits and hardening moduli from the knowledge of the displacement response of the structure. Tin-Loi and Que[100] presented MPEC approaches for indirect parameter identification of cohesive crack properties from a wedge splitting test. Tin-Loi[57] proposed an MPEC formulation with member areas, stresses, nodal displacements, and plastic multipliers as optimization variables for the minimum weight design of path-independent plastic trusses. Tin-Loi[68] also presented the numerical solution of a class of unilateral contact structural optimization problems. The limit analysis of frictional block assemblies was formulated as an MPEC by Ferris and Tin-Loi.[101] The computation of collapse loads of discrete rigid block systems, characterized by frictional and tensionless contact interfaces was formulated. Evgrafov and Patriksson[102] studied stochastic structural topology optimization based on discretization and penalty function approach. The resulting nonsmooth stochastic optimization problem was an instance of stochastic MPEC that was solved using some approximations.

Although many researchers have aimed to simply transform an MPEC into a standard NLP problem and solve it by various parametric, smoothing, relaxation or penalty methods, substantial attention has also been devoted to further understanding and development of theories and efficient algorithms to solve MPECs. A *penalty interior point algorithm* (PIPA), an *implicit programming algorithm* and a *piecewise* SQP were presented by Lou *et al.*[91] Patriksson and Wynter[103] studied stochastic MPECs. Some basic parallel iterative algorithms for discretely distributed stochastic MPEC were discussed. Scholtes and Stöhr[104] studied theoretical and computational aspects of an exact penalization approach to MPECs. A globally convergent trust region method was developed. Complementarity constraint qualifications and simplified B-stationarity conditions (Bouligand first-order optimality conditions) for MPECs were studied by Pang and Fukushima.[105] Andreani and Martinez[106] proved that stationary points of the sum of squares of the constraints were feasible points for the MPEC under reasonable sufficiency conditions. Wan[107] presented some further investigation on feasibility conditions of MPECs. It was demonstrated that these feasibility conditions were also sufficient for quadratic programming subproblems arising from the penalty interior point algorithm (PIPA) and the smooth SQP algorithm for solving MPECs. Birbil *et al.*[108] presented an entropic regularization approach for the MPECs. A three-dimensional null-space approach for the MPECs with steps related to nonlinear inequality constraints, the complementarity conditions and the objective function was proposed by Nie.[109]

8. Optimal Control

The SF-type approaches have also been used to solve open-loop optimal control problems for trajectory design in aerospace engineering,[110,111] robotics or human motion planning,[112,113] and chemical process engineering.[114,115] These problems involve the solution of differential algebraic equations (DAEs), or just differential equations (DEs). The standard optimal control problem is to find the control history $\mathbf{u}(t)$ that minimizes the performance functional in the time interval $[t_0, t_f]$, as:[116]

$$f = \phi(t_f, \mathbf{y}_f) + \int_{t_0}^{t_f} L(t, \mathbf{y}, \mathbf{u}) dt \tag{25}$$

subject to the system dynamics equations

$$\dot{\mathbf{y}} = \ell(t, \mathbf{y}, \mathbf{u}) \tag{26}$$

and the prescribed initial conditions

$$t_0 = t_{0_s}; \quad \mathbf{y}_0 = \mathbf{y}_{0_s} \tag{27}$$

and the prescribed final conditions

$$\varphi(t_f, \mathbf{y}_f) = \mathbf{0} \tag{28}$$

The basic idea of an SF-type approach is to discretize the system of first order differential equations (26), and define a finite dimensional approximations or parametric representation for the state and control variables. The discretized state equations are treated as equality constraints in the optimization process, converting the optimal control problem into an NLP problem. Several viable approaches are available. If the design variables together with the state variables and control variables are all treated as optimization variables, the approach is called the *direct collocation/transcription* method. If the control variables are eliminated from the system (i.e., only the design variables and the state variables are treated as optimization variables), it is called the *differential inclusion* method.[116] Another possibility is the so-called *multiple shooting* technique.[117,118]

Different discretization techniques for the state equations have been studied in the literature. In general there are two classes of methods to transfer the DAEs or DEs to an algebraic system of equations. One is to use some polynomial interpolation between the time grid points, and the other is to use a series expansion in terms of orthogonal polynomials, such as Legendre or Chebyshev polynomials.[119] For the former case, the most common ways are the trapezoidal or Simpson's quadrature schemes based on piecewise quadratic or cubic polynomials.[110,120] Explicit or implicit Runge-Kutta methods were used to discretize the state equations by Biehn *et al.*,[121] and Betts *et al.*[122] Higher degree polynomials for direct collocation were studied by Herman and Conway,[123] and Hu *et al.*[124] If very smooth trajectory is required, B-spline curves can also be used to parameterize the dynamic equations.[125] Seywald,[126] and Kumar and Seywald[127] discussed a technique to eliminate the controls while solving optimal control problems via direct methods. Conway and Larson[128] presented a comparison of collocation and differential inclusion methods in direct trajectory optimization.

Different optimization algorithms have been used for direct collocation or multiple shooting, among which the SQP and interior point algorithms are the popular choices. Hargraves and Paris,[120] Schulz and Bock,[69] Betts,[111] and Itle *et al.*[129] used SQP. The interior point algorithm was employed by Cervantes *et al.*[130] Since the resulting NLP is large and sparse, sparse NLP were extensively discussed.[131,132] Parallel computation was considered by Betts and Huffman.[117] A detailed survey of the numerical methods for simultaneous optimization and control can also be found in Betts.[133]

9. Multidisciplinary Design Optimization (MDO)

The SF formulation is also called the *infeasible path* (IP) approach for aerodynamic design that was pioneered by Rizk.[134] Later, more research was done for this problem.[135-138] Other applications of SAND formulation can be found in heat transfer; e.g., Hrymak *et al.*[139] presented optimization of extended heat transfer surfaces.

The SF formulation has also been demonstrated in many multidisciplinary design optimization (MDO) papers and has been called the *all-at-once* (AAO) formulation.[72,136,140-143] Haftka *et al.* discussed the interdisciplinary optimization of engineering systems from the standpoint of the computational alternatives available to the designers.[140] Optimization of the system could be formulated in several ways, i.e., NAND or SF formulations. Cramer *et al.*[141] presented three MDO formulations, namely multidisciplinary feasible (MDF), AAO, and individual discipline feasible (IDF) formulations. In AAO formulation, the optimization problems were very large and residuals were evaluated in all disciplines. No existing analysis codes were necessary. Though AAO was computationally least expensive, it required a higher degree of software integration. Balling and Wilkinson[143] studied available multidisciplinary design optimization approaches on common test problems. It turned out that the AAO formulation showed the most efficiency among all the approaches for the test problems. Detailed reviews of various MDO formulations can be found in the literature.[142]

10. Recent Evaluations of Sensitivity-Free Formulations

10.1. *Truss Optimization*

Wang and Arora[2,4] have implemented three sensitivity-free formulations with existing analysis software and evaluated them using several truss optimization problems. The cross-sectional areas were the optimization variables in all the formulations. In addition, the first formulation had nodal displacements as additional variables, the second formulation had member forces as variables as well, and the third formulation had member stresses as variables instead of the member forces. The three sensitivity-free formulations and the conventional formulation were analyzed and their advantages and disadvantages were discussed. Existing analysis software was integrated with an optimizer based on a sparse SQP method to solve several numerical examples with known solutions. With the sensitivity-free formulations, only the pre- and post-processing capabilities of the analysis software were used to evaluate the problem functions; however, the conventional formulation also needed the equation solving capabilities for analysis as well as design sensitivity analysis. Numerical performance of all the formulations was evaluated with extensive numerical experiments. The sensitivity-free formulations were compared with the conventional formulation. The three sensitivity-free formulations were also compared to each other. For some example problems, better solutions compared to the known solutions were obtained. Based on those studies, the following points are observed:

- The sensitivity-free formulations are more efficient than the conventional formulations in most cases. More efficient methods to solve large-sparse optimization problems need to be developed.
- In sensitivity-free formulations two and three, the global equilibrium equations in terms of displacements are not needed. Therefore, the global stiffness matrix for the structure need not be assembled.
- Since the variables in the sensitivity-free formulations have different orders of magnitude, scaling of the variables is needed. Appropriate automatic scaling procedures need to be developed.
- Sparsity of the problem functions must be utilized for efficiency and effectiveness of the sensitivity-free formulations.
- Implementation of the sensitivity-free formulations is simpler with the existing analysis software compared to the conventional formulation.

- Use of parallel computations with the sensitivity-free formulations needs to be developed.

10.2. *Framed Structures*

Similar to the evaluations for trusses of the previous section, Wang and Arora[5] have implemented two sensitivity-free formulations with existing analysis software and evaluated them using several frame optimization problems. A restricted form of the frame optimization problem was used where only the cross-sectional areas were treated as the design variables. The moment of inertia of the member was expressed in terms of the cross-sectional area. Also a very simple form of the combined stress constraint was imposed. For practical applications, these limitations need to be relaxed where the design manual constraints can be included in the solution process.

The first sensitivity-free formulation had nodal displacements as additional variables, and the second formulation had member forces as variables as well. The two sensitivity-free formulations and the conventional formulation were analyzed and their advantages and disadvantages were discussed. As for the truss problems, existing analysis software was integrated with an optimizer based on a sparse SQP method to solve several numerical examples with known solutions. For some examples, better solutions were obtained compared to the known solutions. The conclusions drawn from this study were quite similar to the ones for the trusses presented in the previous section.

10.3. *Transient Dynamic Response Optimization*

Three sensitivity-free formulations for transient dynamic response optimization of mechanical systems have been presented, analyzed and evaluated by Wang and Arora.[3] The basic idea of these formulations was to treat various state variables as independent variables in the optimization process (in addition to the real design variables); i.e., generalized displacements, velocities and accelerations. Since all functions of the optimization problem became explicit in terms of the variables in these formulations, their gradients could be calculated easily as compared to the conventional approach. Also, the equations of motion in the second order form were not integrated explicitly; they were directly imposed as equality constraints. The state variables were discretized using standard finite difference methods, making the numerical implementation quite easy and straightforward. For the sensitivity-free formulations, the optimization problem was quite large in terms of the numbers of variables and constraints.

However, the problem functions were quite sparse, which was exploited in the optimization process. Advantages and disadvantages of the formulations were discussed. Several cases of two example problems were solved to study the performance of the formulations. Based on the extensive numerical experiments, it was concluded that the proposed formulations worked well for the example problems, and have potential for further development for practical applications. Further investigation was suggested by studying and evaluating (i) different forms for approximation of velocities and acceleration in terms of the displacements, (ii) use of existing simulation software, and (iii) other optimization algorithms for large-scale optimization problems.

11. Concluding Remarks

Sensitivity-free formulations for optimization of structural and mechanical systems, including configuration and topology design, are described. The formulations are described in the context of continuous variables that use gradient-based optimization algorithms for their solution. Features of various formulations are discussed and their advantages and disadvantages are delineated. These include mathematical programs with equilibrium constraints (MPEC), and partial differential equations-constrained (PDE-constrained) formulations. If design variables and some state variables are combined together in a single and large optimization problem, then the sensitivity-free formulation is obtained.

MPEC is a more general formulation where the equilibrium constraints are defined by variational equalities or inequalities, such as the one for the contact analysis problem.[94,95] In addition, the formulation has complementarity system of equations that makes the problem nondifferentiable. MPEC can also be considered as a special case of the so-called bilevel optimization problems where some of the constraints involve optimization. It is noted that the equilibrium equation (1) is obtained as a necessary condition for minimization of the total potential energy of the structure. Thus the structural optimization problem can also be considered as a special case of the MPEC. This formulation has been studied recently and mathematical foundations for its solution have been presented. First and second order optimality conditions for the formulation have been developed and computational algorithms for its numerical solution have been presented and demonstrated. Some MPEC problems can be reformulated and solved by the standard NLP algorithms.

Another recent formulation is the PDE-constrained optimization of systems. This formulation is similar to the SAND approach except that the equilibrium equations are written as PDEs; i.e., in the distributed parameter form. The formulation offers more flexibility for numerical calculations because any discretization scheme can be used to solve the PDEs. In addition to the foregoing literature, sensitivity-free formulations for the optimal control problem and multidisciplinary optimization problems are briefly reviewed.

Major topics needing further investigation relative to the sensitivity-free formulations are as follows:

- Most of the formulations have focused on use of the displacement-based FEM. Other analysis methods need to be considered, such as the force methods, mixed methods, meshless methods, boundary element methods, and others.
- Implementation aspects with the existing analysis programs have not been adequately discussed; this important aspect needs to be addressed.
- Parallel processing of various steps of the solution process must be considered to solve very large-scale and multidisciplinary problems.
- Since various optimization variables can be of different orders of magnitude, methods for automatic transformation of the variables needs to be developed to improve rate of convergence of optimization algorithms.

Acknowledgments

Support for this work, provided by The University of Iowa under Carver Research Initiation Grant and the Virtual Soldier Research program, is gratefully acknowledged. Parts of this chapter are taken and edited from a recent paper by the authors "Review of formulations for structural and mechanical system optimization" published in *Structural and Multidisciplinary Optimization,* Volume **30**, pages 251-272, 2005.

References

1. J. S. Arora and Q. Wang, *Struct. Multidisc. Optim.*, **30**, 251 (2005).
2. Q. Wang and J. S. Arora, *AIAA J.*, **43**, 2202 (2005).
3. Q. Wang and J. S. Arora, *AIAA J.*, **43**, 2188 (2006).
4. Q. Wang and J. S. Arora, *Int. J. for Num. Meth. In Engrg.*, (2006 in press).
5. Q. Wang and J. S. Arora, *J. of Str. Engrg. ASCE*, (2006 in press).
6. U. Kirsch and G. I. N. Rozvany, *Struct. Multidisc. Optim.*, **7**, 32 (1994).
7. L. A. Schmit and R. L. Fox, *AIAA J.*, **3**, 1104 (1965).
8. R. L. Fox and L. A. Schmit, *J. Spacecraft & Rockets*, **3**, 858 (1966).
9. M. B. Fuchs, *Int. J. Solids Structures*, **18**, 13 (1982).
10. M. B. Fuchs, *Eng. Optim.*, **6**, 213 (1983).

11. R. T. Haftka, *AIAA J.*, **23**, 1099 (1985).
12. R. T. Haftka and M. P. Kamat, *Comp. Mech.*, **4**, 409 (1989).
13. Y. S. Shin, R. T. Haftka and R. H. Plaut, *AIAA J.*, **26**, 738 (1988).
14. U. T. Ringertz, *Eng. Optim.*, **14**, 179 (1989).
15. U. T. Ringertz, *Struct. Optim.*, **4**, 193 (1992).
16. U. T. Ringertz, *Int. J. Num. Meth. Eng.*, **38**, 299 (1995).
17. C. E. Orozco and O. N. Ghattas, *AIAA J.*, **30**, 1877 (1992).
18. C. E. Orozco and O. N. Ghattas, *Int. J. Num. Meth. Eng.*, **40**, 2759 (1997).
19. A. Ben-Tal and G. Roth, *Optim. Meth. & Soft.*, **6**, 283 (1996).
20. A. Ben-Tal and M. Zibulevsky, *SIAM J. Optim.*, **7**, 347 (1997).
21. F. Jarre, M. Kočvara and J. Zowe, *SIAM J. Optim.*, **8**, 1084 (1998).
22. B. Maar and V. Schulz, *Struct. Multidisc. Optim.*, **19**, 214 (2000).
23. J. Herskovits, P. Mappa and L. Juillen, in *Proc. WCSMO4, Fourth World Congress of Structural and Multidisciplinary Optimization*, (held in Dalian, China, 2001).
24. R. H. W. Hoppe, S. I. Petrova and V. Schulz, *Optimiz. Theory Appl.*, **114**, 545 (2002).
25. J. J. McKeown, *Comp. Meth. Appl. Mech. Eng.*, **12**, 155 (1977).
26. J. J. McKeown, *Eng. Optim.*, **14**, 159 (1989).
27. J. J. McKeown, *Struct. Multidisc. Optim.*, **15**, 92 (1998).
28. G.-Y. Wang, Z.-Y. Zhou and D. Huo, *Eng. Optim.*, **8**, 55 (1984).
29. A. G. Striz and J. Sobieszczanski-Sobieski, in *Proc. 6th AIAA/USAF/NASA/ISSMO MAO Symp. on Multidisciplinary Analysis and Optimization* (held in Bellevue, WA, 1996), AIAA-CP-4098.
30. C. Plunkett, A. G. Striz and J. Sobieszczanski-Sobieski, in *Proc. 42th AIAA/ASME/ASCE/AHS/ ASC Structures, Structural Dynamics and Materials Conf.*, (held in Seattle, WA, 2001), AIAA-2001-1552.
31. S. Subramaniyam, H. J. Neeman and A. G. Striz, in *Proc. 10th AIAA/ISSMO MAO Symp. on Multidisciplinary Analysis and Optimization* (held in Albany, NY, 2004), AIAA-2004-4445.
32. S. Missoum and Z. Gürdal, *AIAA J.*, **40**, 154 (2002).
33. W. Gu, Z. Gürdal and S. Missoum, *Comp. Meth. Appl. Mech. Eng.*, **191**, 2907 (2002).
34. S. Missoum, Z. Gürdal and W. Gu, *Struct. Multidisc. Optim.*, **23**, 214 (2002).
35. S. Missoum, Z. Gürdal and L. T. Watson, *Struct. Multidisc. Optim.*, **24**, 195 (2002).
36. E. J. Haug, K. K. Choi and V. Komkov, *Design Sensitivity Analysis of Structural Systems*, (Academic Press, New York, 1986).
37. J. S. Arora, in *Advances in Structural Optimization*, Ed. J. Herskovits (Kluwer Academic Publishers, Boston, 1995), p. 47.
38. R. Sedaghati and E. Esmailzadeh, *Int. J. Mech. Sci.*, **45**, 1369 (2003).
39. N. H. Kim, K. K. Choi and M. E. Botkin, *Struct. Multidisc. Optim.*, **24**, 418 (2003).
40. B. Barthelemy, R. T. Haftka, U. Madapur and S. Sankaranarayanan, *AIAA J.*, **29**, 791 (1991).
41. R. T. Haftka and Z. Gürdal, *Elements of Structural Optimization: Third Revised and Expanded Edition*, (Kluwer Academic Publishers, Dordrecht, 1992).
42. U. Kirsch, *Struct. Multidisc. Optim.*, **20**, 97 (2000).
43. U. Kirsch, *Design-Oriented Analysis of Structures*, (Kluwer Academic Publishers, Dordrecht, 2002).
44. U. Kirsch, M. Kočvara and J. Zowe, *Int. J. Num. Meth. Eng.*, **55**, 233 (2002).
45. U. Kirsch, M. Bogomolni and F. Keulan, *AIAA J.*, **43**, 399 (2004).

46. R. H. Myers and D. C. Montgomery, *Response Surface Methodology, Second Edition*, (Wiley-Interscience, New York, 2002).

47. T. Krishnamurthy, in *Proc. 44th AIAA/ASME/ASCE/AHS/ASC Structures, Structural Dynamics and Materials Conf.*, (held in Norfolk, VA, 2003), AIAA 2003-1748.

48. U. Kirsch, *Optimal Structural Design – Concepts, Methods and Applications*, (McGraw-Hill, New York, 1981).

49. U. Kirsch, *Fundamentals and Applications of Structural Optimization*, (Springer-Verlag, Heidelberg, 1993).

50. O. Canyurt and P. Hajela, in *Proc. 45th AIAA/ASME/ASCE/AHS/ASC Structures, Structural Dynamics and Materials Conf.*, (held in Palm Springs, CA, 2004), AIAA-2004-1913.

51. R. Muralidhar, J. R. J. Rao, K. Badhrinath and A. Kalagatla, *Int. J. Num. Meth. Eng.*, **39**, 2031 (1996).

52. R. Muralidhar and J. R. J. Rao, *Comp. Meth. Appl. Mech. Eng.*, **140**, 109 (1997).

53. W. Achtziger, *Struct. Optim.*, **12**, 63 (1996).

54. W. Achtziger, *Struct. Multidisc. Optim.*, **17**, 235 (1999).

55. W. Achtziger, *Struct. Multidisc. Optim.*, **17**, 247 (1999).

56. W. Achtziger, *Comp. Optim. & Appl.*, **15**, 69 (2000).

57. F. Tin-Loi, *Struct. Multidisc. Optim.*, **17**, 279 (1999).

58. F. Tin-Loi, *Struct. Multidisc. Optim.*, **19**, 130 (2000).

59. M. Stope and K. Svanberg, *Struct. Multidisc. Optim.*, **25**, 62 (2003).

60. M. Stope and K. Svanberg, *Int. J. Num. Meth. Eng.*, **57**, 723 (2003).

61. M. P. Saka, *ASCE J. Struct. Div.*, **106**, 1155 (1980).

62. M. P. Saka, *Comp. Struct.*, **11**, 411 (1980).

63. U. T. Ringertz, *Eng. Optim.*, **10**, 111 (1986).

64. A. Ben-Tal and M. P. Bendsøe, *SIAM J. Optim.*, **3**, 322 (1993).

65. S. Sankaranarayanan, R. T. Haftka and R. K. Kapania, *AIAA J.*, **32**, 420 (1994).

66. H. Smaoui and L. A. Schmit, *Int. J. Num. Meth. Eng.*, **26**, 555 (1988).

67. T. Larsson and M. Rönnqvist, *Struct. Multidisc. Optim.*, **9**, 1 (1995).

68. F. Tin-Loi, *Struct. Optim.*, **17**, 155 (1999).

69. V. Schulz and H. G. Bock, *Nonlin. Anal. Theory Meth. Appl.*, **30**, 4723 (1997).

70. T. Dreyer, B. Maar and V. Schulz, *J. Comp. & Appl. Math.*, **120**, 67 (2000).

71. V. Schulz, *J. Comp. Appl. Math.* **164**, 629 (2004).

72. J. Herskovits, in *Proc. 10th AIAA/ISSMO MAO Symp. on Multidisciplinary Analysis and Optimization*, (held in Albany, NY, 2004), AIAA-2004-4502.

73. R. H. W. Hoppe and S. I. Petrova, *Math. & Comp. Simu.*, **65**, 257 (2004).

74. M. P. Bendsøe, and C. A. Mota Soares, Eds., *Topology Design of Structures, NATO ASI Series. Series E: Applied Sciences – Vol. 227*, (Kluwer Academic Publishers, Dordrecht, 1993).

75. G. I. N. Rozvany, M. P. Bendsøe and U. Kirsch, *Appl. Mech. Rev.*, **48**, 41 (1995).

76. C. C. Swan and I. Kosaka, *Int. J. Num. Meth. Eng.*, **40**, 3033 (1997).

77. H. A. Eschenauer and N. Olhoff, *Appl. Mech. Rev.*, **54**, 331 (2001).

78. M. P. Bendsøe and O. Sigmund, *Topology Optimization, Theory, Methods and Applications*, (Springer-Verlag, Berlin Heidelberg, 2003).

79. M. Beckers and C. Fleury, *Comp. Struct.*, **64**, 77 (1997).

80. M. P. Bendsøe, A. Ben-Tal and J. Zowe, *Struct. Optim.*, **7**, 141 (1994).

81. A. Ben-Tal and A. Nemirovski, *SIAM J. Optim.*, **7**, 991 (1997).
82. M. Kočvara, J. Zowe and A. Nemirovski, *Comp. Struct.*, **76**, 431 (2000).
83. A. Ben-Tal, M. Kočvara, A. Nemirovski and J. Zowe, *SIAM Rev.*, **42**, 695 (2000).
84. W. Achtziger, M. P. Bendsøe, A. Ben-Tal and J. Zowe, *IMPACT of Computing in Sci. & Engrg.*, **4**, 315 (1992).
85. J. M. Oberndorfer, W. Achtziger and H. R. E. M. Hörnlein, *Struct. Optim.*, **11**, 137 (1996).
86. I. E. Grossmann, V. T. Voudouris and O. Ghattas, in *Recent Advances in Global Optimization*, Eds. C. A. Floudas and P. M. Pardalos (Princeton Univ Press, Princeton, N.J., 1992), p. 478.
87. D. Hilding, A. Klarbring and J. Petersson, *App. Mech. Rev.*, **52**, 139 (1999).
88. M. Stope and K. Svanberg, *Struct. Multidisc. Optim.*, **27**, 126 (2004).
89. S. Bollapragada, O. Ghattas and J. N. Hooker, *Operations Research*, **49**, 42 (2001).
90. L. T. Biegler, O. Ghattas, M. Heinkenschloss and B.v. Bloemen Waanders, Eds., *Large-Scale PDE-Constrained Optimization, Lecture Notes in Computational Science and Engineering, Vol. 30,* (Springer-Verlag, Berlin, 2003).
91. Z.-Q. Luo, J.-S. Pang and D. Ralph, *Mathematical Programs with Equilibrium Constraints*, (Cambridge University Press, Cambridge, 1996).
92. J. V. Outrata, M. Kočvara and J. Zowe, *Nonsmooth Approach to Optimization Problems with Equilibrium Constraints: Theory, Applications and Numerical Results*, (Kluwer Academic Publishers, Dordrecht, 1998).
93. M. C. Ferris and J. S. Pang, *SIAM Review*, **39**, 669 (1997).
94. A. R. Mijar and J. S. Arora, *Archives Comp. Meth. Eng.*, **7**, 387 (2000).
95. A. R. Mijar and J. S. Arora, *Struct. Multidisc. Optim.*, **20**, 167 (2000).
96. A. R. Mijar and J. S. Arora, *Struct. Multidisc. Optim.*, **28**, 99 (2004).
97. A. R. Mijar and J. S. Arora, *Struct. Multidisc. Optim.*, **28**, 113 (2004).
98. M. C. Ferris and F. Tin-Loi, *Struct. Eng & Mech.*, **6**, 857 (1998).
99. J.-S. Pang and F. Tin-Loi, *Mech. Struct. & Mach.*, **29**, 85 (2001).
100. F. Tin-Loi and N. S. Que, *Comp. Meth. Appl. Mech. Eng.*, **190**, 5819 (2001).
101. M. C. Ferris and F. Tin-Loi, *Int. J. Mech. Sci.*, **43**, 209 (2001).
102. A. Evgrafov and M. Patriksson, *Struct. Multidisc. Optim.*, **25**, 174 (2003).
103. M. Patriksson and L. Wynter, *Oper. Res. Letters.*, **25**, 159 (1999).
104. S. Scholtes and M. Stöhr, *SIAM J. Contr. & Optim.*, **37**, 617 (1999).
105. J.-S. Pang and M. Fukushima, *Comp. Optim. & Appl.*, **13**, 111 (1999).
106. R. Andreani and J. M. Martinez, *Math. Meth. Oper. Rese.*, **54**, 345 (2001).
107. Z. Wan, *Comp. & Math. with Appl.*, **44**, 7 (2002).
108. S. I. Birbil, S.-C. Fang and J. Han, *Comp. & Oper. Res.*, **31**, 2249 (2004).
109. P.-Y. Nie, *Appl. Math. Comp.*, **149**, 203 (2004).
110. P. J. Enright and B. A. Conway, *J. Guidance, Cont. Dyn.*, **14**, 981 (1991).
111. J. T. Betts, *J. Comp. Appl. Math.*, **120**, 27 (2000).
112. M. L. Kaplan and J. H. Heegaard, *Int. J. Num. Meth. Eng.*, **53**, 2043 (2002).
113. C. Bottasso and A. Croce, *Multibody Sys. Dyn.*, **12**, 17 (2004)
114. J. E. Cuthrell and L. T. Biegler, *Chem. Eng. Res. Des.*, **64**, 341 (1986).
115. L. T. Biegler, *J. Proc. Cont.*, **8**, 301 (1998).
116. D. G. Hull, *Optimal Control Theory for Applications*, (Springer-Verlag, New York, 2003).
117. J. T. Betts and W. P. Huffman, *J. Guidance, Cont. Dyn.*, **14**, 431 (1991).

118. D. B. Leineweber, I. Bauer, H. G. Bock and J. P. Schloder, *Comp. & Chem. Eng.*, **27**, 157 (2003).

119. F. Fahroo and I. M. Ross, *J. Guidance, Cont. Dyn.*, **25**, 160 (2002).

120. C. R. Hargraves and S. W. Paris, *J. Guidance, Cont. Dyn.*, **10**, 338 (1987).

121. N. Biehn, S. L. Campbell, L. Jay and T. Westbrook, *J. Comp. Appl. Math.*, **120**, 109 (2000).

122. J. T. Betts, N. Biehn, S. L. Campbell and W. P. Huffman, *J. Comp. Appl. Math.*, **143**, 237 (2002).

123. A. L. Herman and B. A. Conway, *J. Guidance, Cont. Dyn.*, **19**, 592 (1995).

124. G. S. Hu, C. J. Ong and C. L. Teo, *Eng. Optim.*, **34**, 155 (2002).

125. C. P. Neuman and A. Sen, *Automatica*, **9**, 601 (1973).

126. H. Seywald, *J. Guidance, Cont. Dyn.*, **17**, 480 (1994).

127. R. R. Kumar and H. Seywald, *J. Guidance, Cont. Dyn.*, **19**, 418 (1996).

128. B. A. Conway and K. M. Larson, *J. Guidance, Cont. Dyn.*, **21**, 780 (1998).

129. G. C. Itle, A.G. Salinger, R. P. Pawlowski, J. N. Shadid and L. T. Biegler, *Comp. & Chem. Eng.*, **28**, 291 (2004).

130. A. M. Cervantes, A. Wächter, R. H. Tütüncü and L. T. Biegler, *Comp. & Chem. Eng.*, **24**, 39 (2000).

131. A. M. Cervantes and L. T. Biegler, *AIChE J.*, **44**, 1038 (1998).

132. J. T. Betts and W. P. Huffman, *Comp. Optim. & Appl.*, **14**, 179 (1999).

133. J. T. Betts, *J. Guidance, Cont. Dyn.*, **21**, 193 (1998).

134. M. H. Rizk, *AIAA J.*, **21**, 1640 (1983).

135. P. D. Frank and G. R. Shubin, *J. Comp. Phys.*, **98**, 74 (1992).

136. G. R. Shubin, *J. Comp. Phys.*, **118**, 73 (1995).

137. C. E. Orozco and O. N. Ghattas, *Comp. Systems Eng.*, **3**, 311 (1992).

138. C. E. Orozco and O. N. Ghattas, *AIAA J.*, **34**, 217 (1996).

139. A. N. Hrymak, G. J. McRae and A. W. Westerberg, *J. Heat Transfer*, **107**, 527 (1985).

140. R. T. Haftka, J. Sobieszczanski-Sobieski and S. L. Padula, *Struct. Optim.*, **4**, 65 (1992).

141. E. J. Cramer, J. E., Dennis Jr., P. D. Frank, R. M. Lewis and G. R. Shubin, *SIAM J. Optim.*, **4**, 754 (1994).

142. R. J. Balling and J. Sobieszczanski-Sobieski, *AIAA J.*, **34**, 6 (1996).

143. R. J. Balling and C. A. Wilkinson, *AIAA J.*, **35**, 178 (1997).

CHAPTER 16

KRIGING METAMODEL BASED OPTIMIZATION

Tae Hee Lee and Jae Jun Jung

School of Mechanical Engineering
Hanyang University, Seoul 133-791, Korea
E-mail: thlee@hanyang.ac.kr

Kriging metamodel that is suitable for approximation of highly nonlinear functions is derived systematically and sampling techniques for kriging are summarized and compared. Then optimization of an engineering problem based on kriging metamodel is performed.

1. Introduction

Many design optimization problems can consist of objective functions and constraint equations that require complicated and time-consuming numerical analyses. In this case, conventional optimization technique based on direct integration with computationally expensive high-fidelity simulations may be impractical because of enormous computational cost of many iterative function calls.

One of solutions to overcome the problem is to use approximate responses of the objectives and constraints. The goal of approximation is to provide a continuous function that is inexpensive to evaluate within acceptable fidelity. This approximation model is often referred to metamodel, i.e., 'models of the model'.[1] A wide variety of metamodels have been developed: response surface model,[2] kriging,[3-5] radial basis function[6,7] and multivariate adaptive regression splines.[8] According to how a fitting curve to sampled data is generated, metamodels are classified into regression model and interpolation model. Response surface model is a typical regression model in which the unknown coefficients are estimated by means of least square method, while the others are interpolation models that fitting curve passes exactly through the sampled points. Among these metamodels, kriging model has gained much attention in engineering literatures because it can be useful in predicting uncertainty as well as remarkable in predicting performance for especially highly nonlinear functions. In this chapter, we derive kriging predictor based on notion of statistics.

One of the issues related to development of a kriging model is how to select sample points. In computer experiment, replication is usually meaningless since the response of computer simulation is deterministic–response of a simulation code with the same input always gives identical output. Therefore, sample points should be evenly distributed over entire design space in order to predict the response well at untried points. Sampling techniques based on this concept is referred to as 'space-filling sampling.'[9,10] In addition, 'infill sampling criteria' that are available in approximating various extrema of true function are introduced. They locate sample points in the neighborhood of local optimum and the location at maximum uncertainty.[11-15]

To illustrate the application of kriging model, laterally vibrating circular plate is employed as an engineering optimization problem. We formulate the optimization problem to minimize the mass of vibrating plate satisfying stress and frequency constraints. Responses such as stress and volume, and frequency of vibrating plate are approximated based on various sampling techniques. Accuracy of kriging models is validated and results of kriging-based optimization are compared with those of conventional optimization.

2. Kriging Metamodel

2.1. *Origin of Kriging*

Kriging originally comes from the field of geostatistics as a method to predict geological data, such as the thickness of ore layers. The name of 'Kriging' refers to a South African geologist, D. G. Krige who first used the statistics in analyzing mining data.[16] His work was furthered in the early 1970s by Matheron and formed the fundamental foundation in the field of geostatistics.[17]

At the end of 1980, kriging was taken in a new direction when a group of statisticians applied kriging technique for design of engineering problems.[3,4] Their processes of obtaining experimental design and exploiting kriging as prediction tool for the design were called design and analysis of computer experiments (DACE). Nowadays, kriging model has been widely employed in research area such as multidisciplinary design optimization (MDO) due to powerful applicability.[18]

2.2. *Kriging Theory*

For design variables $\mathbf{x} \in R^{n_d}$, kriging is based on assumption that response function $Y(\mathbf{x})$ is composed of a regression model $\mathbf{f}(\mathbf{x})^T \boldsymbol{\beta}$ and stochastic process $Z(\mathbf{x})$:

$$\bar{Y}(\mathbf{x}) = \mathbf{f}(\mathbf{x})^T \boldsymbol{\beta} + Z(\mathbf{x}) \tag{1}$$

where polynomial terms consist of $(p \times 1)$ vector of regression functions as

$$\mathbf{f}(\mathbf{x}) = [f_1(\mathbf{x}), f_2(\mathbf{x}), \cdots, f_p(\mathbf{x})]^T \tag{2}$$

and the corresponding $(p \times 1)$ vector of the unknown coefficients as follows:

$$\boldsymbol{\beta} = [\beta_1, \beta_2, \cdots, \beta_p]^T \tag{3}$$

Here, polynomials of order 0, 1 and 2, i.e., constant, linear and quadratic, are usually used and $p = 1$, $p = n_d + 1$, $p = (n_d + 1)(n_d + 2)/2$, respectively.

Note that regression model represents global model, i.e., mean model, $E[Y(\mathbf{x})] = \mathbf{f}(\mathbf{x})^T \boldsymbol{\beta}$ while the localized model $Z(\mathbf{x})$ is assumed to be a Gaussian process with zero mean $E[Z(\mathbf{x})] = 0$ and covariance $\mathrm{cov}[Z(\mathbf{x}_i), Z(\mathbf{x}_j)]$, where $E[\cdot]$ denotes expectation.

It is pointed that there is the dependency between deviation terms $Z(\mathbf{x})$ because $Z(\mathbf{x})$ are relatively determined according to estimation of regression model as shown in Fig. 1. This dependency can be expressed in form of covariance.

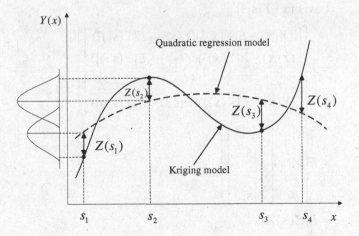

Fig. 1 Visualization for deviation terms of response function

For random responses $Y(\mathbf{s}_i)$ at n sampled points, the vector form of Eq. (1) is written as

$$\mathbf{Y} = \mathbf{F}\boldsymbol{\beta} + \mathbf{Z} \tag{4}$$

$$\mathbf{Y} \equiv \begin{bmatrix} Y(\mathbf{s}_1) \\ Y(\mathbf{s}_2) \\ \vdots \\ Y(\mathbf{s}_n) \end{bmatrix}, \mathbf{F} \equiv \begin{bmatrix} \mathbf{f}(\mathbf{s}_1)^T \\ \mathbf{f}(\mathbf{s}_2)^T \\ \vdots \\ \mathbf{f}(\mathbf{s}_n)^T \end{bmatrix}, \mathbf{Z} \equiv \begin{bmatrix} Z(\mathbf{s}_1) \\ Z(\mathbf{s}_2) \\ \vdots \\ Z(\mathbf{s}_n) \end{bmatrix} \tag{5}$$

where $\mathbf{F}$ is defined as $(n \times p)$ expanded design matrix. Because responses $Y(\mathbf{s}_i)$ are random variable and correlated to each other, $(n \times n)$ covariance matrix for n random responses exists as follows:

$$\mathrm{cov}[Y(\mathbf{s}_i), Y(\mathbf{s}_j)]_{(1 \le i \le n) \times (1 \le j \le n)}$$

$$= E\left[\begin{bmatrix} Y(\mathbf{s}_1) - E[Y(\mathbf{s}_1)] \\ Y(\mathbf{s}_2) - E[Y(\mathbf{s}_2)] \\ \vdots \\ Y(\mathbf{s}_n) - E[Y(\mathbf{s}_n)] \end{bmatrix} \left[Y(\mathbf{s}_1) - E[Y(\mathbf{s}_1)], \quad Y(\mathbf{s}_2) - E[Y(\mathbf{s}_2)], \quad \cdots \quad , Y(\mathbf{s}_n) - E[Y(\mathbf{s}_n)] \right] \right]$$

$$= E\left[\begin{bmatrix} Z(\mathbf{s}_1) \\ Z(\mathbf{s}_2) \\ \vdots \\ Z(\mathbf{s}_n) \end{bmatrix} \left[Z(\mathbf{s}_1), \quad Z(\mathbf{s}_2), \quad \cdots \quad Z(\mathbf{s}_n) \right] \right]$$

$$= E[\mathbf{Z}\mathbf{Z}^T] \tag{6}$$

Similarly, $(n \times 1)$ covariance vector between responses at sampled point $\mathbf{s}_i$ and an untried point $\mathbf{x}$ is represented as

$$\mathrm{cov}[Y(\mathbf{x}), Y(\mathbf{s}_i)]_{(1 \le i \le n) \times 1}$$

$$= E\left[\begin{bmatrix} (Y(\mathbf{x}) - E[Y(\mathbf{x})])(Y(\mathbf{s}_1) - E[Y(\mathbf{s}_1)]) \\ (Y(\mathbf{x}) - E[Y(\mathbf{x})])(Y(\mathbf{s}_2) - E[Y(\mathbf{s}_2)]) \\ \vdots \\ (Y(\mathbf{x}) - E[Y(\mathbf{x})])(Y(\mathbf{s}_n) - E[Y(\mathbf{s}_n)]) \end{bmatrix} \right]$$

$$= E\left[Z(\mathbf{x}) \begin{bmatrix} Z(\mathbf{s}_1) \\ Z(\mathbf{s}_2) \\ \vdots \\ Z(\mathbf{s}_n) \end{bmatrix} \right]$$

$$= E[Z(\mathbf{x})\mathbf{Z}] \tag{7}$$

Notice that $Z(\mathbf{x})$ and $\mathbf{Z}$ are scalar and column vector, respectively. It is interesting that the covariance between responses $Y(\mathbf{s}_i)$ is identical to the covariance between $Z(\mathbf{s}_i)$:

$$
\begin{aligned}
\text{cov}[Z(\mathbf{s}_i), Z(\mathbf{s}_j)]_{(1 \le i \le n) \times (1 \le j \le n)} \\
= E[(\mathbf{Z} - E[\mathbf{Z}])(\mathbf{Z} - E[\mathbf{Z}])^T] (\because E[Z(\mathbf{x})] = 0 \ \forall \mathbf{x}) \\
= E[\mathbf{Z}\mathbf{Z}^T]
\end{aligned}
\tag{8}
$$

Covariance between $Z(\mathbf{x})$ at an untried point and $Z(\mathbf{s}_i)$ at sampled points is given as follows:

$$
\begin{aligned}
\text{cov}[Z(\mathbf{x}), Z(\mathbf{s}_i)]_{(1 \le i \le n) \times 1} \\
= E[(Z(\mathbf{x}) - E[Z(\mathbf{x})])(\mathbf{Z} - E[\mathbf{Z}])] \\
= E[Z(\mathbf{x})\mathbf{Z}]
\end{aligned}
\tag{9}
$$

Covariance between departures $Z(\mathbf{s}_i)$ and $Z(\mathbf{s}_j)$ at sampled points, however, is generally unknown. In kriging, the covariance is assumed to be expressed as the product of process variance σ_z^2 and $(n \times n)$ correlation matrix $\mathbf{R}$:

$$
\text{cov}[Z(\mathbf{s}_i), Z(\mathbf{s}_j)]_{(1 \le i \le n) \times (1 \le j \le n)} \equiv \sigma_z^2 \mathbf{R}
\tag{10}
$$

where σ_z^2 is process variance that is determined by stochastic process. Correlation matrix can be defined by spatial correlation function $R(\boldsymbol{\theta}, \mathbf{s}_i, \mathbf{s}_j)$:

$$
\mathbf{R} =
\begin{bmatrix}
R(\boldsymbol{\theta}, \mathbf{s}_1, \mathbf{s}_1) & R(0, \mathbf{s}_1, \mathbf{s}_2) & \cdots & R(\boldsymbol{\theta}, \mathbf{s}_1, \mathbf{s}_n) \\
R(\boldsymbol{\theta}, \mathbf{s}_2, \mathbf{s}_1) & R(\boldsymbol{\theta}, \mathbf{s}_2, \mathbf{s}_2) & \cdots & R(\boldsymbol{\theta}, \mathbf{s}_2, \mathbf{s}_n) \\
\vdots & \vdots & \ddots & \vdots \\
R(\boldsymbol{\theta}, \mathbf{s}_n, \mathbf{s}_1) & R(\boldsymbol{\theta}, \mathbf{s}_n, \mathbf{s}_2) & \cdots & R(\boldsymbol{\theta}, \mathbf{s}_n, \mathbf{s}_n)
\end{bmatrix}
\tag{11}
$$

where $\boldsymbol{\theta}$ is a vector of unknown correlation parameters.

The form of spatial correlation function is restricted to only correlation that can be expressed as product of one-dimensional correlations [19]:

$$
R(\boldsymbol{\theta}, \mathbf{s}_i, \mathbf{s}_j) = \prod_{k=1}^{n_d} R(\theta_k, d_{ij}^k)
\tag{12}
$$

where θ_k and d_{ij}^k are k-th component of each vector $\boldsymbol{\theta}$ and $\mathbf{d}_{ij}$, respectively. $\mathbf{d}_{ij}$ represents distance between $\mathbf{s}_i$ and $\mathbf{s}_j$.

Covariance between the $Z(\mathbf{x})$ and $Z(\mathbf{s}_i)$ is expressed as product of process variance and correlation vector:

$$\text{cov}[Z(\mathbf{x}), Z(\mathbf{s}_i)]_{(1 \le i \le n) \times 1}$$
$$\equiv \sigma_z^2 \mathbf{r}(\mathbf{x}) \tag{13}$$
$$= \sigma_z^2 [R(\boldsymbol{\theta}, \mathbf{x}, \mathbf{s}_1), R(\boldsymbol{\theta}, \mathbf{x}, \mathbf{s}_2), \cdots, R(\boldsymbol{\theta}, \mathbf{x}, \mathbf{s}_n)]^T$$

Physical meaning of one-dimensional correlation is that the correlation in each dimension is function of only distance in the dimension. Fig. 2 shows geometric expression of the spatial correlation between sampled points $\mathbf{s}_1$ and $\mathbf{s}_2$ in 2-dimensional space. Among types of one-dimensional correlation functions, the commonly used spatial correlation functions are shown in Table 1.

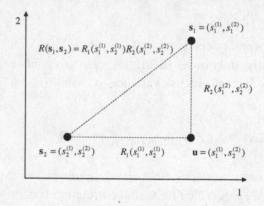

Fig. 2 Calculation of correlation in two-dimensional space

Correlation matrix is positive-definite and the diagonal term is all ones according to definition of correlation function. Thus, the value of spatial correlation function $R(\boldsymbol{\theta}, \mathbf{s}_i, \mathbf{s}_j)$ is specified between 0 and 1. The value approaches zero as the distance d_{ij}^k increases indefinitely, i.e., weak correlation, while the correlation of identical points becomes 1, i.e., strong correlation.

The choice of spatial correlation function plays a critical role in the prediction. For a smooth response function, Gaussian spatial correlation function would be preferred. For non-differentiable response, exponential spatial correlation function is recommended. We need to pay attention to the meaning of correlation parameter in the prediction. Correlation parameter θ_k determines how quickly the correlation decreases as the distance increases (See Fig. 3).

Table 1 Types of spatial correlation functions

Name	$R(\theta_k, d_{ij}^k)$
Exponential	$e^{-\theta_k\left|d_{ij}^k\right|}$
Exponential-Gauss	$e^{-\theta_k\left|d_{ij}^k\right|^h} \quad 0 < h \leq 2$
Gauss	$e^{-\theta_k d_{ij}^{k^2}}$
Linear	$\max\left\{0, 1-\theta_k\left|d_{ij}^k\right|\right\}$
Spherical	$1 - 1.5\xi_{ij}^k + 0.5\xi_{ij}^{k^3}, \quad \xi_{ij}^k = \min\left\{1, \theta_k\left|d_{ij}^k\right|\right\}$
Cubic	$1 - 3\xi_{ij}^{k^2} + 2\xi_{ij}^{k^3}, \quad \xi_{ij}^k = \min\left\{1, \theta_k\left|d_{ij}^k\right|\right\}$

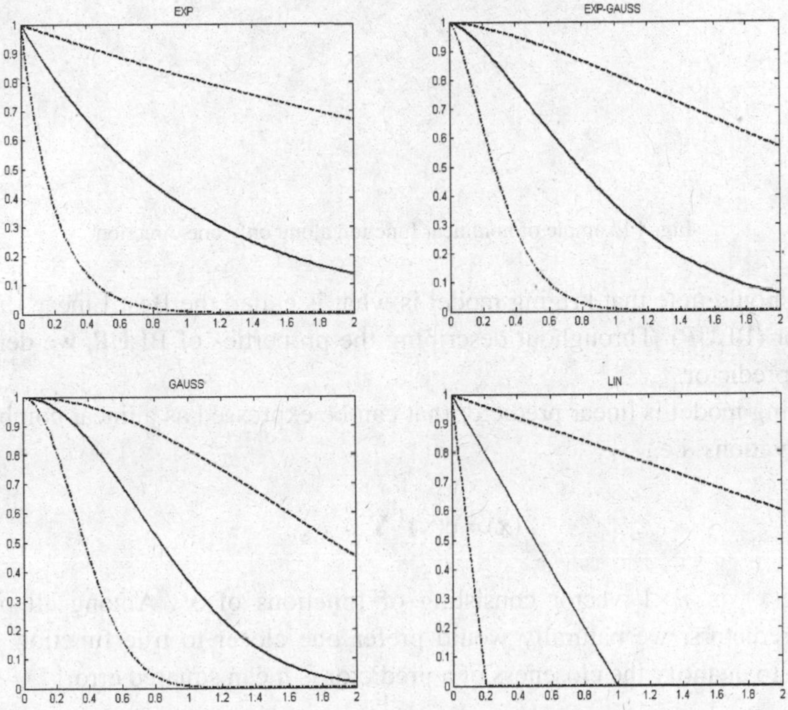

Fig. 3 Various correlation functions for $0 \leq d_k \leq 2$; dashed, full and dash-dotted line for $\theta_k = 0.2, 1, 5$, respectively

Thus, the correlation parameter can reflect the nonlinearity or smoothness of response with respect to design variables. That is, large value of correlation parameter θ_k indicates the low correlation of the response along the axis of k-th design variable. The lower correlation means that response is more nonlinear along the axis because correlation is dominantly influenced by closely located sample point. For example, Fig. 4 shows a function that is highly nonlinear in x_1 dimension but linear in x_2 dimension. When kriging model is reasonably constructed from appropriate sampled points, θ_1 will become even larger than θ_2. It means that the nonlinearity of response is stronger along x_1-axis than along x_2-axis. This is invaluable to infer the curvature information between response and design variables.

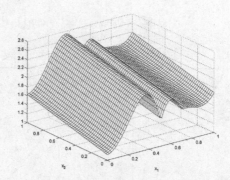

Fig. 4 Example of nonlinear function along only one direction

We should note that kriging model is what is called the Best Linear Unbiased Predictor (BLUP). Throughout describing the properties of BLUP, we derive the kriging predictor.

Kriging model is linear predictor that can be expressed as a linear combination of observations, i.e.,

$$\hat{Y}(\mathbf{x}) = \mathbf{c}(\mathbf{x})^T \mathbf{Y} \tag{14}$$

where $\mathbf{c}(\mathbf{x})$ is $n \times 1$ vector consisting of functions of $\mathbf{x}$. Among all possible linear predictors, we naturally would prefer one closer to true function. Useful measure to quantify the closeness of a predictor is mean squared error:

$$
\begin{aligned}
MSE[\hat{Y}(\mathbf{x})] &= E[(\hat{Y}(\mathbf{x}) - Y(\mathbf{x}))^2] \\
&= \mathrm{var}[Y(\mathbf{x})] + \mathrm{var}[\hat{Y}(\mathbf{x})] + (E[\hat{Y}(\mathbf{x})] - E[Y(\mathbf{x})])^2 \\
&\quad - 2\,\mathrm{cov}[\hat{Y}(\mathbf{x}), Y(\mathbf{x})]
\end{aligned} \tag{15}
$$

Eq. (15) shows how mean squared error is related to variance, variance of predictor, bias of predictor, and covariance between true function and predictor.

In BLUP, 'best' means that we find a predictor with the smallest mean squared error. To do so, we need to look for predictor within the class of unbiased predictors. Unbiasedness condition implies that the expectation of predictor model equals to the expectation of true response:

$$E[\hat{Y}(\mathbf{x})] = E[Y(\mathbf{x})]$$
$$\Leftrightarrow \mathbf{c}(\mathbf{x})^T \mathbf{F}\boldsymbol{\beta} = \mathbf{f}(\mathbf{x})^T \boldsymbol{\beta} \quad \forall \boldsymbol{\beta} \quad (\because \mathbf{c}(\mathbf{x})^T E[\mathbf{Z}] = 0) \tag{16}$$
$$\Leftrightarrow \mathbf{F}^T \mathbf{c}(\mathbf{x}) = \mathbf{f}(\mathbf{x})$$

If we substitute the unbiased condition into Eq. (15), we can eliminate the third term of RHS of Eq. (15) and obtain as

$$MSE[\hat{Y}(\mathbf{x})] = \text{var}[Y(\mathbf{x})] + \text{var}[\hat{Y}(\mathbf{x})] - 2\text{cov}[\hat{Y}(\mathbf{x}), Y(\mathbf{x})] \tag{17}$$

In Eqs. (15) and (17), we can find that mean squared error of unbiased predictor is always less than that of biased predictor because the square of bias of predictor is nonnegative. For this reason, we consider only the unbiased predictor.

Fig. 5 shows a graphical representation of biased predictor and unbiased predictor. Mean of unbiased predictor $\hat{Y}_2(x)$ is identical that of true response. On the other hand, mean of biased predictor $\hat{Y}_1(x)$ stands away from that of true response.

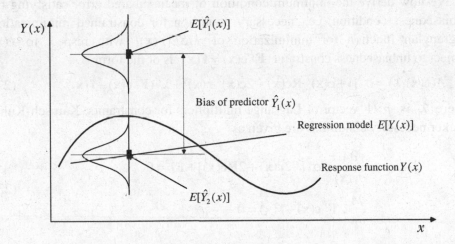

Fig. 5 Illustration of unbiased predictor and biased predictor in one-dimensional function

Let us rewrite Eq. (17) in terms of the notations defined in Eqs. (6), (7), (10) and (13) as follows:

$$
\begin{aligned}
\text{var}[Y(\mathbf{x})] &= E[(Y(\mathbf{x}) - E[Y(\mathbf{x})])^2] \\
&= E[Z(\mathbf{x})^2] \\
&= \sigma_z^2
\end{aligned}
\tag{18}
$$

$$
\begin{aligned}
\text{var}[\hat{Y}(\mathbf{x})] &= E[(\hat{Y}(\mathbf{x}) - E[\hat{Y}(\mathbf{x})])^2] \\
&= E[(\mathbf{c}(\mathbf{x})^T \mathbf{Z}\mathbf{Z}^T \mathbf{c}(\mathbf{x})] \\
&= \sigma_z^2 \mathbf{c}(\mathbf{x})^T \mathbf{R}\mathbf{c}(\mathbf{x})
\end{aligned}
\tag{19}
$$

$$
\begin{aligned}
\text{cov}[Y(\mathbf{x}), \hat{Y}(\mathbf{x})] &= E[(Y(\mathbf{x}) - E[Y(\mathbf{x})])(\hat{Y}(\mathbf{x}) - E[\hat{Y}(\mathbf{x})])] \\
&= E[Z(\mathbf{x})\mathbf{c}(\mathbf{x})^T \mathbf{Z}] \\
&= \mathbf{c}(\mathbf{x})^T E[Z(\mathbf{x})\mathbf{Z}] \\
&= \sigma_z^2 \mathbf{c}(\mathbf{x})^T \mathbf{r}
\end{aligned}
\tag{20}
$$

Thus, mean squared error of kriging predictor becomes

$$
MSE[\hat{Y}(\mathbf{x})] = \sigma_z^2 \left(1 + \mathbf{c}(\mathbf{x})^T \mathbf{R}\mathbf{c}(\mathbf{x}) - 2\mathbf{c}(\mathbf{x})^T \mathbf{r}(\mathbf{x})\right)
\tag{21}
$$

We now derive the minimum condition of mean squared error satisfying the unbiasedness condition, i.e., necessary condition for constrained minimization. Lagrangian function for minimization of $MSE[\hat{Y}(\mathbf{x})]$ with respect to $\mathbf{c}(\mathbf{x})$ subject to unbiasedness constraint $\mathbf{F}^T \mathbf{c}(\mathbf{x}) = \mathbf{f}(\mathbf{x})$ is of the form

$$
A(\mathbf{c}(\mathbf{x}), \lambda) = \sigma_z^2 [1 + \mathbf{c}(\mathbf{x})^T \mathbf{R}\mathbf{c}(\mathbf{x}) - 2\mathbf{c}(\mathbf{x})^T \mathbf{r}(\mathbf{x})] + \lambda^T (\mathbf{F}^T \mathbf{c}(\mathbf{x}) - \mathbf{f}(\mathbf{x}))
\tag{22}
$$

where λ is $p \times 1$ vector of Lagrange multipliers for constraints. Karusch-Kuhn-Tucker necessary conditions are given as

$$
\begin{aligned}
\frac{\partial A}{\partial \mathbf{c}(\mathbf{x})} &: \sigma_z^2 [-2\mathbf{r}(\mathbf{x}) + 2\mathbf{R}\mathbf{c}(\mathbf{x})] + \mathbf{F}\lambda = 0 \\
\frac{\partial A}{\partial \lambda} &: \mathbf{F}^T \mathbf{c}(\mathbf{x}) - \mathbf{f}(\mathbf{x}) = 0
\end{aligned}
\tag{23}
$$

Rewriting Eq. (23) in matrix form yields

$$\begin{bmatrix} \mathbf{0} & \mathbf{F}^T \\ \mathbf{F} & \mathbf{R} \end{bmatrix} \begin{Bmatrix} \lambda/2\sigma_z^2 \\ \mathbf{c(x)} \end{Bmatrix} = \begin{Bmatrix} \mathbf{f(x)} \\ \mathbf{r(x)} \end{Bmatrix} \tag{24}$$

For simplicity of derivation, we replace $\tilde{\lambda} = \lambda/2\sigma_z^2$. Then, we have

$$\begin{bmatrix} \mathbf{0} & \mathbf{F}^T \\ \mathbf{F} & \mathbf{R} \end{bmatrix} \begin{Bmatrix} \tilde{\lambda} \\ \mathbf{c(x)} \end{Bmatrix} = \begin{Bmatrix} \mathbf{f(x)} \\ \mathbf{r(x)} \end{Bmatrix} \tag{25}$$

Solution to Eq. (25) can be found by using inverse theorem of the block matrix (see Appendix) as follows:

$$\begin{bmatrix} \mathbf{0} & \mathbf{F}^T \\ \mathbf{F} & \mathbf{R} \end{bmatrix}^{-1} = \begin{bmatrix} -(\mathbf{F}^T\mathbf{R}^{-1}\mathbf{F})^{-1} & (\mathbf{F}^T\mathbf{R}^{-1}\mathbf{F})^{-1}\mathbf{F}^T\mathbf{R}^{-1} \\ \mathbf{R}^{-1}\mathbf{F}(\mathbf{F}^T\mathbf{R}^{-1}\mathbf{F})^{-1} & \mathbf{R}^{-1}(\mathbf{I}-\mathbf{F}(\mathbf{F}^T\mathbf{R}^{-1}\mathbf{F})^{-1}\mathbf{F}^T\mathbf{R}^{-1}) \end{bmatrix} \tag{26}$$

Finally, $\mathbf{c(x)}$ can be calculated as

$$\mathbf{c(x)} = \mathbf{R}^{-1}[\mathbf{r(x)} - \mathbf{F}(\mathbf{F}^T\mathbf{R}^{-1}\mathbf{F})^{-1}(\mathbf{F}^T\mathbf{R}^{-1}\mathbf{r(x)} - \mathbf{f(x)})] \tag{27}$$

Substituting Eq. (27) into Eq. (14) and simplifying matrix multiplication using symmetry of $\mathbf{R}$, BLUP leads:

$$\hat{Y}(\mathbf{x}) = [\mathbf{r(x)}^T - \{\mathbf{r(x)}^T\mathbf{R}^{-1}\mathbf{F} - \mathbf{f(x)}^T\}(\mathbf{F}^T\mathbf{R}^{-1}\mathbf{F})^{-1}\mathbf{F}^T]\mathbf{R}^{-1}\mathbf{Y} \tag{28}$$

If we introduce generalized least square estimator as

$$\hat{\boldsymbol{\beta}} = (\mathbf{F}^T\mathbf{R}^{-1}\mathbf{F})^{-1}\mathbf{F}^T\mathbf{R}^{-1}\mathbf{Y} \tag{29}$$

Then, the kriging predictor of Eq. (28) becomes

$$\hat{Y}(\mathbf{x}) = \mathbf{f(x)}^T\hat{\boldsymbol{\beta}} + \mathbf{r(x)}^T\mathbf{R}^{-1}(\mathbf{Y} - \mathbf{F}\hat{\boldsymbol{\beta}}) \tag{30}$$

In Eq. (30), first component $\mathbf{f(x)}^T\boldsymbol{\beta}$ is the estimator of global model and denotes a mean model over the entire domain $\mathbf{x}$. And second one is localized model $Z(\mathbf{x})$ and represents the systematic departure from the estimated mean model. From a view of the response surface model, systematic departure $Z(\mathbf{x})$ seems to be the residual. Note that since departures are correlated, generalized least square method using weighting factor is applied.

In kriging model, uncertainty of predictor, i.e., mean squared error, is useful statistics. In Eqs. (25) and (26), $\tilde{\lambda}$ is given follows as

$$\tilde{\lambda} = (\mathbf{F}^T\mathbf{R}^{-1}\mathbf{F})^{-1}(\mathbf{F}^T\mathbf{R}^{-1}\mathbf{r(x)} - \mathbf{f(x)}) \tag{31}$$

Substituting $\mathbf{c(x)}$ and $\tilde{\lambda}$ into Eq. (21) leads to matrix representation of mean squared error as follows:

$$MSE[\hat{Y}(\mathbf{x})] = \sigma_z^2 \left(1 - [\mathbf{f}(\mathbf{x})^T, \mathbf{r}(\mathbf{x})^T] \begin{bmatrix} \mathbf{0} & \mathbf{F}^T \\ \mathbf{F} & \mathbf{R} \end{bmatrix}^{-1} \begin{bmatrix} \mathbf{f}(\mathbf{x}) \\ \mathbf{r}(\mathbf{x}) \end{bmatrix} \right) \tag{32}$$

Mean square error, i.e. uncertainty of prediction, is zero at sample points because kriging model is interpolation model–prediction of kriging always coincides with its true response at sampled point. However, uncertainty of prediction at untried point increase highly as prediction point exits somewhat away from sample points.

Now, we need to estimate correlation parameter $\boldsymbol{\theta}$ because predictor includes parameters that are yet determined in Eqs. (11) and (13). In kriging, maximum likelihood estimation (MLE) is used to estimate correlation parameters $\boldsymbol{\theta}$, regression coefficient $\boldsymbol{\beta}$, and variance σ_z^2. MLE is based on the assumption that we know the probability distribution of sampled responses. We estimate correlation parameters by maximizing the probability that sampled responses can have. To quantify the probability, likelihood function is defined as product of probability density function of each sampled response.

Because we assume that responses follow Gaussian distribution, likelihood function to find correlation parameters is given as

$$L \equiv \frac{1}{(2\pi)^{n/2} \sqrt{|\sigma_z^2 \mathbf{R}|}} e^{-\frac{(\mathbf{Y}-\mathbf{F}\boldsymbol{\beta})^T \mathbf{R}^{-1}(\mathbf{Y}-\mathbf{F}\boldsymbol{\beta})}{2\sigma_z^2}} \tag{33}$$

Recalling the property of determinant, covariance matrix term can be given as

$$\left| \hat{\sigma}_z^2 \mathbf{R} \right| = \left(\hat{\sigma}_z^2 \right)^n |\mathbf{R}| \tag{34}$$

For simplicity, we define logarithm of likelihood function, Eq. (33), as follows:

$$\ln L = -\frac{n}{2} \ln 2\pi - \frac{n}{2} \ln \sigma_z^2 - \frac{1}{2} \ln |\mathbf{R}| - \frac{(\mathbf{Y}-\mathbf{F}\boldsymbol{\beta})^T \mathbf{R}^{-1}(\mathbf{Y}-\mathbf{F}\boldsymbol{\beta})}{2\sigma_z^2} \tag{35}$$

According to the notion of MLE, the estimator of $\boldsymbol{\beta}$ is now obtained by taking the partial derivative of log likelihood function with respect to $\boldsymbol{\beta}$ and setting it to zero:

$$\frac{\partial \ln L}{\partial \boldsymbol{\beta}} = -\mathbf{F}^T \mathbf{R}^{-1}(\mathbf{Y} - \mathbf{F}\boldsymbol{\beta}) = 0, \ \forall \boldsymbol{\beta} \tag{36}$$

Note that solution to Eq. (36) coincides with generalized least square estimator $\hat{\boldsymbol{\beta}}$. Similarly, estimator of process variance σ_z^2 is given by differentiating $\ln L$ with respect to σ_z^2 and setting it to zero:

$$\hat{\sigma}_z^2 = \frac{(\mathbf{Y} - \mathbf{F}\hat{\boldsymbol{\beta}})^T \mathbf{R}^{-1}(\mathbf{Y} - \mathbf{F}\hat{\boldsymbol{\beta}})}{n} \tag{37}$$

For correlation parameters $\boldsymbol{\theta}$, however, the likelihood equation does not lead to a closed form solution. Therefore, a numerical optimization procedure is required. For optimization, Eq. (33) needs be converted into simple equation eliminating the constants. Substituting parameter with estimators into Eq. (35) yields

$$\ln L = -\frac{n}{2}\ln 2\pi - \frac{n}{2}\ln \hat{\sigma}_z^2 - \frac{1}{2}\ln|\mathbf{R}| - \frac{n}{2} \tag{38}$$

Finally, MLE to construct the predictor is written as

$$\underset{\boldsymbol{\theta}}{\text{maximize}} \left(-\frac{n\ln\hat{\sigma}_z^2 + \ln|\mathbf{R}|}{2} \right) \tag{39}$$

Non-logarithm form of likelihood function is often used equally as follows:

$$\underset{\boldsymbol{\theta}}{\text{minimi ze}}\ \hat{\sigma}_z^2 |\mathbf{R}|^{1/n} \tag{40}$$

2.3. *Example of Kriging Modeling*

To clearly understand kriging, we illustrate the process of kriging modeling for 2-dimensional problem as follows:

$$y = x_1 \sin x_2, \quad x_1, x_2 \in [1,8] \tag{41}$$

As shown in Fig. 6, test function is slightly nonlinear in only one dimension.

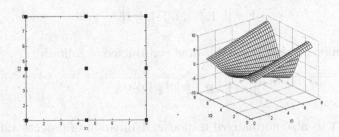

Fig. 6 Sample points and original function of test function

Location of the nine sample points and the corresponding responses is given in Table 2. Sample points and responses are normalized by each means and standard deviations before using the data. This is helpful to alleviate the dimension effect of each design variable and prevent an erratic prediction of the kriging model.

Table 2 Sample data and normalized data

No	x_1	x_2	$N(x_1)^*$	$N(x_2)^*$	Y	$N(Y)^*$
1	1	1	-1.1547	-1.1547	0.8415	-0.0855
2	1	4.5	-1.1547	0	-0.9775	-0.4401
3	1	8	-1.11547	1.1547	0.9894	-0.0566
4	4.5	1	0	-1.11547	3.7866	0.4886
5	4.5	4.5	0	0	-4.3989	-1.1070
6	4.5	8	0	1.1547	4.4521	0.6184
7	8	1	1.1547	-1.1547	6.7318	1.0628
8	8	4.5	1.1547	0	-7.8202	-1.7740
9	8	8	1.1547	1.1547	7.9149	1.2934

Note: $N(\cdot)$=normalized$(\cdot)$

In order to fit a kriging model, we first consider maximum likelihood estimation. Two terms, $\hat{\sigma}_z^2$ and $\mathbf{R}$, that consist of maximum likelihood function, must be constructed. First, correlation matrix with Gaussian correlation function is computed as follows:

$$\mathbf{R} = \begin{bmatrix} 1 & e^{-1.333\theta_2} & e^{-5.333\theta_2} & e^{-1.333\theta_1} & e^{-1.333\theta_1 -1.333\theta_2} & e^{-1.333\theta_1 -5.333\theta_2} & e^{-5.333\theta_1} & e^{-5.333\theta_1 -1.333\theta_2} & e^{-5.333\theta_1 -5.333\theta_2} \\ & 1 & e^{-1.333\theta_2} & e^{-1.333\theta_1 -1.333\theta_2} & e^{-1.333\theta_1} & e^{-1.333\theta_1 -1.333\theta_2} & e^{-5.333\theta_1 -1.333\theta_2} & e^{-5.333\theta_1} & e^{-5.333\theta_1 -1.333\theta_2} \\ & & 1 & e^{-1.333\theta_1 -5.333\theta_2} & e^{-1.333\theta_1 -1.333\theta_2} & e^{-1.333\theta_1} & e^{-5.333\theta_1 -5.333\theta_2} & e^{-5.333\theta_1 -1.333\theta_2} & e^{-5.333\theta_1} \\ & & & 1 & e^{-1.333\theta_2} & e^{-5.333\theta_2} & e^{-1.333\theta_1} & e^{-1.333\theta_1 -1.333\theta_2} & e^{-1.333\theta_1 -5.333\theta_2} \\ & & & & 1 & e^{-1.333\theta_2} & e^{-1.333\theta_1 -1.333\theta_2} & e^{-1.333\theta_1} & e^{-1.333\theta_1 -1.333\theta_2} \\ & & Sym & & & 1 & e^{-1.333\theta_1 -5.333\theta_2} & e^{-1.333\theta_1 -1.333\theta_2} & e^{-1.333\theta_1} \\ & & & & & & 1 & e^{-1.333\theta_2} & e^{-5.333\theta_2} \\ & & & & & & & 1 & e^{-1.333\theta_2} \\ & & & & & & & & 1 \end{bmatrix} \quad (42)$$

Note that the distance is calculated by the normalized data. If we select a constant underlying global model, $\mathbf{F}$ becomes simply a column vector of ones:

$$\mathbf{F} = \begin{bmatrix} 1, 1, 1, 1, 1, 1, 1, 1, 1 \end{bmatrix}^T \quad (43)$$

The estimation of global model is now constructed as follows:

$$\hat{\beta} = (\mathbf{F}^T \mathbf{R}^{-1} \mathbf{F})^{-1} \mathbf{F}^T \mathbf{R}^{-1} \mathbf{Y} \quad (44)$$

Note that $\mathbf{Y}$ is also normalized response. Estimation of process variance is given as

$$\hat{\sigma}_z^2 = \frac{(\mathbf{Y}-\mathbf{F}\hat{\boldsymbol{\beta}})^T \mathbf{R}^{-1}(\mathbf{Y}-\mathbf{F}\hat{\boldsymbol{\beta}})}{9} \tag{45}$$

Maximum likelihood estimation for θ_1 and θ_2 is found by the following expression:

$$\min_{\theta_1,\theta_2} \hat{\sigma}^2 |\mathbf{R}|^{1/9} \tag{46}$$

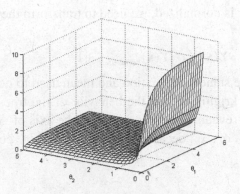

Fig. 7 A plot of likelihood function

Note that the maximum likelihood function is only a function of θ_1 and θ_2. When optimization range of correlation parameters is given in $\theta_1, \theta_2 \in [0.1, 5]$, the "best" kriging model to fit these sample points is obtained as $\theta_1^* = 0.1$ and $\theta_2^* = 5$. These values are the lower-limit and upper-limit, respectively. A plot of maximum likelihood function is given in Fig. 7. As shown in Fig. 7, you can find that the optimum solution exists on the lower left of the plot.

Now, we can predict the response at an untried point within design range. We can substitute the optimum solution into correlation matrix as follows:

$$\mathbf{R} = \begin{bmatrix} 1 & 1.273\times10^{-3} & 2.623\times10^{-12} & 8.752\times10^{-1} & 1.114\times10^{-3} & 2.296\times10^{-12} & 5.867\times10^{-1} & 7.466\times10^{-4} & 1.539\times10^{-12} \\ & 1 & 1.273\times10^{-3} & 1.114\times10^{-3} & 8.752\times10^{-1} & 1.114\times10^{-3} & 7.466\times10^{-4} & 5.867\times10^{-1} & 7.466\times10^{-4} \\ & & 1 & 2.296\times10^{-12} & 1.114\times10^{-3} & 8.752\times10^{-1} & 1.539\times10^{-12} & 7.466\times10^{-4} & 5.867\times10^{-1} \\ & & & 1 & 1.273\times10^{-3} & 2.623\times10^{-12} & 8.752\times10^{-1} & 1.114\times10^{-3} & 2.296\times10^{-12} \\ & & & & 1 & 1.273\times10^{-3} & 1.114\times10^{-3} & 8.752\times10^{-1} & 1.114\times10^{-3} \\ & & Sym & & & 1 & 2.296\times10^{-12} & 1.114\times10^{-3} & 8.752\times10^{-1} \\ & & & & & & 1 & 1.273\times10^{-3} & 2.623\times10^{-12} \\ & & & & & & & 1 & 1.273\times10^{-3} \\ & & & & & & & & 1 \end{bmatrix}$$

$$\tag{47}$$

We need to know the form of the correlation vector as follows:

$$\mathbf{r}(\mathbf{x}) = [R(\mathbf{x}, \mathbf{s}_1), R(\mathbf{x}, \mathbf{s}_2), R(\mathbf{x}, \mathbf{s}_3), R(\mathbf{x}, \mathbf{s}_4), \cdots$$
$$R(\mathbf{x}, \mathbf{s}_5), R(\mathbf{x}, \mathbf{s}_6), R(\mathbf{x}, \mathbf{s}_7), R(\mathbf{x}, \mathbf{s}_8), R(\mathbf{x}, \mathbf{s}_9)]^T \tag{48}$$

It is important to note that the prediction point should be normalized. Correlation vector is also only function of $\mathbf{x}$.

Finally, we can obtain the predictor of kriging as follows:

$$\hat{Y}(\mathbf{x})_{normalized} = \hat{\boldsymbol{\beta}} + r(\mathbf{x})^T \mathbf{R}^{-1}(\mathbf{Y} - \mathbf{F}\hat{\boldsymbol{\beta}}) \tag{49}$$

Because the predictor is normalized, we need to transform the predicted value into real value as follows:

$$\hat{Y}(\mathbf{x}) = \sigma_Y \hat{Y}(\mathbf{x})_{normalized} + \overline{Y} \tag{50}$$

where σ_Y and $\overline{Y}$ are standard deviation and mean used to normalize responses.

Fig. 8 shows that kriging predict the behavior of the true function accurately. In addition, predictor reveals that response along the axis of x_2 is more nonlinear than along the axis of x_1 because the estimation of correlation parameter $\theta_2^* = 5 > \theta_1^* = 0.1$.

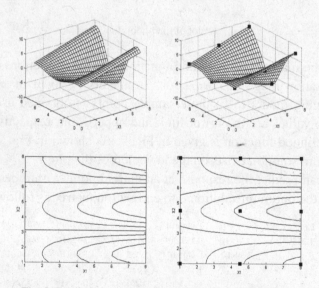

Fig. 8 Contour plots of true function and kriging predictor

3. Sampling Techniques

Prediction performance of metamodel is directly influenced by the nonlinearity of an original function, sampling strategies and types of metamodels. Among these three factors, nonlinearity of functions depends largely on the provided data. Therefore, the other two factors are main concerns in metamodeling. In this section, we consider sampling strategies in order to obtain reasonable metamodels with minimum sampling points.

In computer experiment to construct metamodel, a good sampling strategy is defined as one that provides good performance of the metamodel with minimum number of analyses. In this chapter, however, we only compare the features of various sampling techniques because the subject is beyond the scope of the present chapter.

For systematic taxonomy of a variety of sampling techniques, we classify sampling strategies into 'domain-based sampling' and 'response-based sampling' in according to information type that they use during sampling process. For example, mean squared error (MSE) approach[20] selects sequentially a sample point with maximum uncertainty of kriging model. Therefore MSE belongs to response-based sampling. On the other hand, maximin distance sampling is a domain-based sampling because it uses only information of design domain, the minimum distance between any two sample points. Similarly, sampling techniques using optimal criteria, for example, maximum entropy criterion[21] and maximin eigenvalue criterion [22], are domain-based sampling.

Feature of response-based sampling is to select a sample point or several points by only sequential scheme: Initial sample points are determined and then additional sample point is sequentially selected by specific criteria. This is often referred to as sequential sampling technique. Sequential sampling is originally developed to enhance the efficiency of sampling process by using the information of the existing metamodel.

One of advantages of sequential sampling is that designer can decide systematically the number of sampling points during sampling process. We can monitor accuracy of metamodel during sample process because we add sequentially sample point. Thus, we can decide when the sampling process is terminated as accurate metamodel is obtained. In this case, validation strategy becomes an important research topic.

In addition, sequential sampling technique requires cheap computational cost compared to one-stage optimal sampling. For example, let us consider sampling problem to decide n sample points in n_d dimension. When we use optimality criteria such as entropy, distance and eigenvalue, one-stage sampling problem is to decide coordinates of all sample points, i.e., $(n \times n_d)$ dimensional problem, at once. Moreover, computational cost of the sampling problem is extremely prohibitive as number of sample points and dimension of design variables increase. On the other hand, sequential sampling problem keeps less computational cost because it divides this large-scale optimization problem into small-scale problem to find one or several sample points. Thus, in sequential sampling, computational costs increase linearly with respect to number of sample points and dimensionality.

In category of response-based sampling often used in geo-statistics, infill sampling criteria are noticeable. A variety of infill sampling techniques sequentially locate a sample point to balance the improvement of current best point and global accuracy. To embody the concept, infill sampling criterion adopts the probability of improving upon best sample point. Note that best sampling point is usually local minimum. In this chapter, infill sampling such as Kushner's criterion, expected improvement, WB1 and WB2 are explained.[11-15]

Domain-based sampling is based on the space-filling concept that it evenly fills design domain with sample points. When optimality criterion is used, benefit of the domain-based sampling is to select sample points by both sequential and one-stage scheme. However, it is inefficient to determine all sample points at once, i.e., one-stage sampling because it cannot reflect the nonlinearity of original function, relative importance of design variables, and expected improvement of local optimum. Moreover, it is not easy for designer to decide the number of sample points that is enough to approximate nonlinear response. Thus, domain-based sampling by using all-at-once scheme is suggested as an initial sampling before starting sequential sampling.

To simply distinguish between domain-based sampling and response-based sampling, we need to check whether the result of sampling technique can apply to multi-response. Domain-based sampling evenly spreads out sample points over design domain. This feature can provide average prediction performance of metamodels for all response. However, because response-based sampling uses the metamodel of information of a response to approximate may not guarantee good accuracy of metamodels for other responses.

3.1. *Domain-based Sampling*

3.1.1. *Maximum Entropy Sampling*

Let $\mathbf{X}_s = \{\mathbf{x}_1, \mathbf{x}_2, ..., \mathbf{x}_n\}$ be sampled set with density function $p(\mathbf{X}_s)$. Entropy of sampled set is defined as follows[21]:

$$Ent(\mathbf{X}_s) = E[-\ln p(\mathbf{X}_s)] \tag{51}$$

where $Ent(\cdot)$ denotes entropy.

Assuming that a random vector $\mathbf{x}_i \in \mathbf{X}_s$ has multivariate density function, entropy criterion is derived as follows:

$$Ent(\mathbf{X}_s) = \frac{1}{2}\ln\left((2\pi e)^n |\mathbf{V}|\right) \tag{52}$$

where n and $\mathbf{V}$ represent the number of samples and covariance matrix of samples, respectively. Note that covariance matrix can be defined as the product of process variance and spatial correlation matrix. In sampling problem, the number of sample points is usually pre-determined. It means that n is fixed during optimization. Therefore, by eliminating the constant terms of entropy equation, maximization of the entropy becomes

$$Maximize |\mathbf{R}| \tag{53}$$

where $\mathbf{R}$ is the correlation matrix used in predictor of kriging. In maximum entropy sampling, correlation parameter is user-defined parameter. Design variables of sampling problem are all coordinates of sample points.

Entropy was originally used by Shannon to quantify the "amount of information." Note that maximization of the determinant of correlation matrix can be interpreted as selection of experiment to acquire the maximum amount of information. Intuitively, if the sample points are located in particular design domain, the information obtained from the sampling result will be poor. However, if sample points spread uniformly over design domain, much information can be obtained from the sampling result.

Figs. 9 and 10 show illustration of maximum entropy sampling by Gaussian correlation function. For same number of sample points, sampling results depends severely on the choice of correlation parameter. In entropy sampling, therefore, proper correlation parameter is a prerequisite for arranging sample points uniformly throughout the design domain. It is interesting that the distribution of sample points is point-symmetric with respect to the center of design domain or symmetric with respect to a particular dotted line.

Examples of maximum entropy sampling of 25 points by sequential scheme are given in Figs. 11 and 12. When initial samplings are different each other, sequential sampling may have different distribution of final sample points as shown in Figs. 11 and 12.

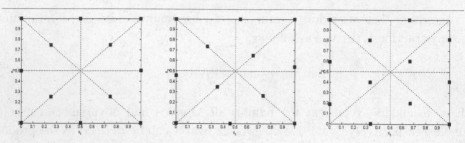

Fig. 9 Maximum entropy designs of 12 points: correlation parameters are assigned as 1, 10 and 50, respectively

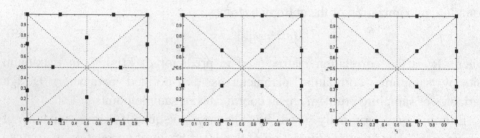

Fig. 10 Maximum entropy sampling design of 16 points: correlation parameters are assigned as 1, 10 and 50, respectively

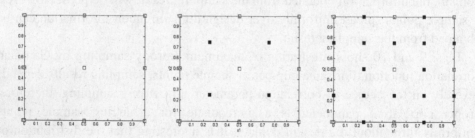

Fig. 11 Influence of 9 initial sample points by maximum entropy sequential sampling

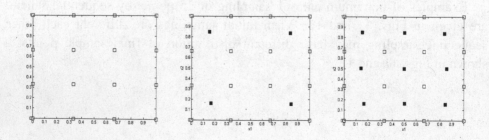

Fig. 12 Influence of 16 initial sampling points by maximum entropy sequential sampling

3.1.2. *Maximin Eigenvalue Sampling*

Maximin eigenvalue sampling[22] is to select sampling points by maximizing minimum eigenvalue of the eigen-problem for correlation matrix, i.e., $(\mathbf{R} - \lambda)\mathbf{u} = \mathbf{0}$:

$$maximize \ \lambda_{min} \tag{54}$$

Here, minimum eigenvalue is theoretically larger than zero because correlation matrix $\mathbf{R}$ is positive-definite.

It is important to note that maximin eigenvalue sampling is a variant of the maximum entropy sampling. Recall that the determinant of correlation matrix is equivalent to the product of all eigenvalues as follows:

$$|\mathbf{R}| = \lambda_1 \lambda_2 \cdots \lambda_n \ (0 < \lambda_1 \leq \lambda_2 \leq \cdots \leq \lambda_n) \tag{55}$$

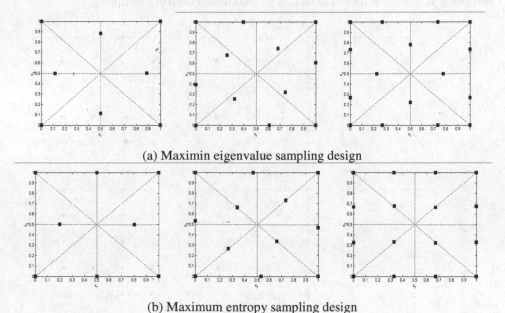

(a) Maximin eigenvalue sampling design

(b) Maximum entropy sampling design

Fig. 13 Comparison of maximum eigenvalue and maximum entropy criteria for 8, 12, and 16 sampling points

Maximin eigenvalue sampling adopts only the first eigenvalue as sampling criterion. Therefore maximization of first eigenvalue, i.e., minimum eigenvalue, leads to effect to maximize entropy. Gaussian function is taken as spatial correlation function and its common correlation parameter is pre-determined as $\theta = 8$. As shown in Fig. 13, however, the sampling results show considerable

466   T. H. Lee and J. J. Jung

dissimilarity in comparison with those of maximum entropy sampling. Distribution of sample points obtained by maximin eigenvalue sampling is geometrically point-symmetric with respect to center point or line-symmetric with respect to a dotted line.

Note that the geometric symmetry can be an evidence of optimal sampling design because the sampling result will be asymmetric when sampling points gather together in particular region or do not converge enough toward optimal solution. It is pointed that the symmetric characteristics doesn't always mean global optimum but the solution is one of various optimum solutions.

Fig. 14 shows the sequential maximin eigenvalue sampling for correlation parameters $\theta = 3$ and $\theta = 10$, respectively. For $\theta = 3$, sampling results reveal slightly irregular distribution compared with those at $\theta = 10$. However, as correlation parameter increases, sample distribution of maximin eigenvalue sampling tends to be identical to that of maximum entropy sampling

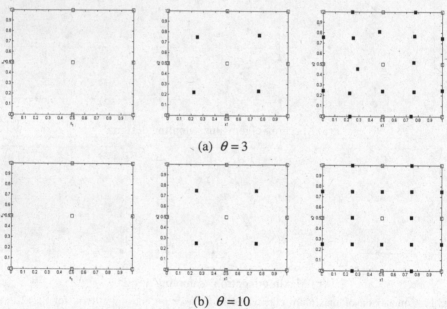

(a) $\theta = 3$

(b) $\theta = 10$

Fig. 14 Sequential maximin eigenvalue sampling design for $\theta = 3$ and $\theta = 10$ from 9 initial sampling points

3.1.3. Maximin Distance Sampling

Maximin distance sampling[23] is to select sampling points by maximizing the minimum distance between any two sample points as follows:

$$Maximize \ \min d(\mathbf{x}_i, \mathbf{x}_j) \tag{56}$$

where $d(\mathbf{x}_i, \mathbf{x}_j)$ denotes the distance between two sample points $\mathbf{x}_i$ and $\mathbf{x}_j$. To balance the magnitudes of the distance in each dimension, design variables are usually transformed into the normalized space $[0,1]^{n_d}$.

Contrast to both maximum entropy sampling and maximin eigenvalue sampling, maximin distance sampling is not required to assign any parameter.

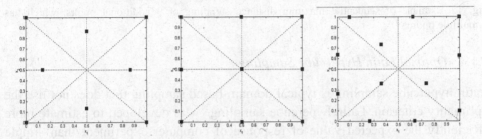

Fig. 15 Maximin distance sampling for 8, 9, and 12 sample points

In some cases, results of sequential maximin distance sampling are identical to those of other sampling techniques such as maximum entropy sampling and maximin eigenvalue sampling as you can see in Fig. 16.

Sequential maximin distance sampling has many multiple optimum solutions during sequential optimization process. Because there are nine optimum solutions as shown in Fig. 17, additional nine sample points can be randomly selected. However, both maximum entropy sampling and maximin eigenvalue sampling tend to always locate sample point sequentially with specific regularity (See Fig. 18): First sample points is filled around boundary and then last point is arranged at the middle of design domain.

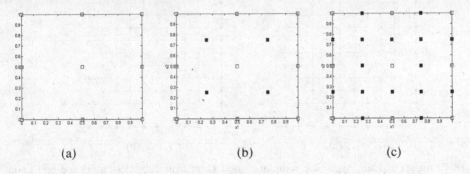

(a)   (b)   (c)

Fig. 16 Sequential maximin distance sampling from 9 initial sample points: initial sampling (a), 13 sampling points (b), and 25 sampling points (c)

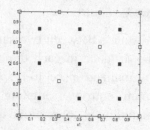

Fig. 17 Example of sequential maximin distance sampling of 9 additional points, which has 9-multiple optima

3.1.4. *Optimal Latin Hypercube Sampling*

Latin hypercube sampling is typical domain-based sampling that does not use the optimality criterion. Latin hypercube sampling[24] was developed to estimate more efficiently the expected value of response of computer experiment than simple random sampling. Feature of this method is to cover the design domain without replicated coordinate values. The number of divisions in each dimension is equivalent to the number of sample points. Fig. 19 shows illustration of nine points by Latin hypercube. As shown in Fig. 19, the projection of sample points to axis of each design variable has the good uniformity.

Moreover, it is computationally cheap to construct. As shown in Fig. 19, however, distribution of Latin hypercube may sometimes be poor because the level of sample points is randomly permuted. As a result, Latin hypercube sampling often produces poor space-filling in spite of good projection properties.

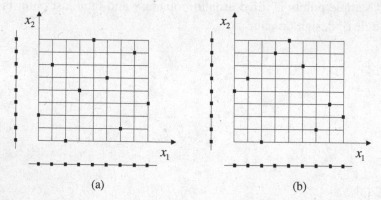

Fig. 18 Examples of Latin hypercube sampling: good Latin hypercube design (a) and bad Latin hypercube design (b)

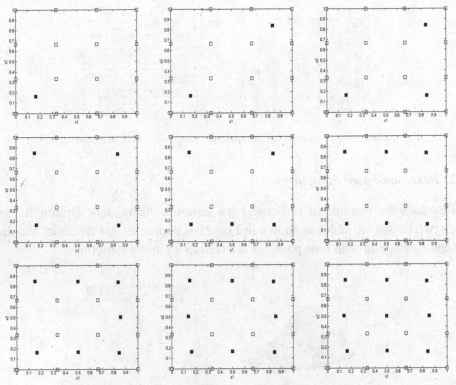

Fig. 19 Example of sequential maximum entropy sampling

Thus, optimal Latin hypercube sampling is suggested to select an optimal design among all possible Latin hypercube samplings. Optimal Latin hypercube sampling is a compromise between optimality criterion and Latin hypercube with the good projection properties.[25] Optimal Latin hypercube sampling has some advantages. From the view of computational cost, optimal Latin hypercube sampling is usually more efficient than other optimal samplings since it evaluates optimal criterion for only candidates of Latin hypercube sampling. In addition, when entropy is adopted as optimal criterion, optimal Latin hypercube sampling does not produce erratic optimum solution that arises from improper choice of correlation parameter. For this reason, we suppose that Latin hypercube sampling scatters sample points enough throughout design domain. Fig. 20 shows the illustrations of 9 and 16 points optimal Latin hypercube sampling. The results represent both good projection and uniformity of sampling points within design domain.

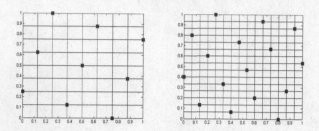

Fig. 20 Examples of optimal Latin hypercube sampling: 9 and 16 sampling points

3.2. *Response-based Sampling*

To explain the features of response-base sampling techniques, Branin function with two design variables as shown in Fig. 21 is chosen as test function. Maximin distance sampling with nine points is selected as an initial sampling.

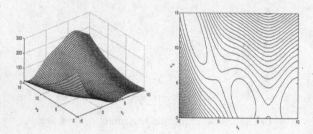

Fig. 21 Branin function

Branin function has three distinct global minimums with same function value as follows:

$$f = (x_2 - 5.1\frac{x_1^2}{4\pi^2} + \frac{5x_1}{\pi} - 6)^2 + 10(1 - \frac{1}{8\pi})\cos x_1 + 10$$

$$-5 \leq x_1 \leq 10 \quad 0 \leq x_2 \leq 15$$

(57)

$$f(-\pi, 12.275) = f(\pi, 2.275) = f(9.42478, 2.475) = 0.397887$$

3.2.1. *Mean Squared Error Approach*

Mean squared error approach is to select a new sample point with the largest uncertainty of prediction.[20] This approach is applicable only when metamodel provides uncertainty of prediction. Criterion of MSE approach is given as

Maximize MSE   (58)

where *MSE* is mean squared error of kriging model from the existing sampled points. As shown in Fig. 22, MSE fills design domain with sample points by adding the point with largest uncertainty. Circle marker denotes initial sampling and x-maker a point sequentially selected by MSE criterion. Because MSE approach employs only stochastic prediction error, the tendency to sample around global optima is not found.

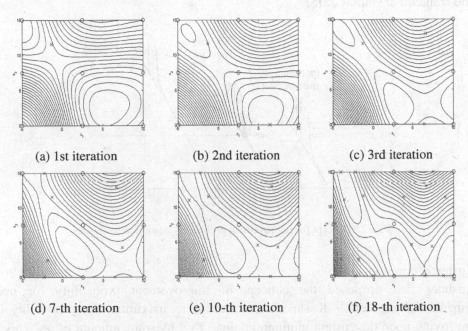

(a) 1st iteration	(b) 2nd iteration	(c) 3rd iteration
(d) 7-th iteration	(e) 10-th iteration	(f) 18-th iteration

Fig. 22 Sampling results of mean square error approach

3.2.2. *Infill Sampling Criterion*

Infill sampling criterion originally came from the branch of geo-statistics. A variety of infill sampling criteria searches for the minimum of the approximate model, the location of maximum uncertainty, or some compromise between these.

To understand the concept of infill sampling, we quantify the probability of improving the current minimum value. Current minimum value is the response that has smallest value among sampled points. Fig. 23 describes the probability that response at the specific points will improve current minimum value. Here, we assume that the random variable is assumed to be normally distribution. We can clarify the improvement probability of current minimum value as follows:

$$\Phi(I) = \Phi\left(\frac{f_{\min} - \hat{Y}(\mathbf{x})}{\sqrt{MSE}}\right) \tag{59}$$

where $\Phi(\cdot)$ is normal cumulative distribution function. Stochastic variable I is assumed to 'be like' the realization of random variable $Y(\mathbf{x})$ with mean $\hat{Y}(\mathbf{x})$ and standard deviation $\sqrt{MSE}$.

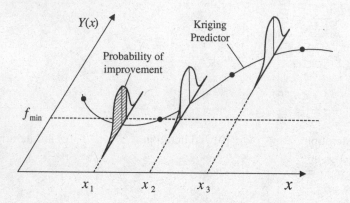

Fig. 23 Concept for improvement probability described by shades

Kushner' criterion

Kushner first proposed the concept of improvement probability for one-dimensional problem.[11] Kushner's criterion is to maximize the probability of improving upon the current minimum point, $f_{\min}$ by some amount of ε. This is quantified as

$$Maximize \ \ \Phi\left(\frac{(f_{\min} - \varepsilon) - \hat{Y}(\mathbf{x})}{\sqrt{MSE}}\right) \tag{60}$$

The value ε is a parameter that may be chosen by the user. Usually, parameter ε is preferred as from 0.1 to 1% of $f_{\min}$. Due to this reason, we can presume that the probability of improvement becomes so large around current minimum point because the uncertainty of prediction in the region approaches very small value. Fig. 24 shows the illustration of Kushner' criterion with 1% margin. As shown from iterations 1 to 3, Kushner's criterion tends to locate a sample point around current minimum point rather than local minimum of metamodel.

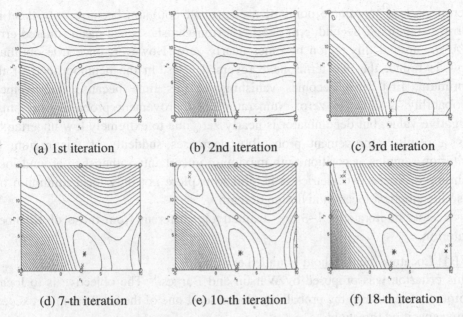

(a) 1st iteration   (b) 2nd iteration   (c) 3rd iteration

(d) 7-th iteration   (e) 10-th iteration   (f) 18-th iteration

Fig. 24 Sampling results by Kushner' criterion: 'o' and 'x' denote initial and sequential sampling points, respectively

Expected improvement

The expected value of the improvement[14,15] is defined as

$$Maximize \ EI = \begin{cases} (f_{min} - \hat{Y})\Phi(I) + \sqrt{MSE}\phi(I) & if \ MSE > 0 \\ 0 & if \ MSE = 0 \end{cases} \tag{61}$$

where $\phi(\cdot)$ denotes the probability density function. Note that mean squared error becomes zero at sampled points.

A glance at the expected improvement reveals two important trends. The first term is the difference between the current minimum and the predicted value multiplied by the probability that $Y(\mathbf{x})$ is smaller than f_{min}. For this reason, it is large where $\hat{Y}(\mathbf{x})$ is likely smaller than f_{min}. The second term tends to be large where high uncertainty occurs. Thus, the expected improvement is large for both regions of likely improvement and regions of high uncertainty. It is helpful to enhance local accuracy of current minimum value and global prediction performance of metamodel over whole design domain.

Let us figure out the principle of expected improvement for exploring both regions. First, assume that the minimum sampled point is closely around the true local minimum, but true local minimum is still unexplored. In this case, several iterations lead to sample points denser around optimum because only the first term is dominant: Second term of EI criterion vanishes since mean squared error $\sqrt{MSE}$ around this region becomes nearly zero. However, once true optimum now is found, the two terms have reversed roles. In the neighborhood of the optimum, first term becomes vanishing at this time because improvement probability is close to zero: Numerator of improvement probability is small negative value, but denominator is nearly zero due to extremely low uncertainty. As a result, improvement probability decreases suddenly at this region. EI criterion samples at position with maximum uncertainty rather than around local minimum. These two metrics is helpful to explore not only local minimum but also the emptiest location in design domain.

Results of iterations 1~3 are distinctively different from those of Kushner' criterion.

WB1: Locating the threshold-bounded extreme

This criterion was proposed by Watson and Barnes.[12] The objective is to locate points that maximize the probability that at least one of the infill samples exceeds some specified threshold.

$$\text{Maximize } WB_1 = \Phi\left(\frac{f_{\min} - \hat{Y}(\mathbf{x})}{\sqrt{MSE}}\right) \tag{62}$$

Note that the criterion is a special case of Kushner's criterion for $\varepsilon = 0$. The behavior of this criterion is illustrated in Fig. 26. The results of WB1 criterion are very similar to those of Kushner's criterion.

WB2: Locating the regional extreme

This criterion is to minimize the expected value of the smallest observation once the infill samples have been added:

$$\text{Minimize } WB_2 = \begin{cases} \hat{Y} + (f_{\min} - \hat{Y})\Phi(I) + \sqrt{MSE}\phi(I) & \text{if } MSE > 0 \\ 0 & \text{if } MSE = 0 \end{cases} \tag{63}$$

The criterion is the form that adds predicted value to EI criterion. Thus, it provides slightly more weight to local search than does the EI criterion. As shown in Fig. 27, this criterion tends to sample excessively many points in the neighborhood of current minimum value. As a result, after 18 iterations, the metamodel does not approximate other two minima exactly.

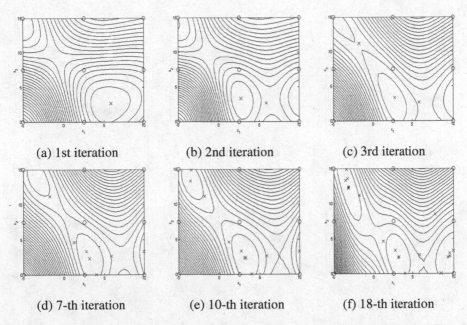

(a) 1st iteration   (b) 2nd iteration   (c) 3rd iteration

(d) 7-th iteration   (e) 10-th iteration   (f) 18-th iteration

Fig. 25 Sampling results by using expected improvement sampling: 'o' and 'x' denote initial and sequential sampling points, respectively

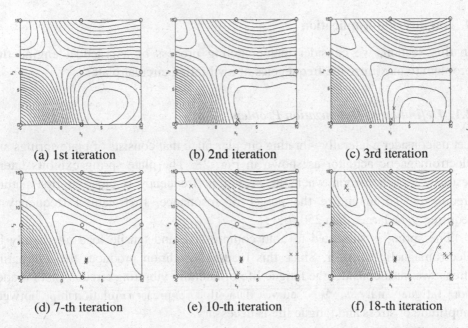

(a) 1st iteration   (b) 2nd iteration   (c) 3rd iteration

(d) 7-th iteration   (e) 10-th iteration   (f) 18-th iteration

Fig. 26 Sampling results by WB1: 'o' and 'x' denote initial and sequential sampling points, respectively

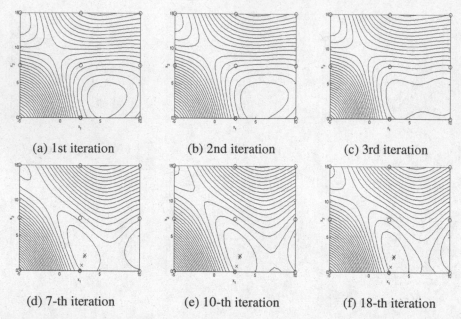

(a) 1st iteration   (b) 2nd iteration   (c) 3rd iteration

(d) 7-th iteration   (e) 10-th iteration   (f) 18-th iteration

Fig. 27 Sampling results by WB2: 'o' and 'x' denote initial and sequential sampling points, respectively

4. Engineering Application

In this section, we consider metamodeling process of a practical engineering problem to maximize the frequencies of a vibrating circular plate.

4.1. *Definition of Optimization Problem*

Let us consider a laterally vibrating circular plate that consists of plate springs and electromagnetic actuator as shown in Fig. 28. The plate spring exhibits lateral resonance phenomenon when the excitation frequency is close to first natural frequency. To maximize the power of resonance, first natural frequency is required to be larger than $250\,(Hz)$.

The spring is subjected to both compressive and tensile load alternately by electromagnetic actuator. Since this loading condition produces the alternating stress, we must consider the fatigue life of laterally vibrating circular plate model. For fatigue analysis, S-N curve data that expresses relationship between amplitude of stress and fatigue life is necessary.

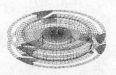

Fig. 28 Finite element model of laterally vibrating circular plate

To facilitate design optimization, however, we simply use static stress as design constraint. This is because many researches demonstrate that fatigue limit is approximately 50% of ultimate strength. As shown in Table 4, the material properties reveal this to be a reasonable assumption. In addition, we adopt safety factor of 1.6. Therefore, allowable stress constraint is specified as 300 MPa. For weight minimization of vibrating circular plate, we will approximate the three responses, i.e. mass, first natural frequency and maximum stress. As boundary condition, all degree of freedom on the edge of spring are fixed. For static analysis, concentrated load of $1(N)$ is applied at the center of BNV model. Material properties of plate spring and magnetic are given in Table 3 and Table 4, respectively.

Table 3 Material properties of vibrating circular plate spring

Material	Beryllium Copper
Modulus of elasticity, E	$128\,(Gpa)$
Poisson' ratio, ν	0.29
Yield strength, σ_Y	$758\,(Mpa)$
Mass density, ρ	$8250\,(kg\,/\,m^3)$
Ultimate strength, σ_{ult}	$966\,(Mpa)$
Endurance limit, σ_e	$430\,(Mpa)$

Table 4 Material properties of the magnetic

Material	Steel, carbon
Modulus of elasticity, E	$210\,(Gpa)$
Poisson' ratio, ν	0.28
Mass density, ρ	$7800\,(kg\,/\,m^3)$

For mass minimization, angle (x_1) and width (x_2) vibrating of circular plate are selected as design variables as shown in Fig. 29.

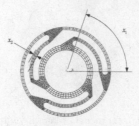

Fig. 29 Description of design variables for optimum design of vibrating circular plate

Finally, we formulate the optimization problem of vibrating circular plate as follows:

$$Min \ mass$$
$$subject \ to \ 250 - f_{1st} \leq 0$$
$$\sigma_{max} \leq 300(Mpa) \quad\quad\quad (64)$$
$$60° \leq x_1 \leq 90°$$
$$0.5 \leq x_2 \leq 1.1(mm)$$

4.2. *Metamodels of Laterally Vibrating Circular Plate*

For metamodel-based design optimization, maximum entropy sampling, maximin distance sampling and maximin eigenvalue sampling are used to approximate the three responses. As response-based sampling, mean squared error and expected improvement sampling are adopted to compare with results of domain-based sampling. All sampling techniques sequentially add sample points. Full factorial design of 3 levels is employed as initial sampling and the number of total sample points is 25.

Fig. 30 shows approximation of three responses by sequential mean square error approach. Because mean squared error approach is response-based sampling, sampling results for three responses are different from each other. Mean square error approach tends to locate many sample points on the boundary of design domain. This is because mean square error generally has large value around the edge of the region. Approximation of mass is linear while those of maximum stress and frequency are slightly nonlinear. In approximation of stress, however, scarce sample points in the middle of design domain bring about erratic approximation.

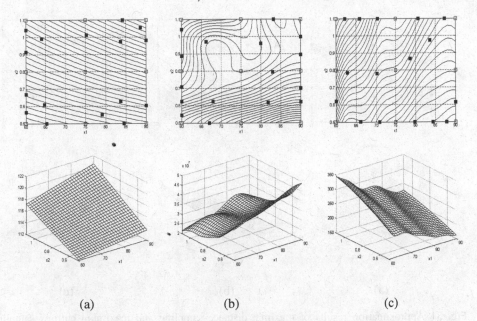

Fig. 30 Approximation results of mean square error approach: (a) mass, (b) stress, and (c) first frequency

Maximin distance sampling and maximum entropy sampling show same sampling results, where distribution of sample points is equivalent to that of grid sampling. Two approaches evenly fill design domain with sample points for approximation of three responses. Two approaches approximate stress and frequency smoothly than the mean square error method.

Expected improvement sampling shows many replications around minimum sample points. However, the approximation does not show inaccurate performance because the behavior of original function itself is nearly linear. If the original function shows high nonlinearity, choice of initial sampling is very important to approximate the response well by the expected improvement.

To verify accuracy of kriging models, R-square is employed:

$$R^2 = 1 - \frac{\sum_{i=1}^{n_p}(Y_i - \hat{Y}_i)^2}{\sum_{i=1}^{n_p}(Y_i - \overline{Y})^2} \tag{65}$$

where $\hat{Y}_i$ is the corresponding predicted value for the observed value Y_i at validation point. $\overline{Y}$ is the mean of the actual value at validation points and n_p is the number of validation points.

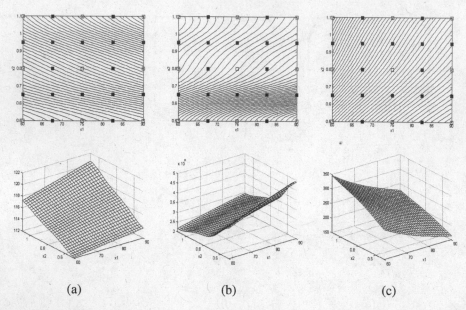

(a)　　　　　　　　(b)　　　　　　　　(c)

Fig. 31 Approximation results of maximin distance sampling and maximum entropy sampling: (a) mass, (b) stress and (c) frequency

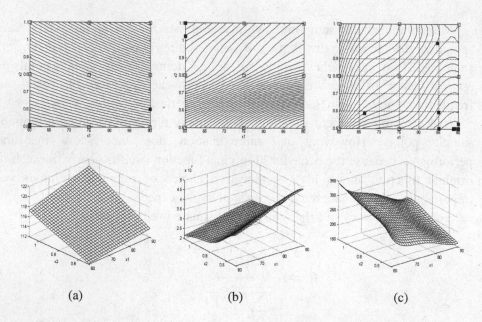

(a)　　　　　　　　(b)　　　　　　　　(c)

Fig. 32 Expected improvement sampling: (a) mass, (b) stress, and (c) first frequency

Table 5 Validation results by R-square

	Mass	Stress	First Frequency
Mean square error	1	0.889	0.967
Maximin distance sampling	0.998	0.879	0.980
Maximum entropy sampling	0.998	0.879	0.980
Expected improvement	0.999	0.8813	0.974

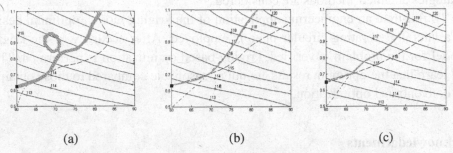

(a)　　　　　　　　　(b)　　　　　　　　　(c)

Fig. 33 Optimum results by various sampling techniques: mean squared error (a), maximin distance and maximum entropy (b), and expected improvement (c)

Validation data are obtained by grid sampling of 20×20. As R-square approaches 1, kriging model become accurate. Table 5 denotes the R-square of various approximations. The results show that approximations for three static responses are quite accurate. However, approximation of stress is relatively inaccurate than mass and frequency.

4.3. Optimum Results Based on Metamodels

Fig. 33 shows contour plots of objective function (solid line) and the boundary of active constraint (dotted line). Feasible region is the upper left portion of design space and optimum point is denoted by square dot.

Table 6 represents optimum solution obtained by conventional optimization based on high-fidelity simulation model. As shown in Fig. 33, all sampling techniques can find optimum solution well for this example.

Table 6 Optimum solution of conventional optimization

x_1(deg)	x_2(mm)	volume(mm^3)	stress(MPa)	frequency(Hz)
60	0.6251	113.223	299.98	276.132

5. Concluding Remarks

To replace complicated simulation models that are time consuming to analyze, we have presented a metamodel, the so-called kriging metamodel, to achieve an optimum solution within reasonable budget. Features of kriging metamodel are systematically described and mathematical derivation is performed to enhance the understanding of kriging metamodel. To pre-determine sampling points that are necessary for metamodel, domain-based sampling and response-based sampling strategies and their methods are considered.

To illustrate an engineering application of the kriging model, optimum design of a laterally vibrating circular plate is employed. After accuracy of the kriging model for the problem is verified, kriging-based optimization is performed. Then we learn that the optimum results are quite reasonable compared to those obtained by conventional optimization.

Acknowledgements

This work was supported by the Center of Innovative Design Optimization Technology (ERC of Korea Science and Engineering Foundation).

References

1. Kleijnen, J. P. C., 1987, *Statistical tools for simulation practitioners*, Marcel Dekker, NY.
2. Myers, R. H. and Montgomery, D. C., 1995, *Response Surface Methodology*, John Wiley & Sons, Inc., New York.
3. Sacks, J., Welch, W. J., Mitchell, T. J. and Wynn, H. P., 1989, "Design and analysis of computer experiments," *Statistical Science*, Vol. 4, pp. 409-435.
4. Sacks J., Schiller, S. B. and Welch, W. J., 1989, "Designs for computer experiments," *Technometics*, Vol. 31, No. 1, pp. 41~47.
5. Martin, J. D. and Simpson, T. W., 2003, "A study on the use of kriging models to approximate deterministic computer models," *ASME 2003 Design Engineering Technical Conference and Computers and Information in Engineering Conference*, Chicago, 2003, DETC2003/DAC-48762.
6. Gutmann, H. M., 2001, "A radial basis function method for global optimization, *Journal of Global optimization*, Vol. 19, pp. 201~227.
7. Hardy, R. L., 1971, "Multiquadratic equations of topography and other irregular surfaces," *Journal of Geophys. Res.*, Vol. 76, pp. 1905~1915.
8. Friedman, J. H., 1991, "Multivariate adaptive regression splines," *The Annals of Statistics*, Vol. 19, pp. 1-141.
9. Koehler, J. R. and Owen, A. B., 1996, "Computer experiments," in Ghosh, S. and Rao, C.R., eds, Handbook of Statistics, 13, pp. 261~308, Elsevier Science, New York.
10. Simpson, T. W., Peplinski, J. D., Koch, P. N. and Allen, J. K., 2001, "Metamodels for computer-based engineering design: survey and recommendations," *Engineering with Computers*, Vol. 17, pp. 129-150.

11. Kushner, H. J., 1964, "A new method of locating the maximum of an arbitrary multi-peak curve in the presence of noise," *Journal of Basic Engineering*, Vol. 86, pp. 97~106.

12. Watson, A. G. and Barnes, R. J., 1995, "Infill sampling criteria to locate extremes," *Mathematical Geology*, Vol. 27, No. 5, pp. 589~608.

13. Sasena, M. J., Papalambros, P. Y. and Goovaerts P., 2000, "Metamodeling sampling criteria in a global optimization framework," *8th AIAA/USAF/NASA/ISSMO Symposium on Multidisciplinary Analysis and Optimization*, Sept. 2000, Long Beach, CA.

14. Jones, D. R., 2001, "A taxonomy of global optimization methods based on response surfaces," *Journal of Global Optimization*, Vol. 21, pp. 345~383.

15. Schonlau, M., 1997, *Computer experiments and global optimization*, Ph.D. dissertation, Waterloo, Ontario, Canada

16. Krige, D. G., 1951, "A statistical approach to some basic mine valuation problems on the Witwatersrand," *Journal of the Chemical, Metallurgical and Mining Society of South Africa*, Vol. 52, pp. 119~139.

17. Matheron, G., 1963, "Principles of geostatistics," *Economic Geology*, Vol. 58, pp. 1246-1266.

18. Simpson, T. W., Mauery, T. M., Korte, J. J. and Mistree, F., 2001, "Kriging models for global approximation in simulation-based multidisciplinary design optimization," *AIAA Journal*, Vol. 39, No. 12, pp. 2233~2241.

19. Mitchell, T.J. and Morris, M.D., 1992, "The spatial correlation function approach to response surface estimation," *Proceedings of the 1992 Winter Simulation Conference*, pp. 565~571.

20. Jin, R., Chen, W. and Sudjianto, A., 2002, "On sequential sampling for global metamodeling in engineering design," *ASME 2002 Design Engineering Technical Conferences And Computers and Information in Engineering Conference*, DETC2002/DAC-34092, Montreal.

21. Shewry, M. C. and Wynn, H. P., 1987, "Maximum entropy sampling," *Journal of applied statistics*, Vol. 14, No. 2, pp. 165~170.

22. Lee, T. H. and Jung, J. J., 2004, "Maximin eigenvalue sampling for kriging model," *10th AIAA/ISSMO Multidisciplinary Analysis and Optimization Conference*, Albany, New York.

23. Johnson, M., Moore, L. and Ylvisaker, D., 1990, "Minimax and maximin distance designs," *Journal of Statistical Planning and Inference*, Vol. 26, pp. 131~148.

24. Mckay, M. D., Beckman, R. J. and Conover, W. J., 1979, "A comparison of three methods for selecting values of input variables in the analysis of output from a computer code," *Technometrics*, Vol. 21, No. 2, pp. 239~245

25. Park, J. S., 1994, "Optimal Latin-hypercube designs for computer experiments," *Journal of Statistical Planning and Inference*, Vol. 39, pp. 95~111.

Appendix

Let $\mathbf{M}$ be $(p+n)\times(p+n)$ block matrix that is defined using smaller matrices

$$\mathbf{M} = \begin{bmatrix} \mathbf{0} & \mathbf{F}^T \\ \mathbf{F} & \mathbf{R} \end{bmatrix} \tag{A.1}$$

Assume that $\mathbf{M}^{-1}$ exists and is expressed in terms of $\mathbf{A}_{p\times p}, \mathbf{B}_{p\times n}, \mathbf{C}_{n\times p}$ and $\mathbf{D}_{n\times n}$ as follows:

$$\mathbf{M}^{-1} = \begin{bmatrix} \mathbf{A} & \mathbf{B} \\ \mathbf{C} & \mathbf{D} \end{bmatrix} \tag{A.2}$$

Since $\mathbf{MM}^{-1} = \mathbf{I}$, we get the following equation

$$\begin{bmatrix} \mathbf{0} & \mathbf{F}^T \\ \mathbf{F} & \mathbf{R} \end{bmatrix} \begin{bmatrix} \mathbf{A} & \mathbf{B} \\ \mathbf{C} & \mathbf{D} \end{bmatrix} = \begin{bmatrix} \mathbf{I} & \mathbf{0} \\ \mathbf{0} & \mathbf{I} \end{bmatrix} \tag{A.3}$$

Multiplication of corresponding matrices in blocks yields

$$\mathbf{F}^T \mathbf{C} = \mathbf{I} \tag{A.4}$$

$$\mathbf{F}^T \mathbf{D} = \mathbf{0} \tag{A.5}$$

$$\mathbf{FA} + \mathbf{RC} = \mathbf{0} \tag{A.6}$$

$$\mathbf{FB} + \mathbf{RD} = \mathbf{I} \tag{A.7}$$

Because square matrix $\mathbf{R}$ is symmetric and positive-definite, the matrix is invertible.

$$\mathbf{C} = -\mathbf{R}^{-1}\mathbf{FA}$$
$$\mathbf{D} = \mathbf{R}^{-1}(\mathbf{I} - \mathbf{FB}) \tag{A.8}$$

If we substitute $\mathbf{C}$ and $\mathbf{D}$ matrices into Eq. (A.4) and Eq. (A.5), we obtain

$$\mathbf{A} = -(\mathbf{F}^T \mathbf{R}^{-1} \mathbf{F})^{-1}$$
$$\mathbf{B} = (\mathbf{F}^T \mathbf{R}^{-1} \mathbf{F})^{-1} \mathbf{F}^T \mathbf{R}^{-1} \tag{A.9}$$

Inserting above equation into Eq. (A.8) is

$$\mathbf{C} = \mathbf{R}^{-1} \mathbf{F} (\mathbf{F}^T \mathbf{R}^{-1} \mathbf{F})^{-1}$$
$$\mathbf{D} = \mathbf{R}^{-1}(\mathbf{I} - \mathbf{F}(\mathbf{F}^T \mathbf{R}^{-1} \mathbf{F})^{-1} \mathbf{F}^T \mathbf{R}^{-1}) \tag{A.10}$$

Therefore, the inverse of block matrix $\mathbf{M}$ becomes

$$\begin{bmatrix} \mathbf{0} & \mathbf{F}^T \\ \mathbf{F} & \mathbf{R} \end{bmatrix}^{-1} = \begin{bmatrix} -(\mathbf{F}^T \mathbf{R}^{-1} \mathbf{F})^{-1} & (\mathbf{F}^T \mathbf{R}^{-1} \mathbf{F})^{-1} \mathbf{F}^T \mathbf{R}^{-1} \\ \mathbf{R}^{-1} \mathbf{F}(\mathbf{F}^T \mathbf{R}^{-1} \mathbf{F})^{-1} & \mathbf{R}^{-1}(\mathbf{I} - \mathbf{F}(\mathbf{F}^T \mathbf{R}^{-1} \mathbf{F})^{-1} \mathbf{F}^T \mathbf{R}^{-1}) \end{bmatrix} \tag{A.11}$$

CHAPTER 17

ROBUST DESIGN BASED ON OPTIMIZATION

Byung Man Kwak

Department of Mechanical Engineering, Korea Advanced Institute of Science and Technology, 373-1 Guseong-dong, Yuseong-gu, Daejeon 305-701, South Korea
E-mail: bmkwak@khp.kaist.ac.kr

The goal of robust design is to make system performance least sensitive to uncertainties. There is, however, no dominant formulation. One approach is to minimize the standard deviation of the performance and the other a reliability-based approach where the reliability in terms of a limit function is maximized or it is imposed as a probabilistic constraint. As methods that do not require probability information, a sensitivity-based formulation using gradient index is introduced with a MEMS example comparing yield rates. A robust design in terms of relative safety index that is defined by a new concept of allowable set in the random variable space finds interesting applications to multi-body mechanism design and human motion trajectory. As a reliability-based approach, a recently developed moment method based on full factorial design of experiments is greatly improved in efficiency by using adaptive response surface constructions. An application of tolerance synthesis has illustrated the method and its practicality for industrial problems.

1. Introduction

In the design of a mechanical or structural system, once a design is obtained, it is manufactured and then used by various users. The design, however, can not be made as specified due to uncertainties in materials and manufacturing processes. Also environmental conditions for operation of the product are constantly changing. Not only loading conditions but many noise factors such as temperature, corrosion, wear and aging effect are involved during the life of a product. All these uncertainties bring in uncontrollable changes in performance and system characteristics. Robust design conceptually refers to a process of obtaining a product performance that is least sensitive to changes of environmental conditions or noise.[1] How can one quantify robustness? How can one formulate the problem?

2. Defining Robustness and Classifying Methods

Insensitiveness of performance to noise may be paraphrased as minimum scatter of performance, that is, standard deviation. So in this sense robust design may be translated as minimizing the standard deviation of a performance. In Fig. 1(a), design B is considered more robust than design A, because the standard deviation, σ_B, of design B is smaller than that of design A. Note that in this case to obtain σ, probability information of noise variables is required. A 100% robustness, although impossible, means that performance does not change ($\sigma = 0$) whatever changes happen in noise factors. Another view of robust design is to maximize the reliability that a performance is within a prescribed acceptance range or a specification. In Fig 1(b), the performance needs to be enclosed within the specified region as much as possible under noisy conditions. This actually implies minimizing the standard deviation, but the approach is different from purely minimizing standard deviation. This second problem belongs to reliability based design optimization (RBDO). Here, an acceptable level of reliability is first prescribed and then the system is designed in such a way that the reliability is larger than this level or simply maximized.

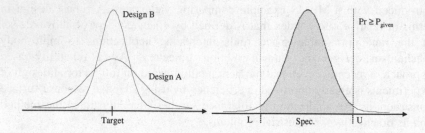

Fig. 1. Definition of robustness: (a) Robust design. (b) Reliability based design.

Both of the formulations need probability information. It is usually the case, however, that probability information is not available or hard to get. Therefore formulations without using detail probability information of random variables are a preferred choice for practical reasons and robust design often refers to this situation. The criteria for the "robust," however, are somewhat obscure in the literature. If one goes to consider constraints, further variations of formulations are possible. For example, a hybrid robust design formulation can be possible where a probability constraint with target reliability is imposed.

For the purpose of classification, robust design in the context of optimal design is arbitrarily categorized depending on whether any data of probability distribution is necessary or not, as summarized in Fig. 2. Since there is no dominant forms yet, the user has to select what he feels best describes his decision goal.

<u>Robust design methods</u>

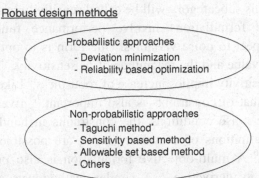

Probabilistic approaches
- Deviation minimization
- Reliability based optimization

Non-probabilistic approaches
- Taguchi method*
- Sensitivity based method
- Allowable set based method
- Others

Fig. 2. Classification of robust design methods. Taguchi method* is not an optimization based method. It may be considered as a probabilistic approach where standard deviation is obtained by DOE using orthogonal arrays.

In this chapter, first non-probabilistic approaches are introduced with a description of some recent ones, and then reliability based approaches. For the non-probabilistic methods, the basic idea is using some approximate quantity that indicates the standard deviation explained above. The most well-known and successful concept in robust design is the Taguchi method. Quality Engineering as Dr. Taguchi called it started in 1950's and it is an established procedure utilized by major companies and a topic of many text books authored from both industrial and academic sectors. [2] Quality of a product is viewed in terms of total loss to the society; occurred due to functional variation and harmful side effect. The Taguchi method is data driven and orthogonal array based design of experiment (DOE) method with the concept of signal-to-noise ratio. The original Taguchi method has been modified to be fit for structural optimization,[3] especially to deal with constraints, which have not been considered in the usual Taguchi method. In this case, an iterative use of the Taguchi method is necessary to translate and shrink the solution range, while constraints are treated as a penalty term in the objective function. The method is not meant for exact optimum, but turns out to be very robust for obtaining near optimum solutions. It is also found that existence and non-existence of a feature in a structure can be taken as levels of a design parameter and this can be utilized for best topology selection under limited topology options.[4] Another development is combining the Taguchi method with response surface method. Although the Taguchi method is a very important tool for design, it will not be covered here because it is not optimization-based per se.

A good approximate measure of standard deviation is using derivatives, often called sensitivity. This subcategory will be called sensitivity based robust design. Here, most of the formulations involve performance functions and their derivatives with respect to noise variables. One form is to optimize a weighted sum of the function value and the magnitude of its sensitivity.[5] The philosophy is that reduction of sensitivity implies increase of robustness.[6] Like robustness of an objective function, that of constraints is also important.[7,8] Even though there is some change in the noise conditions, the constraints in hand still need to be satisfied. Several definitions of sensitivity index are possible in terms of the mean and scatter.[9,10] A multi-objective formulation is also proposed with the function value and its derivative.[11] A similar method was applied to robust optimization of electromechanical device.[12] As a slightly modified approach, the signal-to-noise (S/N) ratio in the Taguchi method is replaced by an objective function with the mean and standard deviations of the quality characteristics.[13] In addition to the usual methods of using design sensitivities for optimization, obtaining a global solution is often necessary, and genetic algorithm is a good choice for example.[14]

Another subcategory of formulations, which is called allowable set based robust design, is not directly related to standard deviation but utilizes a norm called "relative safety index" that is indicative of robustness. This is based on a newly defined concept of allowable set.[15] Initially an allowable load set is considered, where load parameters are considered as uncertainty. Once this concept is given, a relative safety index is defined as the distance from the design load to the boundary of the allowable set. A robust design with this concept is then to maximize the relative safety index. In addition to multi-body mechanism examples, an interesting problem of finding the trajectory of a lifting motion of a human is studied.

The reliability-based approach of robust design is most comprehensive and can be made to be consistent with robust design, by a quantitative measure for robustness using probability. Naturally the resulting formulation falls in the category of general probabilistic optimization, to be treated in other chapters in this book. In the present chapter, however, a general efficient procedure recently established by the author based on moments calculated by design of experiments is presented with an application to geometric tolerance synthesis.[16] This includes an efficient probability analysis and optimization using the previously calculated design of experiment results.

Compared to the Taguchi method, studies on optimization based robust design has a short history, but as market competition becomes keen and robustness or yield rate of new products such as MEMS is a key issue, this topic becomes popular in these years.

There are several others such as robust design using information-gap based method by Ben-Haim,[17] model-based robust design,[18] and stochastic optimization as discussed in a recent survery.[19] It is noted that the classification and methods introduced here are not meant to be exhaustive and thus made at the discretion of the author, mainly covering the topics of his interest.

3. Sensitivity-based Robust Design Formulation

A logical procedure of a robust optimal design is to start with a deterministic optimization, and then to perform robust design. Sensitivity-based methods are very efficient and simple because they do not require probability information. For this purpose, a gradient index (GI), which is a function of gradients of performance functions with respect to uncertain variables, is minimized. The level of constraint feasibility is also enhanced by adding a term determined by a constraint value and its gradient index. A practical formulation is illustrated below with an example of MEMS device where robustness is crucial for high yield rate but information on uncertainties is particularly hard to obtain.

3.1. *Formulations: Sensitivity and Feasibility Robustness*

A general optimization problem to minimize an objective function subject to constraints is expressed as

$$
\begin{aligned}
& Minimize \quad f(\mathbf{x}, \mathbf{z}) \\
& subject\ to \quad g_j(\mathbf{x}, \mathbf{z}) \le 0 \quad j = 1, 2, \ldots, m \\
& \mathbf{x}_L \le \mathbf{x} \le \mathbf{x}_U
\end{aligned} \tag{1}
$$

where $\mathbf{x}$ is an n-dimensional vector of design variables and $\mathbf{z}$ is an l-dimensional system parameter vector. $\mathbf{x}_L$ and $\mathbf{x}_U$ denote the lower and upper bounds of the design variables, respectively. The deterministic optimum, which does not take into account the effect of uncertainties in design variables and system parameters, might be very sensitive to these variations.

A gradient based robust design formulation by Han and Kwak[20,21] is taken as a typical sensitivity based robust design optimization. A gradient index (GI) is defined as the gradients of performance functions with respect to stochastic variables as follows:

$$GI = \max_i |d\Phi/du_i| \qquad i = 1, 2, \ldots, N \qquad (2)$$

where Φ, u_i, and N denote a performance function, uncertain variables, and the number of uncertain variables, respectively. A robust design formulation corresponding to the deterministic one shown above is to use the gradient index and a target value of the objective function as follows:

$$\begin{aligned}
\textit{Minimize} \quad & GI_f \\
\textit{subject to} \quad & g_j(\mathbf{x}, \mathbf{z}) + \Psi_j(g_j(\mathbf{x}, \mathbf{z})) \le 0 \quad j = 1, 2, \ldots, m \\
& f(\mathbf{x}, \mathbf{z}) \cong M \\
\textit{where} \quad & GI_f = \max_i |df(\mathbf{x}, \mathbf{z})/du_i| \qquad i = 1, 2, \ldots, N
\end{aligned} \qquad (3)$$

A penalty term, $\Psi_j(g_j)$, as suggested below, is added in the formulation to enhance the feasibility robustness for each important constraint:

$$\Psi_j(g_j) = \begin{cases} 0 & g_j < CT \\ \dfrac{\kappa_j GI_{g_j}}{CTMIN - CT}(g_j - CT) & CT \le g_j \le CTMIN \\ \kappa_j GI_{g_j} & g_j > CTMIN \end{cases} \qquad (4)$$

$$\textit{where} \quad GI_{g_j} = \max_i |dg_j/du_i| \qquad i = 1, 2, \ldots N$$

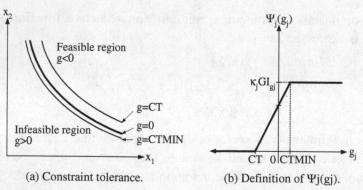

(a) Constraint tolerance.   (b) Definition of $\Psi j(g_j)$.

Fig. 3. Robustness of constraint feasibility.[20]

Figure 3 illustrates a typical shape of $\Psi_j(g_j)$. This is a linear function of g_j with two constants, κ_j and GI_{g_j}, for a critical region defined by a band ($CT \le g_j \le$

CTMIN). A value of κ_j in the range of 0.5-2.0 was used in the following examples.

3.2. Examples: MEMS Applications[20,21]

A resonant-type micro probe[22] to measure surface profile of a specimen, particularly inner region of a high aspect-ratio micro-hole is studied. This device is operated based on a stress-induced frequency shift of resonators which is caused by deflection of the cantilever probe when it scans surfaces of a specimen (Fig. 4(a)). Therefore, the measurement sensitivity of the micro probe is derived as follows:

$$\Delta f / f\,(\%) = \left(f_2' - f_1'\right) / f_1 \times 100 \tag{5}$$

where f_1, f_1', and f_2' are the first resonance frequency of the stress-free probe and the first and second resonance frequency of the deformed state, respectively.

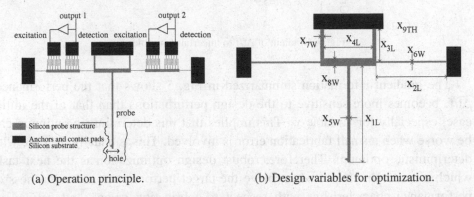

(a) Operation principle.   (b) Design variables for optimization.

Fig. 4. The resonant-type micro probe example.[20]

The overall procedure of a robust design starts with a deterministic optimization problem to maximize the sensitivity with a target value, M, subject to some constraints as formulated below:

$$
\begin{aligned}
Maximize \quad & \Delta f / f\,(\%) \\
subject\ to \quad & f_1 \ge 10000\mathrm{Hz} \\
& f_3 \ge 20000\mathrm{Hz} \\
& \sigma_{max} \le \sigma_{yield} \\
& f_{\Gamma 1i} - f_1 \le 0.001 \\
& \Delta f / f \le M
\end{aligned}
\tag{6}
$$

where f_1, f_3, and σ_{max} denote the first and third natural frequencies and the maximum of von Mises stress, respectively. The fourth constraint implies that the first resonance mode of f_1 must correspond to the in-plane transverse vibration, $f_{\Gamma 1i}$, of the resonator. The nine design variables used are described in Fig. 4(b). A target value, M = 6%, is used in the optimization. The achieved measurement sensitivity is 5.99% while all the constraints are satisfied. The optimized design is twice bigger in width than the initial one.

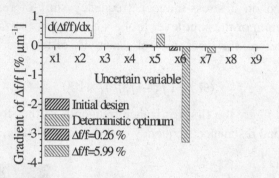

Fig. 5. Gradient of Δf/f to uncertain variables.[20]

The gradient information summarized in Fig. 5 shows that the performance, Δf/f, becomes more sensitive to the design perturbations than that of the initial case, especially for variable x_6. This implies that this deterministic optimum can be worse when a small fabrication error is involved. This is usually the case with deterministic optimum. Therefore, robust design optimization is the next task, which aims to simultaneously achieve the target performance and robustness of performance characteristics with respect to fabrication errors, and defined as follows, from (3) and (6):

$$Minimize \quad \mathrm{GI}_f$$
$$subject\ to \quad g_j + \Psi_j\left(g_j\right) \leq 0 \quad j = 1,2,3$$
$$f_{\Gamma 1i} - f_1 \leq 0.001 \tag{7}$$
$$\Delta f / f \cong M$$

$$\mathrm{GI}_f = \max_k \left| d\left(\Delta f / f\right) / du_k \right| \quad where\ u_k = \left(x_5, x_6, x_7, x_8, x_9\right) \tag{8}$$

Two cases are considered; robust optimum 1 does not take into account the feasibility robustness, but robust optimum 2 does include a penalty term with a factor of $\kappa_1 = \kappa_2 = \kappa_3 = 2.0$ for the robustness of constraints. The gradient index of the measurement sensitivity decreases from 3.25 at the deterministic optimum to about 1.8% μm^{-1} at robust optimums. Due to space limitation, the detail results[20] are suppressed. To investigate the robustness of the deterministic and robust optimums further, yield rate and violation probability of constraints are estimated from Monte Carlo simulations, assuming normal distributions of all variables. A simulation yield is defined as follows to determine the number of designs accepted as satisfactory micro probes.

$$SY(\%) = \frac{N_{ac}}{N_t} \times 100 = \frac{N_t - (N_{g4} + N_S)}{N_t} \times 100(\%) \tag{9}$$

where N_t, N_{g4}, and N_S denote the number of total simulations, violations in constraint g_4 in (6) and designs that satisfy g_4 but $\Delta f/f$ is out of the acceptable range, respectively. The number of samples taken is 10000.

Variations of the design variables taken are $\Delta x_1 = \Delta x_2 = \Delta x_3 = \Delta x_4 = \pm 2.0$, $\Delta x_5 = \Delta x_6 = \Delta x_7 = \Delta x_8 = \pm 1.0$ and $\Delta x_9 = \pm 0.5 \mu m$, where Δ means 3 standard deviations. Acceptance range of the target performance taken is between 5 and 7% as shown in Fig. 6.

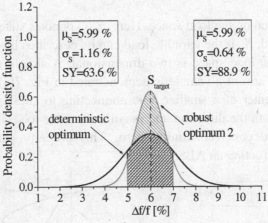

Fig. 6. Comparison of robustness by a Monte Carlo simulation for design variations.[20]

Regardless of the magnitude of the assumed design variations including those not presented here, the three optimums have practically the same mean $\Delta f/f$, and the deterministic optimum shows the largest standard deviation. Robust optimums 1 and 2 have almost the same mean and standard deviation of $\Delta f/f$, but the violation probability of constraint g_2 has decreased from 19.5% to 11.9% when the term $\Psi_j(g_j)$, is added for the feasibility robustness in robust optimization 2. Most importantly, the simulation yield improves from 63.6% at the deterministic optimum, to 89.5% and 88.9% at the robust optimums 1 and 2, respectively. These results illustrate that the sensitivity based robust optimal design method is very simple to use but produces practical solutions as validated in terms of the simulation yield rate. For more information on the design details and other applications refer to Han and Kwak.[20,21]

4. Robust Design by Allowable Set Concept

4.1. *"Allowable Load Set (ALS)" Defined*

Unlike the methods described up to now, Kwak and Kim[15] invented a new concept called "Allowable Set," to deal with the design problems with uncertainties. In this chapter, the "allowable load set" limited to load uncertainty is introduced.

Given a structure and knowing the kind of loads at prescribed locations, one can define the set of allowable loads as:

$$ALS = \{\mathbf{X} : \mathbf{G}(\mathbf{X}, \mathbf{z}(\mathbf{X})) \le 0\} = \bigcap_i \{\mathbf{X} : G_i(\mathbf{X}, \mathbf{z}(\mathbf{X})) \le 0\} \tag{10}$$

This set is defined on the load space. Here, $\mathbf{z}(\mathbf{X})$ denotes state variables which are functions of load, $\mathbf{X}$. If the variable load can be described by two independent parameters, then the load space is two-dimensional. A simple example of ALS for a lower control arm of a vehicle is demonstrated in Fig. 7(a), where the load is applied at the center of a smaller hole connecting to chassis. This is a two-dimensional case with the direction and magnitude of the load. It is noted that an ALS is a truncated cone, containing zero. This property can be utilized in geometrically constructing an ALS.

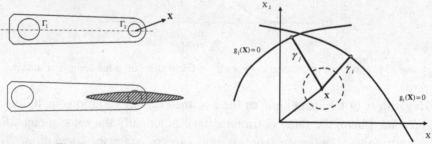

Fig. 7. (a) ALS of the lower control arm. (b) Definition of RSI.

In the ALS analysis problem, we are interested in constructing or visualizing an ALS for a given structure. For linear elastic materials, the construction of ALS for multi-body mechanisms, such as an excavator operating mechanism, is very efficiently done by using element concept and assembling procedure like in finite element method. Since in usual mechanical systems, the load applied has only a few parameters, the ALS can be very easily calculated and visualized for any configuration of the system.

4.2. *Relative Safety Index as a Measure of Robustness*

In the ALS design problem, we want to design the ALS by changing structural design variables. We want to specifically define "relative safety index (RSI)," as a criterion for robustness, because it is our intuition that the larger this value is, the more margin of safety there is. Therefore the robust design problem under this criterion is to maximize RSI. The RSI is the distance from a prescribed mean or design load to the boundary of an ALS. In Fig. 8(b), the RSI is the minimum of the distances, γ_i, to boundary segments. In the case of general uncertainties, one can use an "allowable set" of random variables instead of the ALS. The relative safety index remains the same, the distance to the boundary from the most probable point, that is, the mean point. The RSI for a mechanism may be defined as the minimum distance over all possible operating configurations of the system and this opens a new area of interesting applications as illustrated below.

A very important version of the RSI method is obtained by using a normalized safety index which is a transformed distance in terms of the functional value at the mean load. If there are more than one limit functions, however, there is difficulty in weighting or balancing the constraints. This topic requires more research and is not covered here.

4.3. *Application to Multi-body Mechanism Design*

Robust design of multi-body systems in the context of ALS is a case where the smallest RSI over all possible configurations along all possible trajectories need be calculated. So a global optimization is needed to evaluate the overall RSI. In the mechanism design, the constraints are divided into geometric constraints and constraints related to state variables. For a constraint involving state variables, $G_i(\mathbf{X}, \mathbf{z}(\mathbf{X}))$, RSI γ_i can be calculated by choosing the minimum value among several local solutions of global optimization.

The objective of the robust optimization is then to maximize the overall RSI of the multi-body system:

$$\max b_{n+1}$$
$$subject\ to\quad b_{n+1} \le \gamma_i, \quad i = 1, \cdots, N_1 \tag{11}$$
$$g_{N_1+j} \le 0, \quad j = 1, \cdots, N_2$$

where n is the number of design variables, N_1 is the number of constraints involving state variables, N_2 is the number of geometric constraints, b_{n+1} is an artificial design variable which effectively denotes the overall RSI, g_{N_1+j} are geometric constraints, and γ_i denotes the relative safety index of the i-th constraint. Depending on the problems, the following transformed problem may be used:

$$\max b_{n+1}$$
$$\text{subject to} \quad G_i \leq 0, \quad i=1,\cdots,N_1 \tag{12}$$
$$g_{N_1+j} \leq 0, \quad j=1,\cdots,N_2$$

where

$$G_i = \max_{\mathbf{q}_k} \max_{\mathbf{X} \in D} g_i(\mathbf{X}, \mathbf{q}_k, \mathbf{b}), \quad i=1,\cdots,N_1$$
$$D = \{\mathbf{X} : \|\mathbf{X} - \overline{\mathbf{X}}\| \leq b_{n+1}\} \tag{13}$$

where $\overline{\mathbf{X}}$ is the mean value of loads. The configuration, $\mathbf{q}_k$, which maximizes g_j is called critical. For purpose of numerical implementation, the set of critical and near critical configurations are kept as active constraints, whose number varies from iteration to iteration. The algorithm first finds all critical configurations and then, imposing those constraints at critical configurations, solves (12) for the next improved design. This algorithm, however, does not guarantee convergence to a global solution of (13).

4.4. *Example: Excavator Boom Design*[23]

There are many uncertainties when an excavator is at work. At the tip of the bucket, loading conditions drastically vary as the working configuration changes. The above design methodology is applied to the design of a three-dimensional boom of an excavator.

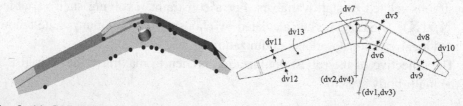

Fig. 8. (a) CAD model of the boom. Twenty potential critical points selected to check stress constraints. (b) The set of design variables selected from CAD parameters representing the shape of the boom.[23]

Assume that all bodies are rigid except the boom. Stress constraints are imposed at twenty points, as marked in Fig. 8(a). A geometric constraint imposed is that the total mass of the boom must not exceed the value of an existing initial design. Design variables are selected to represent the shape of the boom as shown in Fig. 8(b). The number of active constraints varies during optimization. The relative safety index has been improved enormously by 52.5%.

4.5. *Example: Lifting Motion Trajectory of a Human Model*

The concept of the ALS suggests a new hypothesis or principle that can help determine human motions, which look natural. This is an example of "robust path design" of a multi-body mechanism. It is illustrated by an example of finding robust working postures when a biomechanical model lifts an object. The result is compared with that obtained by a deterministic optimization method. This example is cited from the article, [24] and for reasons of space, only the results are introduced.

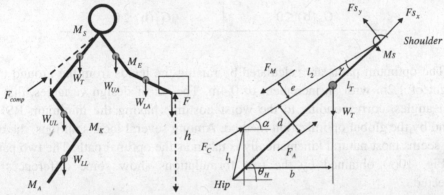

Fig. 9. A biomechanical model. (a) The model has five body segments. (b) Spine model using beam elements.[22,23]

A simplified biomechanical model is used and composed of body-segments namely; lower leg (LL), upper leg (UL), trunk (T), upper arm (UA) and lower arm (LA). All bodies are connected to each other by revolute joints. The spine is of special interest here, to consider low back pain. Most of disc disorders occur at either L5/S1 or L4/L5. The spine is modeled as a two bar linkage (Fig. 9(b)) to consider the disc. The linkages are replaced by beam elements using cylinder elements. The load at L5/S1, for example, can be calculated with the abdominal force, the erector spine muscle force and the resultant force at the shoulder joint.

The model has six independent variables: the angles of bodies and the length of trunk. These are thus six design variables. To make a lifting motion, an equality constraint for the lifting height is given. To evaluate the constraint functions, moments and inter-segmental forces are calculated at the ankle, knee, hip, shoulder and elbow.

Assuming that the man lifts a 20kg object from the ground to 1.2m high, two different optimizations as shown in Table 1 are performed. Formulation 1 is the new approach that maximizes the relative safety index to reduce the effect of loading uncertainties. Formulation 2, which is conventional, maximizes the minimum value of constraints.

Table 1. Human motion formulations.

Formulation 1(RSI)	Formulation 2 (Conventional)
$\max b_{n+1}$	$\min b_{n+1}$
$b_{n+1} \leq \gamma_i$	$b_{n+1} \geq g_i$
$G_j(\mathbf{b}) \leq 0$	$G_j(\mathbf{b}) \leq 0$

The optimum paths are calculated by raising the hands from the ground to a height of 1.2m with an increment of 0.4m. The set of design variables, that is, joint angles, corresponding to the worst postures having the minimum RSI is found by the global optimization method. Among several local solutions, the one that seems most natural kinematically is taken as the optimal path. The two paths in Fig. 10(a) obtained by the two formulations show some differences as expected.

To investigate the results in detail, the ALS are drawn with a scale factor when the height is 0.0m and 0.4m (Fig. 10(b)). The head of the load arrow denotes the location of the mean load. Figure 10(b) reveals that in the posture obtained by Formulation 2 the load is near the borders of constraints 5 and 9, which are related to shoulder and spine. That is, the man with these postures is more liable to hurt than those by Formulation 1. Researchers in ergonomics recommend workers to take the postures like those by RSI. Considering instability and narrow safety margin that might occur when lifting a heavy load, the postures that adopt Formulation 1 seem more natural and safer than those by Formulation 2. It is thus our conclusion that maximizing the RSI is one of the guiding hypotheses suitable for predicting a human motion, comparing with those previously used in the literature. This hypothesis essentially says that a human in natural conditions will move in a way that he is as far off as possible from potential dangers.

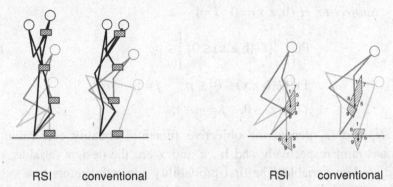

| RSI | conventional | | RSI | conventional |

Fig. 10. (a) Lifting postures by RSI on the left vs. conventional on the right. (b) ALS at two postures by RSI on the left vs. conventional on the right.

To consider the optimum path of a man who has backache, assume that he can endure only 2400N compressive force at L5/S1 instead of 3400N in the normal condition. It is shown that the posture of this man tends to be straightened up (Fig. 11) such that the object becomes closer to the body as should be.

normal   disorder

Fig. 11. Comparing lifting postures of a normal and a man with low back pain.

5. Reliability-based Robust Design Formulation

Reliability refers to the probability of a system or a component to perform its originally specified function during a prescribed period of time. In structural design, the criterion is given by a function, often called a failure function or a limit state function, denoted by $G(\cdot)$. The probability of failure, P_f, is defined as $P_f = \Pr[G(\mathbf{x}) \le 0]$. Then the reliability is $1 - P_f$. A standard form of a problem in structural design is as follows: [26]

$$Minimize\ W(\mathbf{b})$$

$$Subject\ to\ H_i(\mathbf{b},\mathbf{z},\mathbf{x}) = 0 \quad i = 1, \cdots s$$

$$\Pr\left[\bigcup_{j=1}^{m}\{G_j(\mathbf{b},\mathbf{z},\mathbf{x}) \le 0\}\right] \le p_0 \tag{14}$$

$$\Pr[G_j(\mathbf{b},\mathbf{z},\mathbf{x}) \le 0] \le p_j \quad j = 1, \cdots m$$

$$G_j(\mathbf{b},\mathbf{z}) \le 0, \quad j = m+1, ..., m'$$

where W, H_i and G_j denote the objective function, equality constraint and inequality constraint respectively and $\mathbf{b}$, $\mathbf{z}$ and $\mathbf{x}$ are the design variable, state variable and random variable. The first probability constraint refers to a system failure mode, while the second to a component failure mode.

The first task is how to obtain the probability related to an arbitrary function $G(\mathbf{x})$, and the second is how to solve the nonlinear programming problem. The Monte Carlo simulation turns out to be extremely time consuming when a nontrivial problem is considered. Also the accuracy is not good even with a very large sample. Notwithstanding these shortcomings, the Monte Carlo simulation is often used for comparison purpose especially for small size problems, since there is usually no other simple direct method. Next, methods based on design of experiments and first and second order reliability method using a Taylor expansion are popular. In the first order reliability method (FORM), the limit function is linearized, referenced at the most probable failure point (MPP), which is nearest to the origin in the reduced variable space. This space is obtained by transforming the physical random variable space into a standard normal variable space with a mean of zero and a standard deviation of 1. The distance is reliability index, β. One can then obtain the probability from the cumulative distribution function, i.e., $P_f = \Phi(-\beta)$. In terms of the reliability index, a probability constraint in Eq. (14) can be replaced with the following:

$$1 - \Phi(\beta) \le p$$

$$where\ \beta = \min_{\mathbf{u} \in A}(\mathbf{u}^T\mathbf{u})^{1/2} \qquad \text{(RIA)} \tag{15}$$

$$A = \{\mathbf{u} : G(\mathbf{u}) \le 0\}$$

where $\mathbf{u}$ denotes the normalized random variable vector and p is a prescribed level of probability. This can be further transformed into a better type of sub-problem, first developed by Lee and Kwak[26] in 1987. It is noted that this same formulation was later rediscovered by Tu and Choi[27] and called as Performance Measure Approach (PMA). This transformed problem is maximizing the failure function or performance function with a fixed norm constraint, whose value is dependent only on the prescribed failure probability, p:

$$G^* \geq 0$$

$$G^* = \min_{\mathbf{u} \in B} G(\mathbf{x}(\mathbf{u})) \qquad \text{(PMA)} \qquad (16)$$

$$\text{where } B = \left\{ \mathbf{u} \middle| \left(\mathbf{u}^T \mathbf{u} \right)^{\frac{1}{2}} \leq \Phi^{-1} \left(1 - p \right) \right\}$$

This second formulation is sturdy and works better than the RIA, especially for very low probability. The RIA has been most popular in structural design area, but the PMA is getting more attention among researchers. Recently, however, an efficient method of reliability analysis and optimization based on full factorial design-of-experiments has been developed as treated below.[16]

5.1. *Moment and Probability Calculation by Multi-point Information*

In the moment method for reliability analysis, one can obtain structural reliability by fitting the information of a few statistical moments of a system response function in an empirical distribution system, such as the Pearson system, Johnson system, and so on. To obtain the statistical moments, usually the system response function should be evaluated for a set of well-designed calculation points, often called designed experimental points. The methods are based on the numerical integration using Gaussian type quadrature formula. Since the product quadrature rule for a multiple random variable case is very expensive, however, alternative methods have been developed. The location of the evaluation points and the corresponding probability weights can be found as follows.

The k-th order statistical moment of a system response function $g(\mathbf{x})$ is defined as

$$E\left\{ g^k \right\} = \int_{-\infty}^{\infty} \left[g(\mathbf{x}) \right]^k f_{\mathbf{x}}(\mathbf{x}) d\mathbf{x} \qquad (17)$$

where $f_{\mathbf{x}}(\mathbf{x})$ is the joint probability density function of the vector of random variables, $\mathbf{x}$. For a one random variable case, the moment can be calculated as,

$$E\left\{ g^k \right\} \cong \sum_{i=1}^{m} w_i \left[g(\mu + \alpha_i \sigma) \right]^k \qquad (18)$$

where w_i and $\mu + \alpha_i \sigma$ denotes the probability weight and location of evaluation point, respectively, and m the number of integration points. Taguchi[28] proposed a method using 3 point information for normally distributed random variables and later D'Errico and Zaino[29] obtained improved results as $\{w_1, w_2, w_3, \alpha_1, \alpha_2, \alpha_3\} = \{1/6, 4/6, 1/6, -\sqrt{3}, 0, \sqrt{3}\}$. To obtain an accurate approximation up to the 4th moment, at least 3 nodes are necessary. The optimal quadrature points and probability weights, applicable for general distributions, are found by solving the following,[16]

$$M_k = \int_{-\infty}^{\infty} (x - \mu)^k f(x) dx$$
$$= w_1 (\alpha_1 \sigma)^k + w_2 (\alpha_2 \sigma)^k + w_3 (\alpha_3 \sigma)^k \quad (k = 0, 1, 2, 3, 4, and\ 5) \tag{19}$$

where M_k is the k-th order central moment of variable, x. These equations can be arranged by replacing $\mu + \alpha_i \sigma$ with level l_i as follows:

$$w_1 + w_2 + w_3 = 1 \tag{20}$$

$$w_1 l_1 + w_2 l_2 + w_3 l_3 = \mu \tag{21}$$

$$(l_1 - \mu)^2 w_1 + (l_2 - \mu)^2 w_2 + (l_3 - \mu)^2 w_3 = \sigma^2 \tag{22}$$

$$\frac{(l_1 - \mu)^3 w_1 + (l_2 - \mu)^3 w_2 + (l_3 - \mu)^3 w_3}{\sigma^3} = \sqrt{\beta_1} \tag{23}$$

$$\frac{(l_1 - \mu)^4 w_1 + (l_2 - \mu)^4 w_2 + (l_3 - \mu)^4 w_3}{\sigma^4} = \beta_2 \tag{24}$$

$$(l_1 - \mu)^5 w_1 + (l_2 - \mu)^5 w_2 + (l_3 - \mu)^5 w_3 = M_5 \tag{25}$$

where μ, σ, $\sqrt{\beta_1}$, and β_2 denote the mean, standard deviation, skewness and kurtosis, respectively. By solving this system of equations, we can get the levels and weights providing integration order up to 5, but it is hard to find the solution in a simple, closed form. Seo and Kwak[16] replaced the condition on the fifth moment, Eq. (25) with the condition that the mid-level l_2 be located at the mean of the variable and found an explicit form of solutions:

$$\{l_1, l_2, l_3\} = \left[\begin{array}{c} \mu + \dfrac{\sqrt{\beta_1}\,\sigma}{2} - \dfrac{\sigma}{2}\sqrt{4\beta_2 - 3\beta_1} \\ \mu \\ \mu + \dfrac{\sqrt{\beta_1}\,\sigma}{2} + \dfrac{\sigma}{2}\sqrt{4\beta_2 - 3\beta_1} \end{array} \right]^T \tag{26}$$

$$\{w_1, w_2, w_3\} = \begin{bmatrix} \dfrac{(4\beta_2 - 3\beta_1) + \sqrt{\beta_1}\sqrt{4\beta_2 - 3\beta_1}}{2(4\beta_2 - 3\beta_1)(\beta_2 - \beta_1)} \\[2ex] \dfrac{\beta_2 - \beta_1 - 1}{\beta_2 - \beta_1} \\[2ex] \dfrac{(4\beta_2 - 3\beta_1) - \sqrt{\beta_1}\sqrt{4\beta_2 - 3\beta_1}}{2(4\beta_2 - 3\beta_1)(\beta_2 - \beta_1)} \end{bmatrix}^T \tag{27}$$

The levels and weights thus found are very simple and easy to use, and also applicable for non-normal variables. However, it is found that the simplifying assumption putting the mid-level to the mean can cause problems in some cases. First, since the requirement on the fifth moment has been ignored, the integration order provided by the levels and weights is lowered when dissymmetric distributions are considered. Second, for some dissymmetric distributions such as exponential, the first level can be located outside the domain where the random variable is defined. With this reason, an effort to solve the system of equations without the simplifying assumption has been made with numerical methods. Furthermore, the integration order of the method can be increased by adopting more integration points, say, using 5 or 7 levels. These methods, however, require much more function evaluations, which seem a burden for practical consideration.

When there are n random variables, the moments of the response function can be calculated using three point information and the product quadrature rule as follows:

$$\mu_g = \sum_{i_1=1}^{3} w_{i_1} \cdots \sum_{i_n=1}^{3} w_{i_n} g\left(l_{i_1}, \cdots l_{i_n}\right) \tag{28}$$

$$\sigma_g = \left[\sum_{i_1=1}^{3} w_{i_1} \cdots \sum_{i_n=1}^{3} w_{i_n} \left(g\left(l_{i_1}, \cdots l_{i_n}\right) - \mu_g\right)^2\right]^{1/2} \tag{29}$$

$$\sqrt{\beta_{1g}} = \left[\sum_{i_1=1}^{3} w_{i_1} \cdots \sum_{i_n=1}^{3} w_{i_n} \left(g\left(l_{i_1}, \cdots l_{i_n}\right) - \mu_g\right)^3\right] \Big/ \sigma_g^3 \tag{30}$$

$$\beta_{2g} = \left[\sum_{i_1=1}^{3} w_{i_1} \cdots \sum_{i_n=1}^{3} w_{i_n} \left(g\left(l_{i_1}, \cdots l_{i_n}\right) - \mu_g\right)^4\right] \Big/ \sigma_g^4 \tag{31}$$

5.2. *Reliability Calculation and Optimization by Moment Method*

Once the four statistical moments are obtained, the probability density function $f(x)$ of the system response function can be obtained using a suitable empirical distribution system such as the Pearson system[30], as the solution of the differential equation as follows:

$$\frac{1}{f(x)}\frac{df(x)}{dx} = -\frac{\overline{x}+a}{c_0 + c_1\overline{x} + c_2\overline{x}^2} \tag{32}$$

where c_0, c_1, c_2 and a are the coefficients determined from the first four statistical moments of x. The shape of $f(x)$ is classified into seven groups according to the type of the roots of the denominator.

The PDF available now provides very useful information to design engineers, and reliability can be easily calculated. However, one critical disadvantage is that moment-based methods become numerically inefficient as the number of random variables increases. To overcome this problem, a very efficient method, termed RSMM (Response Surface Augmented Moment Method), has been developed.[31] This uses a series of adaptive additions of a new function evaluation to a set of results at core points selected along the axes, and subsequently construct a convergent response surface. This complements all the necessary information necessary for the full factorial analysis described above. The number of function evaluations is almost one order of magnitude smaller than that of the full factorial method, while achieving the accuracy because an adaptive addition is used. By the enormous improvement of efficiency, the RSMM is expected to be a very powerful tool enabling structural optimization manageable.

The DOE-based moment method just described has another good feature. It provides the design sensitivity of the probabilistic performance measure using the data obtained during the reliability analysis process without additional function evaluation. Since the probability of failure P_f is a function of the first four statistical moment of response function $G(\mathbf{x})$, the design sensitivity of the probabilistic constraint can be rewritten as follows:

$$\frac{d\mathrm{P}_f}{d\mathbf{x}} = \frac{\partial\mathrm{P}_f}{\partial\mu_G}\frac{d\mu_G}{d\mathbf{x}} + \frac{\partial\mathrm{P}_f}{\partial\sigma_G}\frac{d\sigma_G}{d\mathbf{x}} + \frac{\partial\mathrm{P}_f}{\partial\sqrt{\beta_{1G}}}\frac{d\sqrt{\beta_{1G}}}{d\mathbf{x}} + \frac{\partial\mathrm{P}_f}{\partial\beta_{2G}}\frac{d\beta_{2G}}{d\mathbf{x}} \tag{33}$$

The terms $\partial P_f/\partial\mu_G$, $\partial P_f/\partial\sigma_G$, $\partial P_f/\partial\sqrt{\beta_{1G}}$, $\partial P_f/\partial\beta_{2G}$ can be calculated by finite difference method and the Pearson system program and the other terms can be calculated via chain rule of differentiation and the data obtained during the reliability calculation. Thus, the design sensitivity of the probabilistic constraint can be evaluated very efficiently without further evaluations of $G(\mathbf{x})$. The design variable **x** can be nominal or mean value in case of RBDO or tolerance of a dimension in case of a tolerance synthesis problem.

5.3. *Tolerance Design Procedure and An Application Case*

As already noted in the introduction, reliability based design requires probability information. Tolerance allocation problem, however, is one important area of application, where probability information is not necessary, because in usual practice, a tolerance of $\pm t$ to a length dimension for example denotes that the manufactured length has a normal distribution with a standard deviation, $\sigma = t/3$. Tolerance is necessary to assure assemblability and quality of a product under randomness of manufacturing errors. For assemblability, a characteristic function, $G(\mathbf{x})$, of the geometric clearances is a function of tolerances. Tolerance analysis is to analyze the probability of successful assemblage, given by an inequality condition such that $G(\mathbf{x}) \geq 0$. A formulation of tolerance synthesis problem may be posed as to minimize the manufacturing cost under assemblability conditions. It is rather hard to get cost information for a specific manufacturing process; Cost as a function of tolerance is usually discrete or at least discontinuous. To simplify this, we often use some kind of continuous inverse function of tolerance. Now it is seen that the tolerance allocation problem proposed above is exactly the same as the reliability problem described earlier, and the same approach can be applied for its solution.

The whole procedure of tolerance synthesis is demonstrated with a block assembly example[32] shown in Fig. 12. The requirements for proper assembly are given by Eqs. (34-39):

$$g_1(\mathbf{x}) = (x_6 - x_5) - (x_8 - x_7) \tag{34}$$

$$g_2(\mathbf{x}) = (x_3 - x_4) - (x_{11} - x_{10}) \tag{35}$$

$$g_3(\mathbf{x}) = (x_8 - x_7)(x_2 - x_3) - (x_6 - x_5)(x_{10} - x_9)$$
$$+ \tan(\pi/180)\{(x_{10} - x_9)(x_2 - x_3) + (x_8 - x_7)(x_6 - x_5)\} \tag{36}$$

$$g_4(\mathbf{x}) = (x_6 - x_5)(x_{10} - x_9) - (x_8 - x_7)(x_2 - x_3)$$
$$+ \tan(\pi/180)\{(x_{10} - x_9)(x_2 - x_3) + (x_8 - x_7)(x_6 - x_5)\} \tag{37}$$

$$g_5(\mathbf{x}) = -x_1 + x_{12} + 0.01 \tag{38}$$

$$g_6(\mathbf{x}) = x_1 - x_{12} + 0.01 \tag{39}$$

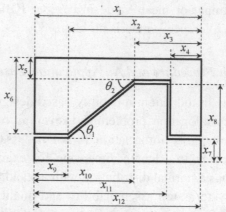

Fig. 12. Illustration of block assembly for nonlinear stack-up tolerance analysis and synthesis.[30]

The dimensions are normally distributed and the distribution parameters are listed in Table 2.

Table 2. Distribution parameters of dimensions in block assembly example. In this example, the tolerance t_i is set as $3\sigma_i$.

i	mean μ_i	tolerance t_i	i	mean μ_i	tolerance t_i
1	50	0.0187	7	10.0	0.0019
2	40.00125	1.2331	8	30.0	0.0015
3	20.05	0.0579	9	10.05	0.1285
4	9.9985	0.0705	10	30.0	0.0714
5	9.9985	0.0019	11	40.0	0.0579
6	30.0	0.0022	12	50.0	0.0168

The tolerance synthesis problem is to minimize the manufacturing cost under assemblability condition as follows:

$$Minimize \quad C_{total}(\mathbf{t}) = \sum_{i=1}^{12} C_i(t_i) = \sum_{i=1}^{12} \frac{a_i \times 10^{-3}}{(t_i)^{b_i}}$$

$$subject \ to \quad 0.95 - \Pr\left[\bigcap_{i=1}^{6}\left[g_i(\mathbf{x}) \geq 0\right]\right] \leq 0, \quad where \ \mathbf{x} = \boldsymbol{\mu} + \mathbf{t} \tag{40}$$

where $a_1 = 0.2$, $a_2 = 1.0$, $a_3 = a_4 = 0.015$, $a_5 = 0.008$, $a_6 = 0.009$, $a_7 = 0.008$, $a_8 = 0.006$, $a_9 = 1.0$, $a_{10} = 0.01$, $a_{11} = 0.015$, $a_{12} = 0.2$, and $b_1 = \ldots = b_{12} = 2.0$. The design variable t_i is set as $3\sigma_i$. This problem is solved with FORM, MCS with 100,000 samples and the full factorial moment method (FFMM) and the results are summarized in Table 3. For this problem, the dimensions are distributed normally, the response functions are close to linear, and the results of FFMM and FORM are almost equal.

Table 3. Results of optimal tolerance synthesis of block assembly.

Variable	Initial tolerance	Optimal tolerance		
		FFMM	FORM	MCS
t_1	0.0050	0.0158	0.0158	0.0147
t_2	0.5000	0.6600	0.6595	0.6988
t_3	0.0100	0.0418	0.0418	0.0212
t_4	0.0100	0.0418	0.0418	0.0217
t_5	0.0005	0.0026	0.0026	0.0027
t_6	0.0005	0.0027	0.0027	0.0029
t_7	0.0005	0.0026	0.0026	0.0029
t_8	0.0005	0.0025	0.0025	0.0022
t_9	0.5000	0.6600	0.6600	0.2235
t_{10}	0.0100	0.0378	0.0377	0.0131
t_{11}	0.0100	0.0418	0.0418	0.0187
t_{12}	0.0050	0.0158	0.0158	0.0144
C_{total}	140.558	6.1341	6.1341	6.4473
Pr_{safe}*	1	0.9501	0.9502	0.9492

6. Summary and Conclusions

Robust design in the area of quality engineering has been practiced for long time; it is rather recent that it has become popular in the optimization sector. In this chapter, after a brief introduction of the Taguchi method, several recently developed formulations are presented. Although the concept of robust design is well-known, no standard formulation is yet available. So the approaches are arbitrarily classified based on whether probability information is required or not. Both are covered with typical and new formulations. For a sensitivity-based approach, a gradient index method is taken showing its simplicity and practicality. Constraint feasibility is also treated using the gradient index. Next introduced is the so-called relative safety index, which is indicative of robustness. This is defined as the distance from the mean point to the boundary of an allowable load set, a new concept developed recently by the author. This new formulation opens a new category of applications in the areas of multi-body mechanism design and provides a new hypothesis in solving human motion trajectories. Lastly presented is an efficient and accurate moment method of probability analysis and design. The first four moments of a performance function are calculated by a numerical integration using full factorial design of experiments. A very efficient method is developed by using adaptive application of response surface instead of the full factorial evaluations. All the three methods are shown to be simple and efficient through nontrivial examples. The choice of approaches is dependent on whether probability data are available or not and what the user's decision goal is.

Acknowledgments

Much of the work was supported by a Korean National Research Laboratory Project and Samsung Chair Professorship. I would like to thank my students, especially, Drs. J.S. Han, J. H. Kim, H. S. Seo, Messrs. J. H. Chang, S. H. Lee, and S. B. Lee for their contributions to the topics cited.

References

1. W. Y. Fowlkes and C. M. Creveling, *Engineering Methods for Robust Product Design*, (1995).
2. M. S. Phadke, *Quality engineering using robust design* (Prentice-Hall, UK, 1989).
3. S. H. Lee and B. M. Kwak, in International Conference on Computational & Experimental Engineering and Science (Reno, Nevada, 2002).
4. S. B. Lee and B. M. Kwak, in Asia-Pacific International Conference on Computational Method in Engineering (JASCOME, Japan, 2003), p. 171.
5. R. J. Balling, J. C. Free and A. R. Parkinson, *J. Mech., Trans., Auto. Design*, 438 (1986).
6. A. D. Belegundu and S. Zhang, *J. Mech. Design*, 213, (1992).
7. A. R. Parkinson, C. Sorensen and N. Pourhassan, *J. Mech. Design*, 74, (1993).
8. J. A. Bennet and R. V. Lust, *AIAA J*, 1491, (1990).
9. S. Sundaresan, K. Ishii and D. R. Houser, *J. Mech. Design*, 318, (1991).
10. Sundaresan, K. Ishii and D. R. Houser, *Eng. Optim.*, 101, (1995).
11. K. W. Lee and G. J. Park, *Comp. & Struc.*, 77, (2001).
12. S. B. Yoon, I. S. Jung and D. S. Hyun, *IEEE trans. magnetics*, 1710, (1999).
13. K. H. Hwang, K. W. Lee and G. J. Park, *Struct. Multidisc. Optim.*, 300, (2001).
14. D. B. Parkinson, *Qual. Reliab. Eng.* I. 201, (2000).
15. B. M. Kwak and J. H. Kim, *Mech. Struct. Mach.*, 213, (2002).
16. H. S. Seo and B. M. Kwak, *Int. J. Prod. Res.*, 40(4), 931-944 (2002).
17. Y. Ben-Haim, *Struct. Saf.*, 269, (1999).
18. K. Al-Widyan, J. Angeles, J. Cervantes-Sanchez, in International Conference on Integrated design and manufacturing in mechanical engineering; Recent advances in integrated design and manufacturing in mechanical engineering (2003, Clermont-Ferrand, France), p. 431.
19. C. Zang, M. I. Friswell and J. E. Mattershead, *Comput. Struct.*, 83, 315-326 (2005).
20. J. S. Han and B. M. Kwak, *Struct. Multidisc. Optim.*, 27, 469-478 (2004).
21. J. S. Han and B. M. Kwak, *J. Micromech. Microeng.* 662, (2001).

22. E. Lebrasseur; T. Bourouina; J. Pourciel; M. Ozaki; T. Masuzawa and H. Fujita, in International Conference on Modeling and Simulation of Microsystems (Computational Publications, San Diego, 2000), p. 285.
23. J. H. Kim, Ph.D. dissertation, KAIST (2002).
24. J. H. Chang, J. H. Kim and B. M. Kwak, in International Conference on Computational Science (Krocow, Poland, 2004).
25. D. Y. Hwang, MS Thesis, KAIST (1993).
26. T. W. Lee and B. M. Kwak, *Mech. Struct. & Mach.*, 15(4), 523-542 (1987-88).
27. J. Tu and K. K. Choi, *ASME J. Mech. Des.*, 557 (1999).
28. G. Taguchi, *Int. J. Prod. Res.*, 521 (1978).
29. J. R. DE'rrico and J. R. Zaino Jr., *Technometrics*, 397 (1988).
30. N. L. Johnson, S. Kotz and N. Balakrishnan, *Continuous univariate distributions*, (John-Wiley & Sons, New York, 1994).
31. S. H. Lee and B. M. Kwak, to appear in *Struct. Safety* (2005). Available online at www.sciencedirect.com.
32. J. Lee and G. E. Johnson, *Comput. aided design*, 601 (1993).

CHAPTER 18

PARALLEL COMPUTATIONS FOR DESIGN OPTIMIZATION

S. D. Rajan and A. Damle[*]

Department of Civil Engineering
Arizona State University, Tempe, AZ 85287

There is increasing evidence that computing clusters created with commodity chips are capable of out-performing traditional supercomputers. The trend of using these commodity computing systems for engineering analysis and design is rapidly gaining momentum. In this chapter we discuss the different parallel processing scenarios and the implementation in the HYI-3D design optimization software system. We examine the hardware and software issues with 32-bit and 64-bit design optimization computations. A scenario for configuring a design engineer's workbench is presented where desktop computations are combined with execution on a computing cluster so as to reduce the design cycle time. Using multi-level parallelism, not only can the function evaluation be carried out in parallel but also other steps in the design optimization algorithm can be computed in parallel – gradients, line search and direction-finding problem. Numerical examples involving sizing, shape and topology optimization show the gains obtained from coarse and fine grain parallelism for both gradient and non-gradient optimization techniques.

Nomenclature

d	Total number of search directions
d_{std}	Standard search direction
d_t	Tangent search direction
f	Number of subdomains the finite element model is split into (also the # of processors involved in parallel finite element analysis)
DF	Direction finding
FE	Function evaluation
GE	Gradient evaluation
l	Number of processors available for parallel LS

---

[*] Design Software Engineer, Hawthorne & York, Intl., Phoenix, AZ.

511

K	System stiffness matrix
LS	Line search
NDV	Number of design variables
NFV_{GE}	Number of function evaluations in GE
NFV_{LS}	Number of function evaluations in LS
NFV_{Total}	Total number of function evaluations (GE + LS + external)
PE	Total # of processors
P_i	Processor $i, 1 \le i \le PE$
$\mathbf{x}$	Design variable vector

1. Introduction

Finite element based design optimization (DO) is now a relatively well-established methodology for engineering design. The use of this methodology involves several areas and techniques such as geometric modeling, finite element model generation, finite element analysis (FEA), and numerical optimization techniques to name a few. Advances in each of these areas have made the overall design process more versatile. A tightly integrated software system would enable an end-to-end solution with minimal designer intervention starting with the conceptual phase of the design and ending with manufacturing of the product or construction of the project. Yet there remain more challenges to meet and hurdles to overcome. As FE models have become more sophisticated and detailed, the execution time has also increased in spite of advances in hardware technology. While FE models with more than 100 million degrees of freedom (DOF) have been analyzed, today it is more common to have engineering models with not more than a few hundred thousand DOF. Similarly, the size of typical design optimization problems has also increased in the last two decades. It is common today to solve design optimization problems with a few hundreds design variables. Compressing the design cycle time so as to reduce the time required for design and redesign process, requires more advances in hardware, software and algorithms. Fueling this growth has been the development of computer hardware at an affordable price level. In small as well as large engineering design firms, a typical design engineer works with a 32-bit or a 64-bit desktop system. This system is used primarily for (a) pre and post-processing, (b) design problems that can be executed on these systems in a reasonable amount of time, and (c) as a gateway to faster computing platforms. When the design engineer needs to solve bigger problems, different type of parallel computing architectures can be used.

In the early 70s, the Intel CPU clock speed was below 1 MHz (Fig. 1). By the early 90s, the clock speed had increased to slightly less than 100 MHz. The early 90s saw a phenomenal increase in clock speed. The 1 GHz mark was reached in early 2000 with the 2 GHz CPU in late 2001. However, this growth has slowed down tremendously. With the recent announcements from Intel it is clear that for the time being the clock speed race will stop at 3.8 GHz and the focus has instead shifted to building multi-core CPUs. This trend is an industry-wide phenomenon with similar multi-core initiatives (at the expense of faster clock speeds) from AMD and IBM, the other two dominant CPU manufacturers. So how does this trend affect software development in general and design optimization in particular? We will examine this question and its answers in greater detail in the rest of this chapter.

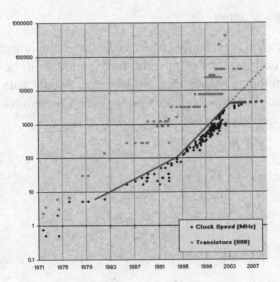

Fig. 1. Intel CPU Clock Speeds

2. Building a Case for Parallel Computations in Design Optimization

A typical single objective design optimization problem can be stated as follows.

Find   **x**   (1a)

to minimize $f(\mathbf{x})$ (1b)

$$g_i(\mathbf{x}) \le 0, \quad i = 1, \ldots, n$$

$$h_j(\mathbf{x}) = 0, \quad j = 1, \ldots, m \tag{1d}$$

$$\mathbf{x}^L \le \mathbf{x} \le \mathbf{x}^U \tag{1e}$$

where $f(\mathbf{x})$ is the objective function, n and m are the numbers of inequality and equality constraints respectively, $g_i(\mathbf{x})$ and $h_j(\mathbf{x})$ are the constraint functions, x is the design variable vector, and x^L and x^U are the lower and upper bounds respectively. The design variables can be continuous, discrete or even 0-1. The objective and constraint functions are usually nonlinear and the problems non-convex.

2.1. *Topology Optimization*

Topology optimization is typically employed at the initial design stage where the form of the structure in terms of material distribution, is to be determined. The dominant methodology is SIMP (Simple Isotropic Model with Penalization) or power-law approach[3,4,5]. The topology optimization problem based on power law approach can be stated as follows.

Find $\qquad \mathbf{x} \in R^k \qquad\qquad\qquad$ (2a)

to minimize $\quad f(\mathbf{x}) = \mathbf{D}^T \mathbf{F} \qquad\qquad$ (2b)

subject to $\quad h = \dfrac{M}{M_0} - F_M = 0 \qquad\quad$ (2c)

$\qquad\qquad \mathbf{KD} = \mathbf{F} \qquad\qquad\qquad$ (2d)

$\qquad\qquad 0 < \mathbf{x}^L < \mathbf{x} < \mathbf{x}^U < 1 \qquad$ (2e)

where $f(\mathbf{x})$, the objective function, is the compliance of the model, h is the mass fraction constraint function (M is the actual mass used in optimization, M_0 is the initial mass, and F_M is the desired mass fraction), $\mathbf{KD} = \mathbf{F}$ are the equilibrium equations, x, the design variable vector, is the material density in each finite element, and x^L and x^U are the design variable lower and upper bounds respectively.

The standard optimality criteria method[6] is used to solve the topology optimization problem. Once the design variable values are known, the checkerboarding problem and mesh dependency issues are handled using a filtering technique[7]. Each iteration of the optimality criteria algorithm involves a complete FEA during which the model's compliance and strain energy are computed using the updated design variables. In other words, evaluation of Eq. (2d) is the most computationally expensive step.

2.2. *Shape and Sizing Optimization*

Shape optimization involves the modification of the parameters that control the shape of the model. There are two widely used approaches – the Natural Design Variable Approach[8,9] (also known as Basis Shape Vectors Approach) and the Geometric Approach[10]. In the former approach, the nodal coordinates are updated as

$$\mathbf{c}^{new} = \mathbf{c}^{old} + \mathbf{Q}\mathbf{x} \tag{3a}$$

$$\mathbf{K}^{aux}\mathbf{Q} = \mathbf{F}^{aux} \tag{3b}$$

where $\mathbf{c}$ is the vector of nodal coordinates and $\mathbf{Q}$ is the velocity field matrix (displacements) that is obtained by solving the equilibrium of the auxiliary structure as shown in Eq. (3b). The velocity field matrix can be computed just once and reused every iteration or can be updated (recomputed) as often as needed. In the Geometric Approach, nodal coordinates are updated either by regenerating the geometric model and then remeshing, or by just updating the mesh without changing the mesh characteristics (number of elements, nodes etc.). Solving Eq. (3b) can be computationally expensive if the mesh updates take place frequently.

Sizing optimization involves the modification of the cross section (of truss and beam elements) or thickness of finite elements (plane elasticity, plate/shell elements) while preserving the shape and topology of the model. It is also possible to combine shape and sizing optimal designs so that the shape of the model and the element properties are modified simultaneously.

To contrast the differences between a typical sequential DO algorithm and the parallel algorithm, we first look at a typical sequential algorithm associated with gradient-based optimizers. The algorithm has eight steps.

(i) Carry out a function evaluation with the initial guess.
(ii) Start design iterations.
(iii) Carry out gradient evaluation at current design point.
(iv) Solve direction-finding problem.
(v) Find optimal step length from line search problem.
(vi) Compute the next design point.

(vii) Converged solution? If no, go to step (iii).
(viii) If yes, carry out the final function evaluation with the optimal values.

The computationally expensive parts of the algorithm are steps involving function evaluations – steps 1, 3, 5 and 8.

Whether we are dealing with topology, shape or sizing optimization, the challenge is not only to carry out the function evaluation (specifically, finite element analysis) efficiently but also to reduce the number of function evaluations. This would imply that apart from parallelizing FE analysis, one would also have to parallelize the design optimization steps (steps 3-6).

3. Hardware for Parallel Processing

Computer systems with multiple processors have been around a long time especially in the form of mainframes and supercomputers. However, in the last decade there has been concerted effort to use commonly available CPUs in a variety of parallel processing architectures starting with Beowulf clusters. Today, the three most popular architectures are clusters, constellations and massively parallel processing (MPP) systems[11]. Clusters are systems made up of several computing nodes. Each node can potentially work in a standalone mode. It has its own CPU and memory. However, the nodes are clustered together in a network to yield faster computing platforms. We have a Beowulf cluster when commodity machines are used as nodes and are networked using commodity networking hardware[12]. One can look at parallelism taking place at two levels. First, increased parallelism is possible by increasing the number of nodes in a cluster. Second, by increasing the number of processors at a node, one can potentially increase the available parallelism. There are specialized hardware manufacturers that provide computing nodes with multiple processors. In the first configuration called uniform memory access SMP (symmetric multiprocessor), they share the same memory (Fig. 2a). The bottleneck is the shared memory bus connecting the CPU to the shared memory. The second configuration is the non-uniform memory access (NUMA) SMP configuration (Fig. 2b). While every processor has its own memory, each process has access to the entire memory. The access is non-uniform since a process cannot access every memory location with the same speed. With these possible architectures, constellations can be thought of as providing "clustering" at two levels – several multi-CPU nodes connected to each other via high-speed switches. Finally, it is more difficult to exactly define MPP systems. They could be a massive distributed memory system, or a large shared memory system employing cache coherence, or a large vector system, and so on. What is perhaps clear is that these MPP systems are much more expensive to buy, build and maintain than commodity machines. The distribution of these three systems in the top 500 computing platforms in the world in shown in Fig. 2(c).

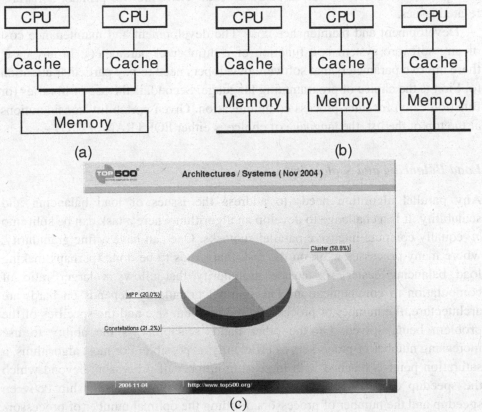

Fig. 2. (a) Uniform memory access (b) Non-uniform memory access (c) Cluster architectures in the top 500 list[i]

4. Software for Parallel Processing

In comparison to the development of faster hardware, little effort has been invested in the development and ready availability of software tools for parallelization. Two widely used approaches in parallelizing computations involve the use of threads (usually on shared memory systems) and some form of message-passing (usually on distributed memory systems). There have been successful efforts at using both threads in the form of OpenMP[13] directives and

---

[i] http://www.top500.org

message-passing in the form of MPI[14] calls to create high-performance software systems. OpenMP is used primarily with shared memory systems while MPI is the choice for distributed memory systems. MPI implementations use TCP/IP protocol or with custom switching hardware, proprietary protocol such as Myrinet, Infiband etc. for increased bandwidth. There are two primary software-related issues.

Development and maintenance cost: The development and maintenance cost of any software system is a function of a number of parameters. In any case, there are two parameters that software developers need to pay particular attention to. First is the choice of programming language. Second, is the use of libraries (or function calls) for inter-process communication. Given portability considerations at the top of the list, the language of choice is either FORTRAN or C/C++.

Load Balancing and Scalability

Any parallel algorithm needs to address the issues of load balancing and scalability. It is a challenge to develop an algorithm where a task can be split into n equally compute intensive parallel subtasks. One can have a fine granularity, where many processes work on the task that needs to be done (perhaps making load balancing easier), or coarse granularity that allows a larger ratio of computation to communication. The optimal granularity depends on hardware architecture, the number of processors, the problem size and the specifics of the problem being solved. On the other hand, scalability is the ability to use increasing number of processors as efficiently as possible. For most algorithms, a saturation point is reached with increasing number of processors beyond which the speedup drops. In other words, there is a nonlinear relationship between speedup and the number of processors. Finding the optimal number of processors for a given task is a demanding job.

The software development challenge is to design a system that will scale with available hardware. To make the most of the system, one should know the topology of the system and the desired topology for the application's threads and/or processes. Software development under this scenario offers challenges that are unique to parallel computations.

5. Methodologies for Parallel Design Optimization

In this section we look at what steps in the DO algorithm can be parallelized and how.

5.1. *Parallel Finite Element Analysis*

Function evaluation (FE) usually implies the use of a numerical simulation scheme such as finite element analysis that is both compute and resource intensive. For example, for a function evaluation using static finite element analysis for one load case, the system equations are of the form

$$\mathbf{K}_{n \times n} \mathbf{D}_{n \times 1} = \mathbf{F}_{n \times 1} \tag{4}$$

where $\mathbf{K}$ is the system stiffness matrix, $\mathbf{D}$ is the vector of nodal displacements and $\mathbf{F}$ is the vector of nodal forces.

One of the approaches to parallelize the FE computations is to use the well-known domain decomposition (DD) idea[15]. We develop the relevant system equations for a typical domain or substructure, k, using the notation (I) for interior and (B) for boundary, e.g. the degrees-of-freedom are identified as $\mathbf{D}_I^k$ and $\mathbf{D}_B^k$. Partitioning the system equations into interior and boundary degrees-of-freedom, we have

$$\begin{bmatrix} \mathbf{K}_{II}^k & \mathbf{K}_{IB}^k \\ \mathbf{K}_{BI}^k & \mathbf{K}_{BB}^k \end{bmatrix} \begin{Bmatrix} \mathbf{D}_I^k \\ \mathbf{D}_B^k \end{Bmatrix} = \begin{Bmatrix} \mathbf{F}_{II}^k \\ \mathbf{F}_{BB}^k \end{Bmatrix} \tag{5}$$

The equations can be expanded and rewritten as follows.

$$\mathbf{K}_{II}^k \mathbf{D}_I^k + \mathbf{K}_{IB}^k \mathbf{D}_B^k = \mathbf{F}_{II}^k \tag{6}$$

$$\mathbf{K}_{BI}^k \mathbf{D}_I^k + \mathbf{K}_{BB}^k \mathbf{D}_B^k = \mathbf{F}_{BB}^k \tag{7}$$

Using the above equations and with algebraic manipulations, we can derive two important relationships as follows.

$$\mathbf{D}_I^k = \left[\mathbf{K}_{II}^k \right]^{-1} \left[\mathbf{F}_{II}^k - \mathbf{K}_{IB}^k \mathbf{D}_B^k \right] \tag{8}$$

and

$$\left[\mathbf{K}_{BB}^k - \mathbf{K}_{BI}^k \left(\mathbf{K}_{II}^k \right)^{-1} \mathbf{K}_{IB}^k \right] \mathbf{D}_B^k = \mathbf{F}_{BB}^k - \mathbf{K}_{BI}^k \left(\mathbf{K}_{II}^k \right)^{-1} \mathbf{F}_{II}^k \tag{9}$$

Or, simplifying the notation, we can write the effective equations as

$$\mathbf{K}_B^k \mathbf{D}_B^k = \mathbf{F}_B^k \tag{10}$$

where

$$\mathbf{K}_B^k = \mathbf{K}_{BB}^k - \mathbf{K}_{BI}^k \left(\mathbf{K}_{II}^k \right)^{-1} \mathbf{K}_{IB}^k \tag{11a}$$

$$\mathbf{F}_B^k = \mathbf{F}_{BB}^k - \mathbf{K}_{BI}^k \left(\mathbf{K}_{II}^k \right)^{-1} \mathbf{F}_{II}^k \tag{11b}$$

are the effective boundary stiffness and the effective load vector. Once these effective equations are constructed for each substructure (using a direct solver), the system equations can be assembled as

$$\sum_k \mathbf{K}_B^k \mathbf{D}_B^k = \mathbf{F}_B^k \Rightarrow \mathbf{K}_B \mathbf{D}_B = \mathbf{F}_B \tag{12}$$

and can be solved for $\mathbf{D}_B$ using an iterative solver. The boundary conditions are typically imposed at the substructure level. After solving Eqs. (12), one can then go back to Eq. (8) to recover $\mathbf{D}_I^k$ for each substructure (using a direct solver) and compute the secondary quantities.

The performance of this methodology is a function of (a) how well the original problem is split into subdomains, (b) the underlying data structure to support the computations, (c) the efficiency of the sparse matrix algorithms, (d) the efficiency of the iterative solver, and (e) the efficiency of the algorithm to minimize the communications between the processes as the number of subdomains increases.

5.2. *Parallelization in Gradient-based Optimization Techniques*

There has been considerable attention paid to coarse-grain, single level parallelization of optimization algorithms[16,17,18,19,20]. We will examine steps where parallel computations can take place.

5.2.1. *Parallel Gradient Evaluation (GE)*

If gradients are evaluated using the forward difference method, the derivative of a function $f(\mathbf{x})$ with respect to the ith design variable is given by

$$\frac{\partial f(x_1, x_2, \ldots x_n)}{\partial x_i} = \frac{f(x_1, x_2, \ldots x_i + \Delta x_i, \ldots x_n) - f(x_1, x_2, \ldots x_n)}{\Delta x_i} \tag{13}$$

The number of FEAs required during gradient evaluation is equal to the number of design variables. When multiple processors are available, gradient evaluation can be parallelized such that the number of FEAs is divided equally among the available processes. With the right combination of processors, this is an embarrassingly parallel problem.

5.2.2. *Parallel Line Search (LS)*

Parallel line search is implemented using a combination of the multi-section scheme[21] and parallel Avriel search[22]. The overall algorithm can be split into three steps.

Step 1: Bracketing the minimum

In this step, the idea is to bracket the interval whose end points have maximum constraint values of opposite sign.

Step 2: Zero-finding

Once the minimum has been bracketed, the next step is to find a feasible point that is as close to the constraint surface within the constraint thickness tolerance.

Step 3: Function minimization using parallel Avriel search

Finally, using information from Step 2, we can compute the lowest function value $f(\alpha_j^i)$ in parallel until the size of the final uncertainty interval falls below a predefined tolerance value.

Both an analysis of the algorithm and numerical experimentation show that a decent speedup is possible with only a few number of processors (between 4 and 16)[18].

5.2.3. *Parallel Direction Finding (DF)*

One of the most popular nonlinear programming techniques is the Method of Feasible Directions (MFD). It should be noted that line search involves searching along a direction vector that lies within the usable feasible cone (Fig. 3). It is possible to compute more than one search direction and conduct a line search along each of those directions so as to find the best possible solution. Numerical results using MFD show that these parallel DF and LS approaches are extremely promising[23]. The search directions can be found in three steps (Fig. 4).

Step 1: Compute the standard search direction, $\mathbf{d}_{std}$.

Step 2: Compute the tangent search direction, $\mathbf{d}_t$.

Step 3: Compute the intermediate search directions that lie in the cone defined by A and B. These search directions are called the γ search directions. Upon obtaining $\mathbf{d}_{std}$ and $\mathbf{d}_t$, the search directions $\mathbf{d} = \mathbf{d}_{\gamma i}$, $i = 1,...,n_\gamma$ are obtained from the relaxed LP subproblem as follows.

minimize β

subject to $\nabla f^{\mathrm{T}}\mathbf{d} \le \gamma\,(\nabla f^{\mathrm{T}}\mathbf{d}_t)\qquad 0<\gamma<1$

$\nabla g_j^{\mathrm{T}}\mathbf{d} \le \beta \qquad\qquad j\in J_{NL}$ (14)

$\nabla g_j^{\mathrm{T}}\mathbf{d} \le -g_j^0 \qquad\qquad j\in J_B$

$-1\le d_i \le 1 \qquad\qquad i=1,...,NDV$

where J_{NL} and J_B are the sets of active nonlinear and bound constraints, respectively.

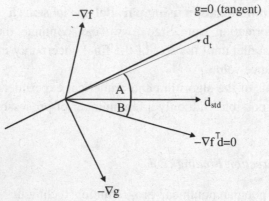

Fig. 3. The usable-feasible cone formed by A and B

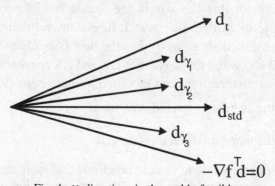

Fig. 4. γ-directions in the usable-feasible cone

As we see from these discussions, almost every step in the design optimization algorithm can be parallelized. An efficient implementation can lead to a decent speedup with a few or a large number of processors depending on the problem size. As we will show later, a strict master-slave implementation is really not necessary. Instead, three types of master processors are defined whose job is to coordinate the activities during GE, DF and LS computations and consolidate the final answer. The GE Manager is the master processor during the parallel gradient computations and constructs the final gradient vectors of the objective function and the active constraints. Similarly, there are DF and LS managers. Figure 5 shows a three-level implementation involving 8 processors, 4 search directions with 2 processors used for parallel FEA.

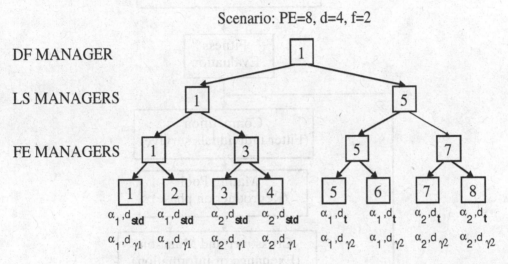

Fig. 5. Example scenario depicting 3-level parallelism

5.3. *Non-gradient Based Techniques*

Research into parallelizing non-gradient based optimizers has taken place over the years[24,25,26,27]. In this section we present a specific implementation of a simple Genetic Algorithm (GA).

5.3.1. *Parallel GA[28]*

The overall algorithm used in a GA-based design optimization problem is quite simple (Fig. 6). There are different ways one could devise the algorithm for evaluating the fitness function in parallel. We present two approaches.

Load-Balanced Approach: Initially, the master process instructs each slave process to evaluate the fitness associated with one member of the population by sending the values of the design variables for that member. Then it waits to receive the values of the objective function and the maximum violation from any of the slave processes (until the fitness evaluations of all the members of the population are completed). If there are more evaluations to be done, it passes the values of the design variables to that process. If all the evaluations are completed, it sends a message to that process that no more fitness evaluations are necessary.

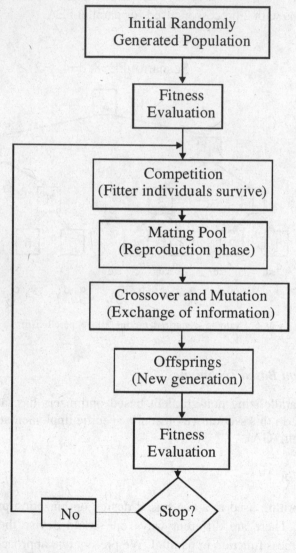

Fig. 6. Flow in a Simple Genetic Algorithm (SGA)

Master Process

(i) Set number of evaluations completed, $n_{eval} = 0$.

(ii) Loop through $j = 1, ..., \min(n_{pop}, n_p)$.

(iii) Generate the vector of design variables, **x**. Pass this vector to process j. Increment n_{eval}.

(iv) End loop.

(v) Loop through all members of the population, $i = 1, 2, ..., n_{pop}$.

(vi) Receive the objective function value and the maximum constraint violation from process j.

(vii) If $n_{eval} < n_{pop}$, generate the vector of design variables, **x** for member i. Pass this vector to process j. Increment n_{eval}. Else send "no-more-evaluation" message to process j.

(viii) End loop.

Slave Process (Valid only for process $j < n_{pop}$)

(i) Loop until "no-more-evaluation" message is received.

(ii) Receive the vector of design variables, compute and send the objective function value and the maximum constraint violation to the master process.

(iii) End loop.

It should be noted that in the abovementioned approach, only the master process executes the GA. Assuming that one integer word is 4 bytes, one double precision word is 8 bytes and the objective function and maximum constraint violation are designated as double precision, we can compute the total number of send and receive bytes as follows for every generation (n_b, n_d and n_s are the number of binary, discrete and continuous design variables) as follows.

$$n_{send} = n_{receive} = 4n_{pop}\left(n_b + n_d + 2n_s\right) + 8\left(2n_{pop}\right) \tag{15}$$

Do-All Load-Balanced Approach: In the previous approach, the GA is implemented and executed in the master process with only the fitness evaluations taking place in the slave processes. This approach requires that the values of the design variables be sent from the master process to the slaves. It should be noted that as a fraction of the total program time, the time taken to execute the GA steps is a very tiny fraction. In this DLB approach, all the processes execute exactly the same program statements except for the part where the entire population is evaluated. As a byproduct, program maintenance is much easier since there is a single block of the program where process-related logic needs to be used.

Master Process

(i) Set number of evaluations completed, $n_{eval} = 0$

(ii) Loop through $j = 1, ..., \min(n_{pop}, n_p)$.

(iii) Increment n_{eval}. Ask process j to evaluate member n_{eval} of the population.

(iv) End loop.

(v) Loop through all members of the population, $i = 1, 2, ..., n_{pop}$.

(vi) Receive the objective function value and the maximum constraint violation from process j.

(vii) If $n_{eval} < n_{pop}$, increment n_{eval}, ask process j to evaluate member n_{eval} of the population. Else send "no-more-evaluation" message to process j.

(viii) End loop.

(ix) Broadcast the objective function and maximum constraint violation values for all the members of the population.

Slave Process (Valid only for process $j < n_{pop}$)

(i) Loop until "no-more-evaluation" message is received.

(ii) Receive index of the member of the population whose objective function and constraints must be evaluated. Compute and send the objective function value and the maximum constraint violation to the master process.

(iii) End loop.

(iv) Receive the objective function and maximum constraint violation values for all the members of the population.

With this approach, the number of point-to-point send and receive bytes per generation is as follows

$$n_{send} = n_{receive} = 20n_{pop} \tag{16a}$$

and the number of broadcast bytes is

$$n_{broadcast} = 16n_{pop} \tag{16b}$$

With this approach, the communication traffic is independent of the number and type of design variables.

In the next section, we look at numerical examples that are solved using the methodologies discussed in this chapter.

6. Numerical Examples

The numerical examples in this section are solved using HYI-3D[29,30] software system that is developed using object-oriented concepts in C++ and uses MPI for message passing. The software suite is a collection of independent modules that can be plugged together to create the required set of finite element analysis and design optimization capabilities and can be executed either sequentially or in parallel. Solutions to three example problems that are solved on the following cluster configurations are discussed in this section.

Beowulf Cluster Information: (a) Number of machines in the cluster = 2 (b) Computer 1 (C1): Intel Dual P3-866 MHz, 768 MB RDRAM, Ultra 7200 rpm IDE Drive, 3COM 3C920 NIC. Computer 2 (C2): AMD 1.2 GHz Athlon,

512 MB SDRAM, Ultra 7200 rpm IDE Drive, 3COM 3C920 NIC, (c) Windows 2000, MPI-Softtech 1.6.4, Linksys BEFSR41 10/100 Router.

ASU FEM Cluster Information: (a) Number of machines in the ˙cluster = 8 (b) Typical machine: Intel P4 1.7 GHz Dual Xeon, 1 GB RDRAM, Ultra 7200 rpm IDE Drive, Intel PRO/1000 T NIC, (c) Windows 2000, MPI-Softtech 1.6.4, Cisco Catalyst 3550-12T switch.

ASU CML Cluster Information: (a) Number of machines in the cluster = 24 (b) Typical machine: Intel P4-3.06 GHz Xeon with 2 GB RAM, SCSI hard disk, Intel PRO/1000 T NIC, (c) Red Hat Linux and MPI (mpich), Dell PowerConnect 2624 Unmanaged Switch.

Maui High-Performance Computing Cluster (MPCC) Information: (a) Number of machines in the cluster = 260 (b) Typical machine: Intel Pentium 3-933 MHz Dual Xeon, 1 GB RAM and SCSI hard disk, (c) Red Hat Linux and mpich, Myrinet switch capable of 200 MB/second sustained bandwidth.

TCP/IP communication protocol is used in all the clusters. All timings are wall clock timings. The term 1PN denotes execution when one process is launched per node in a cluster. Similarly, the term 2PN denotes execution when two processes are launched per node in a cluster. For a homogenous system, we compute the speedup as follows.

$$\text{Speedup obtained for } n \text{ processes} = \frac{\text{Time for one process}}{\text{Time for n processes}} \tag{17}$$

For a heterogeneous system, we compute the speedup as follows. Let the relative speed of the n_p processes with respect to the slowest process be denoted as $s_1, s_2, ..., 1, ..., s_{n_p}$ (with 1 corresponding to the slowest process), the time taken for each single process run be denoted as $t_1, t_2, ..., t_i, ..., t_{n_p}$, and t be the total time taken when n processes are used. Then

$$s_j = \frac{\max(t_1, t_2, ..., t_{n_p})}{t_j} \tag{18}$$

$$\text{Speedup obtained for } n \text{ processes} = \frac{\frac{1}{\sum_{j=1}^{n_p} s_j} \left(\sum_{i=1}^{n_p} s_i t_i \right)}{t} \tag{19}$$

The following nomenclature is used when gradient-based optimization results are discussed.

```
-ge      Parallel GE is carried out using all available processors
-df:d    Parallel DF is carried out using d directions
-ls:l    Parallel LS is carried out using l processors
-fea:f   Parallel FEA is carried out using f processors
```

Example 1: Parallel FEA

Figure 7 shows the FE model of a thick-support, steel platform fixed at both ends and loaded uniformly at the center. The FE model has 955,430 4-noded tetrahedral elements, 169,143 nodes for a total of 507,429 degrees-of-freedom. The FE analysis is carried out to compute the nodal displacements, and element strains and stresses.

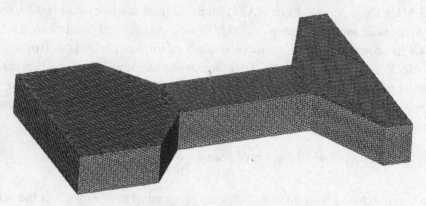

Fig. 7. FE Model of Support Platform

The accuracy of the results from the outer loop in the FE analysis (Eq. (12)) can be controlled by using an appropriate value for the convergence tolerance in the iterative solver. Considerable savings in computational time are possible by using a larger tolerance value without losing accuracy. Two convergence criteria are used as follows.

$$\|\mathbf{KD} \text{-} \mathbf{F}\| \le \delta_1 \text{ and } \frac{\|\mathbf{KD} \text{-} \mathbf{F}\|}{\|\mathbf{F}\|} \le \delta_2 \tag{20}$$

where δ_1 and δ_2 are user-defined convergence tolerances. The iterations are terminated when at least one of the above criteria is satisfied. The timing results of the FE analysis are shown in Table 1.

Table 1. Timing Results (MHPCC Cluster)

# of processors	Wall Clock time (sec)	Wall Clock time (sec)	% savings
Convergence Tolerance	10^{-9}	10^{-6}	
8	2,878	2,675	7.1
16	1,348	1,000	25.8
32	828	589	28.8
64	629	579	7.9

The speedup obtained in this example is typical of the performance gains obtained from the domain decomposition sparse direct plus iterative solver strategy. However, the biggest challenge that still remains after all these years, is obtaining a good pre-conditioner for the iterative solver.

Example 2: Parallel GA

The structural system that is designed is shown in Fig. 8. The planar truss is described in terms of two parameters – the number of bays and the number of stories. The truss members are grouped into three groups per storey – horizontal members, vertical members and diagonal members. Hence, the number of design variables (cross-sectional areas) is equal to three times the number of stories. The optimization problem is to find the optimal values of the cross-sectional areas of the members so as to minimize the mass of the truss subject to axial stress constraint. The specific values used in the following examples are as follows.

Mass density, $\rho = 0.00881448\ \text{lbm/in}^3$
Allowable stress, $\sigma_a = 10000\ psi$
$x_j^L = 0.1\,in^2$ and $x_j^U = 20\,in^2$ and precision is taken as $0.1\,in^2$.
Bay width = 240 in
Story height = 120 in
Applied load, P=10000 lb

The problem-specific data are as follows.

Number of bays	30
Number of storeys	50
Nodes	1581
Elements	4550
Number of design variables	150
Chromosome Length	1200

# of Generations   50
# of function evals/generation   2400

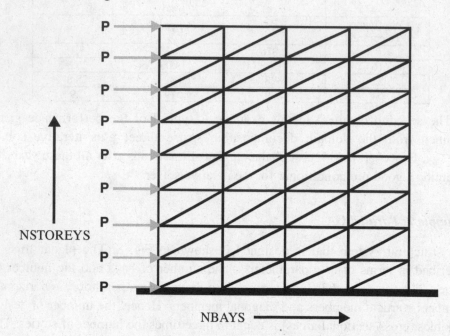

Fig. 8. Layout of the Planar Truss

The initial population is randomly generated and the GA is terminated after 50 generations. The results are shown in Tables 2(a) and 2(b). The initial objective function value is 149,290 lbm and the final objective function value after 50 generations is 65,116 lbm.

Table 2(a). Results of Truss Problem (FEM Cluster)

PE (Machines)	LB Version		DLB Version	
	Time (s)	Speedup	Time (s)	Speedup
1 (1)	6762	1.0	6762	1.0
3 (2)	3366	2.0	3370	2.0
4 (3)	2269	2.98	2260	2.99
5 (4)	1709	3.96	1701	3.98
6 (5)	1382	4.89	1376	4.91
7 (6)	1155	5.85	1148	5.89
11 (5)	716	9.44	690	9.8
15 (7)	524	12.9	510	13.25

Table 2(b). Results of Truss Problem (Beowulf Cluster)

# of Processes (Machines)	LB Version		DLB Version	
	Time (s) (C1/C2)	Speedup	Time (s) (C1/C2)	Speedup
1 (1)	18491/11930	1.0	18491/11930	1.0
3 (2)	7473	1.94	7301	1.99

With the LB approach, a total of 5,836,800 bytes are sent and received every generation. With the DLB approach, 48,000 point-to-point bytes are sent and received every generation, and 38,400 broadcast bytes are sent and received every generation. The amount of physical memory appears to be adequate for the problem being solved. When the system performance is monitored, the hard page faults are shown to be minimal. The Beowulf cluster shows an almost linear speedup. The DLB approach is more efficient showing an ideal speedup because the communication traffic is much less compared to the LB approach. On the other hand with the FEM cluster, the DLB approach becomes more effective with increasing number of processes. The design history is shown in Fig. 9.

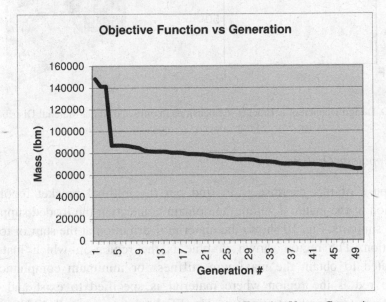

Fig. 9. Design History Showing Objective Function Versus Generation

Example 3: Parallel Gradient-based Design Optimization

An L-bracket is initially modeled as shown in Fig. 10. The thickness of the bracket is 10 mm. The top face of the bracket (AB) is completely constrained. A load of 1000 N is applied along the negative y-direction at the bottom right corner and is distributed over the edge along the thickness. The optimal design takes place in two steps. In the first step we carry out topology optimization. In the next step, the shape optimal design procedure is applied to obtain the final shape and form.

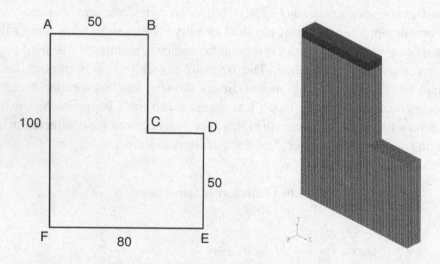

Fig. 10. Design Domain of L-Bracket (Dimensions in mm) and Initial Material Distribution

Step 1: Topology Optimization

The purpose of this exercise is to find out the optimal bracket topology or distribution of the material within the domain connecting the loaded support and the fixed supports. Fig. 10 shows the material distribution at the start of topology optimization. The region shown in blue is the region in which material is redistributed to obtain the maximum stiffness or minimum compliance. The region in red is the region where material is specified to exist and do not participate in topology optimization. The FE model consists of 47790 8-noded hexahedral elements and 54740 nodes. The topology optimization is carried out with the material density, ρ restricted to $0.005 \leq \rho \leq 0.995$ with the initial guess as 0.3. The mass fraction constraint is taken as 0.3. The convergence criteria are taken as (a) 0.001 for the absolute change in the objective function value for 3 consecutive iterations and (b) 0.01 as the change in the Lagrange multiplier value for 3 consecutive iterations.

The final material density plot is shown in Fig. 11. The final results are obtained in 20 iterations involving 21 finite element analyses. The compliance reduces from 0.0058 N-mm to 0.001 N-mm. The results seem to indicate the presence of one rectangular hole and two triangular holes. The optimization timing results are shown in Table 3.

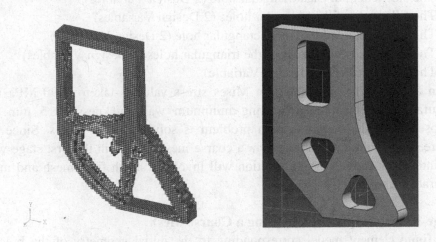

Fig. 11. Final Material Distribution and Subsequent Geometric Model

Table 3. Performance of Parallel Topology Optimization (FEM Cluster)

PE	Topology Optimization Time (seconds)	Speedup (Optimization)	Parallel FEA Time (seconds)	Speedup (FEA)
1	68040	1.0	1300	1.0
4 (1PN)	3330	20.4	134	9.7
8 (1PN)	1605	42.4	64	20.3

Clearly the speedup is superlinear. This is because of two reasons. First, the problem size is memory bound when using a smaller number of processors, and memory becomes less of an issue with increasing number of processors (and subdomains). While the communication traffic increases with increasing number of subdomains, this is offset by the speedup obtained from the iterative solver. Second, the single processor runs are made with the sparse direct solver that is less efficient than the iterative solver for this problem.

Step 2: Shape Optimization

With the optimal material distribution obtained from topology optimization, a Unigraphics model of the L-bracket was created as shown in Fig. 10. The model used for shape optimization has 2 triangular cutouts and 1 rectangular cutout with filleted corners to prevent stress concentrations. There are 11 shape-related parameters or design variables.

 (i) The side length of each triangular hole (2 Design Variables)
 (ii) The fillet radii of the triangular holes (2 Design Variables)
(iii) The length and width of the rectangular hole (2 Design Variables)
(iv) The centroidal coordinates of the triangular holes (4 Design Variables)
 (v) The plate thickness (1 Design Variable)

In addition, the allowable von Mises stress value is taken as 90 MPa and manufacturing constraints requiring minimum wall thickness of 5 mm are imposed. The shape optimization problem is solved in two stages. Since the nature of the solution is not known, a coarse mesh is used in the first stage with the intent that the next stage solution will involve a much finer mesh and more accurate FE solution.

Stage 1. Shape Optimization using a Coarse Mesh

The finite element model corresponding to the initial geometry of the bracket consists of 4932 nodes and 3060 8-noded hexahedral elements. The volume at the beginning of shape optimization is 40839 mm^3 and is reduced to 10714 mm^3. The dimensions of the cutouts increase. A 74% savings in volume from shape optimization and a total saving of 84% in volume as compared to the pre-topology optimization volume of the bracket are obtained. The stress distribution of the final shape is shown in Fig. 12(a). The stress constraint is active at the optimum – the maximum stress is 84 MPa close to the allowable value. The final geometric model is shown in Fig. 12(b). This model is then used to create a much finer mesh that is used in Stage 2.

Stage 2. Shape Optimization using a Fine Mesh

There are 48870 nodes and 39416 8-noded hexahedral elements. The optimization is run for 2 iterations after which the maximum stress increases to 87 MPa but still less than the allowable value of 90 MPa. The volume reduces from 11390 mm^3 to 10933 mm^3. The stress distribution of the final shape is shown in Fig. 13(a). The final geometric model is shown in Fig. 13(b). The computational results for Stage 2 are shown in Table 4.

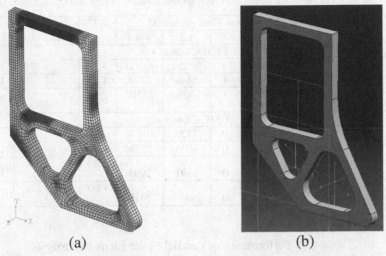

(a)   (b)

Fig. 12. (a) Final Stress Distribution (b) Final Geometric Model

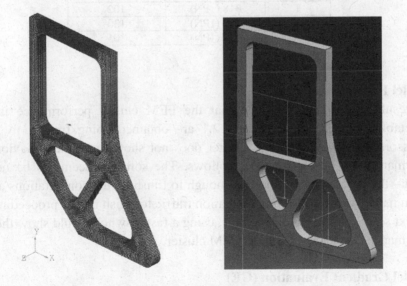

Fig. 13. (a) Final Stress Distribution (b) Final Geometric Model

Table 4(a). Performance of Parallel Shape Optimization (FEM and CML Cluster)

PE	Parallel Computations	Total Time (seconds)				NT[ii]	N_{GE} /PE	N_{LS} /PE	F^{*}[iii] (mm³)
		GE	DF	LS	Total				
FEM Cluster									
1	NA	6924	4	26255	35253	3.24	22	76	10933
8	-ge -df:4 -ls:2	2386	1	5441	10890	1.00	6	9	10937
16	-ge -df:4 -ls:2 -fea:4	11051	1	14080	32500	2.98	6	8	10937
CML Cluster									
1	NA	2236	0	7728	10879	4.51	22	76	10933
8	-ge -df:4 -ls:2	623	0	1034	2578	1.07	6	10	10937
16	-ge -df:4 -ls:2 -fea:2	796	0	1191	2993	1.24	6	9	10937
24	-ge -df:4 -ls:2 -fea:3	629	0	849	2410	1.00	6	8	10913

Table 4(b). Performance of Parallel Finite Element Analysis

PE	Analysis Time (seconds)	
	FEM	CML
1	318 (1PN)	102
2	189 (1PN)	130
3	152 (1PN)	105

Parallel FE Analysis

The results from Table 4 show that the FEM cluster performance is more predictable. A speedup of 1.7 and 2.1 are obtained going from 1 to 2 to 3 processors. However, the CML cluster does not show the same behavior. This performance can be explained as follows. The software seems to be network bound – the processor speed is fast enough to finish all the computations and the system has to wait for the communication traffic to finish before proceeding with the next set of computations. Perhaps, using a faster switch would show the same performance characteristics as the FEM cluster.

Parallel Gradient Evaluation (GE)

Consider the scenario with 16 processors on the CML cluster (Table 3). Since there are 11 design variables and 16/2 = 8 GE managers, the max number of design variables handled by a GE manager = 11/8+1+1 = 3. Thus the expected

---

[ii] NT: Normalized Timing
[iii] F^{*}: Final Objective

speedup for GE is $(11/3)(102/130) = 2.88$. The obtained speedup is $(2236/796) = 2.81$ which agrees very well with the expected speedup.

Again, the speedup for combined parallel GE and FE depends on the reduction in number of design variables or FEs to be performed by the GE manager as well as the speedup due to parallel FE analysis. However for this problem, there is no significant gain using more processors for FEA (Table 4). Thus the maximum speedup obtained for GE is with 8 processors = $(2236/623) = 3.59$.

Combined Parallel Direction Finding (DF), Line Search (LS) and Function Evaluation (FE)

Here we discuss the performance of the scenario described in Fig. 5. When parallel DF is combined with parallel LS and parallel FEA, in addition to the number of directions to be computed in each iteration, it is necessary to specify the number of processors to be used for each parallel FE and the number of LS managers per direction. Consider the scenario with 24 processors on the CML cluster (Table 3). In this case, LS is performed along 4 directions simultaneously. 2 LS processes are specified per direction indicating that 2 steps will be taken simultaneously along each direction. The benefit of parallel LS is the reduction in the number of interval reductions needed to find the step size along a particular direction vector. Thus, parallel DF and parallel LS results in the reduction of NFV_{LS}/PE from 76 to 8. Parallel DF contribution would be approximately $(76/4) = 19$. This assertion in problem specific – NFV_{LS} is almost the same along each direction. Hence the speedup in terms of reduction in FEs due to parallel LS is $19/8 = 2.38$. Hence total expected speedup due to parallel DF and LS is $(76/8) = 9.5$. Additionally, 3 processors are specified for parallel FE. This is expected to further reduce computation time by reducing FE time. However for this problem, there is no reduction in FE time using 3 processors for parallel FEA. Thus the expected speedup is $(76/8)(102/105) = 9.23$ while the obtained speedup is $(7728/849) = 9.1$.

The direction-finding time is extremely small since the number of active constraints is small. The FEM cluster's performance for this problem deteriorates when both the processors per node are used – total time increases from 2386 seconds (8 processors) to 11051 seconds (16 processors). The problem is memory bound and the amount of hard page faults increases appreciably when both processors per node are used.

7. Conclusions

Two major changes bode well for the development and use of parallel processing software for design optimization. First, the requirements for design optimization are increasing for a number of reasons. The finite element analyses now require more computational effort. The design models have larger number of design variables and constraints. Second, the growth in computational speed of CPUs has slowed down and the new focus is on multicore CPUs. Both these changes require the use of parallel processing techniques to obtain the computational gains required to obtain solution to bigger problems.

As the discussions show in this chapter, developments in parallel computations have just begun to yield the expected computational gains. Depending on the type of problem being solved and the solution technique, once can use a small 2-computer heterogeneous cluster to as large as a 260-node homogenous cluster. Much more effort is needed in a multitude of areas – hardware development, compiler technology, parallel sparse direct and iterative solvers, design optimization algorithm development and fault-tolerant, optimized message passing libraries.

References

1. ANSYS Breaks Engineering Simulation Solution Barrier, 2004 International ANSYS Conference.
2. Sutter, H., "The Free Lunch Is Over: A Fundamental Turn Toward Concurrency in Software", Dr. Dobb's Journal, Vol. 30, No. 3, 2005.
3. Bendsoe, M.P., "Optimal Shape Design as a Material Distribution Problem", Structural Optimization, Vol. 1, pp. 193-202, 1989.
4. Bendsoe, M.P. and Sigmund, O., "Material Interpolations in Topology Optimization", *Arch. Applied Mechanics*, Vol. 69, pp. 635-654, 1999.
5. Zhou, M. and Rozvany, G.I.N., "The COC Algorithm, Part II: Topological, Geometry and Generalized Shape Optimization", *Computer Methods in Applied Mechanical Engineering*, Vol. 89, pp. 197-224, 1991.
6. Sigmund, O., "A 99 Line Topology Optimization Code Written in Matlab", *Structural Multidisciplinary Optimization*, Vol. 21, pp. 120-127, 2001.
7. Sigmund, O., "On the Design of Compliant Mechanisms using Topology Optimization", *Mech. Struct. Mach.*, Vol. 25, pp. 495-526, 1997.
8. Belegundu, A.D. and Rajan, S.D., "A Shape Optimization Approach Based on Natural Design Variables and Shape Functions", *Computer Methods in Applied Mechanics and Engineering*, Vol. 66, pp. 87-106, 1988.
9. Rajan, S.D., Chin, S.W. and Gani, L., "Toward a Practical Design Optimization Tool", *Microcomputers in Civil Engineering*, Vol. 11, pp. 259-274, 1996.
10. Briabant, V. and Fluery, C., "Shape Optimum Design using B-splines," *Computer Methods in Applied Mechanics and Engineering*, Vol. 44(3), pp. 247-267, 1984.
11. Dongarra, J., Sterling, T., Simon, H. and Strohmaier, E., "High Performance Computing. Clusters, Constellations, MPPs, and Future Directions", IEEE Computing in Science and Engineering, Vol. 7, No. 2, pp.51-59, 2005.

12. Sloan, J., High Performance Linux Clusters with OSCAR, Rocks, openMosix and MPI, O'Reilly, 2004.
13. Chandra, R., Dagum, L., Kohr, D., Maydan, D., McDonald, J. and Menon, R., Parallel Programming in OpenMP, Academic Press, 2001.
14. Gropp, Lusk and Skjellum, Using MPI: Portable Parallel Programming with the Message-Passing Interface, 2nd Edition, The MIT Press, 1999.
15. Moayyad, M.A. and Nguyen, D.T., "An Algorithm for Domain Decomposition in Finite Element Analysis", *Journal of Computers and Structures*, Vol. 39, No. 1-4, pp. 227-290, 1991.
16. Benson, S., McInnes, L., More, J., and Sarich, J., "The Toolkit for Advanced Optimization for Large-Scale Minimization Problems", *10th AIAA/ISSMO Multidisciplinary Analysis and Optimization Conference*, Albany, NY, 2004.
17. Burgee, S., Giunta, A.A., Narducci, R., Watson, L.T., Grossman, B. and Haftka, R.T., "A Coarse Grained Parallel Variable-Complexity Multidisciplinary Optimization Paradigm," *The International Journal of Supercomputer Applications and High Performance Computing*, Vol. 10, No. 4, pp. 269-299, 1996.
18. Grauer, M. and Barth, T., "Grid Based Computing for Multidisciplinary Analysis and Optimization", *10th AIAA/ISSMO Multidisciplinary Analysis and Optimization Conference*, Albany, NY, 2004.
19. Padula, S.L. and Stone, S.C., "Parallel implementation of large-scale structural optimization", *Structural Optimization*, Vol. 16, No. 2-3, pp. 176-185, 1998.
20. Venter, G. and Watson, B., "Efficient Optimization Algorithms for Parallel Applications", *8th AIAA/USAF/NASA/ISSMO Symposium at Multidisciplinary Analysis and Optimization*, Long Beach, CA, Sept. 6-8, 2000.
21. Belegundu, A.D., Damle, A., Rajan, S.D., Dattaguru, B. and St. Ville, J., "Parallel Line Search in Method of Feasible Directions", *Optimization and Engineering Journal, Optimization and Engineering Journal*, 5, 379-288, 2004.
22. Nguyen, D.T., "Nonlinear constrained optimization and parallel processing for golden block line search", *Computer Assisted Mechanics and Engineering Sciences*, Vol. 6, No. 3, 8th International Conference on Numerical Mathematics and Computational Mechanics (NMCM98), Miskolc, Hungary, pp. 469-477, 1998.
23. Rajan, S.D., Belegundu, A.D., Damle, A.S., Lim, H. and St. Ville, J., "General Implementation of Multilevel Parallelization in a Gradient-based Design Optimization Algorithm", submitted to *AIAA Journal*, 2005.
24. Miki, M., Hiroyasu, T. and Hatanaka, K., "Parallel Genetic Algorithms with Distributed-Environment Multiple Population Scheme", *3rd WCSMO World Congress of Structural and Multidisciplinary Optimization*, Niagara Falls, NY, 1999.
25. Chipperfield, A. and Fleming, P., "Parallel Genetic Algorithms", Chapter 39, *Parallel and Distributed Computing Handbook*, Ed. Albert Y. Zomaya, McGraw-Hill, 1996.
26. Sarma, K. and Adeli, H., "Bilevel Parallel Genetic Algorithms for Optimization of Large Steel Structures", *Computer-Aided Civil and Infrastructure Engineering*, 16, 295-304, 2001.
27. Scott, S. D., Samal, A. and Seth, S., "HGA: A Hardware-Based Genetic Algorithm", *Proc. of the 1995 ACM/SIGDA Third International Symposium on Field-Programmable Gate Arrays*, 53-59, Monterey, CA, February 1995.
28. Rajan, S. D. and Nguyen, D.T., "Design Optimization of Discrete Structural Systems Using MPI-enabled Genetic Algorithm", *Journal of Structural and Multidisciplinary Optimization*, 27, 1-9, 2004.
29. HYI, Inc., HYI-3D, Software System for Volumetrically Controlled Manufacturing, Version 1.8, Phoenix, AZ, 2004.
30. Rajan, S.D., Nguyen, D.T., Yu, S., Lau, D. and St. Ville, J., "Improving the Performance of Finite Element Computations through Parallel Processing", *Maui High-Performance Computing Cluster Application Briefs*, pp. 2-3, 2003.

CHAPTER 19

SEMIDEFINITE PROGRAMMING FOR STRUCTURAL OPTIMIZATION

Makoto Ohsaki

Department of Architecture and Architectural Engineering, Kyoto University,
Kyotodaigaku-Katsura, Nishikyo, Kyoto 615-8540, Japan
E-mail: ohsaki@archi.kyoto-u.ac.jp

Yoshihiro Kanno

Department of Urban and Environmental Engineering, Kyoto University,
Sakyo, Kyoto 606-8501, Japan
E-mail: kanno@archi.kyoto-u.ac.jp

Basic formulations of SemiDefinite Programming (SDP) are presented, and the properties including optimality conditions and duality are summarized. It is shown that structural optimization problems considering compliance, eigenvalue of vibration, etc. can be formulated by SDP problems. Interior point methods for solving SDP are briefly introduced. Finally, it is shown that the analysis problem of a cable network can be formulated as a Second-Order Cone Programming (SOCP) problem that is a particular case of SDP.

1. Introduction

SemiDefinite Programming (SDP) problem is a minimization problem of a linear function under linear constraints and requirement such that variable matrices are positive semidefinite. Since the positive semidefinite constraints include linear and convex quadratic constraints, the SDP is an extension of Linear Programming (LP), and unifies several nonlinear convex optimization problems such as convex Quadratic Programming (convex QP), Second-Order Cone Programming (SOCP), etc. Also SDP is a special case of more general problems of minimizing a linear function over convex cones.[49]

SDP has been extensively studied in the past ten years, because it has wide areas of application in mathematical and engineering fields.[68,40,41] For a combinatorial optimization problem, a better bound can be obtained by SDP relaxation than by LP relaxation. A graph partitioning problem such as maximum cut problem can also be relaxed to an SDP problem.[19] In the field of control theory, semidefinite constraints have been often used in the form of Linear Matrix Inequality (LMI).[10]

Recently, some applications have been found in the field of structural optimization. Ben-Tal and Nemirovski[7] presented an SDP formulation for robust truss topology design under compliance constraints. Ohsaki *et al.*[54] demonstrated that optimal trusses with multiple eigenvalues of vibration can be obtained without any difficulty by SDP, and extended the method to linear buckling constraints.[30]

Another reason for extensive research on SDP is that a polynomial-time algorithm called primal-dual interior-point method has been developed, and large practical problems can be solved within reasonable computational time. The convergence and quality of the obtained solution defined by the duality gap are guaranteed by the strong duality theory.

In this chapter, outline of SDP and its application to structural optimization are summarized. The details of mathematical background and algorithms for SDP may be consulted to the review papers[22,63,61] and the handbook.[65]

2. Mathematical Preliminaries

Fundamental properties of positive semidefinite matrices are listed as follows for references to derivations of equations in the subsequent sections. Throughout this chapter, I^n denotes an identity matrix in $\mathbf{R}^{n \times n}$.

Let $\mathcal{S}^n \subset \mathbf{R}^{n \times n}$ denote the set of all $n \times n$ real symmetric matrices. We write $A \succeq O$ if $A \in \mathcal{S}^n$ is positive semidefinite; i.e., if all the eigenvalues of A are nonnegative. $A \bullet B$ is the inner product of $A = (A_{ij}) \in \mathcal{S}^n$ and $B = (B_{ij}) \in \mathcal{S}^n$ defined by

$$A \bullet B = \sum_{i=1}^{n} \sum_{j=1}^{n} A_{ij} B_{ij}.$$

Fundamental properties:

(P1) For $A \in \mathcal{S}^n$, the following three properties are equivalent:

    (a) A is positive semidefinite.

    (b) $u^\top A u \geq 0$ for all $u \in \mathbf{R}^n$.

    (c) All eigenvalues of A are nonnegative.

(P2) Suppose $A \in \mathcal{S}^m$ is positive definite and $B \in \mathcal{S}^n$. Then the matrix $D \in \mathcal{S}^{m+n}$ defined by

$$D = \begin{pmatrix} A & C \\ C^\top & B \end{pmatrix}$$

    is positive definite (semidefinite) if and only if $E = B - C^\top A^{-1} C$ is positive definite (semidefinite). Here, the matrix $E \in \mathcal{S}^n$ is called *Schur complement* of A in D.

(P3) $A \in \mathcal{S}^n$ and $B \in \mathcal{S}^n$ satisfy $A \bullet B = \text{trace}(AB)$.

(P4) Let $\mathcal{S}^n \ni A \succeq O$.

(a) There exists the unique matrix $\mathcal{S}^n \ni A^{1/2} \succeq O$ satisfying $A^{1/2}A^{1/2} = A$. Here, $A^{1/2}$ is often called *square root* of A.

(b) For $\mathcal{S}^n \ni B \succeq O$, we obtain $A \bullet B = \text{trace}(A^{1/2}A^{1/2}B) = \text{trace}(A^{1/2}BA^{1/2}) \geq 0$, because the matrix $A^{1/2}BA^{1/2}$ is positive semidefinite.

(P5) Let λ_1 denote the minimum eigenvalue of $\mathcal{S}^n \ni A \succeq O$. $\lambda_1 \geq \bar{\lambda}$ holds if and only if $A - \bar{\lambda}I^n \succeq O$, i.e., the matrix $(A - \bar{\lambda}I^n)$ is positive semidefinite.

(P6) $A \in \mathcal{S}^n$ and $u \in \mathbf{R}^n$ satisfy $u^\top A u = A \bullet (uu^\top)$.

(P7) For $\mathcal{S}^n \ni A \succeq O$ and $\mathcal{S}^n \ni B \succeq O$, the condition $A \bullet B = 0$ is equivalent to $AB = O$.

(P8) For $q \in \mathbf{R}^n$, $\gamma \in \mathbf{R}$, $\mathcal{S}^n \ni Q \succeq O$, and $L \in \mathbf{R}^{n \times n}$ satisfying $Q = LL^\top$, we have

$$x^\top Q x + q^\top x + \gamma \leq 0 \iff \begin{pmatrix} I^n & L^\top x \\ x^\top L & -q^\top x - \gamma \end{pmatrix} \succeq O.$$

3. Problem Formulation

Let $A_i \in \mathcal{S}^n$ $(i = 1, \ldots, m)$, $C \in \mathcal{S}^n$, and $b = (b_i) \in \mathbf{R}^m$ denote constant (data) matrices and vector. $X \in \mathcal{S}^n$ and $Z \in \mathcal{S}^n$ are the variable matrices and $y = (y_i) \in \mathbf{R}^m$ is a variable vector. In the following, all the vectors are column vectors, and the components are given by the subscripts.

The standard form of the SDP problem is formulated as

$$\left. \begin{aligned} \mathcal{P} : \min_{X} \quad & C \bullet X \\ \text{s.t.} \quad & A_i \bullet X = b_i, \quad (i = 1, \ldots, m) \\ & X \succeq O. \end{aligned} \right\} \tag{1}$$

The dual of $\mathcal{P}$, which is formulated by using the same data matrices and vector as $\mathcal{P}$, refers to the optimization problem having the form of

$$\left. \begin{aligned} \mathcal{D} : \max_{y, Z} \quad & b^\top y \\ \text{s.t.} \quad & \sum_{i=1}^{m} A_i y_i + Z = C \\ & Z \succeq O. \end{aligned} \right\} \tag{2}$$

Here, $\mathcal{D}$ is also an SDP problem, since it can be embedded into the form of $\mathcal{P}$.[22] Moreover, the dual problem of $\mathcal{D}$ coincides with $\mathcal{P}$ again. Note that most of the engineering problems are formulated in the form of $\mathcal{D}$. However, we define 'primal' and 'dual' as shown above along with the mathematical tradition.

Example 1: Consider a problem of minimizing $y_1 + y_2$ under constraints of $y_1 \geq 0$ and $y_1 y_2 \geq 1$. This problem can be formulated in the dual form of SDP as

$$\left. \begin{aligned} \max \quad & -y_1 - y_2 \\ \text{s.t.} \quad & Z = \begin{pmatrix} y_1 & 1 \\ 1 & y_2 \end{pmatrix} \succeq O. \end{aligned} \right\} \tag{3}$$

Here, we see

$$b = (-1, -1)^\top, \quad y = (y_1, y_2), \quad A_1 = \begin{pmatrix} -1 & 0 \\ 0 & 0 \end{pmatrix}, \quad A_2 = \begin{pmatrix} 0 & 0 \\ 0 & -1 \end{pmatrix}, \quad C = \begin{pmatrix} 0 & 1 \\ 1 & 0 \end{pmatrix}$$

and $m = n = 2$ in $\mathcal{D}$. Hence the primal form $\mathcal{P}$ is written by $X \in \mathcal{S}^2$ as

$$\left. \begin{array}{ll} \min & \begin{pmatrix} 0 & 1 \\ 1 & 0 \end{pmatrix} \bullet X \\ \text{s.t.} & \begin{pmatrix} -1 & 0 \\ 0 & 0 \end{pmatrix} \bullet X = -1, \quad \begin{pmatrix} 0 & 0 \\ 0 & -1 \end{pmatrix} \bullet X = -1, \\ & X \succeq O \end{array} \right\} \tag{4}$$

which is reduced to

$$\left. \begin{array}{ll} \min & 2X_{21} \\ \text{s.t.} & X_{11} = 1, \quad X_{22} = 1 \\ & X \succeq O \end{array} \right\} \tag{5}$$

where $X_{12} = X_{21}$ has been used.   □

Example 2: The problem with several positive semidefinite variable matrices can be converted into the standard form of SDP problem. Let $A_{ij} \in \mathcal{S}^n$ ($i = 1, \ldots, m$; $j = 1, \ldots, k$). The problem

$$\max_{y, Z_1, \ldots, Z_k} \quad b^\top y$$

$$\text{s.t.} \quad \sum_{i=1}^{m} A_{ij} y_i + Z_j = C_j, \quad Z_j \succeq O, \quad (j = 1, \ldots, k)$$

where $Z_j \in \mathcal{S}^{n_j}$ ($j = 1, \ldots, k$), can be formulated equivalently as $\mathcal{D}$ in (2), because

$$Z_j \succeq O, \quad (j = 1, \ldots, k) \quad \Longleftrightarrow \quad \begin{pmatrix} Z_1 & O & \cdots & O \\ O & Z_2 & \cdots & O \\ \vdots & \vdots & \ddots & \vdots \\ O & O & \cdots & Z_k \end{pmatrix} \succeq O.$$

In a particular case of $n_j = 1$ ($j = 1, \ldots, k$), we see that a linear inequality $\mathbf{R}^k \ni z \geq 0$ is equivalent to $\mathrm{diag}(z) \succeq O$, which shows that LP is included in SDP.   □

Example 3: Any convex quadratic programming problem is formulated as an SDP problem. Indeed, a convex quadratic inequality can be expressed as a linear matrix inequality. See property (P8).   □

Example 4: Let $\lambda_1(F)$ denote the minimum eigenvalue of $F \in \mathcal{S}^n$. Suppose that F depends linearly on $p \in \mathbf{R}^k$ as

$$F = F_0 + \sum_{i=1}^{k} F_i p_i$$

where $F_i \in S^n$ $(i = 0, 1, \ldots, k)$ are constant matrices. Consider the problem of maximizing the minimum eigenvalue of $F(p)$, namely, $\max_p \lambda_1(F(p))$. From property (P5), this problem is equivalent to the following problem in variables p and t:

$$\max_{p,t} \quad t$$

$$\text{s.t.} \quad F_0 + \sum_{i=1}^{k} F_i p_i - tI^n \succeq O$$

which can be embedded into the form of $\mathcal{D}$ with

$$y = (p^\top, t)^\top, \quad (m = k + 1),$$
$$A_i = -F_i, \ (i = 1, \ldots, k), \quad A_{k+1} = I^n, \quad b = (0^\top, 1)^\top, \quad C = F_0.$$

See Alizadeh[1] for more examples of eigenvalue optimization. □

Example 5: Recall that the dual SDP problem $\mathcal{D}$ has linear equality constraints. However, there exist many problems with nonlinear equality constraints that can be represented as SDP problems. Letting $Q_1, \ldots, Q_m \in S^n$ denote constant positive semidefinite matrices, consider the following problem:

$$\left. \begin{array}{rl} \max\limits_{y,Z} & b^\top y \\ \text{s.t.} & -\sum\limits_{i=1}^{m} Q_i y_i^2 + \sum\limits_{i=1}^{m} A_i y_i + Z = C \\ & Z \succeq O. \end{array} \right\} \tag{6}$$

For each $i = 1, \ldots, m$, there exists an L_i satisfying $Q_i = L_i L_i^\top$, e.g., we can choose L_i as the Cholesky factor of Q_i. By using such L_i and property (P2), we see that the problem (6) is equivalent to

$$\max_{y,\widehat{Z}} \quad b^\top y$$

$$\text{s.t.} \quad \sum_{i=1}^{m} \begin{pmatrix} O & -L_i \\ -L_i^\top & A_i \end{pmatrix} y_i + \widehat{Z} = \begin{pmatrix} I^n & O \\ O & C \end{pmatrix}$$
$$\widehat{Z} \succeq O$$

which is in the form of $\mathcal{D}$. □

The readers may find a comprehensive list of constraints that can be converted to SDP constraints in Ben-Tal and Nemirovski[8] (Section 4.2).

4. Duality

The *duality gap* refers to the gap between the value $C \bullet X$ of the primal objective and $b^\top y$ of the dual objective. The following relation of *weak duality* holds between $\mathcal{P}$ and $\mathcal{D}$:

Proposition 6: (weak duality). *If X and (y, Z) are feasible in $\mathcal{P}$ and $\mathcal{D}$, respectively, then the duality gap is nonnegative, i.e., the inequality*

$$C \bullet X - b^\top y = X \bullet Z \geq 0$$

holds.

The weak duality implies that $\mathcal{D}$ provides a lower bound of $\mathcal{P}$, and $\mathcal{D}$ provides an upper bound of $\mathcal{P}$. It follows immediately from the weak duality that feasible X and (y, Z) satisfy $X \bullet Z = 0$ only if they are optimal solutions. The converse is guaranteed by the following *strong duality*, which requires a certain constraint qualification:

Proposition 7: (strong duality). *Assume that $\mathcal{P}$ and $\mathcal{D}$ have feasible solutions such that X and Z are positive definite. Then $\mathcal{P}$ and $\mathcal{D}$ have optimal solutions $\widetilde{X}$ and $(\widetilde{y}, \widetilde{Z})$, respectively, and the duality gap becomes zero, i.e.,*

$$C \bullet \widetilde{X} - b^\top \widetilde{y} = \widetilde{X} \bullet \widetilde{Z} = 0.$$

Assume that the assumption of Proposition 7 holds. If $\widetilde{X}$ and $(\widetilde{y}, \widetilde{Z})$ are optimal solutions, then $\widetilde{X}\widetilde{Z} = O$ is obtained from the strong duality and property (P7). Conversely, suppose that feasible $\widetilde{X}$ and $(\widetilde{y}, \widetilde{Z})$ satisfy $\widetilde{X}\widetilde{Z} = O$. By using property (P7), we obtain $\widetilde{X} \bullet \widetilde{Z} = 0$, from which and the weak duality it follows that $\widetilde{X}$ and $(\widetilde{y}, \widetilde{Z})$ are optimal solutions. As a consequence, under the same assumption as Proposition 7, the necessary and sufficient conditions for optimality of $\mathcal{P}$ and $\mathcal{D}$ are obtained as

$$A_i \bullet X = b_i, \ (i = 1, \ldots, m), \quad X \succeq O \tag{7a}$$

$$\sum_{i=1}^{m} A_i y_i + Z = C, \quad Z \succeq O \tag{7b}$$

$$XZ = O \tag{7c}$$

Here, (7a) and (7b) correspond to the primal and dual feasibility conditions, respectively. Eq. (7c) is called a *complementarity condition*. Note that (7a)–(7c) are called Karush–Kuhn–Tucker (KKT) conditions which are extended from those of LP. The condition (7) plays a crucial role in developing the primal-dual interior-point method that solves $\mathcal{P}$ and $\mathcal{D}$ simultaneously. See Section 6.

Example 8: Consider the pair of primal and dual problems in Example 1. $X_{11} = 1$ and $X_{22} = 1$ hold from the constraints in the primal problem (5), and by using $X_{12} = X_{21}$, the complementarity condition (7c) is written as

$$XZ = \begin{pmatrix} X_{21} + y_1 & X_{21}y_2 + 1 \\ X_{21}y_1 + 1 & X_{21} + y_2 \end{pmatrix} = O,$$

which leads to the optimal solutions $(y_1, y_2) = (1, 1)$ and $X_{21} = -1$. □

5. Structural Optimization by SDP

Some structural optimization problems are shown in this section in the forms of SDPs, where the dual form $\mathcal{D}$ in (2) that has linear objective function is usually used. The restriction of linear objective function is not very serious, because any objective function can be linearized by introducing an auxiliary variable and additional inequality constraints. However, the constraints should be converted in the form of $\mathcal{D}$, i.e., only positive semidefinite constraints of matrices are allowed in $\mathcal{D}$, and the matrices should linearly depend on the variables.

For structures such as trusses, plates with in-plane deformation, frames and plates with sandwich cross-sections, etc., both the stiffness and mass matrices are linear functions of the design variables, and the optimization problems under constraints on compliance or eigenvalues of vibration can be formulated by SDP. Even if the dependence of the matrices on the variables is not linear, there are many cases where nonlinear convex constraints can be converted into a form of positive semidefinite constraints as demonstrated in Ref.5. For a more general case where the matrices nonlinearly depend on the variable, local optimal solution can be found by successively solving linearized SDPs as illustrated in the example of linear buckling constraints in Section 5.3.

In this section, we consider a finite dimensional linear elastic structure subjected to nodal loads $p \in \mathbf{R}^n$, where n is the total number of degrees of freedom. Small displacements as well as small strains are assumed. Let $x = (x_i) \in \mathbf{R}^m$ denote the vector of design variables such as the cross-sectional areas of a truss and the thickness of a plate discretized into finite elements. The stiffness matrix is denoted by $K(x) \in \mathcal{S}^n$. The displacement vector $u \in \mathbf{R}^n$ is found from the stiffness (equilibrium) equation

$$K(x)u = p$$

where p is assumed, for brevity, to be independent of x, and the argument x is omitted in the following.

5.1. *Compliance optimization*

The compliance W is defined as the external work that is equivalent to twice of the strain energy; i.e.

$$W = p^\top u = u^\top K u = 2 \left(p^\top u - \frac{1}{2} u^\top K u \right) = p^\top K^{-1} p.$$

Note that W is a measure of flexibility, and the stiffness of a structure is to be increased by reducing W. Since the compliance is a simple global performance measure of a structure, it is often used for formulating structural optimization problems.

The structural volume for unit value of x_i is denoted by s_i; e.g., s_i is a member length of a truss, area of a plate element, etc. The minimization problem of compliance under the constraint of structural volume can be formulated as

$$\min_{\boldsymbol{x}} \quad W(\boldsymbol{x}) \atop \text{s.t.} \quad \boldsymbol{x} \geq \overline{\boldsymbol{x}}, \quad \boldsymbol{s}^\top \boldsymbol{x} \leq V \Bigg\} \tag{8}$$

where V is the specified total volume, and the nonnegative lower bound $\overline{\boldsymbol{x}}$ is given for $\boldsymbol{x}$. Observe that $\tau - \boldsymbol{p}^\top \boldsymbol{K}^{-1} \boldsymbol{p}$ is the Schur complement of $\boldsymbol{K}$ in the matrix

$$\begin{pmatrix} \tau & \boldsymbol{p}^\top \\ \boldsymbol{p} & \boldsymbol{K} \end{pmatrix}.$$

Hence, by using property (P2), we see

$$\tau \geq \boldsymbol{p}^\top \boldsymbol{K}^{-1} \boldsymbol{p} \quad \Longleftrightarrow \quad \begin{pmatrix} \tau & \boldsymbol{p}^\top \\ \boldsymbol{p} & \boldsymbol{K} \end{pmatrix} \succeq \boldsymbol{O}$$

from which it follows that (8) is equivalent to

$$\mathcal{D}_{COMP}: \quad \max_{\tau, \boldsymbol{x}} \quad -\tau$$
$$\text{s.t.} \quad \left(\begin{array}{c|c} \begin{matrix} \tau & \boldsymbol{p}^\top \\ \boldsymbol{p} & \boldsymbol{K}(\boldsymbol{x}) \end{matrix} & \\ \hline \text{diag}(\boldsymbol{x} - \overline{\boldsymbol{x}}) & \\ \hline & (V - \boldsymbol{s}^\top \boldsymbol{x}) \end{array} \right) \succeq \boldsymbol{O}$$

where τ is an auxiliary variable. Note that maximizing $-\tau$ is equivalent to minimizing τ that coincides with $\boldsymbol{p}^\top \boldsymbol{K}^{-1} \boldsymbol{p}$ at an optimal solution of $\mathcal{D}_{COMP}$. If $\boldsymbol{K}$ is a linear function of $\boldsymbol{x}$, then $\mathcal{D}_{COMP}$ has a linear objective function and a semidefinite constraint of a matrix that is a linear function of the variables $\boldsymbol{x}$; i.e. $\mathcal{D}_{COMP}$ can be embedded in a dual form $\mathcal{D}$ of SDP with $\boldsymbol{y} = (\tau, \boldsymbol{x}^\top)^\top$.

If we consider multiple load cases, a robust design can be obtained by minimizing the maximum compliance among the load cases.[37] Let l denote the number of load cases, and the values corresponding to the ith load case is denoted by the superscript i. The robust design problem is formulated as

$$\min_{\boldsymbol{x}} \sup_{i \in \{1, \dots, l\}} \left(\boldsymbol{p}^{i\top} \boldsymbol{u}^i - \frac{1}{2} \boldsymbol{u}^{i\top} \boldsymbol{K}(\boldsymbol{x}) \boldsymbol{u}^i \right)$$

which can be converted to an SDP as

$$\max_{\tau, \boldsymbol{x}} \quad -\tau$$
$$\text{s.t.} \quad \begin{pmatrix} \tau & (\boldsymbol{p}^i)^\top \\ \boldsymbol{p}^i & \boldsymbol{K}(\boldsymbol{x}) \end{pmatrix} \succeq \boldsymbol{O}, \quad (i = 1, \dots, l) \atop \boldsymbol{x} - \overline{\boldsymbol{x}} \geq \boldsymbol{0}, \quad V - \boldsymbol{s}^\top \boldsymbol{x} \geq 0, \Bigg\}$$

where τ is an auxiliary variable, and the constraints can be easily converted to the positive semidefinite form as in $\mathcal{D}_{COMP}$.

Let a convex set $\mathcal{M}$ of the load vectors be defined as $\mathcal{M} = \{ \boldsymbol{Q}\boldsymbol{e} \,|\, \boldsymbol{e} \in \mathbf{R}^l, \, \|\boldsymbol{e}\| \leq 1 \}$, where $\boldsymbol{Q} = (\boldsymbol{p}^1, \dots, \boldsymbol{p}^l) \in \mathbf{R}^{n \times l}$, $\boldsymbol{e}$ is a coefficient vector, and $\|\cdot\|$ is the standard

Euclidean norm. The problem of minimizing the worst-case value of the compliance against the possible load vectors $p \in \mathcal{M}$ can be formulated as

$$\min_{x} \sup_{p \in \mathcal{M}} p^{\top} K(x)^{-1} p$$

which can be converted to an SDP as[7,37]

$$\max_{\tau, x} \quad -\tau$$
$$\text{s.t.} \quad \begin{pmatrix} \tau I^l & Q^{\top} \\ Q & K(x) \end{pmatrix} \succeq O$$
$$x - \overline{x} \geq 0, \quad V - s^{\top} x \geq 0$$

5.2. *Eigenvalue of free vibration*

It is shown in this section that optimization problem under constraints on eigenvalues of free vibration can be formulated as an SDP problem.[54]

The structural and nonstructural mass matrices are denoted by $M^{S}(x) \in S^n$ and $M^0 \in S^n$, respectively. Note that $M^{S}(x)$ and the stiffness matrix $K(x)$ are linear functions of design variables x, whereas M^0 is a constant matrix. Let Ω_r and $\Phi_r \in \mathbf{R}^n$ denote the rth eigenvalue and eigenvector, respectively. The eigenvalue problem of vibration is formulated as

$$[K - \Omega_r(M^{S} + M^0)]\Phi_r = 0, \quad (r = 1, \ldots, n) \tag{9}$$

where Φ_r is normalized as $\Phi_r^{\top}(M^{S} + M^0)\Phi_r = 1$.

Consider a problem of minimizing the total structural volume $s^{\top} x$ under the lower-bound constraints on the eigenvalues. In a usual nonlinear programming formulation, the optimization problem is formulated as

$$\mathrm{P}_{EIG} : \min_{x} \quad s^{\top} x$$
$$\text{s.t.} \quad \Omega_r(x) \geq \overline{\Omega}, \quad (r = 1, 2, \ldots, n)$$
$$x_i \geq \overline{x}_i, \quad (i = 1, 2, \ldots, m)$$

where $\overline{\Omega} > 0$ is the specified lower bound for the eigenvalues, and the nonnegative lower bound $\overline{x}_i$ is given for x_i.

It is well known that an optimal solution of P_{EIG} often has multiple eigenvalues for which the sensitivity coefficients are discontinuous functions of the design variables.[21] Therefore, it is very difficult to obtain the optimal solution of P_{EIG} by using a gradient-based nonlinear programming algorithm for a large structure especially for topology optimization problem with $\overline{x} = 0$.

In the following, we convert P_{EIG} to an SDP problem. From the eigenvalue constraint and Rayleigh's principle, the following inequality should be satisfied by any vector $\Psi \in \mathbf{R}^n$:

$$\Psi^{\top}[K - \overline{\Omega}(M^{S} + M^0)]\Psi \geq 0$$

which is equivalent to

$$K - \overline{\Omega}(M^S + M^0) \succeq O. \tag{10}$$

Note that the condition (10) can be obtained also by transforming the generalized eigenvalue problem (9) to a standard form.

Suppose the case where K and M^S are defined as follows as linear functions of x without constant terms:

$$K = \sum_{i=1}^m x_i K_i, \quad M^S = \sum_{i=1}^m x_i M_i$$

where $K_i \in \mathcal{S}^n$ and $M_i \in \mathcal{S}^n$ are constant matrices. Hence, (10) is written as

$$\sum_{i=1}^m (K_i - \overline{\Omega} M_i) x_i - \overline{\Omega} M^0 \succeq O.$$

Finally, the optimization problem is formulated as the dual SDP form $\mathcal{D}$ introduced in (2) as

$$\mathcal{D}_{EIG} : \min_x \quad s^\top x$$

$$\text{s.t.} \quad \sum_{i=1}^m (K_i - \overline{\Omega} M_i) x_i - \overline{\Omega} M^0 \succeq O, \quad x - \overline{x} \geq 0.$$

The dual problem of $\mathcal{D}_{EIG}$ is formulated in the form of $\mathcal{P}$ in (1) as

$$\mathcal{P}_{EIG} : \max_{Y, \eta} \quad \overline{\Omega} M^0 \bullet Y$$

$$\text{s.t.} \quad (K_i - \overline{\Omega} M_i) \bullet Y + \eta_i = s_i, \quad \eta_i \geq 0, \quad (i = 1, \ldots, m)$$

$$Y \succeq O$$

where η_i is a slack variable.

Let x^* denote a feasible solution of P_{EIG}, i.e., of $\mathcal{D}_{EIG}$. Assume that, at x^*, the lowest eigenvalue Ω_1 in (9) is equal to $\overline{\Omega}$. Let t denote the multiplicity of the lowest eigenvalues. The eigenvectors corresponding to $\Omega_r = \overline{\Omega}$ are denoted by $\Phi_1, \ldots, \Phi_t$, which are chosen so as to be linearly independent. It was shown by Kanno and Ohsaki[27] that x^* is an optimal solution of P_{EIG} if and only if there exists an $H = (H_{pq}) \in \mathcal{S}^t$ satisfying

$$\sum_{p=1}^t \sum_{q=1}^t H_{pq} \Phi_p^\top (K_i - \overline{\Omega} M_i) \Phi_q \leq s_i, \quad \text{if } x_i^* = \overline{x}_i, \tag{11a}$$

$$\sum_{p=1}^t \sum_{q=1}^t H_{pq} \Phi_p^\top (K_i - \overline{\Omega} M_i) \Phi_q = s_i, \quad \text{if } x_i^* > \overline{x}_i, \tag{11b}$$

$$H \succeq O. \tag{11c}$$

Here, H plays a role similar to the Lagrange multiplier.

By using the SDP formulation, optimal solutions with many multiplicity of the lowest eigenvalues can be found without any difficulty. The possible number of

Fig. 1.   A spring-mass model.

Table 1.   Symmetry of the fundamental eigenmodes of the optimal double-layer grid.

	xz-plane	yz-plane
Mode 1	S	S
Mode 2	S	A
Mode 3	A	S
Mode 4	S	A
Mode 5	A	S

multiplicity can be investigated from the number of rank deficiencies of the matrices at the optimal solutions.[3]

Example 9: As an illustrative example, consider a spring-mass model as shown in Fig. 1. The extensional stiffness of the bars 1 and 2 are denoted by x_1 and x_2, respectively, which are the design variables. Each node has the lumped mass q, and the structural mass is neglected; i.e. the mass matrix is independent of stiffness. Let $q = 1$, $\overline{\Omega} = 1$, and $\overline{x} = 0$, for brevity, and the length s_i of member i is equal to 1. The dual and primal problems are written as

$$
\left.
\begin{aligned}
\mathcal{D}_{EX} : \min_{x} \quad & x_1 + x_2 \\
\text{s.t.} \quad & X = \begin{pmatrix} 1 & 0 \\ 0 & 0 \end{pmatrix} x_1 + \begin{pmatrix} 1 & -1 \\ -1 & 1 \end{pmatrix} x_2 - \begin{pmatrix} 1 & 0 \\ 0 & 1 \end{pmatrix} \succeq O \\
& x_1 \geq 0, \quad x_2 \geq 0
\end{aligned}
\right\}
\tag{12}
$$

$$
\left.
\begin{aligned}
\mathcal{P}_{EX} : \max_{Y, \eta} \quad & Y_{11} + Y_{22} \\
\text{s.t.} \quad & Y_{11} + \eta_1 = 1, \quad Y_{11} + Y_{22} - 2Y_{12} + \eta_2 = 1, \quad \eta_1 \geq 0, \quad \eta_2 \geq 0 \\
& \begin{pmatrix} Y_{11} & Y_{12} \\ Y_{12} & Y_{22} \end{pmatrix} \succeq O
\end{aligned}
\right\}
\tag{13}
$$

$\square$

Example 10: Consider a double-layer grid truss illustrated in Fig. 2, where $m = 288$ and $n = 243$. A nonstructural mass of 1.0×10^4 kg is located at each node on the upper layer. The elastic modulus and the mass density of each member are 200 GPa and 7.86×10^3 kg/m^3, respectively. The lengths of the members in x- and y-directions are 2.0 m and 1.6 m, respectively, and the distance between the

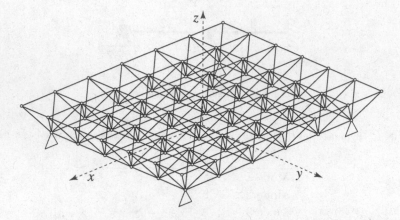

Fig. 2. A double-layer grid.

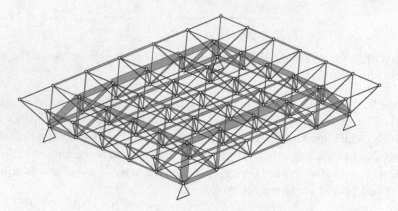

Fig. 3. Optimal design of the double-layer grid.

upper and lower layers is 1.3 m. The truss is symmetric with respect to xz- and yz-planes. We set the lower bounds of eigenvalues and member cross-sectional areas, respectively, as $\overline{\Omega} = 1000.0 \text{ rad}^2/\text{s}^2$ and $\overline{x} = 0$.

We solve the primal-dual pair of SDP problems $\mathcal{P}_{EIG}$ and $\mathcal{D}_{EIG}$ by using Se-DuMi Ver. 1.05[59], which implements the primal-dual interior-point method, under the environment of MATLAB.[43] The obtained optimal design is shown in Fig. 3, where the width of each member is proportional to its cross-sectional area. At the optimal solution, the total volume of members is 1.1394 m^3. The five lowest eigenvalues are all equal to $\overline{\Omega}$, i.e, the multiplicity of fundamental eigenvalues is five. Symmetry properties of the five fundamental eigenmodes are as listed in Table 1, where 'S' and 'A' stand for *symmetric* and *antisymmetric*, respectively.

The primal-dual interior-point method has no difficulty in finding an optimal solution with many multiplicity of eigenvalues. An extra benefit of using this method is that a strictly symmetric optimal solution can be automatically found even if we incorporate neither any additional constraint on symmetry nor a variable linking technique. See, for more details, Kanno *et al.*[32] □

5.3. *Linear buckling*

The linear buckling load factor is a basic performance measure of stability of structures mainly for thin-walled or slender structures. It is well known that optimum designs often have multiple buckling load factors.[55] In this case, similarly to the case of eigenvalue of vibration[57], the buckling load factor is not continuously differentiable.

However, contrary to the vibration problem, the geometrical stiffness matrix for formulating linear buckling problem depends on the internal forces that are usually nonlinear functions of the design variables. Therefore, in this section, an optimization method is presented based on successive solution of linearized SDPs.

5.3.1. *Formulation with positive semidefinite constraints*

Consider a finite dimensional structure with fixed locations of nodes and members (elements). The optimal design variables such as cross-sectional areas and thickness are found so as to minimize the total volume of the structure with a specified linear buckling load factor. The quasi-static proportional loads are defined by the load factor Λ and the specified vector $\boldsymbol{p}_0 \in \mathbf{R}^n$ of load pattern as $\boldsymbol{p} = \Lambda \boldsymbol{p}_0$. Let $\boldsymbol{K}^{\mathrm{G}} \in \mathcal{S}^n$ denote the geometrical stiffness matrix. See, e.g., Gallagher[18] for details of the matrix. The rth linear buckling load factor Λ_r and the corresponding buckling mode $\boldsymbol{\Phi}_r \in \mathbf{R}^n$ are defined by

$$(\boldsymbol{K} + \Lambda_r \boldsymbol{K}^{\mathrm{G}})\boldsymbol{\Phi}_r = \mathbf{0}, \quad (r = 1, \ldots, n). \tag{14}$$

For frames and curved structures such as arches and shells, $\boldsymbol{K}^{\mathrm{G}}$ is not negative definite and generally there exist positive and negative buckling load factors. However, we are concerned with the lowest positive value of Λ_r, because the directions of the loads are fixed. Therefore, the reciprocal values of load factors $1/\Lambda_r$ are numbered in nonascending order; i.e., Λ_1 is the lowest positive buckling load factor.

The positive lower-bound $\overline{\Lambda}$ is specified for positive Λ_r, but no constraint is given for the negative values of Λ_r; i.e., Λ_r should satisfy

$$\Lambda_r \geq \overline{\Lambda} \text{ or } \Lambda_r < 0, \quad (r = 1, \ldots, n). \tag{15}$$

Observe that (15) is alternatively written as

$$\frac{1}{\Lambda_r} \leq \frac{1}{\overline{\Lambda}}, \quad (r = 1, \ldots, n). \tag{16}$$

Hence, the optimization problem for minimizing the total structural volume under linear buckling constraints is formulated as

$$P_{LB} : \min_{x} \quad s^\top x$$
$$\text{s.t.} \quad \frac{1}{\Lambda_r(x)} \leq \frac{1}{\overline{\Lambda}}, \quad (r = 1, \ldots, n) \tag{17}$$
$$x \geq \overline{x}$$

where $\overline{x}$ is the specified nonnegative lower bound for x.

By using Rayleigh's principle, the condition (16) is reduced to

$$-\frac{\Psi^\top K^G \Psi}{\Psi^\top K \Psi} \leq \frac{1}{\overline{\Lambda}}, \quad \text{for any } \Psi \in \mathbf{R}^n. \tag{18}$$

Eq. (18) is then rewritten as the following positive semidefinite condition of a matrix:

$$\left(\frac{1}{\overline{\Lambda}} K + K^G\right) \succeq O.$$

The displacement vector u_0 for p_0 is found from

$$K u_0 = p_0$$

Let $f_i \in \mathbf{R}^\nu$ denote the vector of generalized stress of the ith element or member. For a truss and a frame, $\nu = 1$ and f_i is the axial force. For a plate subjected to in-plane deformation, ν is three times the number of integration points if a numerical integration is used. For simplicity, $f \in \mathbf{R}^{\nu m}$ is defined by

$$f = (f_1^\top, \ldots, f_m^\top)^\top$$

and f_j denotes the jth component of f, which is a linear function of u_0.

K^G is written as a linear function of f as

$$K^G = \sum_{j=1}^{\nu m} f_j K_j^G$$

where $K_j^G \in \mathcal{S}^n$ is a symmetric constant matrix. Consequently, the problem P_{LB} in (17) can be equivalently reformulated as

$$P'_{LB} : \quad \min_{x, u_0} \quad s^\top x \tag{19a}$$

$$\text{s.t.} \quad \frac{1}{\overline{\Lambda}} \sum_{i=1}^{m} x_i K_i + \sum_{j=1}^{\nu m} f_j(u_0) K_j^G \succeq O, \tag{19b}$$

$$\sum_{i=1}^{m} x_i K_i u_0 = p_0, \tag{19c}$$

$$x - \overline{x} \geq 0. \tag{19d}$$

P'_{LB} cannot be formulated as an SDP problem, because the equality constraints (19c), i.e., the equilibrium equations, are nonconvex with respect to x and u_0 for a general statically indeterminate structure. Ben-Tal *et al.*[5] proposed a successive SDP algorithm solving P'_{LB} based on the semidefinite relaxation of (19c). In the next section, the authors' approach of successive linearization is briefly presented.

5.3.2. *Sequential SDP algorithm*

An optimal solution of nonlinear problem P'_{LB} with positive semidefinite constraints can be found by sequentially solving SDP problems. Let the superscript (k) denote the iteration counter for the sequential SDP. We approximate $f_j(\boldsymbol{x})$ $(j = 1, \ldots, \nu m)$ by $\widetilde{f}_j^{(k)}(\boldsymbol{x})$ defined by the values at $\boldsymbol{x} = \boldsymbol{x}^{(k)}$ as

$$\text{Method (a):} \quad \widetilde{f}_j^{(k)}(\boldsymbol{x}) = f_j(\boldsymbol{x}^{(k)}) + \nabla f_j(\boldsymbol{x}^{(k)})^\top (\boldsymbol{x} - \boldsymbol{x}^{(k)});$$

$$\text{Method (b):} \quad \widetilde{f}_j^{(k)}(\boldsymbol{x}) = f_i(\boldsymbol{x}^{(k)}).$$

In Method (a), we approximate f_j as a linear function of the design variables $\boldsymbol{x}$ by using the sensitivity coefficients of $\boldsymbol{f}$. Method (b) simply assumes that f_j does not change from the value at the current solution $\boldsymbol{x}^{(k)}$.

By using either Method (a) or (b), the problem P'_{LB} is approximated as the following dual SDP form:

$$\mathcal{D}_{LB}^{(k)}: \quad \min_{\boldsymbol{x}} \quad \boldsymbol{s}^\top \boldsymbol{x}$$

$$\text{s.t.} \quad \boldsymbol{Z} = \sum_{i=1}^{m} \boldsymbol{F}_i x_i + \boldsymbol{F}_0$$

$$\boldsymbol{Z} \succeq \boldsymbol{O}, \quad \boldsymbol{x} - \overline{\boldsymbol{x}} \geq \boldsymbol{0}$$

where $\boldsymbol{F}_i \in \mathcal{S}^n$ $(i = 0, 1, \ldots, m)$ are the constant matrices defined as

$$\text{Method (a):} \quad \boldsymbol{F}_i = \frac{1}{\Lambda} \boldsymbol{K}_i + \sum_{j=1}^{\nu m} \frac{\partial f_j(\boldsymbol{x}^{(k)})}{\partial x_i} \boldsymbol{K}_j^{\mathrm{G}}, \quad (i = 1, \ldots, m)$$

$$\boldsymbol{F}_0 = \sum_{j=1}^{\nu m} \left[f_j(\boldsymbol{x}^{(k)}) - \nabla f_j(\boldsymbol{x}^{(k)})^\top \boldsymbol{x}^{(k)} \right] \boldsymbol{K}_j^{\mathrm{G}}; \tag{20}$$

$$\text{Method (b):} \quad \boldsymbol{F}_i = \frac{1}{\Lambda} \boldsymbol{K}_i, \quad (i = 1, \ldots, m)$$

$$\boldsymbol{F}_0 = \sum_{j=1}^{\nu m} f_j(\boldsymbol{x}^{(k)}) \boldsymbol{K}_j^{\mathrm{G}}. \tag{21}$$

The original problem P'_{LB} can be solved by the following algorithm:

Algorithm 11:

Step 0: Choose $\boldsymbol{x}^{(0)} \geq \overline{\boldsymbol{x}}$, $c^{(0)} > 0$, and the tolerance $\epsilon > 0$. Set $k := 0$.

Step 1: Find an optimal solution $(\boldsymbol{x}^*, t^*)$ of the SDP problem

$$\left. \begin{aligned} \min_{\boldsymbol{\tau}, t} \quad & \boldsymbol{s}^\top \boldsymbol{x} + c^{(k)} t \\ \text{s.t.} \quad & \sum_{i=1}^{m} \boldsymbol{F}_i x_i - \boldsymbol{F}_0 \succeq \boldsymbol{O} \\ & t \geq \| \boldsymbol{x} - \boldsymbol{x}^{(k)} \| \\ & \boldsymbol{x} - \overline{\boldsymbol{x}} \geq \boldsymbol{0}. \end{aligned} \right\} \tag{22}$$

If $t^* \leq \epsilon$ and $\Lambda_1(\boldsymbol{x}^*) \geq \overline{\Lambda}$, then STOP.

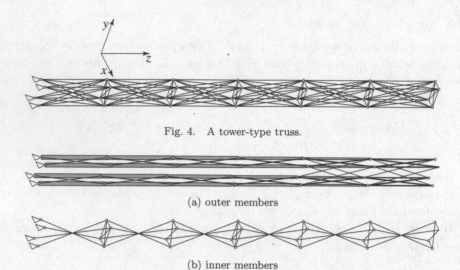

Fig. 4.   A tower-type truss.

(a) outer members

(b) inner members

Fig. 5.   Optimal design of the tower-type truss.

Step 2: Set $x^{(k+1)} := x^*$.

Step 3: Choose $c^{(k+1)} > 0$. Set $k \leftarrow k + 1$, and go to Step 1.

In the problem (22), the penalty term $c^{(k)}t$ with the second constraint for the norm of $x - x^{(k)}$ plays a role similar to the conventional move limit on x, and $t^* = \|x^* - x^{(k)}\|$ is satisfied at an optimal solution of (22).

Note that the sensitivity coefficients of buckling load factors are not used in Algorithm 11. Therefore, the optimal design with a large number of multiplicity of buckling load factors can be found without any difficulty.

It should be emphasized that a solution obtained by Method (a) satisfies the first-order optimality conditions of P'_{LB}. Therefore, the solution is guaranteed to be a stationary point of P'_{LB}. On the contrary, the optimality conditions are not generally satisfied by the solution obtained by Method (b).

Some software packages based on the primal-dual interior-point method incorporate efficient methods for computation when the coefficient matrices F_i of $\mathcal{D}_{LB}^{(k)}$ are sparse; i.e., when these matrices have large numbers of zero elements. See, e.g., Fujisawa *et al.*[15] Note that K_i and K_j^G are very sparse matrices. Consequently, if we use Method (b), then F_i in $\mathcal{D}_{LB}^{(k)}$, and hence in (22), have larger sparsity than those by Method (a). It follows that the computational cost for solving each $\mathcal{D}_{LB}^{(k)}$ with Method (b) is smaller than that with Method (a) if an SDP software exploiting sparsity is used.

Example 12: Consider a tower-type truss illustrated in Fig. 4, where $m = 132$ and $n = 72$. The lengths of members in x-, y- and z-directions are 50.0 cm, 100.0 cm and 300.0 cm, respectively. The external loads of 1000.0 kN are applied at the four nodes

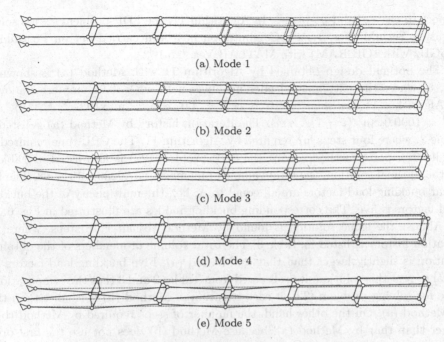

(a) Mode 1

(b) Mode 2

(c) Mode 3

(d) Mode 4

(e) Mode 5

Fig. 6. Buckling modes of the optimal tower-type truss.

Table 2. Iteration history of the tower-type truss by Method (a)

k	Volume (10^5 cm^3)	Λ_1	Λ_2	Λ_3	Λ_4	Λ_5	CPU (sec.)	$\|x - x^k\|$
0	329.142	17.1360	73.7784	146.785	369.231	622.829		
1	7.91929	0.19873	0.41141	0.54588	0.81982	0.85576	19.4	1.1
2	7.92204	0.99938	0.99984	1.00000	1.02025	1.04674	15.1	5.3×10^{-5}
3	7.92300	0.99917	0.99980	0.99980	1.00000	1.00000	18.2	1.1×10^{-5}
4	7.92301	1.00000	1.00000	1.00000	1.00000	1.00000	14.4	1.8×10^{-7}

Table 3. Iteration history of the tower-type truss by Method (b)

k	Volume (10^5 cm^3)	Λ_1	Λ_2	Λ_3	Λ_4	Λ_5	CPU (sec.)	$\|x - x^k\|$
0	329.142	17.1360	73.7784	146.785	369.231	622.829		
1	7.91919	0.17782	0.39808	0.53972	0.81368	0.84645	3.4	1.1
2	7.92335	1.00002	1.00025	1.00085	1.00800	1.01612	2.9	2.4×10^{-5}
3	7.92315	0.99982	0.99992	0.99998	1.00000	1.00398	3.1	2.7×10^{-6}
4	7.92315	0.99976	0.99999	1.00000	1.00000	1.00020	3.0	4.6×10^{-7}
10	7.92315	1.00000	1.00000	1.00000	1.00000	1.00000	3.1	1.6×10^{-12}

on the top layer (right layer in Fig. 4) in the negative direction of the z-axis. We set the lower bounds of positive linear buckling factors and member cross-sectional areas, respectively, as $\overline{\Lambda} = 1.0$ and $\overline{x} = 0$. Hence, the model is symmetric with respect to the xz- and yz-planes. In Algorithm 11, the SDP problem (22) is solved by using SeDuMi Ver. 1.05.[59] Computation has been carried out on Pentium M (1.5GHz with 1GB RAM) with MATLAB Ver. 6.5.1.[43]

The optimal design obtained by Algorithm 11 with Method (a) is shown in Fig. 5, where the extremely slender members with $x_i < 0.01$ cm^2 are removed. In Algorithm 11, we set $\epsilon = 10^{-6}$, $c^{(0)} = 4.0$, $c^{(k)} = \max\{0.6c^{(k-1)}, 0.8\}$, and $x_i^{(0)} = 1000.0$ cm^2 $(i = 1, \ldots, m)$. The iteration history by Method (a) is listed in Table 2, where four steps are required by Algorithm 11. The CPU time required by the interior-point method solving the SDP problem (22) is also listed in Table 2. At the optimal solution, the total volume of members is 7.9230×10^5 cm^3. The five linear buckling load factors are all equal to $\overline{\Lambda}$, i.e., the multiplicity of the buckling load factors is five. The corresponding buckling modes are illustrated in Fig. 6.

Almost the same solution is found by Method (b) after 16 steps, where the iteration history is listed in Table 3. The total volume of members of the obtained solution is slightly larger than that by Method (a). Five buckling load factors are equal to $\overline{\Lambda}$ at this solution. It is observed from Tables 2 and 3 that the computational time for solving problem (22) at each iteration of Method (b) is smaller than that of Method (a). On the other hand, the number of steps required by Method (b) is larger than that by Method (a), because Method (b) does not use the first-order information of axial forces. □

6. Interior-Point Algorithm

Interior-point method originally referred the penalty approach with interior penalty function such as SUMT.[14] The new generation interior-point method is first developed for LP[66] by Karmarkar[34], which has been extended to SDP in 1990s. It is theoretically guaranteed[39] that the primal-dual interior-point method converges to the global optimal solution of the pair $\mathcal{P}$ and $\mathcal{D}$ within polynomial time of m and n, which demonstrates excellent performance even in large scale applications if sparsity of the matrices is utilized.

Recall that the SDP problem $\mathcal{P}$ has the nonlinear constraint $\mathcal{S}^n \ni \boldsymbol{X} \succeq \boldsymbol{O}$. For this constraint, we first introduce the barrier term $-\mu \ln(\det \boldsymbol{X})$ called the *logarithmic barrier function*, where $\mu > 0$ is a parameter. By adding this barrier function to the objective function of $\mathcal{P}$, we obtain the following family of problems parameterized by $\mu > 0$:

$$\mathcal{P}(\mu): \min_{\boldsymbol{X}} \quad \boldsymbol{C} \bullet \boldsymbol{X} - \mu \ln(\det \boldsymbol{X})$$
$$\text{s.t.} \quad \boldsymbol{A}_i \bullet \boldsymbol{X} = b_i, \quad (i = 1, \ldots, m), \quad \boldsymbol{X} \succ \boldsymbol{O}$$

where $\boldsymbol{X} \succ \boldsymbol{O}$ means that $\boldsymbol{X}$ is positive definite. $\boldsymbol{X}$ is an optimal solution of $\mathcal{P}(\mu)$

if and only if there exist $y \in \mathbf{R}^m$ and $Z \in \mathcal{S}^n$ satisfying

$$\mathrm{CP}(\mu): \left.\begin{array}{c} A_i \bullet X = b_i, \ (i = 1, \ldots, m), \quad X \succ O \\ \sum_{i=1}^{m} A_i y_i + Z = C, \quad Z \succ O \\ XY = \mu I^n \end{array}\right\} \tag{23}$$

where $y_1, \ldots, y_m$ and Z can be regarded as the Lagrange multipliers and slack variables. It is known that the system (23) has the unique solution (X, y, Z) for any $\mu > 0$. Moreover, (y, Z) satisfying (23) is also an optimal solution of the following barrier problem of the dual form SDP:

$$\mathcal{D}(\mu): \max_{y, Z} \quad b^\top y + \mu \ln(\det Z)$$

$$\text{s.t.} \quad \sum_{i=1}^{m} A_i y_i + Z = C, \quad Z \succ O.$$

Notice here that the optimality conditions (7) of the SDP problems $\mathcal{P}$ and $\mathcal{D}$ can be obtained from (23) by taking a limit $\mu \to 0$ and replacing $X \succ O$ and $Z \succ O$ with $X \succeq O$ and $Z \succeq O$, respectively. This implies that a sequence of solutions to (23), so-called *central path*[61], converges to a pair of optimal solutions of $\mathcal{P}$ and $\mathcal{D}$ with $\mu \to 0$. It is known that the central path is a unique and smooth curve. The primal-dual path-following interior-point method finds a pair of optimal solutions of $\mathcal{P}$ and $\mathcal{D}$ by numerically tracing the central path $\mathrm{CP}(\mu)$ by decreasing $\mu \to 0$.

Let (X^k, y^k, Z^k) denote a current solution obtained in iteration procedure. The fundamental idea of primal-dual interior-point methods is summarized as follows: first, the search direction $(\Delta X, \Delta y, \Delta Z)$ is determined by solving the following Newton equations of the system of central path (23) with an appropriately chosen $\mu > 0$:

$$A_i \bullet \Delta X = b_i - A_i \bullet X^k, \quad (i = 1, \ldots, m), \tag{24a}$$

$$\sum_{i=1}^{m} A_i \Delta y_i + \Delta Z = C - \sum_{i=1}^{m} A_i y_i^k - Z^k, \tag{24b}$$

$$X^k(\Delta Y) + (\Delta X) Y^k = \mu I^n - X^k Z^k. \tag{24c}$$

Next, the step size is determined so that X and Z remain positive definite, and we move to the new solution $(X^{k+1}, y^{k+1}, Z^{k+1})$. Then we decrease μ with an appropriate criterion, and solve (24a)–(24c) at $(X^{k+1}, y^{k+1}, Z^{k+1})$ again to find a new search direction.

Unfortunately, ΔX obtained by solving (24a)–(24c) is not necessarily symmetric. However, an optimal solution X of $\mathcal{P}$ should be symmetric, which implies that some additional scheme is required to symmetrize ΔX. Various symmetrizing methods were proposed that lead to different search directions[61], e.g., AHO[4], HRVW[24]/KSH[39]/M[46], and NT[51]. However, we should not go into details here. The following is a prototype of typical primal-dual path-following interior-point method solving $\mathcal{P}$ and $\mathcal{D}$:

Algorithm 13: (prototype of interior-point method)

Step 0: Choose (X^0, y^0, Z^0) satisfying $X^0 \succ O$, $Z^0 \succ O$ and some additional conditions. Set $k := 0$.

Step 1: If (X^k, y^k, Z^k) is feasible and $\mu^k := X^k \bullet Z^k / n$ is small enough, then STOP.

Step 2: Choose $\mu \in (0, \mu^k)$, and compute $(\Delta X, \Delta y, \Delta Z)$ by solving the system of (24a), (24b) and a variant of (24c).

Step 3: Choose $\alpha_\mathrm{p}, \alpha_\mathrm{d} \in (0, 1]$ so that $X^k + \alpha_\mathrm{p} \Delta X \succ O$ and $Z^k + \alpha_\mathrm{d} \Delta Z \succ O$.

Step 4: Set $(X^{k+1}, y^{k+1}, Z^{k+1}) := (X^k + \alpha_\mathrm{p} \Delta X, y^k + \alpha_\mathrm{d} \Delta y, Z^k + \alpha_\mathrm{d} \Delta Z)$.

Step 5: Set $k \leftarrow k + 1$, and go to Step 1.

Here, we shall not discuss how to update μ, α_p and α_d. By taking an appropriate updating scheme of these parameters, it is shown that an optimal pair of $\mathcal{P}$ and $\mathcal{D}$ can be found with the number of arithmetic operations bounded by a polynomial of m and n. The interested readers may consult Refs.[39,46,51] The other types of interior-point methods than path-following methods, e.g., potential-reduction methods[61], self-dual embedding methods[67], have been also developed. Several software packages have been developed based on the interior-point methods[45], e.g., SDPA[16], SeDuMi[59], and SDPT3.[62]

7. Second-Order Cone Programming

Second-Order Cone Programming (SOCP) is a convex programming that minimizes a linear function under constraints that the variable vectors should be contained in the second-order (Lorenz) cones.[2] SOCP includes LP, convex QP, etc., and, conversely, is included in SDP as a particular case. Hence, SOCP can be solved by using an interior-point method for SDP. However, special polynomial-time interior point methods have been developed for SOCP[47] that should be used if an SDP is converted into SOCP. In this section, the outline of SOCP and its application to analysis of cable networks are briefly presented. The details of mathematical background and algorithms for SOCP may be found in Alizadeh and Goldfarb[2] and Ben-Tal and Nemirovski[8] (Chap. 3).

Let $x_i \in \mathbf{R}^{n_i}$ $(i = 1, \ldots, r)$ denote a set of column vectors. For each $i = 1, \ldots, r$, the components of the vector x_i are divided into the first and remaining components as

$$x_i = (x_{i0}, x_{i1}^\top)^\top, \quad x_{i0} \in \mathbf{R}, \quad x_{i1} \in \mathbf{R}^{n_i - 1}$$

The second-order cone $\mathcal{K}(n_i)$ refers to a convex cone in $\mathbf{R}^{n_i}$ defined as

$$\mathcal{K}(n_i) = \{x_i \in \mathbf{R}^{n_i} \,|\, x_{i0} \geq \|x_{i1}\|\} \tag{25}$$

For simple presentation, the following notation is used for a vector $\boldsymbol{x}_i$:

$$\boldsymbol{x}_i \succeq \boldsymbol{0} \quad \Longleftrightarrow \quad \boldsymbol{x}_i \in \mathcal{K}(n_i)$$

Also, we write $\boldsymbol{x}_i \succ \boldsymbol{0}$ if $x_{i0} > \|\boldsymbol{x}_{i1}\|$, i.e., if $\boldsymbol{x}_i$ is in the interior of $\mathcal{K}(n_i)$. The dual cone $\mathcal{K}^*(n_i)$ of $\mathcal{K}(n_i)$ is defined by

$$\mathcal{K}^*(n_i) = \{\boldsymbol{s}_i \in \mathbf{R}^{n_i} \,|\, \boldsymbol{x}_i^\top \boldsymbol{s}_i \geq 0, \ \forall \boldsymbol{x}_i \in \mathcal{K}(n_i)\}.$$

The second-order cone is known to be *self-dual*, i.e., $\mathcal{K}(p) = \mathcal{K}^*(p)$.

Let $\boldsymbol{b} \in \mathbf{R}^m$, $\boldsymbol{c}_i \in \mathbf{R}^{n_i}$ and $\boldsymbol{A}_i \in \mathbf{R}^{m \times n_i}$ $(i = 1, \ldots, r)$ denote constant vectors and matrices. The primal SOCP problem is formulated in variables $\boldsymbol{x}_i \in \mathbf{R}^{n_i}$ $(i = 1, \ldots, r)$ as

$$\mathcal{P}_{SOCP} : \min_{\boldsymbol{x}_1, \ldots, \boldsymbol{x}_r} \quad \sum_{i=1}^{r} \boldsymbol{c}_i^\top \boldsymbol{x}_i$$
$$\text{s.t.} \quad \sum_{i=1}^{r} \boldsymbol{A}_i \boldsymbol{x}_i = \boldsymbol{b}, \quad \boldsymbol{x}_i \succeq \boldsymbol{0}, \quad (i = 1, \ldots, r)$$

The dual of $\mathcal{P}_{SOCP}$ is formulated with $\boldsymbol{y} \in \mathbf{R}^m$ and $\boldsymbol{z}_i \in \mathbf{R}^{n_i}$ $(i = 1, \ldots, r)$ as

$$\mathcal{D}_{SOCP} : \max_{\boldsymbol{y}, \boldsymbol{z}_1, \ldots, \boldsymbol{z}_r} \quad \boldsymbol{b}^\top \boldsymbol{y}$$
$$\text{s.t.} \quad \boldsymbol{A}_i^\top \boldsymbol{y} + \boldsymbol{z}_i = \boldsymbol{c}_i, \quad \boldsymbol{z}_i \succeq \boldsymbol{0}, \quad (i = 1, \ldots, r)$$

In a manner similar to the case of SDP duality, $\mathcal{D}_{SOCP}$ is also an SOCP problem, and the dual problem of $\mathcal{D}_{SOCP}$ coincides with $\mathcal{P}_{SOCP}$. Moreover, the following strong duality holds:

Proposition 14: (strong duality of SOCP). *Assume that $\mathcal{P}_{SOCP}$ and $\mathcal{D}_{SOCP}$ have feasible solutions satisfying $\boldsymbol{x}_i \succ \boldsymbol{0}$ and $\boldsymbol{z}_i \succ \boldsymbol{0}$ $(i = 1, \ldots, r)$. Then $\mathcal{P}_{SOCP}$ and $\mathcal{D}_{SOCP}$ have optimal solutions $(\widetilde{\boldsymbol{x}}_1, \ldots, \widetilde{\boldsymbol{x}}_r)$ and $(\widetilde{\boldsymbol{y}}, \widetilde{\boldsymbol{z}}_1, \ldots, \widetilde{\boldsymbol{z}}_r)$, respectively, and the duality gap becomes zero, i.e.,*

$$\sum_{i=1}^{r} \boldsymbol{c}_i^\top \widetilde{\boldsymbol{x}}_i - \boldsymbol{b}^\top \widetilde{\boldsymbol{y}} = \sum_{i=1}^{r} \widetilde{\boldsymbol{x}}_i^\top \widetilde{\boldsymbol{z}}_i = 0.$$

It should be noted that Proposition 14 is very analogous to the strong duality theorem of SDP (Proposition 7). This similarity enables us to develop theory and algorithms, especially the primal-dual interior-point methods, of SOCP[2] in a way similar to SDP. For example, the necessary and sufficient conditions for optimality of $\mathcal{P}_{SOCP}$ and $\mathcal{D}_{SOCP}$ are given as

$$\sum_{i=1}^{r} \boldsymbol{A}_i \boldsymbol{x}_i = \boldsymbol{b}, \quad \boldsymbol{x}_i \succeq \boldsymbol{0}, \quad (i = 1, \ldots, r)$$
$$\boldsymbol{A}_i^\top \boldsymbol{y} + \boldsymbol{z}_i = \boldsymbol{c}_i, \quad \boldsymbol{z}_i \succeq \boldsymbol{0}, \quad (i = 1, \ldots, r)$$
$$\boldsymbol{x}_i^\top \boldsymbol{z}_i = 0, \quad (i = 1, \ldots, r)$$

which should be compared with (7) of SDP.

Example 15: For a vector $x \in \mathbf{R}^n$ and the scalars u and v, the quadratic constraint

$$x^\top x \leq uv, \quad u \geq 0, \quad v \geq 0$$

can be written as the following second-order cone constraint:

$$u + v \geq \left\| \begin{pmatrix} 2x \\ u - v \end{pmatrix} \right\|.$$

$\square$

In engineering field, there are several studies on application of SOCP. Jarre *et al.*[26] formulated the problem of minimizing the compliance of trusses by SOCP. Sasakawa and Tsuchiya[58] solved the optimal magnetic shielding design problem, which arises in the development of a superconducting magnetically levitated vehicle. Other applications of SOCP can be found in Lobo *et al.*[42]

In the following, we summarize the authors' SOCP formulation for large-deformation analysis of cable networks.[29] Consider an elastic pin-jointed cable network in three-dimensional space discretized into m members, that cannot transmit compression forces. Let $x \in \mathbf{R}^n$ denote the nodal location vector after deformation. The unstressed length of the ith member is denoted by l_i^0. The extension c_i of the ith member is defined by

$$c_i = \|B_i x - d_i\| - l_i^0, \quad (i = 1, \ldots, m).$$

Here, for each $i = 1, \ldots, m$, $B_i \in \mathbf{R}^{3 \times n}$ defines the connectivity of the member, $d_i \in \mathbf{R}^3$ is a constant vector related to support conditions.

The strain energy w_i of the ith member is defined by

$$w_i(c_i) = \begin{cases} \frac{1}{2} k_i c_i^2 & (c_i \geq 0) \\ 0 & (c_i < 0) \end{cases}$$

where k_i is the extensional stiffness. Thus, the definition of w_i, and consequently the definition of analysis problem, depends on the sign of c_i.

Kanno *et al.*[29] showed that the minimization problem of the total potential energy of a cable network can be formulated as

$$\mathcal{P}_{CN} : \min_{t_1, \ldots, t_m, x} \quad \sum_{i=1}^{m} \frac{1}{2} k_i t_i^2 - p^\top x$$

$$\text{s.t.} \quad t_i \geq \|B_i x - d_i\| - l_i^0, \quad (i = 1, \ldots, m)$$

where $p \in \mathbf{R}^n$ is the vector of external loads, and k_i is the extensional stiffness of the ith member. Note that Problem $\mathcal{P}_{CN}$ can easily be converted to a primal form of SOCP by using the following relation:

$$\tau_i \geq \frac{1}{2} k_i t_i^2 \quad \Longleftrightarrow \quad \frac{\tau_i}{2k_i} + 1 \geq \left\| \begin{pmatrix} \frac{\tau_i}{2k_i} - 1 \\ t_i \end{pmatrix} \right\|$$

where τ_i is an auxiliary variable.

Kanno and Ohsaki[28] showed that the minimization problem of complementary energy of a cable net can be formulated as the dual form $\mathcal{D}_{CN}$ of $\mathcal{P}_{CN}$ as follows that includes the force variables only, which is referred to as truly complementary form:

$$\mathcal{D}_{CN}: \quad \max_{f_1,\ldots,f_m,\boldsymbol{\pi}_1,\ldots,\boldsymbol{\pi}_m} \quad \sum_{i=1}^{m}\left(\frac{f_i^2}{2k_i} + l_i^0 f_i + \boldsymbol{d}_i^\top \boldsymbol{\pi}_i\right)$$

$$\text{s.t.} \quad \sum_{i=1}^{m} \boldsymbol{B}_i^\top \boldsymbol{\pi}_i = \boldsymbol{p},$$

$$f_i \geq \|\boldsymbol{\pi}_i\|, \quad (i = 1,\ldots,m)$$

where $(f_1,\ldots,f_m,\boldsymbol{\pi}_1,\ldots,\boldsymbol{\pi}_m)$ with $\boldsymbol{\pi}_i \in \mathbf{R}^3$ are the force variables. At an optimal solution of $\mathcal{D}_{CN}$, f_i and $\boldsymbol{\pi}_i$ coincide with the axial force and member force vector of the ith member, respectively.

8. Related Topics and Outlook

Various interior-point methods, especially the primal-dual path-following interior-point algorithms, have been extensively studied for SDP and SOCP. Several software packages[16,59,62] have been already well developed based on them, which have been proven to be efficient and robust for medium-sized SDP and SOCP problems.[45] Most of these packages take full advantage of sparsity of data matrices $\boldsymbol{A}_1,\ldots,\boldsymbol{A}_m$ and $\boldsymbol{C}$ in (1) and (2), which leads to significant reduction in computational time and required memory. Some other computational methods have also been proposed and intensively studied for solving large scale SDP problems: the spectral bundle method[23], nonlinear programming reformulation of SDP[64,11], the Lagrangian dual interior-point method[17], etc.

Recall that the primal and dual SDP problems (1) and (2) have linear constraints other than the positive semidefinite constraints on variable matrices, i.e., only $\boldsymbol{X} \succeq \boldsymbol{O}$ and $\boldsymbol{Z} \succeq \boldsymbol{O}$ are nonlinear constraints. We may extend SDP naturally by introducing some additional nonlinear and possibly nonconvex constraints on $\boldsymbol{X}$ and/or $(\boldsymbol{y}, \boldsymbol{Z})$. The resulting problem is sometimes called *nonlinear SDP*.[33] Note that, in some literature, problems (1) and (2) are called *linear SDP* problems in spite of the fact that they are nonlinear optimization problems. Some algorithms, including a sequential SDP method, have been proposed for various nonlinear SDP problems.[33,25,5]

For various classes of convex optimization problems including SDP, a unified methodology of *robust optimization*, or *robust counterpart* scheme, was developed by Ben-Tal and Nemirovski[9], where the data in optimization problems are assumed to be unknown but bounded.

Suppose that a symmetric matrix $\boldsymbol{W}$ is given as a linear function of $\boldsymbol{y}$. It can be easily seen that the minimization problem of maximum eigenvalue of $\boldsymbol{W}(\boldsymbol{y})$ is formulated as an SDP problem.[63] See also Example 5. Burke *et al.*[12] assumed that

$W(y)$ is nonsymmetric, and attempted to minimize the maximum real parts of its eigenvalues.

Another interest of extension of SDP is increasing in the field of complementarity problems. Linear complementarity problem can be obtained as a natural extension of the linear programming problem. Analogously, SDP has been extended to *semidefinite complementarity problem* $(SDCP)$[48], which has application in the control theory.[13]

References

1. F. Alizadeh, "Interior point methods in semidefinite programming with applications to combinatorial optimizations," *SIAM Journal on Optimization,* **5**, pp. 13–51, 1995.
2. F. Alizadeh and D. Goldfarb, "Second-order cone programming", *Mathematical Programming,* **B95**, pp. 3–51, 2003.
3. F. Alizadeh, J.A. Haeberly and M.L. Overton, "Complementarity and nondegeneracy in semidefinite programming", *Mathematical Programming,* **77**, pp. 111–128, 1997.
4. F. Alizadeh, J.-P.A. Haeberly and M.L. Overton, "Primal-dual interior-point methods for semidefinite programming: Convergence rates, stability and numerical results", *SIAM Journal on Optimization,* **8**, pp. 746–68, 1998.
5. A. Ben-Tal, F. Jarre, M. Kočvara, A. Nemirovski and J. Zowe, "Optimal design of trusses under a nonconvex global buckling constraint", *Optimization and Engineering,* **1**, pp. 189-213, 2000.
6. A. Ben-Tal, M. Kočvara, A. Nemirovski and J. Zowe, "Free material design via semidefinite programming: the multiload case with contact conditions", *SIAM Journal on Optimization,* **9**, pp. 813–832, 1999.
7. A. Ben-Tal and A. Nemirovski, "Robust truss topology optimization via semidefinite programming", *SIAM Journal on Optimization,* **7**, pp. 991–1016, 1997.
8. A. Ben-Tal and A. Nemirovski, *Lectures on Modern Convex Optimization: Analysis, Algorithms, and Engineering Applications,* SIAM, Philadelphia, PA, 2001.
9. A. Ben-Tal and A. Nemirovski, "Robust Optimization — Methodology and Applications", *Mathematical Programming,* **B92**, pp. 453–480, 2002.
10. S. Boyd, L. El Ghaoui, E. Feron and V. Balakrishnan, *Linear Matrix Inequalities in System and Control Theory,* SIAM, Philadelphia, PA, 1994.
11. S. Burer, R.D.C. Monteiro and Y. Zhang, "Interior-point algorithms for semidefinite programming based on a nonlinear formulation", *Computational Optimization and Applications,* **22**, pp. 49–79, 2002.
12. J.V. Burke, A.S. Lewis and M.L. Overton, "Two numerical methods for optimizing matrix stability", *Linear Algebra and its Applications,* **351&352**, pp. 117–145, 2002.
13. L. El Ghaoui, F. Oustry, M. AitRami, "A cone complementarity linearization algorithm for static output-feedback and related problems", *IEEE Transactions on Automatic Control,* **42**, pp. 1171 - 1176, 1997.
14. A.V. Fiacco and G.P. McCormick, *Nonlinear Programming: Sequential Unconstrained Minimization Techniques,* John Wiley & Sons, New York, 1968; also: Classics in Applied Mathematics, SIAM, Philadelphia, PA, 1990.
15. K. Fujisawa, M. Kojima and K. Nakata, Exploiting sparsity in primal-dual interior-point methods for semidefinite programming, *Mathematical Programming,* **79**, pp. 235-253, 1997.
16. K. Fujisawa, M. Kojima, K. Nakata and M. Yamashita, *SDPA (Semidefinite Programming Algorithm) User's Manual—Version 6.00,* Research Report B-308, Depart-

ment of Mathematical and Computing Sciences, Tokyo Institute of Technology, Tokyo, Japan, December 1995 (revised July 2002).

17. M. Fukuda, M. Kojima and M. Shida, Lagrangian dual interior-point methods for semidefinite programs, *SIAM Journal on Optimization,* **12**, pp. 1007–1031, 2002.

18. R.H. Gallagher, *Finite Element Analysis: Fundamentals,* Prentice-Hall, New Jersey, 1975.

19. M.X. Goemans, Semidefinite programming in combinatorial optimization, *Mathematical Programming,* **79** (1997) 143-161.

20. D. Goldfarb and K. Scheinberg, "Interior point trajectories in semidefinite programming," *SIAM J. Optim.,* **8**, pp. 871–886, 1998.

21. E.J. Haug, K.K. Choi and V. Komkov, *Design Sensitivity Analysis of Structural Systems,* Academic Press, 1986.

22. C. Helmberg, "Semidefinite programming", *European Journal of Operational Research,* **137**, pp. 461–482, 2002.

23. C. Helmberg and F. Rendl, "A spectral bundle method for semidefinite programming", *SIAM Journal on Optimization,* **10**, pp. 673–696, 2000.

24. C. Helmberg, F. Rendl, R. J. Vanderbei and H. Wolkowicz, "An interior-point method for semidefinite programming," *SIAM Journal on Optimization,* **6**, pp. 342–361, 1996.

25. F. Jarre, An interior method for nonconvex semidefinite programs, *Optimization and Engineering,* **1**, pp. 347–372, 2001.

26. F. Jarre, M. Kočvara and J. Zowe, "Optimal truss topology design by interior-point methods", *SIAM Journal on Optimization,* **8**, pp. 1084–1107, 1998.

27. Y. Kanno and M. Ohsaki, "Necessary and sufficient conditions for global optimality of eigenvalue optimization problems", *Struct. Multidisc. Optim.,* **22**, pp. 248-252, 2001.

28. Y. Kanno and M. Ohsaki, "Minimum principle of complementary energy of cable networks by using second-order cone programming", *Int. J. Solids and Struct.,* **40**, pp. 4437–4460, 2003.

29. Y. Kanno, M. Ohsaki and J. Ito, "Large-deformation and friction analysis of non-linear elastic cable networks by second-order cone programming", *Int. J. Numer. Meth. Engng.,* **55**, pp. 1079–1114, 2002.

30. Y. Kanno, M. Ohsaki and N. Katoh, "Sequential semidefinite programming for optimization of framed structures under multimodal buckling constraints", *Int. J. Structural Stability and Dynamics,* **1**, pp. 585-602, 2001.

31. Y. Kanno, M. Ohsaki and N. Katoh, "Symmetricity of the solution of semidefinite programming", *Struct. Multidisc. Optim.,* **24**, pp. 225-232, 2002.

32. Y. Kanno, M. Ohsaki, K. Murota and N. Katoh, "Group symmetry in interior-point methods for semi-definite program", *Optimization and Engineering,* **2**, pp. 293–320, 2001.

33. C. Kanzow, C. Nagel, H. Kato and M. Fukushima, "Successive linearization methods for nonlinear semidefinite programs", to appear in *Computational Optimization and Applications.*

34. N.K. Karmarkar, "A new polynomial-time algorithm for linear programming", *Combinatorica,* **4**, pp. 373–395, 1984.

35. M. Kočvara, "On the modeling and solving of the truss design problem with global stability constraints", *Struct. Multidisc. Optim.,* **23**, pp. 189-203, 2002.

36. M. Kočvara and M. Stingl, "Solving nonconvex SDP problems of structural optimization with stability control", *Optimization Methods and Software,* **19**, pp. 595-609, 2004.

37. M. Kočvara, J. Zowe and A. Nemirovski, "Cascading — an approach to robust material optimization", *Computers & Structures,* **76**, pp. 431–442, 2000.

38. M. Kojima, S. Mizuno and A. Yoshise, "A primal-dual interior point algorithm for linear programming", in *Progress in Mathematical Programming: Interior-Point Algorithms and Related Methods*, N. Megiddo (ed.), Springer-Verlag, New York, pp. 29–47, 1989.

39. M. Kojima, S. Shindoh and S. Hara, "Interior-point methods for the monotone semidefinite linear complementarity problems", *SIAM Journal on Optimization*, **7**, pp. 86–125, 1997.

40. M. Kojima and L. Tunçel, "Discretization and localization in successive convex relaxation methods for nonconvex quadratic optimization problems," *Math. Programming*, **89**, pp. 79–111, 2000.

41. A.S. Lewis and M.L. Overton, "Eigenvalue optimization," *Acta Numerica*, **5**, pp. 149–190, 1996.

42. M.S. Lobo, L. Vandenberghe, S. Boyd and H. Lebret, "Applications of second order cone programming", *Linear Algebra and its Applications* **284**, pp. 193–228, 1998.

43. The MathWorks, Inc., *Using MATLAB*, The MathWorks, Inc., Natick, MA, 2002.

44. S. Mehrotra, "On the implementation of a primal-dual interior point method," *SIAM J. Optim.*, **2**, pp. 575–601, 1992.

45. H.D. Mittelmann, "An independent benchmarking of SDP and SOCP solvers", *Mathematical Programming*, **B95**, pp. 407-430, 2003.

46. R.D.C. Monteiro, "Primal-dual path-following algorithms for semidefinite programming", *SIAM Journal on Optimization*, **7**, pp. 663–78, 1997.

47. R.D.C. Monteiro and T. Tsuchiya, "Polynomial convergence of primal-dual algorithms for the second-order cone program based on the MZ-family of directions", *Mathematical Programming*, **88** pp. 61–83, 2000.

48. R.D.C. Monteiro and P.R. Zanjácomo, "General interior-point maps and existence of weighted paths for nonlinear semidefinite complementarity problems", *Mathematics of Operations Research*, **25**, pp. 381–399, 2000.

49. Y.E. Nesterov and A.S. Nemirovski, *Interior Point Polynomial Methods in Convex Programming: Theory and Applications*, SIAM: Philadelphia, PA, 1994.

50. Y.E. Nesterov and M.J. Todd, "Self-scaled barriers and interior-point methods in convex programming," *Mathematics of Operations Research*, **22**, pp. 1–42, 1997.

51. Y.E. Nesterov and M.J. Todd, "Primal-dual interior-point methods for self-scaled cones", *SIAM Journal on Optimization*, **8**, pp. 324–364, 1998.

52. M. Ohsaki, "Genetic algorithm for topology optimization of trusses," *Comput. & Struct*, vol. 57, pp. 219–225, 1995.

53. M. Ohsaki, "Optimization of geometrically non-linear symmetric systems with coincident critical points", *Int. J. Numer. Methods Engng*, **48**, pp. 1345–1357, 2000.

54. M. Ohsaki, K. Fujisawa, N. Katoh and Y. Kanno, "Semi-definite programming for topology optimization of trusses under multiple eigenvalue constraints", *Comput. Meth. Appl. Mech. Engrg.*, **180**, pp. 203–217, 1999.

55. N. Olhoff and S.H. Rasmussen, "On single and bimodal optimum buckling loads of clamped columns", *Int. J. Solids Structures*, **13**, pp. 605–614, 1977.

56. H.C. Rodrigues, J.M. Guedes and M.P. Bendsøe, "Necessary conditions for optimal design of structures with a nonsmooth eigenvalue based criterion", *Struct. Optim.*, **9**, pp. 52–56, 1995.

57. A.P. Seyranian, E. Lund and N. Olhoff, "Multiple eigenvalues in structural optimization problems", *Struct. Optim.*, **8**, pp. 207–227, 1994.

58. T. Sasakawa and T. Tsuchiya, "Optimal magnetic shield design with second-order cone programming", *SIAM Journal on Scientific Computing*, **24**, pp. 1930–1950, 2003.

59. J.F. Sturm, "Using SeDuMi 1.02, a MATLAB toolbox for optimization over symmetric cones", *Optimization Methods and Software,* **11/12**, pp. 625–653, 1999.
60. J.F. Sturm and S. Zhang, "Symmetric primal-dual path following algorithms for semidefinite programming," *Applied Numerical Mathematics,* **29**, pp. 301–315, 1999.
61. M.J. Todd, "Semidefinite optimization", *Acta Numerica,* **10**, pp. 515–560, 2001.
62. R.H. Tütüncü, K.C. Toh and M.J. Todd, "Solving semidefinite-quadratic-linear programs using SDPT3", *Mathematical Programming,* **B95**, pp. 189-217, 2003.
63. L. Vandenberghe and S. Boyd, "Semidefinite programming," *SIAM Review,* **38**, pp. 49–95, 1996.
64. R.J. Vanderbei and H.Y. Benson, "Solving problems with semidefinite and related constraints using interior-point methods for nonlinear programming", *Mathematical Programming,* **95**, pp. 279–302, 2003.
65. H. Wolkowicz, R. Saigal and L. Vandenberghe (eds.), *Handbook of Semidefinite Programming — Theory, Algorithms, and Applications,* Kluwer, Dordrecht, The Netherlands, 2000.
66. S.J. Wright, *Primal-Dual Interior-Point Methods,* SIAM, Philadelphia, PA, 1997.
67. S. Zhang, "A new self-dual embedding method for convex programming", *Journal of Global Optimization,* **29**, pp. 479–496, 2004.
68. Z. Zhao, B. Braams, M. Fukuda, M.L. Overton and J.K. Percus, "The reduced density matrix method for electronic structure calculations and the role of three-index representability conditions", *Journal of Chemical Physics,* **120**, pp. 2095–2104, 2004.

CHAPTER 20

NONLINEAR OPTIMAL CONTROL

Soura Dasgupta

Department of Electrical and Computer Engineering
University of Iowa, Iowa City, Iowa 52242, USA
E-mail: dasgupta@engineering.uiowa.edu

This chapter considers the rich field of nonlinear optimal control. After providing a brief history of feedback control theory, we provide some preliminaries and some general feedback control practices for both linear and nonlinear systems. We use these to set up a review of both linear and nonlinear optimal control.

1. Introduction

The field of feedback control has been part of the engineering lexicon dating as far back in history as to the ancient Egyptians. In its more modern incarnation its applications can be found in the 19th century governors. In the early 20th century engineers discovered one of its major attraction: A well designed feedback system renders a system insensitive to modeling imperfections. This discovery is often attributed to Black, an engineer working for Bell Labs. Each day Black would have to spend hours retuning his amplifiers, because of changing ambient conditions from one day to the next. Legend has it that frustrated by this daily occurrence, on his way to work on the ferry, Black literally perfomed a set of back of the envelope calculations and concluded that amplifiers designed using feedback would not have such annoying sensitivity. This feature of feedback is often under appreciated. Thus many believe that to design a good controller one must have access to a highly accurate and complicated model. Quite to the contrary, the beauty of a good controller is that it can be designed on the basis of simple models that capture only the dominant features of the system that work against achieving control performance.

The basic feedback control setting in its broadest generality is depicted in Fig. 1. Here P is the system, often known as *plant* to be controlled, and the C_i are the controllers: C_1 is often called the feed forward controller, C_3 the feedback controller, and C_2 the compensator. Further $r(t)$, $y(t)$ and $u(t)$ are respectively known as the reference input, output, and the control or plant input respectively. In *Single Input Single Output (SISO)* each of these signals is a scalar function of time. In *Multiple Input Multiple Output (MIMO)* systems they are vector functions.

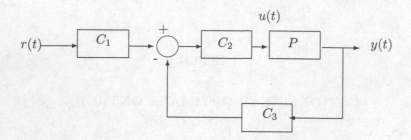

Fig. 1.   A general feedback control setting.

Nyquist,[1] Bode,[2] and Black[3] were early pioneers who in the period spanning the 1930's and the 1950's developed a preliminary mathematical framework of control theory. Their work focused on meeting relatively simple objectives such as forcing $y(t)$ to track a prespecified trajectory, while maintaining system stability with certain margins. The plants and controllers they considered were linear time invariant (LTI) plants described by ordinary differential equations, which they treated in a *transfer function* framework using the *Laplace Transform*. The looming ubiquity of digital computers prompted researchers such as Tsypkin,[4] and Jury,[5] to consider *sampled data control*, where the controllers process *discrete time* samples of the plant output and generate a discrete time control input that is converted to a continuous time signal via *digital to analog convertors*. The underlying mathematical framework involves the plant being modeled by an equivalent difference equation that governs the behavior of the system relating the discrete time control input to the output samples. The controllers are also modeled by difference equations. A transfer function approach in the LTI case uses the *z-Transform*.

Soon there was growing interest in not just meeting simple performance objectives but to do so *optimally*, e.g. by keeping the control input as small as possible. This was facilitated by the pioneering work of Kalman, who among other things started using a *state variable representation (SVR)* of the underlying plants. These developments married such strands from the optimization literature as *calculus of variations*,[6] *Pontryagin's Maximum Principle*[7] and the *Hamilton-Jacobi-Bellman (HJB) equations*[8] in continuous time and *Dynamic Programming*[8] in discrete time. Other approaches used *game theory*[9] and *Risk Sensitive Control*.[10]

Early work focussed on *Linear Optimal Control* and led to the famous *Linear Quadratic Regulator (LQR)*[11] and in the *stochastic framework* to the solution of the *Linear Quadratic Gaussian (LQG)*[12] problem. In the 1980's primarily motivated by robustness issues, these approaches were augmented by linear $\mathcal{H}_\infty$ and $\mathcal{H}_2$ optimal control.[14]

The foregoing describes briefly the history of linear control in general and linear optimal control in particular. Over the years many of these approaches have been extended in bites to *nonlinear control* and *nonlinear optimal control*. A more concerted effort in these directions commenced in the 80's, though these fields are by no means totally mature.

In Section 2 presents preliminaries. Section 3 reviews certain basic non-optimal control techniques. Section 4 focuses on optimal control.

2. Preliminaries

In this section we provide preliminary concepts, focusing primarily though not exclusively on LTI systems.

2.1. *Models*

We describe below various models used in control design.

2.1.1. *State Variable Representations (SVR)*

Most dynamical systems can be described by an SVR. In continuous time these comprise a pair of ordinary differential equations:

$$\dot{x}(t) = f(t, x, u) \tag{1}$$

$$y(t) = h(t, x, u). \tag{2}$$

In discrete time with k the sample index one has:

$$x(k+1) = f(k, x, u) \tag{3}$$

$$y(k) = h(k, x, u). \tag{4}$$

In both cases for a q-input, p-output system, referred to henceforth as a $p \times q$ system, $u(\cdot)$ and $y(\cdot)$ are q and p-dimensional column vectors respectively. The n-vector $x(.)$ is called the state vector; n is the degree or dimension of the system. The state vector contains the history of the system: e.g. in continuous time for any t_0, to deterimine $y(t)$ for all $t \geq t_0$, it suffices to know $x(t_0)$ and $u(t)$ for all $t \geq t_0$. Note (1) and (3) are known as the *state equations* and (2) and (4) are known as the *output equations*.

These are general nonlinear time varying (NLTV) systems. For time invariant systems f and h do not have explicit dependence on time. In the case of general Linear Time Varying (LTV) systems these reduce to

$$\dot{x}(t) = A(t)x(t) + B(t)u(t) \tag{5}$$

$$y(t) = C(t)x(t) + D(t)u(t), \tag{6}$$

where A, B, C and D are $n \times n$, $n \times q$, $p \times n$ and $p \times q$ dimensional matrices respectively. In discrete time with k the sample index one has:

$$x(k+1) = A(k)x(k) + B(k)u(k) \tag{7}$$

$$y(k) = C(k)x(k) + D(k)u(k) \tag{8}$$

Henceforth we will focus mostly on the time invariant cases:

$$\dot{x}(t) = f(x, u) \tag{9}$$

$$y(t) = h(x, u). \tag{10}$$

In discrete time with k the sample index one has:

$$x(k + 1) = f(x, u) \tag{11}$$

$$y(k) = h(x, u). \tag{12}$$

In the LTI case the quadruple $\{A, B, C, D\}$ are constant matrices, i.e. we have in continuous time

$$\dot{x}(t) = Ax(t) + Bu(t) \tag{13}$$

$$y(t) = Cx(t) + Du(t), \tag{14}$$

and in discrete time

$$x(k + 1) = Ax(k) + Bu(k) \tag{15}$$

$$y(k) = Cx(k) + Du(k). \tag{16}$$

We now provide some key definitions pertaining to SVR's. The first is the notion of reachability that is fundamental to the ability to control the state by appropriately choosing the input.

Definition 1. The system (1, 2) (3, 4) is completely reachable (cr) if for every x_0, x_f and t_0 (k_0) there exists a finite time t_f (k_f) and an input $u(t)$ $(u(k))$, defined on $t_0 \leq t \leq t_f$ $(k_0 \leq k \leq k_f$) such that with $x(t_0) = x_0$ $(x(k_0) = x_0$) one achieves $x(t_f) = x_f$ $(x(k_f) = x_f$). It is uniformly completely reachable (ucr) if t_f (k_f) is independent of t_0 (k_0).

Thus a cr system can be driven from any intial state to any final state in a finite time. For time invariant systems ucr is equivalent to cr. Unsurprisingly for an LTV system the cr/ucr property is determined entirely by the pair $[A(t), B(t)]$. Thus we often say this pair is cr/ucr. The second property is that of observability.

Definition 2. The system (1, 2) (3, 4) is completely observable (co) if for every x_0 and t_0 (k_0) there exists a finite time t_f (k_f) such that the knowledge of the input $u(t)$ $(u(k))$, and output $y(t)$ $(y(k))$, on $t_0 \leq t \leq t_f$ $(k_0 \leq k \leq k_f)$ suffices to estimate $x(t_0) = x_0$ $(x(k_0) = x_0)$ uniquely. It is uniformly completely observable (uco) if t_f (k_f) is independent of t_0 (k_0).

Again uco is equivalent to co for time invariant systems, and for LTV systems it is entirely associated with the pair $[A(t), C(t)]$ which is thus called an observable pair. Further $[A(t), C(t)]$ is co/uco iff $[A^T(t), C^T(t)]$ is cr/ucr. This property enables one to uniquely estimate the state of the system from its input and output signals. If a system is either not cr and or not co then its state cannot be completely controlled using a feedback scheme as depicted in Fig. 1. Further, the violation of either property imples that the representation (1, 2) (3, 4) are nonminimal, i.e. the input/output behavior is captured by lower dimensional SVR's.

2.1.2. *Input/Output Models*

A second class of models assume that the $x(-\infty) = 0$, i.e. the system is at *initial rest*, and focuses on the input to output mapping. Typically, one uses the notation:

$$y = Gu, \tag{17}$$

G being the underlying operator. A key difference between linear and nonlinear operators is that while for linear operators for a constant scalar α, $G\{\alpha u\} = \alpha Gu$, and $G\{u_1 + u_2\} = Gu_1 + Gu_2$, one or other of these will be violated for nonlinear operators. For discrete and continuous time linear time invariant systems, one uses the notion of *transfer function* to characterize operators. In continuous time this is the ratio of the Laplace Transform[15] of the output and input and in discrete time the ratio of the z-Transform[15] of the output and input. Specifically for the continuous and discrete time systems (13) and (15) it is respectively given by the first and second equality below:

$$G(s) = D + C(sI - A)^{-1}B \text{ and } G(z) = D + C(zI - A)^{-1}B. \tag{18}$$

Observe that for a $p \times q$ system, $G(s)$ and $G(z)$ are $p \times q$ rational matrices in s, and z, i.e. each element of these matrices are ratios of two polynomials. The values of s, z where these matrices become infinity are called the *poles* of the system. Values at which they become singular are called *zeros*. Lack of either co or cr leads to a transfer function where a zero coincides with a pole.

2.2. *Stability*

Stability is often described as the *sine qua non* of control. There are many definitions if stability. In broad terms one involves zero input stability and the second zero state stability. The latter requires that bounded inputs generate bounded outputs with zero intial state values. The former examines state behavior under zero inputs. Below we briefly reprise certain definitions and criteria of checking stability.

We first define certain common notions of zero input stability. We will define this for continuous time systems. The definition extends to discrete time systems with t replacing k. Further in the sequel we will assume that $x = 0$ is an equilibrium point of all systems, e.g. in (1)

$$f(t, 0, 0) = 0 \text{ for all t.} \tag{19}$$

Further the vector p-norm of a vector $x = [x_1, \cdots, x_n]^T$ is defined by

$$\|x\|_p = \left(\sum_{i=1}^{n} |x_i|^p \right)^{1/p}.$$

If we drop the subscript p we would mean that the concept applies to all p-norms.

Definition 3.[13] The system (1) is:

- stable if for all $\epsilon > 0$ and t_0 there is a $\delta(\epsilon, t_0)$, such that whenever $\|x(t_0)\| \leq \delta(\epsilon, t_0)$, one has,

$$\|x(t)\| \leq \epsilon$$

  for all $t \geq t_0$.
- asymptotically stable (a.s.) if it is stable and there is a $\delta(\epsilon, t_0)$, such that whenever $\|x(t_0)\| \leq \delta(\epsilon, t_0)$,

$$\lim_{t \to \infty} x(t) = 0.$$

- uniformly stable (u.s.) if it is stable and $\delta(\epsilon, t_0)$ is independent of t_0.
- uniformly asymptotically stable (uas) if it is a.s. and $\delta(\epsilon, t_0)$ is independent of t_0.
- globally uniformly asymptotically stable (guas) if δ in the definition of uas is infinity.
- exponentially asymptotically zero (eas) in continuous time if there exists a c_1 and $\lambda_1 > 0$ such that for all t_0 and $t \geq t_0$

$$\|x(t)\| \leq c_1 \|x(t_0)\| e^{-\lambda(t-t_0)}. \tag{20}$$

In discrete time it is exponentially asymptotically zero (eas) if there exists a c_1 and $|\lambda_1| < 1$ such that for all k_0 and $k \geq k_0$

$$\|x(k)\| \leq c_1 \|x(k_0)\| \lambda_1^{k-k_0}. \tag{21}$$

We make some observations. Uniformity in all definitions is automatic for all time invariant systems. A uas linear system is both guas and eas. An LTI continuous time system is eas iff all the eigenvalues of A have negative real parts or for that matter all poles of the system have negative real parts, i.e. they lie in the *open left half plane* (OLHP). An LTI discrete time system is eas iff all the eigenvalues of A have magnitude less than one or for that matter all poles have magnitude less than one, i.e. they lie in the *open unit circle*.

A key device for checking stability is Lyapunov theory. In the simplest form suppose one has a scalar function $V(x, t)$ such that:

- For some $c_i > 0$, all t and x

$$c_1 \|x(t)\|^2 \leq V(x, t) \leq c_2 \|x(t)\|^2. \tag{22}$$

- For all t

$$V(0, t) = 0. \tag{23}$$

- When $u(t) = 0$, along the trajectories of (1), for some $c_3 > 0$,

$$\dot{V}(x, t) \leq -c_3 \|x(t)\|^2. \tag{24}$$

Then observe and that with zero input

$$\lim_{t \to \infty} V(x, t) = 0;$$

i.e. because of (23)

$$\lim_{t \to \infty} x(t) = 0.$$

Thus the system is a.s. In fact under these conditions it is eas. The function $V(x, t)$ is known as a *Lyapunov function*. As stated Lyapunov theory provides sufiicient conditions for stability of various kinds, and most methods for selection such a function, except in the linear case, are *ad hoc*. Nonetheless there are converse Lyapunov theorems that demonstrate that stability of various types imply the existence of various forms of Lyapunov functions.[13] In discrete time (24) is replaced by

$$V(x(k + 1)) - V(x(k)) \leq -c_3 \|x(k)\|^2.$$

It is useful to consider specialization to the linear case. In continuous time it is known that (5) is eas iff there is a symmetric matrix $0 \leq c_1 I \leq P(t) = P^T(t) \leq c_2 I$ and a matrix $Q(t)$, such that (a) the pair $[A(t), Q(t)]^{16}$ is uco and there holds for all t:

$$\dot{P}(t) + A^T(t)P(t) + P(t)A(t) = -Q(t)Q^T(t). \tag{25}$$

In fact under these conditions $V(t) = x^T(t)P(t)x(t)$ serves as a Lyapunov function. In discrete time (25) is replaced by

$$P(k + 1) - P(k) + A^T(k)P(k)A(k) = -Q(k)Q^T(k). \tag{26}$$

The equations (25) and (26) are respectively known as the Lyapunov Differential and Lyapunov Difference Equations. In the LTI case P and Q are constants and one has the two Lyapunov equations

$$A^T P + P A = -QQ^T \tag{27}$$

and,

$$-P + A^T P A = -QQ^T. \tag{28}$$

In both cases P is called a *Lyapunov Matrix* for A.

The second kind of stability is zero-state stability. Without going into too many nuances, it essentially requires that with zero intial state, all bounded inputs generate bounded outputs. Such stability is known as Bounded Input Bounded Output

(BIBO) stability. For uco and ucr linear systems eas is equivalent to BIBO stability. For continuous time LTI systems BIBO stability is equivalent to a transfer function having all poles in the OLHP *after pole-zero cancellations* if any. In discrete time they must be inside the unit circle again *after pole-zero cancellations* if any.

3. Nonoptimal Control Strategies

In this section we provide insights into some standard nonoptimal control techniques. Section 3.1 describes some standard control tasks. Section 3.2 and 3.3 respectively review some linear and nonlinear control techniques.

3.1. *Control Objectives*

Controllers are designed to meet performance objectives in a *robust* way. Robustness refers to acceptable behavior in the face of modeling imperfections. At all times a minimum requirement in the setting of Fig. 1 is that the *closed loop* systems relating $r(t)$ to $y(t)$ and $u(t)$ is BIBO stable.

By and large all concern inducing the right trajectory of the system output. Thus for example one may require $y(t)$ to follow a constant value, a ramp, a sinusoid, or something more complicated. Some times one may simply want the output to be forced to zero, or as a stricter measure, force the state to be zero. This latter task is known as *regulation*, and will be treated in some detail in the sequel.

Additionally one often wishes to minimize the impact on the output of disturbances appearing at various places in the closed loop of Fig. 1.

3.2. *Linear Control Techniques*

One approach to linear control systems is to use the transfer function approach.[17] They largely rely on a frequency domain approach that recognizes that in continuous time if a BIBO stable LTI system is excited by a sinusoid then the output at steady state, i.e. after the decay of initial condition effects, is also a sinusoid. For example for a SISO continuous time system with transfer function $G(s)$, the input $cos(\omega t)$ generates the output

$$|G(j\omega)|cos(\omega t + \angle G(j\omega)).$$

This leads to the use of a slew of techniques using devices such as Root Locus, Nyquist and Bode plots and Nichols chart to SISO controller design. In particular if in Fig. 1, the transfer functions of the controllers and plant are $C_i(s)$ and $P(s)$ respectively, then the transfer function relating $r(t)$ to $y(t)$ is:

$$(I + P(s)C_2(s)C_3(s))^{-1}P(s)C_2(s)C_1(s). \tag{29}$$

The poles of the system are thus the zeros of

$$I + P(s)C_2(s)C_3(s).$$

These expressions form the basis for various design rules for stabilization etc. The Internal Model Principle[18] is used to effect robust tracking and disturbance rejection.

The second approach involves SVR techniques. Since SVR methods are directly used in optimal control, we spend some time explaining certain salient features of SVR techniques. Consider for example state regulation by *pole placement*. Work with (13) and (14) and assume full access to the system state $x(t)$. Consider the control law

$$u(t) = r(t) - Kx(t), \tag{30}$$

where K is a $q \times n$ feedback gain matrix. Under (30), (13, 14) become:

$$\dot{x}(t) = (A - BK)x(t) + Br(t),$$

and

$$y(t) = (C - BK)x(t) + Br(t).$$

Thus the closed loop transfer function from $r(t)$ to $y(t)$ is

$$(C - BK)(sI - (A - BK))^{-1}B + D. \tag{31}$$

Observe now that effectively $x(t)$ remains the state vector of the closed loop system and that the poles are the eigenvalues of $A - BK$. The pole placement problem becomes selecting the gain matrix K so that the eigenvalues of $A - BK$ have prescribed values. The feedback law (30) (with t replaced by k), also leads to an identical design issue. If in continuous time the eigenvalues are in OLHP, then with zero $r(t)$, $x(t)$ goes to zero. In discrete time the eigenvalues must be inside the unit circle.

It turns out,[18] that as long as $[A, B]$ is cr, K can be found to arbitrarily place the eigenvalues of $A - BK$. The celebrated Ackermann's formula provides K.

In many settings however, due to e.g. limited sensing infrastructure the full state $x(t)$ cannot be directly measured, and one needs to work with an output feedback law. To do this one must seek to estimate the system state, a task performed by an *observer* described below:

$$\dot{\hat{x}}(t) = A\hat{x}(t) + Bu(t) + L(y(t) - C\hat{x}(t) - Du(t)), \tag{32}$$

where $\hat{x}(t)$ is the state estimate and L is an $n \times p$ observer gain matrix.

Without the term $L(y(t) - C\hat{x}(t) - Du(t))$, (32) simply mimics (30). The term $L(y(t) - C\hat{x}(t) - Du(t))$, acts as a correction to counteract the disparity between $x(t)$ and $\hat{x}(t)$. Notice in particular that $y(t) - C\hat{x}(t) - Du(t)$ is a gauge of how well the estimate $\hat{x}(t)$ matches $x(t)$. It is readily seen that with

$$\tilde{x}(t) = x(t) - \hat{x}(t),$$

$$\dot{\tilde{x}}(t) = (A - LC)\tilde{x}(t). \tag{33}$$

Thus if $A - LC$ has all eigenvalues inside the OLHP, $\tilde{x}(t)$ will converge to zero. The eigenvalues of $A - LC$ are called the *observer poles*. One can choose them arbitrarily by fixing L iff $[A, C]$ is co. Again a modest variation of Ackermann's formula provides L.

If the state is not directly available, the question arises as to what happens if (30) is replaced by

$$u(t) = r(t) - K\hat{x}(t),\qquad(34)$$

i.e. one uses the state estimate in the stead of $x(t)$. The closed loop system relating $r(t)$ to $y(t)$ now has the higher dimensional SVR:

$$\begin{bmatrix} \dot{x}(t) \\ \dot{\tilde{x}}(t) \end{bmatrix} = \begin{bmatrix} A - BK & BK \\ 0 & A - LC \end{bmatrix} \begin{bmatrix} x(t) \\ \tilde{x}(t) \end{bmatrix} + \begin{bmatrix} B \\ 0 \end{bmatrix} r(t)\qquad(35)$$

and

$$y(t) = \begin{bmatrix} C - DK & DK \end{bmatrix} \begin{bmatrix} x(t) \\ \tilde{x}(t) \end{bmatrix} + Dr(t)\qquad(36)$$

First observe that the closed loop poles are the eigenvalues of:

$$\begin{bmatrix} A - BK & BK \\ 0 & A - LC \end{bmatrix}$$

or equivalently the eigenvalues of $A - BK$ and $A - LC$. Thus one can design K and L independently as above. This is the celebrated *separation principle or certainty equivalence control*. A second apect of the separation principle is that after cancelling common factors, the closed loop transfer function is identical to (31). Thus with this design procedure one can use a state estimate to effect the control law in the control law of (30). The placement of the observer and other poles affects the closed loop dynamics. *Optimum control and the celebrated Kalman Filter*[19] provide one set of optimum solutions for the selection of K and L.

3.3. *Nonlinear Control Techniques*

We now briefly reprise some related techniques for controlling nonlinear systems. Perhaps the most rudimentary yet still the most widely employed is that of *linearization* around a state trajectory. In particular, suppose $x(t)$ and $u(t)$ in (9) remain "close" to nominal values of x^* and u^*. Suppose also without loss of generality (through a simple redefintion of $x(t)$ and $u(t)$ involving a translation) that

$$f(x^*, u^*) = 0.$$

Then to a first order approximation, one has that with

$$\Delta x(t) = x(t) - x^*$$

and

$$\Delta u(t) = u(t) - u^*,$$

$$f(x,u) \approx \left.\frac{\partial f(x,u)}{\partial x}\right|_{x=x^*,u=u^*} \Delta x(t) + \left.\frac{\partial f(x,u)}{\partial u}\right|_{x=x^*,u=u^*} \Delta u(t). \qquad (37)$$

Consequently (9) can be approximated as

$$\Delta \dot{x}(t) \approx A\Delta x(t) + B\Delta u(t), \qquad (38)$$

with

$$A = \left.\frac{\partial f(x,u)}{\partial x}\right|_{x=x^*,u=u^*}$$

and

$$B = \left.\frac{\partial f(x,u)}{\partial u}\right|_{x=x^*,u=u^*}.$$

Consequently as long as $x(t)$ and $u(t)$ in (9) remain "close" to nominal values of x^* and u^*, one can employ linear control techniques.

A more sophisticated version of this approach is *gain scheduling*. Simply put, it uses a host of set points like (x^*, u^*), and whenever the trajectories approach any of these set points, uses a model linearized around the corresponding set point.

Much more can be done in the special case of *input affine systems*. These are represented by the SVR

$$\dot{x} = f(x) + g(x)u. \qquad (39)$$

In this case under the right conditions (whose characterization involves differential geometric concepts) one can in fact perform *exact* feedback linearization.[13] Specifically under these conditions there exists a $n \times 1$ vector function $h(x)$, and a state transformation with $z = h(x)$ which achieves the following: There exist functions $\gamma(x)$ $(q \times q)$ and $\alpha(x)$ $(q \times 1)$ with $\gamma(x)$ nonsingular every where so that (39) becomes

$$\dot{z} = A_c z + B_c \gamma(x)[u - \alpha(x)].$$

Then the feedback law

$$u = [\gamma(x)]^{-1} r(t) + \alpha(x)$$

results in the linear state equation

$$\dot{z} = A_c z + B_c r.$$

These techniques involve full state linearization as above, or when only the r to y map is linear. After either type of linearization has been achieved, one can then again employ linear control techniques. A related technique is that of backstepping.[13] Other control strategies for (39) are sliding mode control, introduction of nonlinear damping and Lyapunov redesign.[13] The latter in particular, permits one to design a control law using a judiciously chosen Lyapunov function, so that closed stability is determined using this Lyapunov function. One should also note for general nonlinear systems there exist frequency domain methods using the so called describing functions.[20]

4. Optimal Control

This section reviews basic concepts and techniques from the optimal control literature. In Section 4.1 we present the nonlinear optimal control problem in its broadest generality, and provide two distinct solution approaches. Section 4.2 sets the stage for more specialized approaches by looking at the linear case. Section 4.3 uses this as a spring board to consider nonlinear approaches.

4.1. *The Bolza Problem*

A very general optimal control, known as the Bolza problem[21,22] is to find a $u(t)$ that minimizes

$$J(u, x, t) = \phi[x(T)] + \int_t^T l(x, u, \tau)d\tau \tag{40}$$

subject to (1). We will assume here that the loss function $l(x, u, t) \geq 0$ as is the terminal function $\phi[\cdot]$ although the problem in its broadest generality does not mandate this. We will also assume that $l(0, 0, t) = m(0) = 0$.

There are two basic approaches to solving this problem. The first uses the *Hamilton-Jacobi-Bellman (HJB) equation*, that provides a necessary condition for optimality. Specifically, with $J^*(x, t)$ the optimal value of $J(u, x, t)$, HJB equation requires that

$$-\frac{\partial J^*(x, t)}{\partial t} = \min_{u(t)} \left\{ l(x, u, t) + \left[\frac{\partial J^*(x, t)}{\partial t} \right]^T f(t, x, u) \right\}$$

and

$$J^*(x, T) = \phi(x) \text{ for all } x.$$

In general this provides a state feedback control law that must be then checked to see if it corresponds to the optimum. A key difficulty is that the HJB equation requires that a differential equation be solved *backwards in time* making an online solution in general impossible. A second approach that uses Calculus of variations and Pontryagin's Minimum Principle, also involves a similar difficulty. Nonetheless one can circumvent this in many situations, some that are presented in the sequel.

4.2. *Linear Quadratic Control*

See Refs. 21 and 22 for details of results presented here. A considerable body of work exists on a special case of the Bolza problem known as the Linear Quadratic Regulator (LQR). In this case the problem becomes: Find $u(t)$ to minimize

$$J(u) = x^T(T)Mx(T) + \int_0^T \left[x^T(t)Q(t)x(t) + u^T(t)R(t)u(t) \right] dt, \tag{41}$$

subject to the LTV system equation (5). In this case $M = M^T \geq 0$, $Q(t) = Q^T(t) \geq 0$, and for some $\alpha > 0$, $R(t) = R^T(t) \geq \alpha I$. Consequently, $J(u)$ penalizes large state and input values. When T is finite this is called the *finite horizon* problem. Otherwise it is called the *infinite horizon problem*. In the latter case $M = 0$. It turns out that in this case the optimal control law is

$$u(t) = -R^{-1}(t)B^T(t)P(t)x(t), \tag{42}$$

where $P(t)$ is the solution of the celebrated *Riccati Differential Equation (RDE)*,

$$\dot{P}(t) = P(t)B(t)R^{-1}(t)B^T(t)P(t) - A^T(t)P(t) - P(t)A(t) - Q(t); \quad P(T) = M. \tag{43}$$

Several features of this solution are noteworthy. First the solution is (42) is a linear state feedback law. Second, while the RDE must be solved backwards in time, if the time profiles of $A(t), B(t), Q(t)$ and $R(t)$ are available *a priori*, the RDE can be solved off line, and the control law (42) can be implemented on line.

In the special case where $M = 0$, and A, B, Q, R are all constant, and $[A, B]$ cr, then P itself is a constant and the unique positive definite solution to the *Algebraic Riccati Equation (ARE)*

$$PBR^{-1}B^TP - A^TP - PA = Q \tag{44}$$

Observe, in this case (42) becomes a state feedback law of the type described earlier. As long as $[A, B]$ is cr and $[A, Q^{1/2}]$ is co, the closed loop is always stable and further more the optimum cost is simply $x^T(0)Px(0)$.

It is particularly interesting to note that as long as the $[A, B]$ is cr and $[A, Q^{1/2}]$ is co, a unique positive definite P solving the ARE exists. Even more interestingly the closed loop looks like:

$$\dot{x}(t) = (A - BR^{-1}B^TP)x(t).$$

It turns out that this same P acts as a Lyapunov matrix for this closed loop, in particular

$$(A - BR^{-1}B^TP)^TP + P(A - BR^{-1}B^TP) = -Q. \tag{45}$$

Thus $V(x) = x^TPx$ acts as a Lyapunov function for proving the stability of the closed loop. Recall also that $V(x(0))$ is the optimal value of the LQR cost function. This has implications to nonlinear optimal control to be explained later.

There are disctrete time versions of both the finite and infinite horizon LQR. Specifically, with T integer they seek $u(k)$ to minimize

$$J(u) = x^T(T)Mx(T) + \sum_{k=0}^{T} \left[x^T(k)Q(k)x(k) + u^T(k)R(k)u(k) \right] \tag{46}$$

subject to the LTV system equation (7). As before $M = M^T \geq 0$, $Q(k) = Q^T(k) \geq 0$, and for some $\alpha > 0$, $R(k) = R^T(k) \geq \alpha I$, and in the infinite horizon $(T = \infty)$ case $M = 0$. The sloution in this case is obtained using *Dynamic Programming*, and has a similar structure to the continuous time solution. In particular the optimal control law is again linear state feedback:

$$u(k) = -\left(R(k) + B^T(k)P(k+1)B(k) \right)^{-1} Mx(k)$$

where $P(k)$ is the solution of the *Riccati Difference Equation*,

$$P(k) = Q(k) + A^T(k)P(k+1)A(k) - (M + B^T(k)P(k+1)B(k))^T$$
$$(R(k) + B^T(k)P(k+1)A(k))^{-1}(M + B^T(k)P(k+1)B(k)); \quad P(T) = M.$$

In the infinite horizon case with all matrices constant, $[A, B]$ is cr P itself is a constant and the unique positive definite solution to the *Discrete Time Algebraic Riccati Equation*

$$P = Q + A^T PA - (M + B^T PB)^T(R + B^T PA)^{-1}(M + B^T PB).$$

Observe the LQR solution requires the availability of the full state. In many instances this is not available. This gives rise to the *Linear Quadratic Gaussian (LQG)* problem that applies optimal state estimation to LQR. We will consider here only the case where the system is LTI and the cost function has constant matrices. The SVR in this case is

$$\dot{x}(t) = Ax(t) + Bu(t) + w(t) \tag{47}$$

and

$$y(t) = Cx(t) + v(t), \tag{48}$$

where $w(t)$ and $v(t)$ are uncorrelated, zero mean, Gaussian white noise vectors with covariance matrices $W = W^T > 0$ and $V = V^T > 0$, respectively. They respectively reprent noise at the control input and the measured output. The objective now is to find $u(t)$, that minimizes for fixed $Q = Q^T \geq 0$ and $R + R^T > 0$,

$$J = \lim_{t \to \infty} E\left(x^T Q x + u^T R u\right),$$

subject to (47), (48) and on the assumption that $u(t)$ must be chosen simply on the basis of observations of $y(t)$ and $u(t)$, and without access to $x(t)$.

To this end recall that in the output feedback pole placement problem a certainty equivalence principle held, in that one could take the state estimate provided by the state observer and apply it to the feedback law that one would have used in the case where full state feedback is used. Now consider the optimal state estimate for (47), (48) in the Maximum Liklihood sense. It is provided by the celebrated Kalman Filter[19] as $\hat{x}(t)$ below:

$$\dot{\hat{x}}(t) = A\hat{x}(t) + Bu(t) + L_f\left(y(t) - C\hat{x}(t) - Du(t)\right), \tag{49}$$

where L_f

$$L_f = SC^T V^{-1}$$

where assuming that $[A, C]$ is co, S is the unique symmetric positive definite solution of the alternative ARE,

$$0 = AS + SA^T + W - SC^T V^{-1} CS. \tag{50}$$

Observe under the observability assumption $A - L_f C$ has all eigenvalues in OLHP, that (49) is exactly of the same form as (32) and that L_f is an optimal observer gain, known as the *Kalman Gain*. Surprisingly the optimal $u(t)$ for the LQG problem, is simply

$$u(t) = -R^{-1}B^T P\hat{x}(t),\tag{51}$$

where as with LQR, P is the unique symmetric positive definite matrix, (as long as $[A, B]$ is cr) that solves the ARE (44). Observe that (51) is precisely (41) with the actual state replaced by the optimal state extimate provided by the Kalman filter. This is another instance of certainty equivalence control and an elegant example of a separation principle. Further two ARE's (44, 50) are key to the solution. Further the closed loop is BIBO stable as long as $[A, B]$ is cr and $[A, C]$ is co.

We observe that LQG control had a profound impact on the development of the US space program. Yet elegant as the LQG solution is, and influential as it was to a generation of space technology, it has surprisingly bad robustness margins, in that the slightest deviation from the idealizing assumptions may cause closed loop instability.[23] A proposed amelioration is the so called method of Loop Transfer Recovery (LTR)[24] that improves robustness for systems all whose zeros are in the OLHP. Related robustness enhancements come from the literature of gauaranteed costs.[25] Another influential stream is in $\mathcal{H}_\infty$-control. These are alternative classes of optimal control problems, that admit optimizing solutions involving two Riccati equations,[14] and assist in achieving certain margins of robustness.

4.3. *Nonlinear Optimal Control*

The foregoing strives to explain the key difficulty with nonlinear optimal control: That the solution to the HJB equation is not in general amenable to online control, the linear quadratic case being a notable exception. Consequently much of the nonlinear control literature is devoted to conjuring suboptimal but near optimal control. To illustrate these approaches we reconsider (39) with the objective of choosing $u(t)$ to minimize

$$\int_0^\infty (q(x) + u^T u)dt.\tag{52}$$

Then the HJB equation leads to Refs. 29 and 26 the optimal control law

$$u^* = -\frac{1}{2}g^T\frac{\partial V^T}{\partial x}$$

where $V(x)$ is positive definite solution of the equation

$$\frac{\partial V}{\partial x}f - \frac{1}{4}\frac{\partial V}{\partial x}gg^T\frac{\partial V^T}{\partial x} + q = 0\tag{53}$$

and $V(x(0))$ is the minimum value of (52). Further, $V(x)$ serves as a Lyapunov function for proving closed loop stability. Observe that this mirrors precisely the LQR case with (53) specializing to (44) and $V(x) = x^T P x$.

The solution of (53) is in general difficult. Several approaches for approximate solutions exist. The first Refs. 27, 30 and 31 is to solve a linearized version of (53). Another approach is to obtain iterative, numerical solutions.[32]

Another approach uses the so called *Frozen Riccati Equation* approach.[26] Specifically this notes that one can represent (39) by the SVR

$$\dot{x} = A(x)x + B(x)u$$

and (52) by

$$J(u) = \int_0^\infty \left\{ x^T Q(x)x + u^T R(x)u \right\} dt \tag{54}$$

where $Q(x) = Q^T(x) \geq 0$ and $R(x) = R^T(x) > 0$ for all x. Then one suboptimal solution is to use the feedback controller

$$u(x) = -R^{-1}(x)B^T(x)P(x)x$$

where $P(x)$ solves the equation:

$$P(x)B(x)R^{-1}(x)B^T(x)P(x) - A^T(x)P(x) - P(x)A(x) = Q(x).$$

Note the similarity to the LQR solution. This is of course a suboptimal solution that can nonetheless be implemented on line. On the other hand observe that the $A(x)$ and $B(x)$ matrices are in general nonuinque. This is exploited in Ref. 26 to show that under mild assumptions there always exist $A(x)$ and $B(x)$ such that the frozen solutions provide the optimum. Finding the right representation remains a challenge.

Other approches could be to work with linearized models, approximate or exact. Another approach advocated by Ref. 27 is to linearize the exact solution to the HJB equations, which leads to a nonlinear control law that is optained by solving a classical ARE.

Yet another approach involves the use of Control Lyapunov Functions (CLF),[29] that exploits the fact that $V(x)$ acts as a Lyapunov function for the optimal closed loop, and searches over spaces of such Lyapunov functions to obtain approximate optima. It has been noted in Ref. 29 that each such law is a solution of a particular HJB equation. Of course the HJB equation that it is optimal for may not correspond to the actual requirements of the application.

A promising and yet practical approach used in Process Control and recently in alleviating internet traffic congestion is that of Receding Horizon Control also known as Model Predictive Control (MPC).[28] We illustrate this approach using a discrete time paradigm. Consider the optimization of

$$J(u) = \sum_{i=k+1}^{k+T} l(x(i), u(i)) \tag{55}$$

under (11), and potentially additional constraints, such as those prohibiting excusrsions of $x(t)$ beyond certain domains. Then at each k MPC, obtains a $u(k)$, by performing an *open loop* optimization of (55) in terms of $x(k)$. This leads to a state feedback law that can be implemented online. When the state is not accessible, an additional state estimation procedure is incorporated along with a certainty equivalence approach. The CLF appraoch is useful for MPC.

5. Conclusion

We have provided but a glimpse of the rich field of nonlinear optimal control in this all too brief review. The reader who wishes to delve deeper into this rapidly developing field should study the references we have provided.

References

1. Nyquist, H., "Regeneration Theory", *Bell Syst. Tech. J.*, p. 126, 1932.
2. Bode, H. W., *Network Analysis and Feedback Amplifier Design*, Von Nostrand, 1945.
3. Black, H. S., "Stabilized Feedback Amplifiers", *Electrical Engineering*, p. 1311, September 1934.
4. Bissell, C. C., Ed., "Yakov Tsypkin — A Russian Life in Control", *IEE Review*, London, pp. 313-316, U.K.: IEE Press, 199.
5. Jury, E. I., 1964. *Theory and Applications of the z-Transform Method*, Wiley, New York.
6. Kirk, D. E., *Optimal Control Theory*, Prentice Hall, Englewood Cliffs, 1970.
7. Anderson, B. D. O. and Moore, J. B., *Linear Optimal Control*, Prentice Hall, Englewood Cliffs, 1971.
8. Bellman, R. E., "The Theory of Dynamic Programming", *Proc. Nat. Acad. Sci. USA*, pp. 716-719, 1952.
9. Basar, T. and Bernhard, P., *Differential Games and Applications*, Lecture Notes in Control and information Sciences, Berlin: Springer-Verlag, 1989.
10. Whittle, P., *Risk-Sensitive Optimal Control*, John Wiley-Interscience Series in Systems and Optimization, 1990.
11. Kalman, R. E., "Contributions to the theory of optimal control", "Sensor network localization with imprecise distance measurements", *Bol. Soc. Matem. Mex. 5* pp. 102-119, 1960.
12. Athans, M., "The role and use of the stochastic linear quadratic gaussian problem", *IEEE Transactions on Automatic Control*, pp. 529-552, Dec. 1971.
13. Khalil, H. K., *Nonlinear Systems*, Prentice Hall, 2002.
14. Doyle, J. C., Glover, K., Khargonekar, P. P. and Francis, B. A., "State space solutions to standard $\mathcal{H}_\infty$ and $\mathcal{H}_2$ problems", *IEEE Transactions on Automatic Control*, pp. 831-847, August 1989.
15. Oppenheim, A. V. and Willsky, A. S., with Nawab, S. H., *Signals and Systems*, Prentice Hall, 1997.
16. Anderson, B. D. O. and Moore, J. B., "Detectability and stabilizability of time-varying discrete-time linear systems", *SIAM J. Contr. Optim.*, vol. 19, no. 1, pp. 20-32, Jan. 1981.
17. Kuo, B. C., *Automatic Control Systems*, Prentice-Hall, Inc. Upper Saddle River, NJ, USA, 1987.
18. Chen, C. T., *Linear System Theory and Design*, 3rd Ed. Oxford University Press, 1998.

19. Kalman, R. E. and Bucy, R. S., "New results in linear filtering and prediction theory", *Trans. ASME Ser. D. (J. Basic Engr.)*, pp. 95-107, 1961.
20. Mees, A. I. *Dynamics of Feedback Systems*, Wiley, New York, 1981.
21. Barnett, S. and Cameron, R. G., *Introduction to Mathematical Control Theory*, Clarendon Press, Oxford, 1985.
22. Dorato, P., Abdallah, C. and Cerone, V., *Linear Quadratic Contro: An Introduction*, Prentice Hall, Englewood Cliffs, 1995.
23. Doyle, J. C., "Guaranteed margins for LQG regulators", *IEEE Transactions on Automatic Control*, pp. 756-757, Dec. 1978.
24. Stein, G. and Athans, M., "The LQG/LTR procedure for multivariable feedback control design", *IEEE Transactions on Automatic Control*, pp. 105-114, Aug. 1987.
25. Chang, S. and Peng, T., "Adaptive guaranteed cost control of systems with uncertain parameters", *IEEE Transactions on Automatic Control*, pp. 474-483, Aug. 1972.
26. Huang, Y. and Lu, W., "Nonlinear optimal control: Alternatives to the Hamilton-Jacobi Equation", in *Proc. 35th IEEE CDC*, pp. 3942-3947, 1996.
27. Tigrek, T., Dasgupta, S. and Smith, T. F., "Adaptive optimal control of HVAC systems", in *Preprints of IFAC World Congress*, Barcelona, Spain, 2002.
28. Michalska, H. and Mayne, D. Q., "Robust receding horizon control of constrained nonlinear systems", *IEEE Transactions on Automatic Control*, pp. 1623-1633, Nov. 1993.
29. Primbs, J. A., *Nonlinear Optimal Control: A Receding Hprizon Approach*, Ph. D. Thesis, Caltech, 1999.
30. Lukes, D. L., "Optimal regulation of nonlinear dynamical systems", *SIAM J. of Cont*, pp. 75-100, 1969.
31. Huang, Y. and Lu, W., "A numerical approach to computing nonlinear $\mathcal{H}_\infty$ control laws", J. Guidance Contr. and Dynamics, 1996.
32. Cranda, M. and Lions, P., "Two approximations of solutions to the Hamilton-Jacobi Equation", *Mathematics of Computations*, pp. 1-19, 1984.

INDEX